T0186299

GPU Parallel Program Development Using CUDA

Chapman & Hall/CRC
Computational Science Series

SERIES EDITOR

Horst Simon
Deputy Director
Lawrence Berkeley National Laboratory
Berkeley, California, U.S.A.

PUBLISHED TITLES

COMBINATORIAL SCIENTIFIC COMPUTING
Edited by Uwe Naumann and Olaf Schenk

CONTEMPORARY HIGH PERFORMANCE COMPUTING: FROM PETASCALE
TOWARD EXASCALE
Edited by Jeffrey S. Vetter

CONTEMPORARY HIGH PERFORMANCE COMPUTING: FROM PETASCALE
TOWARD EXASCALE, VOLUME TWO
Edited by Jeffrey S. Vetter

DATA-INTENSIVE SCIENCE
Edited by Terence Critchlow and Kerstin Kleese van Dam

ELEMENTS OF PARALLEL COMPUTING
Eric Aubanel

THE END OF ERROR: UNUM COMPUTING
John L. Gustafson

EXASCALE SCIENTIFIC APPLICATIONS: SCALABILITY AND
PERFORMANCE PORTABILITY
Edited by Tjerk P. Straatsma, Katerina B. Antypas, and Timothy J. Williams

FROM ACTION SYSTEMS TO DISTRIBUTED SYSTEMS: THE REFINEMENT APPROACH
Edited by Luigia Petre and Emil Sekerinski

FUNDAMENTALS OF MULTICORE SOFTWARE DEVELOPMENT
Edited by Victor Pankratius, Ali-Reza Adl-Tabatabai, and Walter Tichy

FUNDAMENTALS OF PARALLEL MULTICORE ARCHITECTURE
Yan Solihin

THE GREEN COMPUTING BOOK: TACKLING ENERGY EFFICIENCY AT LARGE SCALE
Edited by Wu-chun Feng

GRID COMPUTING: TECHNIQUES AND APPLICATIONS
Barry Wilkinson

GPU PARALLEL PROGRAM DEVELOPMENT USING CUDA
Tolga Soyata

PUBLISHED TITLES CONTINUED

HIGH PERFORMANCE COMPUTING: PROGRAMMING AND APPLICATIONS
John Levesque with Gene Wagenbreth

HIGH PERFORMANCE PARALLEL I/O
Prabhat and Quincey Koziol

HIGH PERFORMANCE VISUALIZATION:
ENABLING EXTREME-SCALE SCIENTIFIC INSIGHT
Edited by E. Wes Bethel, Hank Childs, and Charles Hansen

INDUSTRIAL APPLICATIONS OF HIGH-PERFORMANCE COMPUTING:
BEST GLOBAL PRACTICES
Edited by Anwar Osseyran and Merle Giles

INTRODUCTION TO COMPUTATIONAL MODELING USING C AND
OPEN-SOURCE TOOLS
José M Garrido

INTRODUCTION TO CONCURRENCY IN PROGRAMMING LANGUAGES
Matthew J. Sottile, Timothy G. Mattson, and Craig E Rasmussen

INTRODUCTION TO ELEMENTARY COMPUTATIONAL MODELING: ESSENTIAL
CONCEPTS, PRINCIPLES, AND PROBLEM SOLVING
José M. Garrido

INTRODUCTION TO HIGH PERFORMANCE COMPUTING FOR SCIENTISTS
AND ENGINEERS
Georg Hager and Gerhard Wellein

INTRODUCTION TO MODELING AND SIMULATION WITH MATLAB® AND PYTHON
Steven I. Gordon and Brian Guilfoos

INTRODUCTION TO REVERSIBLE COMPUTING
Kalyan S. Perumalla

INTRODUCTION TO SCHEDULING
Yves Robert and Frédéric Vivien

INTRODUCTION TO THE SIMULATION OF DYNAMICS USING SIMULINK®
Michael A. Gray

PEER-TO-PEER COMPUTING: APPLICATIONS, ARCHITECTURE, PROTOCOLS,
AND CHALLENGES
Yu-Kwong Ricky Kwok

PERFORMANCE TUNING OF SCIENTIFIC APPLICATIONS
Edited by David Bailey, Robert Lucas, and Samuel Williams

PETASCALE COMPUTING: ALGORITHMS AND APPLICATIONS
Edited by David A. Bader

PROCESS ALGEBRA FOR PARALLEL AND DISTRIBUTED PROCESSING
Edited by Michael Alexander and William Gardner

PUBLISHED TITLES CONTINUED

PROGRAMMING FOR HYBRID MULTI/MANY-CORE MPP SYSTEMS
John Levesque and Aaron Vose

SCIENTIFIC DATA MANAGEMENT: CHALLENGES, TECHNOLOGY, AND DEPLOYMENT
Edited by Arie Shoshani and Doron Rotem

SOFTWARE ENGINEERING FOR SCIENCE
Edited by Jeffrey C. Carver, Neil P. Chue Hong, and George K. Thiruvathukal

GPU Parallel Program Development Using CUDA

Tolga Soyata

CRC Press
Taylor & Francis Group
Boca Raton   London   New York

CRC Press is an imprint of the
Taylor & Francis Group, an **informa** business

A CHAPMAN & HALL BOOK

CRC Press
Taylor & Francis Group
6000 Broken Sound Parkway NW, Suite 300
Boca Raton, FL 33487-2742

International Standard Book Number-13: 978-1-4987-5075-2 (Hardback)

Library of Congress Cataloging-in-Publication Data

Names: Soyata, Tolga, 1967- author.
Title: GPU parallel program development using CUDA
/ by Tolga Soyata.
Description: Boca Raton, Florida : CRC Press, [2018] | Includes bibliographical
references and index.
Identifiers: LCCN 2017043292 | ISBN 9781498750752 (hardback) |
ISBN 9781315368290 (e-book)
Subjects: LCSH: Parallel programming (Computer science) | CUDA (Computer architecture) |
Graphics processing units–Programming.
Classification: LCC QA76.642.S67 2018 | DDC 005.2/75–dc23
LC record available at https://lccn.loc.gov/2017043292

Visit the Taylor & Francis Web site at
http://www.taylorandfrancis.com

and the CRC Press Web site at
http://www.crcpress.com

Printed and bound in Great Britain by
TJ International Ltd, Padstow, Cornwall

To my wife Eileen
and my step-children Katherine, Andrew, and Eric.

Contents

List of Figures   xxiii

List of Tables   xxix

Preface   xxxiii

About the Author   xxxv

PART I   Understanding CPU Parallelism

CHAPTER   1 ▪ Introduction to CPU Parallel Programming   3

1.1   EVOLUTION OF PARALLEL PROGRAMMING   3

1.2   MORE CORES, MORE PARALLELISM   4

1.3   CORES VERSUS THREADS   5

    1.3.1   More Threads or More Cores to Parallelize?   5

    1.3.2   Influence of Core Resource Sharing   7

    1.3.3   Influence of Memory Resource Sharing   7

1.4   OUR FIRST SERIAL PROGRAM   8

    1.4.1   Understanding Data Transfer Speeds   8

    1.4.2   The main() Function in imflip.c   10

    1.4.3   Flipping Rows Vertically: FlipImageV()   11

    1.4.4   Flipping Columns Horizontally: FlipImageH()   12

1.5   WRITING, COMPILING, RUNNING OUR PROGRAMS   13

    1.5.1   Choosing an Editor and a Compiler   13

    1.5.2   Developing in Windows 7, 8, and Windows 10
          Platforms   13

    1.5.3   Developing in a Mac Platform   15

    1.5.4   Developing in a Unix Platform   15

1.6   CRASH COURSE ON UNIX   15

    1.6.1   Unix Directory-Related Commands   15

    1.6.2   Unix File-Related Commands   16

1.7   DEBUGGING YOUR PROGRAMS   19

    1.7.1   gdb   20

    1.7.2   Old School Debugging   21

    1.7.3   valgrind   22

1.8	PERFORMANCE OF OUR FIRST SERIAL PROGRAM	23
1.8.1	Can We Estimate the Execution Time?	24
1.8.2	What Does the OS Do When Our Code Is Executing?	24
1.8.3	How Do We Parallelize It?	25
1.8.4	Thinking About the Resources	25

CHAPTER   2 ▪ Developing Our First Parallel CPU Program   27

2.1	OUR FIRST PARALLEL PROGRAM	27
2.1.1	The main() Function in imflipP.c	28
2.1.2	Timing the Execution	29
2.1.3	Split Code Listing for main() in imflipP.c	29
2.1.4	Thread Initialization	32
2.1.5	Thread Creation	32
2.1.6	Thread Launch/Execution	34
2.1.7	Thread Termination (Join)	35
2.1.8	Thread Task and Data Splitting	35
2.2	WORKING WITH BITMAP (BMP) FILES	37
2.2.1	BMP is a Non-Lossy/Uncompressed File Format	37
2.2.2	BMP Image File Format	38
2.2.3	Header File ImageStuff.h	39
2.2.4	Image Manipulation Routines in ImageStuff.c	40
2.3	TASK EXECUTION BY THREADS	42
2.3.1	Launching a Thread	43
2.3.2	Multithreaded Vertical Flip: MTFlipV()	45
2.3.3	Comparing FlipImageV() and MTFlipV()	48
2.3.4	Multithreaded Horizontal Flip: MTFlipH()	50
2.4	TESTING/TIMING THE MULTITHREADED CODE	51

CHAPTER   3 ▪ Improving Our First Parallel CPU Program   53

3.1	EFFECT OF THE "PROGRAMMER" ON PERFORMANCE	53
3.2	EFFECT OF THE "CPU" ON PERFORMANCE	54
3.2.1	In-Order versus Out-Of-Order Cores	55
3.2.2	Thin versus Thick Threads	57
3.3	PERFORMANCE OF IMFLIPP	57
3.4	EFFECT OF THE "OS" ON PERFORMANCE	58
3.4.1	Thread Creation	59
3.4.2	Thread Launch and Execution	59
3.4.3	Thread Status	60

	3.4.4	Mapping Software Threads to Hardware Threads	61
	3.4.5	Program Performance versus Launched Pthreads	62
3.5	IMPROVING IMFLIPP		63
	3.5.1	Analyzing Memory Access Patterns in MTFlipH()	64
	3.5.2	Multithreaded Memory Access of MTFlipH()	64
	3.5.3	DRAM Access Rules of Thumb	66
3.6	IMFLIPPM: OBEYING DRAM RULES OF THUMB		67
	3.6.1	Chaotic Memory Access Patterns of imflipP	67
	3.6.2	Improving Memory Access Patterns of imflipP	68
	3.6.3	MTFlipHM(): The Memory Friendly MTFlipH()	69
	3.6.4	MTFlipVM(): The Memory Friendly MTFlipV()	71
3.7	PERFORMANCE OF IMFLIPPM.C		72
	3.7.1	Comparing Performances of imflipP.c and imflipPM.c	72
	3.7.2	Speed Improvement: MTFlipV() versus MTFlipVM()	73
	3.7.3	Speed Improvement: MTFlipH() versus MTFlipHM()	73
	3.7.4	Understanding the Speedup: MTFlipH() versus MTFlipHM()	73
3.8	PROCESS MEMORY MAP		74
3.9	INTEL MIC ARCHITECTURE: XEON PHI		76
3.10	WHAT ABOUT THE GPU?		77
3.11	CHAPTER SUMMARY		78
CHAPTER 4 ■ Understanding the Cores and Memory			79
4.1	ONCE UPON A TIME ... INTEL ...		79
4.2	CPU AND MEMORY MANUFACTURERS		80
4.3	DYNAMIC (DRAM) VERSUS STATIC (SRAM) MEMORY		81
	4.3.1	Static Random Access Memory (SRAM)	81
	4.3.2	Dynamic Random Access Memory (DRAM)	81
	4.3.3	DRAM Interface Standards	81
	4.3.4	Influence of DRAM on our Program Performance	82
	4.3.5	Influence of SRAM (Cache) on our Program Performance	83
4.4	IMAGE ROTATION PROGRAM: IMROTATE.C		83
	4.4.1	Description of the imrotate.c	84
	4.4.2	imrotate.c: Parametric Restrictions and Simplifications	84
	4.4.3	imrotate.c: Theory of Operation	85
4.5	PERFORMANCE OF IMROTATE		89
	4.5.1	Qualitative Analysis of Threading Efficiency	89
	4.5.2	Quantitative Analysis: Defining Threading Efficiency	89

4.6   THE ARCHITECTURE OF THE COMPUTER   91
    4.6.1   The Cores, L1$ and L2$ 91
    4.6.2   Internal Core Resources   92
    4.6.3   The Shared L3 Cache Memory (L3$)   94
    4.6.4   The Memory Controller   94
    4.6.5   The Main Memory   95
    4.6.6   Queue, Uncore, and I/O   96
4.7   IMROTATEMC: MAKING IMROTATE MORE EFFICIENT   97
    4.7.1   Rotate2(): How Bad is Square Root and FP Division? 99
    4.7.2   Rotate3() and Rotate4(): How Bad Is sin() and cos()? 100
    4.7.3   Rotate5(): How Bad Is Integer Division/Multiplication? 102
    4.7.4   Rotate6(): Consolidating Computations   102
    4.7.5   Rotate7(): Consolidating More Computations   104
    4.7.6   Overall Performance of imrotateMC   104
4.8   CHAPTER SUMMARY   106

CHAPTER   5 ▪ Thread Management and Synchronization   107

5.1   EDGE DETECTION PROGRAM: IMEDGE.C   107
    5.1.1   Description of the imedge.c   108
    5.1.2   imedge.c: Parametric Restrictions and
        Simplifications   108
    5.1.3   imedge.c: Theory of Operation   109
5.2   IMEDGE.C : IMPLEMENTATION   111
    5.2.1   Initialization and Time-Stamping   112
    5.2.2   Initialization Functions for Different Image
        Representations   113
    5.2.3   Launching and Terminating Threads   114
    5.2.4   Gaussian Filter   115
    5.2.5   Sobel   116
    5.2.6   Threshold   117
5.3   PERFORMANCE OF IMEDGE   118
5.4   IMEDGEMC: MAKING IMEDGE MORE EFFICIENT   118
    5.4.1   Using Precomputation to Reduce Bandwidth   119
    5.4.2   Storing the Precomputed Pixel Values   120
    5.4.3   Precomputing Pixel Values   121
    5.4.4   Reading the Image and Precomputing Pixel
        Values   122
    5.4.5   PrGaussianFilter   123
    5.4.6   PrSobel   124
    5.4.7   PrThreshold   125
5.5   PERFORMANCE OF IMEDGEMC   126

5.6   IMEDGEMCT: SYNCHRONIZING THREADS EFFICIENTLY   127
    5.6.1   Barrier Synchronization   128
    5.6.2   MUTEX Structure for Data Sharing   129
5.7   IMEDGEMCT: IMPLEMENTATION   130
    5.7.1   Using a MUTEX: Read Image, Precompute   132
    5.7.2   Precomputing One Row at a Time   133
5.8   PERFORMANCE OF IMEDGEMCT   134

Part II   GPU Programming Using CUDA

Chapter   6 ▪ Introduction to GPU Parallelism and CUDA   137

6.1   ONCE UPON A TIME ... NVIDIA ...   137
    6.1.1   The Birth of the GPU   137
    6.1.2   Early GPU Architectures   138
    6.1.3   The Birth of the GPGPU   140
    6.1.4   Nvidia, ATI Technologies, and Intel   141
6.2   COMPUTE-UNIFIED DEVICE ARCHITECTURE (CUDA)   143
    6.2.1   CUDA, OpenCL, and Other GPU Languages   143
    6.2.2   Device Side versus Host Side Code   143
6.3   UNDERSTANDING GPU PARALLELISM   144
    6.3.1   How Does the GPU Achieve High Performance?   145
    6.3.2   CPU versus GPU Architectural Differences   146
6.4   CUDA VERSION OF THE IMAGE FLIPPER: IMFLIPG.CU   147
    6.4.1   imflipG.cu: Read the Image into a CPU-Side Array   149
    6.4.2   Initialize and Query the GPUs   151
    6.4.3   GPU-Side Time-Stamping   153
    6.4.4   GPU-Side Memory Allocation   155
    6.4.5   GPU Drivers and Nvidia Runtime Engine   155
    6.4.6   CPU→GPU Data Transfer   156
    6.4.7   Error Reporting Using Wrapper Functions   157
    6.4.8   GPU Kernel Execution   157
    6.4.9   Finish Executing the GPU Kernel   160
    6.4.10   Transfer GPU Results Back to the CPU   161
    6.4.11   Complete Time-Stamping   161
    6.4.12   Report the Results and Cleanup   162
    6.4.13   Reading and Writing the BMP File   163
    6.4.14   Vflip(): The GPU Kernel for Vertical Flipping   164
    6.4.15   What Is My Thread ID, Block ID, and Block Dimension?   166
    6.4.16   Hflip(): The GPU Kernel for Horizontal Flipping   169

6.4.17   Hardware Parameters: threadIDx.x, blockIdx.x, blockDim.x   169

6.4.18   PixCopy(): The GPU Kernel for Copying an Image   169

6.4.19   CUDA Keywords   170

6.5   CUDA PROGRAM DEVELOPMENT IN WINDOWS   170

6.5.1   Installing MS Visual Studio 2015 and CUDA Toolkit 8.0   171

6.5.2   Creating Project imflipG.cu in Visual Studio 2015   172

6.5.3   Compiling Project imflipG.cu in Visual Studio 2015   174

6.5.4   Running Our First CUDA Application: imflipG.exe   177

6.5.5   Ensuring Your Program's Correctness   178

6.6   CUDA PROGRAM DEVELOPMENT ON A MAC PLATFORM   179

6.6.1   Installing XCode on Your Mac   179

6.6.2   Installing the CUDA Driver and CUDA Toolkit   180

6.6.3   Compiling and Running CUDA Applications on a Mac   180

6.7   CUDA PROGRAM DEVELOPMENT IN A UNIX PLATFORM   181

6.7.1   Installing Eclipse and CUDA Toolkit   181

6.7.2   ssh into a Cluster   182

6.7.3   Compiling and Executing Your CUDA Code   182

CHAPTER   7 ▪ CUDA Host/Device Programming Model   185

7.1   DESIGNING YOUR PROGRAM'S PARALLELISM   185

7.1.1   Conceptually Parallelizing a Task   186

7.1.2   What Is a Good Block Size for Vflip()?   187

7.1.3   imflipG.cu: Interpreting the Program Output   187

7.1.4   imflipG.cu: Performance Impact of Block and Image Size   188

7.2   KERNEL LAUNCH COMPONENTS   189

7.2.1   Grids   189

7.2.2   Blocks   190

7.2.3   Threads   191

7.2.4   Warps and Lanes   192

7.3   IMFLIPG.CU: UNDERSTANDING THE KERNEL DETAILS   193

7.3.1   Launching Kernels in main() and Passing Arguments to Them   193

7.3.2   Thread Execution Steps   194

7.3.3   Vflip() Kernel Details   195

7.3.4   Comparing Vflip() and MTFlipV()   196

7.3.5   Hflip() Kernel Details   197

7.3.6   PixCopy() Kernel Details   197

7.4   DEPENDENCE OF PCI EXPRESS SPEED ON THE CPU   199

7.5   PERFORMANCE IMPACT OF PCI EXPRESS BUS   200

    7.5.1   Data Transfer Time, Speed, Latency, Throughput, and
        Bandwidth   200

    7.5.2   PCIe Throughput Achieved with imflipG.cu   201

7.6   PERFORMANCE IMPACT OF GLOBAL MEMORY BUS   204

7.7   PERFORMANCE IMPACT OF COMPUTE CAPABILITY   206

    7.7.1   Fermi, Kepler, Maxwell, Pascal, and Volta Families   207

    7.7.2   Relative Bandwidth Achieved in Different Families   207

    7.7.3   imflipG2.cu: Compute Capability 2.0 Version of imflipG.cu   208

    7.7.4   imflipG2.cu: Changes in main()   210

    7.7.5   The PxCC20() Kernel   211

    7.7.6   The VfCC20() Kernel   212

7.8   PERFORMANCE OF IMFLIPG2.CU   214

7.9   OLD-SCHOOL CUDA DEBUGGING   214

    7.9.1   Common CUDA Bugs   216

    7.9.2   return Debugging   218

    7.9.3   Comment-Based Debugging   220

    7.9.4   printf() Debugging   220

7.10   BIOLOGICAL REASONS FOR SOFTWARE BUGS   221

    7.10.1   How Is Our Brain Involved in Writing/Debugging Code?   222

    7.10.2   Do We Write Buggy Code When We Are Tired?   222

        7.10.2.1 Attention   223

        7.10.2.2 Physical Tiredness   223

        7.10.2.3 Tiredness Due to Heavy Physical Activity   223

        7.10.2.4 Tiredness Due to Needing Sleep   223

        7.10.2.5 Mental Tiredness   224

CHAPTER   8 ■ Understanding GPU Hardware Architecture   225

8.1   GPU HARDWARE ARCHITECTURE   226

8.2   GPU HARDWARE COMPONENTS   226

    8.2.1   SM: Streaming Multiprocessor   226

    8.2.2   GPU Cores   227

    8.2.3   Giga-Thread Scheduler   227

    8.2.4   Memory Controllers   229

    8.2.5   Shared Cache Memory (L2$)   229

    8.2.6   Host Interface   229

8.3   NVIDIA GPU ARCHITECTURES   230

    8.3.1   Fermi Architecture   231

    8.3.2   GT, GTX, and Compute Accelerators   231

    8.3.3   Kepler Architecture   232

	8.3.4	Maxwell Architecture	232
	8.3.5	Pascal Architecture and NVLink	233
8.4		CUDA EDGE DETECTION: IMEDGEG.CU	233
	8.4.1	Variables to Store the Image in CPU, GPU Memory	233
		8.4.1.1 `TheImage` and `CopyImage`	233
		8.4.1.2 `GPUImg`	234
		8.4.1.3 `GPUBWImg`	234
		8.4.1.4 `GPUGaussImg`	234
		8.4.1.5 `GPUGradient` and `GPUTheta`	234
		8.4.1.6 `GPUResultImg`	235
	8.4.2	Allocating Memory for the GPU Variables	235
	8.4.3	Calling the Kernels and Time-Stamping Their Execution	238
	8.4.4	Computing the Kernel Performance	239
	8.4.5	Computing the Amount of Kernel Data Movement	239
	8.4.6	Reporting the Kernel Performance	242
8.5		IMEDGEG: KERNELS	242
	8.5.1	BWKernel()	242
	8.5.2	GaussKernel()	244
	8.5.3	SobelKernel()	246
	8.5.4	ThresholdKernel()	249
8.6		PERFORMANCE OF IMEDGEG.CU	249
	8.6.1	imedgeG.cu: PCIe Bus Utilization	250
	8.6.2	imedgeG.cu: Runtime Results	250
	8.6.3	imedgeG.cu: Kernel Performance Comparison	252
8.7		GPU CODE: COMPILE TIME	253
	8.7.1	Designing CUDA Code	253
	8.7.2	Compiling CUDA Code	255
	8.7.3	GPU Assembly: PTX, CUBIN	255
8.8		GPU CODE: LAUNCH	255
	8.8.1	OS Involvement and CUDA DLL File	255
	8.8.2	GPU Graphics Driver	256
	8.8.3	CPU⟷GPU Memory Transfers	256
8.9		GPU CODE: EXECUTION (RUN TIME)	257
	8.9.1	Getting the Data	257
	8.9.2	Getting the Code and Parameters	257
	8.9.3	Launching Grids of Blocks	258
	8.9.4	Giga Thread Scheduler (GTS)	258
	8.9.5	Scheduling Blocks	259
	8.9.6	Executing Blocks	260
	8.9.7	Transparent Scalability	261

CHAPTER 9 ▪ Understanding GPU Cores 263

9.1 GPU ARCHITECTURE FAMILIES 263
   9.1.1 Fermi Architecture 263
   9.1.2 Fermi SM Structure 264
   9.1.3 Kepler Architecture 266
   9.1.4 Kepler SMX Structure 267
   9.1.5 Maxwell Architecture 268
   9.1.6 Maxwell SMM Structure 268
   9.1.7 Pascal GP100 Architecture 270
   9.1.8 Pascal GP100 SM Structure 271
   9.1.9 Family Comparison: Peak GFLOPS and Peak DGFLOPS 272
   9.1.10 GPU Boost 273
   9.1.11 GPU Power Consumption 274
   9.1.12 Computer Power Supply 274
9.2 STREAMING MULTIPROCESSOR (SM) BUILDING BLOCKS 275
   9.2.1 GPU Cores 275
   9.2.2 Double Precision Units (DPU) 276
   9.2.3 Special Function Units (SFU) 276
   9.2.4 Register File (RF) 276
   9.2.5 Load/Store Queues (LDST) 277
   9.2.6 L1$ and Texture Cache 277
   9.2.7 Shared Memory 278
   9.2.8 Constant Cache 278
   9.2.9 Instruction Cache 278
   9.2.10 Instruction Buffer 278
   9.2.11 Warp Schedulers 278
   9.2.12 Dispatch Units 279
9.3 PARALLEL THREAD EXECUTION (PTX) DATA TYPES 279
   9.3.1 *INT8*: 8-bit Integer 280
   9.3.2 *INT16*: 16-bit Integer 280
   9.3.3 24-bit Integer 280
   9.3.4 *INT32*: 32-bit Integer 281
   9.3.5 Predicate Registers (32-bit) 281
   9.3.6 *INT64*: 64-bit Integer 282
   9.3.7 128-bit Integer 282
   9.3.8 FP32: Single Precision Floating Point (`float`) 282
   9.3.9 FP64: Double Precision Floating Point (`double`) 283
   9.3.10 FP16: Half Precision Floating Point (`half`) 284
   9.3.11 What is a FLOP? 284

9.3.12 Fused Multiply-Accumulate (FMA) versus Multiply-Add (MAD)   285

9.3.13 Quad and Octo Precision Floating Point   285

9.3.14 Pascal GP104 Engine SM Structure   285

9.4 IMFLIPGC.CU: CORE-FRIENDLY IMFLIPG   286

9.4.1 Hflip2(): Precomputing Kernel Parameters   288

9.4.2 Vflip2(): Precomputing Kernel Parameters   290

9.4.3 Computing Image Coordinates by a Thread   290

9.4.4 Block ID versus Image Row Mapping   291

9.4.5 Hflip3(): Using a 2D Launch Grid   292

9.4.6 Vflip3(): Using a 2D Launch Grid   293

9.4.7 Hflip4(): Computing Two Consecutive Pixels   294

9.4.8 Vflip4(): Computing Two Consecutive Pixels   295

9.4.9 Hflip5(): Computing Four Consecutive Pixels   296

9.4.10 Vflip5(): Computing Four Consecutive Pixels   297

9.4.11 PixCopy2(), PixCopy3(): Copying 2,4 Consecutive Pixels at a Time   298

9.5 IMEDGEGC.CU: CORE-FRIENDLY IMEDGEG   299

9.5.1 BWKernel2(): Using Precomputed Values and 2D Blocks   299

9.5.2 GaussKernel2(): Using Precomputed Values and 2D Blocks   300

CHAPTER 10 ■ Understanding GPU Memory   303

10.1 GLOBAL MEMORY   303

10.2 L2 CACHE   304

10.3 TEXTURE/L1 CACHE   304

10.4 SHARED MEMORY   305

10.4.1 Split versus Dedicated Shared Memory   305

10.4.2 Memory Resources Available Per Core   306

10.4.3 Using Shared Memory as Software Cache   306

10.4.4 Allocating Shared Memory in an SM   307

10.5 INSTRUCTION CACHE   307

10.6 CONSTANT MEMORY   307

10.7 IMFLIPGCM.CU: CORE AND MEMORY FRIENDLY IMFLIPG   308

10.7.1 Hflip6(),Vflip6(): Using Shared Memory as Buffer   308

10.7.2 Hflip7(): Consecutive Swap Operations in Shared Memory   310

10.7.3 Hflip8(): Using Registers to Swap Four Pixels   312

10.7.4 Vflip7(): Copying 4 Bytes (int) at a Time   314

10.7.5 Aligned versus Unaligned Data Access in Memory   314

10.7.6 Vflip8(): Copying 8 Bytes at a Time   315

10.7.7 Vflip9(): Using Only Global Memory, 8 Bytes at a Time   316

10.7.8 PixCopy4(), PixCopy5(): Copying One versus 4 Bytes Using Shared Memory   317

10.7.9 PixCopy6(), PixCopy7(): Copying One/Two Integers Using Global Memory   318

10.8 IMEDGEGCM.CU: CORE- & MEMORY-FRIENDLY IMEDGEG   319

10.8.1 BWKernel3(): Using Byte Manipulation to Extract RGB   319

10.8.2 GaussKernel3(): Using Constant Memory   321

10.8.3 Ways to Handle Constant Values   321

10.8.4 GaussKernel4(): Buffering Neighbors of 1 Pixel in Shared Memory   323

10.8.5 GaussKernel5(): Buffering Neighbors of 4 Pixels in Shared Memory   325

10.8.6 GaussKernel6(): Reading 5 Vertical Pixels into Shared Memory   327

10.8.7 GaussKernel7(): Eliminating the Need to Account for Edge Pixels   329

10.8.8 GaussKernel8(): Computing 8 Vertical Pixels   331

10.9 CUDA OCCUPANCY CALCULATOR   333

10.9.1 Choosing the Optimum Threads/Block   334

10.9.2 SM-Level Resource Limitations   335

10.9.3 What is "Occupancy"?   336

10.9.4 CUDA Occupancy Calculator: Resource Computation   336

10.9.5 Case Study: GaussKernel7()   340

10.9.6 Case Study: GaussKernel8()   343

CHAPTER 11 ■ CUDA Streams   345

11.1 WHAT IS PIPELINING?   347

11.1.1 Execution Overlapping   347

11.1.2 Exposed versus Coalesced Runtime   348

11.2 MEMORY ALLOCATION   349

11.2.1 Physical versus Virtual Memory   349

11.2.2 Physical to Virtual Address Translation   350

11.2.3 Pinned Memory   350

11.2.4 Allocating Pinned Memory with cudaMallocHost()   351

11.3 FAST CPU⟷GPU DATA TRANSFERS   351

11.3.1 Synchronous Data Transfers   351

11.3.2 Asynchronous Data Transfers   351

11.4 CUDA STREAMS   352

11.4.1 CPU→GPU Transfer, Kernel Exec, GPU→CPUTransfer   352

11.4.2 Implementing Streaming in CUDA   353

11.4.3 Copy Engine 353

11.4.4 Kernel Execution Engine 353

11.4.5 Concurrent Upstream and Downstream PCIe Transfers 354

11.4.6 Creating CUDA Streams 355

11.4.7 Destroying CUDA Streams 355

11.4.8 Synchronizing CUDA Streams 355

11.5 IMGSTR.CU: STREAMING IMAGE PROCESSING 356

11.5.1 Reading the Image into Pinned Memory 356

11.5.2 Synchronous versus Single Stream 358

11.5.3 Multiple Streams 359

11.5.4 Data Dependence Across Multiple Streams 361

11.5.4.1 Horizontal Flip: No Data Dependence 362

11.5.4.2 Edge Detection: Data Dependence 363

11.5.4.3 Preprocessing Overlapping Rows Synchronously 363

11.5.4.4 Asynchronous Processing the Non-Overlapping Rows 364

11.6 STREAMING HORIZONTAL FLIP KERNEL 366

11.7 IMGSTR.CU: STREAMING EDGE DETECTION 367

11.8 PERFORMANCE COMPARISON: IMGSTR.CU 371

11.8.1 Synchronous versus Asynchronous Results 371

11.8.2 Randomness in the Results 372

11.8.3 Optimum Queuing 372

11.8.4 Best Case Streaming Results 373

11.8.5 Worst Case Streaming Results 374

11.9 NVIDIA VISUAL PROFILER: NVVP 375

11.9.1 Installing nvvp and nvprof 375

11.9.2 Using nvvp 376

11.9.3 Using nvprof 377

11.9.4 imGStr Synchronous and Single-Stream Results 377

11.9.5 imGStr 2- and 4-Stream Results 378

PART III More To Know

CHAPTER 12 ▪ CUDA Libraries 383

MOHAMADHADI HABIBZADEH, OMID RAJABI SHISHVAN, and TOLGA SOYATA

12.1 cuBLAS 383

12.1.1 BLAS Levels 383

12.1.2 cuBLAS Datatypes 384

12.1.3 Installing cuBLAS 385

12.1.4 Variable Declaration and Initialization 385

12.1.5 Device Memory Allocation   386

12.1.6 Creating Context   386

12.1.7 Transferring Data to the Device   386

12.1.8 Calling cuBLAS Functions   387

12.1.9 Transfer Data Back to the Host   388

12.1.10 Deallocating Memory   388

12.1.11 Example cuBLAS Program: Matrix Scalar   388

12.2 CUFFT   390

12.2.1 cuFFT Library Characteristics   390

12.2.2 A Sample Complex-to-Complex Transform   390

12.2.3 A Sample Real-to-Complex Transform   391

12.3 NVIDIA PERFORMANCE PRIMITIVES (NPP)   392

12.4 THRUST LIBRARY   393

CHAPTER 13 ■ Introduction to OpenCL   397

CHASE CONKLIN and TOLGA SOYATA

13.1 WHAT IS OpenCL?   397

13.1.1 Multiplatform   397

13.1.2 Queue-Based   397

13.2 IMAGE FLIP KERNEL IN OPENCL   398

13.3 RUNNING OUR KERNEL   399

13.3.1 Selecting a Device   400

13.3.2 Running the Kernel   401

13.3.2.1 Creating a Compute Context   401

13.3.2.2 Creating a Command Queue   401

13.3.2.3 Loading Kernel File   402

13.3.2.4 Setting Up Kernel Invocation   403

13.3.3 Runtimes of Our OpenCL Program   405

13.4 EDGE DETECTION IN OpenCL   406

CHAPTER 14 ■ Other GPU Programming Languages   413

SAM MILLER, ANDREW BOGGIO-DANDRY, and TOLGA SOYATA

14.1 GPU PROGRAMMING WITH PYTHON   413

14.1.1 PyOpenCL Version of imflip   414

14.1.2 PyOpenCL Element-Wise Kernel   418

14.2 OPENGL   420

14.3 OPENGL ES: OPENGL FOR EMBEDDED SYSTEMS   420

14.4 VULKAN   421

14.5 MICROSOFT'S HIGH-LEVEL SHADING LANGUAGE (HLSL)   421

14.5.1 Shading   421

14.5.2 Microsoft HLSL   422

14.6 APPLE'S METAL API 422

14.7 APPLE'S SWIFT PROGRAMMING LANGUAGE 423

14.8 OPENCV 423

    14.8.1 Installing OpenCV and Face Recognition 423

    14.8.2 Mobile-Cloudlet-Cloud Real-Time Face Recognition 423

    14.8.3 Acceleration as a Service (AXaas) 423

CHAPTER 15 ▪ Deep Learning Using CUDA 425

OMID RAJABI SHISHVAN and TOLGA SOYATA

15.1 ARTIFICIAL NEURAL NETWORKS (ANNS) 425

    15.1.1 Neurons 425

    15.1.2 Activation Functions 425

15.2 FULLY CONNECTED NEURAL NETWORKS 425

15.3 DEEP NETWORKS/CONVOLUTIONAL NEURAL NETWORKS 427

15.4 TRAINING A NETWORK 428

15.5 CUDNN LIBRARY FOR DEEP LEARNING 428

    15.5.1 Creating a Layer 429

    15.5.2 Creating a Network 430

    15.5.3 Forward Propagation 431

    15.5.4 Backpropagation 431

    15.5.5 Using cuBLAS in the Network 431

15.6 KERAS 432

Bibliography 435

Index 439

List of Figures

1.1   Harvesting each coconut requires two consecutive 30-second tasks (threads). Thread 1: get a coconut. Thread 2: crack (process) that coconut using the hammer.   4

1.2   Simultaneously executing Thread 1 ("1") and Thread 2 ("2"). Accessing shared resources will cause a thread to wait ("-").   6

1.3   Serial (single-threaded) program imflip.c flips a 640×480 dog picture (left) horizontally (middle) or vertically (right).   8

1.4   Running gdb to catch a segmentation fault.   20

1.5   Running valgrind to catch a memory access error.   23

2.1   Windows Task Manager, showing 1499 threads, however, there is 0% CPU utilization.   33

3.1   The life cycle of a thread. From the creation to its termination, a thread is cycled through many different statuses, assigned by the OS.   60

3.2   Memory access patterns of MTFlipH() in Code 2.8. A total of 3200 pixels' RGB values (9600 Bytes) are flipped for each row.   65

3.3   The memory map of a process when only a single thread is running within the process (left) or multiple threads are running in it (right).   75

4.1   Inside a computer containing an i7-5930K CPU [10] (CPU5 in Table 3.1), and 64 GB of DDR4 memory. This PC has a GTX Titan Z GPU that will be used to test a lot of the programs in Part II.   80

4.2   The imrotate.c program rotates a picture by a specified angle. Original dog (top left), rotated $+10°$ (top right), $+45°$ (bottom left), and $-75°$ (bottom right) clockwise. Scaling is done to avoid cropping of the original image area.   84

4.3   The architecture of one core of the i7-5930K CPU (the PC in Figure 4.1). This core is capable of executing two threads (hyper-threading, as defined by Intel). These two threads share most of the core resources, but have their own register files.   92

4.4   Architecture of the i7-5930K CPU (6C/12T). This CPU connects to the GPUs through an external PCI express bus and memory through the memory bus.   94

5.1   The imedge.c program is used to detect edges in the original image astronaut.bmp (top left). Intermediate processing steps are: GaussianFilter() (top right), Sobel() (bottom left), and finally Threshold() (bottom right).   108

5.2   Example barrier synchronization for 4 threads. Serial runtime is 7281 ms and the 4-threaded runtime is 2246 ms. The speedup of 3.24× is close to the best-expected 4×, but not equal due to the imbalance of each thread's runtime.   128

5.3   Using a MUTEXdata structure to access shared variables.   129

6.1   Turning the dog picture into a 3D wire frame. Triangles are used to represent the object, rather than pixels. This representation allows us to map a texture to each triangle. When the object moves, so does each triangle, along with their associated textures. To increase the resolution of this kind of an object representation, we can divide triangles into smaller triangles in a process called *tesselation*.   139

6.2   Steps to move triangulated 3D objects. Triangles contain two attributes: their *location* and their *texture*. Objects are moved by performing mathematical operations only on their coordinates. A final texture mapping places the texture back on the moved object coordinates, while a 3D-to-2D transformation allows the resulting image to be displayed on a regular 2D computer monitor.   140

6.3   Three farmer teams compete in Analogy 6.1: (1) Arnold competes alone with his 2× bigger tractor and "the strongest farmer" reputation, (2) Fred and Jim compete together in a much smaller tractor than Arnold. (3) Tolga, along with 32 boy and girl scouts, compete together using a bus. Who wins?   145

6.4   Nvidia Runtime Engine is built into your GPU drivers, shown in your Windows 10 Pro SysTray. When you click the Nvidia symbol, you can open the Nvidia control panel to see the driver version as well as the parameters of your GPU(s).   156

6.5   Creating a Visual Studio 2015 CUDA project named imflipG.cu. Assume that the code will be in a directory named Z:\code\imflipG in this example.   172

6.6   Visual Studio 2015 source files are in the Z:\code\imflipG\imflipG directory. In this specific example, we will remove the default file, kernel.cu, that VS 2015 creates. After this, we will add an existing file, imflipG.cu, to the project.   173

6.7   The default CPU platform is x86. We will change it to x64. We will also remove the GPU debugging option.   174

6.8   The default Compute Capability is 2.0. This is too old. We will change it to Compute Capability 3.0, which is done by editing *Code Generation* under *Device* and changing it to compute_30, sm_30.   175

6.9   Compiling imflipG.cu to get the executable file imflipG.exe in the Z:\code\imflipG\x64\Debug directory.   176

6.10   Running imflipG.exe from a CMD command line window.   177

6.11   The /usr/local directory in Unix contains your CUDA directories.   181

6.12   Creating a new CUDA project using the Eclipse IDE in Unix.   183

7.1   The PCIe bus connects for the host (CPU) and the device(s) (GPUs). The host and each device have their own I/O controllers to allow transfers through the PCIe bus, while both the host and the device have their own memory, with a dedicated bus to it; in the GPU this memory is called *global memory*.   205

8.1 Analogy 8.1 for executing a massively parallel program using a significant number of GPU cores, which receive their instructions and data from different sources. Melissa (*Memory controller*) is solely responsible for bringing the coconuts from the jungle and dumping them into the big barrel (*L2$*). Larry (*L2$ controller*) is responsible for distributing these coconuts into the smaller barrels (L1$) of Laura, Linda, Lilly, and Libby; eventually, these four folks distribute the coconuts (*data*) to the scouts (*GPU cores*). On the right side, Gina (*Giga-Thread Scheduler*) has the big list of tasks (*list of blocks to be executed*); she assigns each block to a school bus (*SM* or *streaming multiprocessor*). Inside the bus, one person — Tolga, Tony, Tom, and Tim — is responsible to assign them to the scouts (*instruction schedulers*).   228

8.2 The internal architecture of the GTX550Ti GPU. A total of 192 GPU cores are organized into six streaming multiprocessor (SM) groups of 32 GPU cores. A single L2$ is shared among all 192 cores, while each SM has its own L1$. A dedicated memory controller is responsible for bringing data in and out of the GDDR5 global memory and dumping it into the shared L2$, while a dedicated host interface is responsible for shuttling data (and code) between the CPU and GPU over the PCIe bus.   230

8.3 A sample output of the imedgeG.cu program executed on the astronaut.bmp image using a GTX Titan Z GPU. Kernel execution times and the amount of data movement for each kernel is clearly shown.   242

9.1 GF110 Fermi architecture with 16 SMs, where each SM houses 32 cores, 16 LD/ST units, and 4 Special Function Units (SFUs). The highest end Fermi GPU contains 512 cores (e.g., GTX 580).   264

9.2 GF110 Fermi SM structure. Each SM has a 128 KB register file that contains 32,768 (32 K) registers, where each register is 32-bits. This register file feeds operands to the 32 cores and 4 Special Function Units (SFU). 16 Load/Store (LD/ST) units are used to queue memory load/store requests. A 64 KB total cache memory is used for L1$ and shared memory.   265

9.3 GK110 Kepler architecture with 15 SMXs, where each SMX houses 192 cores, 48 double precision units (DPU), 32 LD/ST units, and 32 Special Function Units (SFU). The highest end Kepler GPU contains 2880 cores (e.g., GTX Titan Black); its "double" version GTX Titan Z contains 5760 cores.   266

9.4 GK110 Kepler SMX structure. A 256 KB (64 K-register) register file feeds 192 cores, 64 Double-Precision Units (DPU), 32 Load/Store units, and 32 SFUs. Four warp schedulers can schedule four warps, which are dispatched as 8 half-warps. Read-only cache is used to hold constants.   267

9.5 GM200 Maxwell architecture with 24 SMMs, housed inside 6 larger GPC units; each SMM houses 128 cores, 32 LD/ST units, and 32 Special Function Units (SFU), *does not contain* double-precision units (DPUs). The highest end Maxwell GPU contains 3072 cores (e.g., GTX Titan X).   268

9.6 GM200 Maxwell SMM structure consists of 4 identical sub-structures with 32 cores, 8 LD/ST units, 8 SFUs, and 16 K registers. Two of these sub-structures share an L1$, while four of them share a 96 KB shared memory.   269

9.7 GP100 Pascal architecture with 60 SMs, housed inside 6 larger GPC units, each containing 10 SMs. The highest end Pascal GPU contains 3840 cores (e.g., P100 compute accelerator). NVLink and High Bandwidth Memory

(HBM2) allow significantly faster memory bandwidths as compared to previous generations. 270

9.8   GP100 Pascal SM structure consists of two identical sub-structures that contain 32 cores, 16 DPUs, 8 LD/ST units, 8 SFUs, and 32 K registers. They share an instruction cache, however, they have their own instruction buffer. 271

9.9   IEEE 754-2008 floating point standard and the supported floating point data types by CUDA. `half` data type is supported in Compute Capability 5.3 and above, while `float` has seen support from the first day of the introduction of CUDA. Support for `double` types started in Compute Capability 1.3. 284

10.1 CUDA Occupancy Calculator: Choosing the Compute Capability, max. shared memory size, registers/kernel, and kernel shared memory usage. In this specific case, the occupancy is 24 warps per SM (out of a total of 64), translating to an occupancy of $24 \div 64 = 38\%$. 337

10.2 Analyzing the occupancy of a case with (1) registers/thread=16, (2) shared memory/kernel=8192 (8 KB), and (3) threads/block=128 (4 warps). CUDA Occupancy Calculator plots the occupancy when each kernel contains more registers (top) and as we launch more blocks (bottom), each requiring an additional 8 KB. With 8 KB/block, the limitation is 24 warps/SM; however, it would go up to 32 warps/block, if each block only required 6 KB of shared memory (6144 Bytes), as shown in the shared memory plot (below). 338

10.3 Analyzing the occupancy of a case with (1) registers/thread=16, (2) shared memory/kernel=8192 (8 KB), and (3) threads/block=128 (4 warps). CUDA Occupancy Calculator plots the occupancy when we launch our blocks with more threads/block (top) and provides a summary of which one of the three resources will be exposed to the limitation before the others (bottom). In this specific case, the limited amount of shared memory (48 KB) limits the total number of blocks we can launch to 6. Alternatively, the number of registers or the maximum number of blocks per SM does not become a limitation. 339

10.4 Analyzing the GaussKernel7(), which uses (1) registers/thread ≈ 16, (2) shared memory/kernel=40,960 (40 KB), and (3) threads/block=256. It is clear that the shared memory limitation does not allow us to launch more than a single block with 256 threads (8 warps). If you could reduce the shared memory down to 24 KB by redesigning your kernel, you could launch at least 2 blocks (16 warps, as shown in the plot) and double the occupancy. 341

10.5 Analyzing the GaussKernel7() with (1) registers/thread=16, (2) shared memory/kernel=40,960, and (3) threads/block=256. 342

10.6 Analyzing the GaussKernel8() with (1) registers/thread=16, (2) shared memory/kernel=24,576, and (3) threads/block=256. 343

10.7 Analyzing the GaussKernel8() with (1) registers/thread=16, (2) shared memory/kernel=24,576, and (3) threads/block=256. 344

11.1 Nvidia visual profiler. 376

11.2 Nvidia profiler, command line version. 377

11.3 Nvidia NVVP results with no streaming and using a single stream, on the K80 GPU. 378

11.4 Nvidia NVVP results with 2 and 4 streams, on the K80 GPU. 379

14.1 imflip.py kernel runtimes on different devices. 417

15.1 Generalized architecture of a fully connected artificial neural network with n inputs, k hidden layers, and m outputs. 426

15.2 Inner structure of a neuron used in ANNs. ω_{ij} are the weights by which inputs to the neuron $(x_1, x_2, ..., x_n)$ are multiplied before they are summed. "Bias" is a value by which this sum is augmented, and $f()$ is the activation function, which is used to introduce a non-linear component to the output. 426

List of Tables

1.1   A list of common gdb commands and functionality.   21

2.1   Serial and multithreaded execution time of imflipP.c, both for vertical flip and horizontal flip, on an i7-960 (4C/8T) CPU.   51

3.1   Different CPUs used in testing the imflipP.c program.   55
3.2   imflipP.c execution times (ms) for the CPUs listed in Table 3.1.   58
3.3   Rules of thumb for achieving good DRAM performance.   67
3.4   imflipPM.c execution times (ms) for the CPUs listed in Table 3.1.   72
3.5   Comparing imflipP.c execution times (H, V type flips in Table 3.2) to imflipPM.c execution times (I, W type flips in Table 3.4).   73
3.6   Comparing imflipP.c execution times (H, V type flips in Table 3.2) to imflipPM.c execution times (I, W type flips in Table 3.4) for Xeon Phi 5110P.   77

4.1   imrotate.c execution times for the CPUs in Table 3.1 ($+45°$ rotation).   89
4.2   imrotate.c threading efficiency (η) and parallelization overhead ($1-\eta$) for CPU3, CPU5. The last column reports the speedup achieved by using CPU5 that has more cores/threads, although there is no speedup up to 6 launched SW threads.   90
4.3   imrotateMC.c execution times for the CPUs in Table 3.1.   105

5.1   Array variables and their types, used during edge detection.   111
5.2   imedge.c execution times for the W3690 CPU (6C/12T).   118
5.3   imedgeMC.c execution times for the W3690 CPU (6C/12T) in ms for a varying number of threads (above). For comparison, execution times of imedge.c are repeated from Table 5.2 (below).   126
5.4   imedgeMCT.c execution times (in ms) for the W3690 CPU (6C/12T), using the Astronaut.bmp image file (top) and Xeon Phi 5110P (60C/240T) using the dogL.bmp file (bottom).   134

6.1   CUDA keyword and symbols that we learned in this chapter.   170

7.1   Vflip() kernel execution times (ms) for different size images on a GTX TITAN Z GPU.   188
7.2   Variables available to a kernel upon launch.   190
7.3   Specifications of different computers used in testing the imflipG.cu program, along with the execution results, compiled using *Compute Capability 3.0*.   202

7.4   Introduction date and peak bandwidth of different bus types.   203

7.5   Introduction date and peak throughput of different CPU and GPU memory types.   206

7.6   Results of the imflipG2.cu program, which uses the VfCC20() and PxCC20() kernels and works in Compute Capability 2.0.   215

8.1   Nvidia microarchitecture families and their hardware features.   233

8.2   Kernels used in imedgeG.cu, along with their source array name and type.   235

8.3   PCIe bandwidth results of imedgeG.cu on six different computer configurations   249

8.4   imedgeG.cu kernel runtime results; red numbers are the best option for the *number of threads* and blue are fairly close to the best option (see ebook for color version).   251

8.5   Summarized imedgeG.cu kernel runtime results; runtime is reported for 256 threads/block for every case.   252

9.1   Nvidia microarchitecture families and their peak computational power for single precision (GFLOPS) and double-precision floating point (DGFLOPS).   273

9.2   Comparison of kernel performances between (Hflip() and Hflip2()) as well as (Vflip() and HVflip2()).   289

9.3   Kernel performances: Hflip(),···,Hflip3(), and Vflip(),···,Vflip3().   293

9.4   Kernel performances: Hflip(),···,Hflip4(), and Vflip(),···,Vflip4().   295

9.5   Kernel performances: Hflip(),···,Hflip5(), and Vflip(),···,Vflip5().   297

9.6   Kernel performances: PixCopy(), PixCopy2(), and PixCopy3().   298

9.7   Kernel performances: BWKernel() and BWKernel2().   300

9.8   Kernel performances: GaussKernel() and GaussKernel2().   301

10.1   Nvidia microarchitecture families and the size of global memory, L1\$, L2\$ and shared memory in each one of them.   305

10.2   Kernel performances: Hflip() vs. Hflip6() and Vflip() vs. Vflip6().   309

10.3   Kernel performances: Hflip(), Hflip6(), and Hflip7() using mars.bmp.   311

10.4   Kernel performances: Hflip6(), Hflip7(), Hflip8() using mars.bmp.   313

10.5   Kernel performances: Vflip(), Vflip6(), Vflip7(), and Vflip8().   315

10.6   Kernel performances: Vflip(), Vflip6(), Vflip7(), Vflip8(), and Vflip9().   316

10.7   Kernel performances: PixCopy(), PixCopy2(),..., PixCopy5().   317

10.8   Kernel performances: PixCopy(), PixCopy4(),..., PixCopy7().   318

10.9   Kernel performances: BWKernel(), BWKernel2(), and BWKernel3().   320

10.10   Kernel performances: GaussKernel(), GaussKernel2(), GaussKernel3()   322

10.11   Kernel performances: GaussKernel(),..., GaussKernel4().   324

10.12   Kernel performances: GaussKernel1(),..., GaussKernel5().   325

10.13   Kernel performances: GaussKernel3(),..., GaussKernel6().   327

10.14   Kernel performances: GaussKernel3(),..., GaussKernel7().   330

10.15   Kernel performances: GaussKernel3(),..., GaussKernel8().   331

10.16 Kernel performances for GaussKernel6(), GaussKernel7(), and GaussKernel8() for Box IV, under varying threads/block choices.   334

11.1 Runtime for edge detection and horizontal flip for astronaut.bmp (in ms).   346

11.2 Execution timeline for the second team in Analogy 11.1.   347

11.3 Streaming performance results (in ms) for imGStr, on the astronaut.bmp image.   371

13.1 Comparable terms for CUDA and OpenCL.   399

13.2 Runtimes for imflip, in ms.   405

13.3 Runtimes for imedge, in ms.   411

15.1 Common activation functions used in neurons to introduce a non-linear component to the final output.   427

Preface

I am from the days when computer engineers and scientists had to write assembly language on IBM mainframes to develop high-performance programs. Programs were written on punch cards and compilation was a one-day process; you dropped off your punch-code written program and picked up the results the next day. If there was an error, you did it again. In those days, a good programmer had to understand the underlying machine hardware to produce good code. I get a little nervous when I see computer science students being taught only at a high abstraction level and languages like Ruby. Although abstraction is a beautiful thing to develop things without getting bogged down with unnecessary details, it is a bad thing when you are trying to develop super high performance code.

Since the introduction of the first CPU, computer architects added incredible features into CPU hardware to "forgive" bad programming skills; while you had to order the sequence of machine code instructions by hand two decades ago, CPUs do that in hardware for you today (e.g., out of order processing). A similar trend is clearly visible in the GPU world. Most of the techniques that were taught as *performance improvement techniques* in GPU programming five years ago (e.g., thread divergence, shared memory bank conflicts, and reduced usage of atomics) are becoming less relevant with the improved GPU architectures because GPU architects are adding hardware features that are improving these previous inefficiencies so much that it won't even matter if a programmer is sloppy about it within another 5–10 years. However, this is just a guess. What GPU architects can do depends on their (i) *transistor budget*, as well as (ii) their customers' demands. When I say *transistor budget*, I am referring to how many transistors the GPU manufacturers can cram into an Integrated Circuit (IC), aka a "chip." When I say *customer demands*, I mean that even if they can implement a feature, the applications that their customers are using might not benefit from it, which will mean a wasted transistor budget.

From the standpoint of writing a book, I took all of these facts to heart and decided that the best way to teach GPU programming is to show the differences among different families of GPUs (e.g., Fermi, Kepler, Maxwell, and Pascal) and point out the *trend*, which lets the reader be prepared about the upcoming advances in the next generation GPUs, and the next, and the next...I put a lot of emphasis on concepts that will stay relevant for a long period of time, rather than concepts that are platform-specific. That being said, GPU programming is all about performance and you can get a lot higher performance if you know the exact architecture you are running on, i.e., if you write platform-dependent code. So, providing platform-dependent explanations are as valuable as generalized GPU concepts. I engineered this book in such a way so that the later the chapter, the more platform-specific it gets.

I believe that the most unique feature of this book is the fact that it starts explaining parallelism by using CPU multi-threading in Part I. GPU massive parallelism (which differs from CPU parallelism) is introduced in Part II. Due to the way the CPU parallelism is explained in Part I, there is a smooth transition into understanding GPU parallelism in Part II. I devised this methodology within the past six years of teaching GPU programming; I realized that the concept of massive parallelism was not clear with students who have never taken a parallel programming class. Understanding the concept of "parallelizing a task" is a lot easier to understand in a CPU architecture, as compared to a GPU.

The book is organized as follows. In Part I (Chapters 1 through 5), a few simple programs are used to demonstrate the concept of dividing a large task into multiple parallel sub-tasks and map them to CPU threads. Multiple ways of parallelizing the same task are analyzed and their pros/cons are studied in terms of both core and memory operation. In Part II (Chapters 6 through 11) of the book, the same programs are parallelized on multiple Nvidia GPU platforms (Fermi, Kepler, Maxwell, and Pascal) and the same performance analysis is repeated. Because the core and memory structures of CPUs and GPUs are different, the results differ in interesting—and sometimes counterintuitive—ways; these differences are pointed out and detailed discussions are provided as to what would make a GPU program work faster. The end goal is to make the programmer aware of all of the good ideas—as well as the bad ideas—so he or she can apply the good ideas and avoid the bad ideas to his or her programs.

Although Part I and Part II totally cover what is needed to write successful CUDA programs, there is always more to know. In Part III of the book, pointers are provided for readers who want to expand their horizons. Part III is *not* meant to be a complete reference to the topics that are covered in it; rather, it is meant to provide the initial introduction, from which the reader can build a momentum toward understanding the entire topic. Included topics in this part are an introduction to some of the popular CUDA libraries, such as cuBLAS, cuFFT, Nvidia Performance Primitives, and Thrust (Chapter 12), an introduction to the OpenCL programming language (Chapter 13), and an overview of GPU programming using other programming languages and API libraries such as Python, Metal, Swift, OpenGL, OpenGL ES, OpenCV, and Microsoft HLSL (Chapter 14), and finally the deep learning library cuDNN (Chapter 15).

To download the code go to: https://www.crcpress.com/GPU-Parallel-Program-Development-Using-CUDA/Soyata/p/book/9781498750752.

Tolga Soyata

About the Author

Tolga Soyata received his BS degree from Istanbul Technical University, Department of Electronics and Communications Engineering in 1988. He came to the United States to pursue his graduate studies in 1990; he received his MS degree from Johns Hopkins University, Department of Electrical and Computer Engineering (ECE), Baltimore, MD in 1992 and PhD degree from University of Rochester, Department of ECE in 2000. Between 2000 and 2015, he owned an IT outsourcing and copier sales/service company. While operating his company, he came back to academia, joining University of Rochester (UR) ECE as a research scientist. Later he became an Assistant Processor - Research and continued serving as a research faculty member at UR ECE until 2016. During his tenure at UR ECE, he supervised three PhD students. Two of them received their PhD degrees under his supervision and one stayed at UR ECE when he joined State University of New York - Albany (SUNY Albany) as an Associate Professor of ECE in 2016. Soyata's teaching portfolio includes VLSI, circuits, and parallel programming using FPGA and GPUs. His research interests are in the field of cyber physical systems, digital health, and high-performance medical mobile-cloud computing systems.

His entry into teaching GPU programming dates back to 2009, when he contacted Nvidia to certify University of Rochester (UR) as a CUDA Teaching Center (CTC). Upon Nvidia's certification of UR as a CTC, he became the primary contact (PI). Later Nvidia also certified UR as a CUDA Research Center (CRC), with him as the PI. He served as the PI for these programs at UR until he left to join SUNY Albany in 2016. These programs were later named GPU Education Center and GPU Research Center by Nvidia. While at UR, he taught GPU programming and advanced GPU project development for five years, which were cross-listed between the ECE and CS departments. He has been teaching similar courses at SUNY Albany since he joined the department in 2016. This book is a product of the experiences he gained in the past seven years, while teaching GPU courses at two different institutions.

PART I

Understanding CPU Parallelism

Introduction to CPU Parallel Programming

THis book is a self-sufficient GPU and CUDA programming textbook. I can imagine the surprise of somebody who purchased a GPU programming book and the first chapter is named "Introduction to CPU Parallel Programming." The idea is that this book expects the readers to be sufficiently good at a low-level programming language, like C, but not in CPU parallel programming. To make this book a self-sufficient GPU programming resource for somebody that meets this criteria, any prior CPU parallel programming experience cannot be expected from the readers, yet it is not difficult to gain sufficient CPU parallel programming skills within a few weeks with an introduction such as Part I of this book.

No worries, in these few weeks of learning CPU parallel programming, no time will be wasted toward our eventual goal of learning GPU programming, since almost every concept that I introduce here in the CPU world will be applicable to the GPU world. If you are skeptical, here is one example for you: The *thread ID*, or, as we will call it *tid*, is the identifier of an executing thread in a multi-threaded program, whether it is a CPU or GPU thread. All of the CPU parallel programs we write will use the *tid* concept, which will make the programs directly transportable to the GPU environment. Don't worry if the term *thread* is not familiar to you. Half the book is about threads, as it is the backbone of how CPUs or GPUs execute multiple tasks simultaneously.

1.1 EVOLUTION OF PARALLEL PROGRAMMING

A natural question that comes to one's mind is: why even bother with parallel programming? In the 1970s, 1980s, even part of the 1990s, we were perfectly happy with *single-threaded* programming, or, as one might call it, serial programming. You wrote a program to accomplish one task. When done, it gave you an answer. Task is done ... Everybody was happy ... Although the task was done, if you were, say, doing a particle simulation that required millions, or billions of computations per second, or any other image processing computation that works on thousands of pixels, you wanted your program to work much faster, which meant that you needed a faster CPU.

Up until the year 2004, the CPU makers IBM, Intel, and AMD gave you a faster processor by making it work at a higher speed, 16 MHz, 20 MHz, 66 MHz, 100 MHz, and eventually 200, 333, 466 MHz ... It looked like they could keep increasing the CPU speeds and provide higher performance every year. But, in 2004, it was obvious that continuously increasing the CPU speeds couldn't go on forever due to technological limitations. Something else was needed to continuously deliver higher performances. The answer of the CPU makers was to put two CPUs inside one CPU, even if each CPU worked at a lower speed than a single one would. For example, two CPUs (*cores*, as they called them) working at 200 MHz could

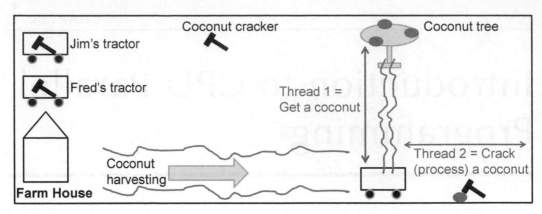

FIGURE 1.1 Harvesting each coconut requires two consecutive 30-second tasks (threads). Thread 1: get a coconut. Thread 2: crack (process) that coconut using the hammer.

do more computations per second cumulatively, as compared to a single core working at 300 MHz (i.e., $2 \times 200 > 300$, intuitively).

Even if the story of "multiple cores within a single CPU" sounded like a dream come true, it meant that the programmers would now have to learn the *parallel programming* methods to take advantage of both of these cores. If a CPU could execute two programs at the same time, this automatically implied that a programmer had to write those two programs. But, could this translate to twice the program speed? If not, then our $2 \times 200 > 300$ thinking is flawed. What if there wasn't enough work for one core? So, only truly a single core was busy, while the other one was doing nothing? Then, we are better off with a single core at 300 MHz. Numerous similar questions highlighted the biggest problem with introducing multiple cores, which is the *programming* that can allow utilizing those cores efficiently.

1.2 MORE CORES, MORE PARALLELISM

Programmers couldn't simply ignore the additional cores that the CPU makers introduced every year. By 2015, INTEL had an 8-core desktop processor, i7-5960X [11], and 10-core workstation processors such as Xeon E7-8870 [14] in the market. Obviously, this multiple-core frenzy continued and will continue in the foreseeable future. Parallel programming turned from an exotic programming model in early 2000 to the only acceptable programming model as of 2015. The story doesn't stop at desktop computers either. On the mobile processor side, iPhones and Android phones all have two or four cores. Expect to see an ever-increasing number of cores in the mobile arena in the coming years.

So, what is a *thread*? To answer this, let's take a look at the 8-core INTEL CPU i7-5960X [11] one more time. The INTEL archive says that this is indeed an 8C/16T CPU. In other words, it has 8 cores, but can execute 16 threads. You also hear parallel programming being *incorrectly* referred to as multi-core programming. The correct terminology is *multi-threaded* programming. This is because when the CPU makers started designing multi-core architectures, they quickly realized that it wasn't difficult to add the capability to execute two tasks within one core by sharing some of the core resources, such as cache memory.

ANALOGY 1.1: *Cores versus Threads.*

Figure 1.1 shows two brothers, Fred and Jim, who are farmers that own two tractors. They drive from their farmhouse to where the coconut trees are every day. They harvest the coconuts and bring them back to their farmhouse. To harvest (process) the coconuts, they use the hammer inside their tractor. The *harvesting* process requires two separate consecutive tasks, each taking 30 seconds: Task 1 go from the tractor to the tree, bringing one coconut at a time, and Task 2 crack (process) them by using the hammer, and store them in the tractor. Fred alone can process one coconut per minute, and Jim can also process one coconut per minute. Combined, they can process two coconuts per minute.

One day, Fred's tractor breaks down. He leaves the tractor with the repair shop, forgetting that the coconut cracker is inside his tractor. It is too late by the time he gets to the farmhouse. But, they still have work to do. With only Jim's tractor, and a single coconut cracker inside it, can they still process two coconuts per minute?

1.3 CORES VERSUS THREADS

Let's look at our Analogy 1.1, which is depicted in Figure 1.1. If harvesting a coconut requires the completion of two consecutive tasks (we will call them *threads*): Thread 1 picking a coconut from the tree and bringing it back to the tractor in 30 seconds, and Thread 2 cracking (i.e., *processing*) that coconut using the hammer inside the tractor within 30 seconds, then each coconut can be harvested in 60 seconds (one coconut per minute). If Jim and Fred each have their own tractors, they can simply harvest twice as many coconuts (two coconuts per minute), since during the harvesting of each coconut, they can share the road from the tractor to the tree, and they have their own hammer.

In this analogy, each tractor is a **core**, and harvesting one coconut is the **program execution** using one data element. Coconuts are **data elements**, and each person (Jim, Fred) is an **executing thread**, using the coconut cracker. Coconut cracker is the **execution unit**, like **ALU** within the core. This program consists of two dependent threads: You cannot execute Thread 2 before you execute Thread 1. The number of coconuts harvested is equivalent to the **program performance**. The higher the performance, the more money Jim and Fred make selling coconuts. The coconut tree is the **memory**, from which you get data elements (coconuts), so the process of getting a coconut during Thread 1 is analogous to **reading data elements from the memory**.

1.3.1 More Threads or More Cores to Parallelize?

Now, let's see what happens if Fred's tractor breaks down. They used to be able to harvest two coconuts per minute, but, now, they only have one tractor and only one coconut cracker. They drive to the trees and park the tractor. To harvest each coconut, they have to execute Thread 1 (Th1) and Th2 consecutively. They both get out of the tractor and walk to the trees in 30 seconds, thereby completing Th1. They bring back the coconut they picked, and now, they have to crack their coconut. However, they cannot execute Th2 simultaneously, since there is only one coconut cracker. Fred has to wait for Jim to crack the coconut, and he cracks his after Fred is done using the cracker. This takes 30+30 more seconds, and they finish harvesting two coconuts in 90 seconds total. Although not as good as two

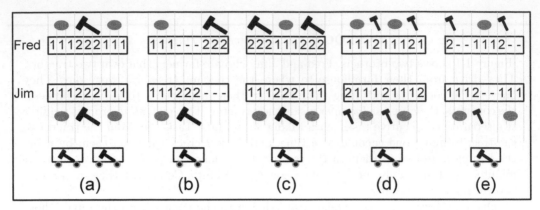

FIGURE 1.2 Simultaneously executing Thread 1 ("1") and Thread 2 ("2"). Accessing shared resources will cause a thread to wait ("-").

coconuts per minute, they still have a performance improvement from 1 to 1.5 coconuts per minute.

After harvesting a few coconuts, Jim asks himself the question: *"Why do I have to wait for Fred to crack the coconut? When he is cracking the coconut, I can immediately walk to the tree, and get the next coconut. Since Th1 and Th2 take exactly the same amount of time, we never have to be in a situation where they are waiting for the cracker to be free. Exactly the time Fred is back from picking the next coconut, I will be done cracking my coconut and we will both be 100% busy."* This genius idea brings them back to the 2 coconuts/minute speed without even needing an extra tractor. The big deal was that Jim re-engineered the **program**, which is the sequence of the threads to execute, so the threads are never caught in a situation where they are waiting for the *shared resources* inside the core, like the cracker inside the tractor. As we will see very shortly, a shared resource inside a core is an ALU, FPU, cache memory, and more ... For now, don't worry about these.

The two scenarios I described in this analogy are having two cores (2C), each executing a single thread (1T) versus having a single core (1C) that is capable of executing two threads (2T). In the CPU world, they are called 2C/2T versus 1C/2T. In other words, there are two ways to give a program the capability to execute two simultaneous threads: 2C/2T (2 cores, which are capable of executing a single thread each—just like two separate tractors for Jim and Fred) or 1C/2T (a single core, capable of executing two threads—just like a single tractor shared by Jim and Fred). Although, from the programmer's standpoint, both of them mean the ability to execute two threads, they are very different options from the hardware standpoint, and they require the programmer to be highly aware of the implications of the threads that share resources. Otherwise, the performance advantages of the extra threads could vanish. Just to remind again: our almighty INTEL i7-5960X [11] CPU is an 8C/16T, which has eight cores, each capable of executing two threads.

Three options are shown in Figure 1.2: (a) is the 2C/2T option with two separate cores. (b) is the 1C/2T option with bad programming, yielding only 1.5 coconuts per minute, and (c) is the sequence-corrected version, where the access to the cracker is never simultaneous, yielding 2 coconuts per minute.

1.3.2 Influence of Core Resource Sharing

Being so proud of his discovery which brought their speed back to 2 coconuts per minute, Jim wants to continue inventing ways to use a single tractor to do more work. One day, he goes to Fred and says "*I bought this new automatic coconut cracker which cracks a coconut in 10 seconds.*" Extremely happy with this discovery, they hit the road and park the tractor next to the trees. This time they know that they have to do some **planning**, before harvesting ...

Fred asks: "*If our Th1 takes 30 seconds, and Th2 takes 10 seconds, and the only task for which we are sharing resources is Th2 (cracker), how should we harvest the coconuts?*" The answer is very clear to them: The only thing that matters is the sequence of execution of the threads (i.e., the design of the program), so that they are never caught in a situation where they are executing Th2 together and needing the only cracker they have (i.e., shared core resources). To rephrase, their program consists of two dependent threads: Th1 is 30 seconds, and does not require shared (memory) resources, since two people can walk to the trees simultaneously, and Th2 is 10 seconds and cannot be executed simultaneously, since it requires the shared (core) resource: the cracker. Since each coconut requires a 30+10=40 seconds of total execution time, the best they can hope for is 40 seconds to harvest two coconuts, shown in Figure 1.2d. This would happen if everybody executed Th1 and Th2 sequentially, without waiting for any shared resource. So, their speed will be an average of 3 coconuts per minute (i.e., average 20 seconds per coconut).

1.3.3 Influence of Memory Resource Sharing

After harvesting 3 coconuts per minute using the new cracker, Jim and Fred come back the next day and see something terrible. The road from the tractor to the tree could only be used by one person today, since a heavy rain last night blocked half of the road. So, they plan again ... Now, they have two threads that each require resources that cannot be shared. Th1 (30 seconds — denoted as 30 s) can only be executed by one person, and Th2 (10 s) can only be executed by one person. Now what?

After contemplating multiple options, they realize that the *limiting factor* in their speed is Th1; the best they can hope for is harvesting each coconut in 30 s. When the Th1 could be executed together (shared memory access), each person could execute 10+30 s sequentially and both of them could continue without ever needing to access shared resources. But now, there is no way to sequence the threads to do so. The best they can hope for is to execute 10+30 s and wait for 20 s during which both need to access the memory. Their speed is back to 2 coconuts per minute on average, as depicted in Figure 1.2e.

This heavy rain reduced their speed back to 2 coconuts per minute. Th2 no longer matters, since somebody could easily crack a coconut while the other is on the road to pick a coconut. Fred comes up with the idea that they should bring the second (slower) hammer from the farmhouse to help. However, this would absolutely not help anything in this case, since the limiting factor to harvesting is Th1. This concept of a limiting factor by a resource is called *resource contention*. This example shows what happens when our access to memory is the limiting factor for our program speed. It simply doesn't matter how fast we can process the data (i.e., core execution speed). We will be limited by the speed at which we can *get* the data. Even if Fred had a cracker that could crack a coconut in 1 second, they would still be limited to 2 coconuts per second if there is a resource contention during memory access. In this book, we will start making a distinction between two different programs: ones that are *core intensive*, which do not depend so much on the

FIGURE 1.3 Serial (single-threaded) program imflip.c flips a 640×480 dog picture (left) horizontally (middle) or vertically (right).

memory access speed, and ones that are *memory intensive*, which are highly sensitive to the memory access speed, as I have just shown.

1.4 OUR FIRST SERIAL PROGRAM

Now that we understand parallel programming in the coconut world, it is time to apply this knowledge to real computer programming. I will start by introducing our first serial (i.e., single-threaded) program, and we will parallelize it next. Our first serial program imflip.c takes the dog picture in Figure 1.3 (left) and flips it horizontally (middle) or vertically (right). For simplicity in explaining the program, we will use Bitmap (BMP) images and will write the result in BMP format too. This is an extremely easy to understand image format and will allow us to focus on the program itself. Do not worry about the details in this chapter. They will become clear soon. For now, just focus on high-level functionality.

The imflip.c source file can be compiled and executed from a Unix prompt as follows:

```
gcc imflip.c -o imflip

./imflip dogL.bmp dogh.bmp V
```

"H" is specified at the command line to flip the image *horizontally* (Figure 1.3 middle), while "V" specifies a *vertical* flip (Figure 1.3 right). You will get an output that looks like this (numbers might be different, based on the speed of your computer):

Input BMP File name : dogL.bmp (3200×2400)

Output BMP File name : dogh.bmp (3200×2400)

Total execution time : 81.0233 ms (10.550 ns per pixel)

The CPU that this program is running on is so fast that I had to artificially expand the original 640×480 image dog.bmp to 3200×2400 dogL.bmp, so, it could run in an amount of time that can be measured; dogL.bmp is 5× bigger in each dimension, thereby making it 25× bigger than dog.bmp. To time the program, we have to record the clock at the beginning of image flipping and at the end.

1.4.1 Understanding Data Transfer Speeds

It is very important to understand that the process of reading the image from the disk (whether it is an SSD or hard drive) should be excluded from the execution time. In other words, we read the image from the disk, and make sure that it is in memory (in our array), and then measure only the amount of time that we spend flipping the image. Due to the drastic differences in the data transfer speeds of different hardware components, we need to analyze the amount of time spent in the disk, memory, and CPU separately.

In many of the parallel programs we will develop in this book, our focus is CPU time and memory access time, because we can influence them; disk access time (which we will call I/O time) typically saturates even with a single thread, thereby seeing almost no benefit from multithreaded programming. Also, make a mental note that this slow I/O speed will haunt us when we start GPU programming; since I/O is the slowest part of a computer and the data from the CPU to the GPU is transferred through the PCI express bus, which is a part of the I/O subsystem, we will have a challenge in feeding data to the GPU fast enough. But, nobody said that GPU programming was easy! To give you an idea about the magnitudes of transfer speeds of different hardware components, let me now itemize them:

- A typical network interface card (NIC) has a transfer speed of 1 Gbps (Giga-bits-per-second or billion-bits-per-second). These cards are called "Gigabit network cards" or "Gig NICs" colloquially. Note that 1 Gbps is only the amount of "raw data," which includes a significant amount of error correction coding and other synchronization signals. The amount of *actual* data that is transferred is less than half of that. Since my goal is to give the reader a *rough idea* for comparison, this detail is not that important for us.

- A typical hard disk drive (HDD) can barely reach transfer speeds of 1–2 Gbps, even if connected to a SATA3 cable that has a peak 6 Gbps transfer speed. The mechanical read-write nature of the HDDs simply do not allow them to access the data that fast. The transfer speed isn't even the worst problem with an HDD, but the *seek time* is; it takes the mechanical head of an HDD some time to locate the data on the spinning metal cylinder, therefore forcing it to wait until the rotating head reaches the position where the data resides. This could take milli-seconds (ms) if the data is distributed in an irregular fashion (i.e., fragmented). Therefore, HDDs could have transfer speeds that are far less than the peak speed of the SATA3 cable that they are connected to.

- A flash disk that is hooked up to a USB 2.0 port has a peak transfer speed of 480 Mbps (Mega-bits-per-second or million-bits-per-second). However, the USB 3.0 standard has a faster 5 Gbps transfer speed. The newer USB 3.1 can reach around 10 Gbps transfer rates. Since flash disks are built using flash memory, there is no seek time, as they are directly accessible by simply providing the data address.

- A typical solid state disk (SSD) can be read from a SATA3 cable at speeds close to 4–5 Gbps. Therefore, an SSD is really the only device that can saturate a SATA3 cable, i.e., deliver data at its intended peak rate of 6 Gbps.

- Once the data is transferred from I/O (SDD, HDD, or flash disk) into the memory of the CPU, transfer speeds are drastically higher. Core i7 family all the way up to the sixth generation (i7-6xxx) and the higher-end Xeon CPUs use DDR2, DDR3, and DDR4 memory technologies and have memory-to-CPU transfer speeds of 20–60 GBps (Giga-Bytes-per-second). Notice that this speed is **Giga-Bytes**; a Byte is 8 bits, thereby translating to memory transfer speeds of 160–480 Gbps (Giga-bits-per-second) just to compare readily to the other slower devices.

- As we will see in Part II and beyond, transfer speeds of GPU internal memory subsystems can reach 100–1000 GBps. The new Pascal series GPUs, for example, have an internal memory transfer rate, which is close to the latter number. This translates to 8000 Gbps, which is an order-of-magnitude faster than the CPU internal memory and three orders-of-magnitude faster than a flash disk, and almost four orders-of-magnitude faster than an HDD.

CODE 1.1:   imflip.c   main() {...}

The main() function of the imflip.c reads three command line parameters to determine the input and output BMP images and also the flip direction (horizontal or vertical). Operation is repeated multiple times (REPS) to improve the accuracy of the timing.

```
#define REPS   129
...
int main(int argc, char** argv)
{
    double timer;       unsigned int a;     clock_t start,stop;
    if(argc != 4){ printf("\n\nUsage: imflip [input][output][h/v]");
                   printf("\n\nExample: imflip square.bmp square_h.bmp h\n\n");
                   return 0;   }
    unsigned char** data = ReadBMP(argv[1]);
    start = clock();            // Start timing the code without the I/O part
    switch (argv[3][0]){
      case 'v' :
      case 'V' : for(a=0; a<REPS; a++) data = FlipImageV(data); break;
      case 'h' :
      case 'H' : for(a=0; a<REPS; a++) data = FlipImageH(data); break;
      default : printf("\nINVALID OPTION\n"); return 0;
    }
    stop = clock();
    timer = 1000*((double)(stop-start))/(double)CLOCKS_PER_SEC/(double)REPS;
    //merge with header and write to file
    WriteBMP(data, argv[2]);
    // free() the allocated memory for the image
    for(int i = 0; i < ip.Vpixels; i++) { free(data[i]); }
    free(data);
    printf("\n\nTotal execution time: %9.4f ms",timer);
    printf(" (%7.3f ns/pixel)\n", 1000000*timer/(double)(ip.Hpixels*ip.Vpixels));
}
```

1.4.2 The main() Function in imflip.c

Our program, shown in Code 1.1, reads a few command line parameters and flips an input image either vertically or horizontally, as specified by the command line. The command line arguments are placed by C into the argv array.

The clock() function reports time in ms intervals; this crude reporting resolution is improved by repeating the same operations an odd number of times (e.g., 129), specified in the "#define REPS 129" line. This number can be changed based on your system.

The ReadBMP() function reads the source image from the disk and WriteBMP() writes the processed (i.e., flipped) image back to the disk. The amount of time spent reading and writing the image from/to the disk is defined as *I/O time* and we will exclude it from the *processing* time. This is why I placed the "start = clock()" and "stop = clock()" lines between the actual code that flips the image that is in memory and intentionally excluded the I/O time.

Before reporting the elapsed time, the imflip.c program de-allocates all of the memory that was allocated by ReadBMP() using a bunch of free() functions to avoid memory leaks.

CODE 1.2:   imflip.c   ... FlipImageV() {...}

To flip the rows vertically, each pixel is read and replaced with the corresponding pixel of the mirroring row.

```c
#include <stdlib.h>
#include <stdio.h>
#include <time.h>
#include "ImageStuff.h"
#define REPS   129
struct ImgProp ip;

unsigned char** FlipImageV(unsigned char** img)
{
    struct Pixel pix; //temp swap pixel
    int row, col;
    for(col=0; col<ip.Hbytes; col+=3){ //go through the columns
        row = 0;
        while(row<ip.Vpixels/2){    // go through the rows
            pix.B = img[row][col];          pix.G = img[row][col+1];
            pix.R = img[row][col+2];

            img[row][col]   = img[ip.Vpixels-(row+1)][col];
            img[row][col+1] = img[ip.Vpixels-(row+1)][col+1];
            img[row][col+2] = img[ip.Vpixels-(row+1)][col+2];

            img[ip.Vpixels-(row+1)][col]   = pix.B;
            img[ip.Vpixels-(row+1)][col+1] = pix.G;
            img[ip.Vpixels-(row+1)][col+2] = pix.R;

            row++;
        }
    }
    return img;
}
```

1.4.3 Flipping Rows Vertically: FlipImageV()

The FlipImageV() in Code 1.2 goes through every column and swaps each vertical pixel with its mirroring vertical pixel for that column. The Bitmap (BMP) image functions are placed in another file named ImageStuff.c and ImageStuff.h is the associated header file. They will be explained in detail in the next chapter. Each pixel of the image is stored as type "struct Pixel," which contains the R, G, and B color components of that pixel in unsigned char type; since unsigned char takes up one byte, each image pixel requires 3 bytes to store.

The ReadBMP() places the image width and height in two variables `ip.Hpixels` and `ip.Vpixels`, respectively. The number of bytes that we need to store each row of the image is placed in `ip.Hbytes`. FlipImageV() function has two loops: The outer loop goes through all `ip.Hbytes` of the image and for every column, it swaps the corresponding vertical mirror pixels one at a time in the inner loop.

CODE 1.3:   imflip.c   FlipImageH() {...}

To flip the columns horizontally, each pixel is read and replaced with the corresponding pixel of the mirroring column.

```c
unsigned char** FlipImageH(unsigned char** img)
{
    struct Pixel pix; //temp swap pixel
    int row, col;

    //horizontal flip
    for(row=0; row<ip.Vpixels; row++){   // go through the rows
        col = 0;
        while(col<(ip.Hpixels*3)/2){        // go through the columns
            pix.B = img[row][col];
            pix.G = img[row][col+1];
            pix.R = img[row][col+2];

            img[row][col]   = img[row][ip.Hpixels*3-(col+3)];
            img[row][col+1] = img[row][ip.Hpixels*3-(col+2)];
            img[row][col+2] = img[row][ip.Hpixels*3-(col+1)];

            img[row][ip.Hpixels*3-(col+3)] = pix.B;
            img[row][ip.Hpixels*3-(col+2)] = pix.G;
            img[row][ip.Hpixels*3-(col+1)] = pix.R;

            col+=3;
        }
    }
    return img;
}
```

1.4.4 Flipping Columns Horizontally: FlipImageH()

The FlipImageH() function of imflip.c flips the image horizontally, as shown in Code 1.3. This function is identical to the vertical counterpart, except the inner loops are opposite. Each swap uses the temporary pixel variable `pix`, which is "struct Pixel" type.

Since the pixels are stored as 3 bytes in a row using the RGB, RGB, RGB, ... format, accessing consecutive pixels requires reading 3 bytes at a time. This will be detailed in the next section. For now, all we need to know is that the following lines

---

```c
pix.B = img[row][col];
pix.G = img[row][col+1];
pix.R = img[row][col+2];
```

---

are simply reading one pixel at the vertical row and horizontal column col; the blue color components of the pixel are at address $img[row][col]$, the green component is at address $img[row][col + 1]$, and the red components are at $img[row][col + 2]$. The pointer to the beginning address of the image, img, was passed onto the FlipImageH() function by main() after the ReadBMP() allocated space for it, as we will see in the next chapter.

1.5 WRITING, COMPILING, RUNNING OUR PROGRAMS

In this section, we will learn how to develop our program in one of these platforms: Windows, Mac, or a Unix box running one of the Unix clones like Fedora, CentOS, or Ubuntu. The readers are supposed to pick one of these platforms as their favorite and will be able to follow the rest of Part I using that platform, both for serial and parallel program development.

1.5.1 Choosing an Editor and a Compiler

To *develop* a program, you need to be able to *write, compile*, and *execute* that program. I am using the plain-and-simple C programming language in this book instead of C++, since this is all we need to demonstrate CPU parallelism, or GPU programming. I do not want unnecessary distracting complexity to take us away from the CPU and GPU and parallelization concepts.

To *write* a C program, the easiest thing to do is to use an editor such as Notepad++ [17]. It is a free download and it works in every platform. It also colors the keywords in the C programming language. There are, of course, more sophisticated integrated development environments (IDEs) such as Microsoft Visual Studio. But, we will favor simplicity in Part I. The results in Part I will be displayed in a Unix command line. This will work even if you have Windows 7, as you will see in a second. To *compile* a C program, we will be using the g++ compiler in Part I which also works in every platform. I will be providing a Makefile which will allow us to compile our programs with the appropriate command line arguments. To *execute* the compiled binary code, you simply have to run it within that same platform that you compiled it in.

1.5.2 Developing in Windows 7, 8, and Windows 10 Platforms

The freely available download that allows you to emulate Unix in Windows is Cygwin64 [5]. Simply put, Cygwin64 is "Unix in Windows." Be careful to get the 64-bit version of Cygwin (called Cygwin64), since it has the newest packages. Your PC must be capable of 64-bit x86. You can skip this section if you have a Mac or a Unix box. If you have a Windows PC (preferably Windows x Professional), then your best bet is to install Cygwin64, which has a built-in g++ compiler. To install Cygwin64 [5], go to http://www.cygwin.com and choose the 64-bit installation. This process takes hours if your Internet connection is slow. So, I strongly suggest that you download everything into a temporary directory, and start installation from that local directory. If you do a direct Internet install, you are highly likely to get interrupted and start over. DO NOT install the 32-bit version of Cygwin, since it is heavily outdated. None of the code in this book will work properly without Cygwin64. Also, the newest GPU programs we are running will require 64-bit execution.

In Cygwin64, you will have two different types of shells: The first type is a plain simple command line (text) shell, which is called "Cygwin64 Terminal." The second type is an "xterm" which means "X Windows terminal," and is capable of displaying graphics. For maximum compatibility across every type of computer, I will use the first kind: a strictly text terminal, "Cygwin64 Terminal," which is a Unix **bash** shell. Using a text-only shell has a few implications:

1. Since every single program we are developing operates on an image, we need a way to display our images outside the terminal. In Cygwin64, since each directory you are browsing on the Cygwin64 terminal corresponds to an actual Windows directory, all you have to do is to find that Windows directory and display the input and output images using a simple program like mspaint or an Internet Explorer browser. Both programs will allow you to resize the monster 3200×2400 image to any size you want and display it comfortably.

2. Cygwin commands ls, md, and cd are all indeed working on a Windows directory. The cygwin64-Windows directory mapping is:

   ~/Cyg64dir ⟷ **C:\cygwin64\home\Tolga\Cyg64dir**

   where Tolga is my login, and, hence, my Cygwin64 root directory name. Every cyg-win64 user's home directory will be under the same C:\cygwin64\home directory. In many cases, there will be only one user, which is your name.

3. We need to run Notepad++ outside the Cygwin64 terminal (i.e., from Windows) by drag-and-dropping the C source files inside Notepad++ to edit them. Once edited, we compile them in the Cygwin64 terminal, and display them outside the terminal.

4. There is another way to run Notepad++ and display the images in Cygwin64, without going to Windows. Type the following command lines:

   ```
   cygstart notepad++ imflip.c
   gcc imflip.c -o imflip
   ./imflip dogL.bmp dogh.bmp V
   cygstart dogh.bmp
   ```

   Command line cygstart notepad++ imflip.c is as if you double-clicked the Notepad++ icon and ran it to edit your file named imflip.c. The second line will compile the imflip.c program and the third line will run it and display the execution time, etc. The last line will run the default Windows program to display the image. This **cygstart** command in Cygwin64 is basically equivalent to "double click in Windows." The result of this last command line is as if you double-clicked on an image dogh.bmp in Windows, which would tell Windows to open a photo viewer. You can change this default viewer by changing "file associations" in Windows Explorer.

One thing might look mysterious to you: Why did I precede our program's name with ./ and didn't do the same thing for **cygstart**? Type this:

   ```
   echo $PATH
   ```

and you will not have ./ in the current PATH environment variable after an initial Cygwin64 install. Therefore, Cygwin64 won't know to search the *current directory* for any command you type. If you already have the ./ in your PATH, you do not have to worry about this. If you don't, you can add that to your PATH within your .bash_profile file, and now it will start recognizing it. This file is in your home directory and the line to add is:

   ```
   export PATH=$PATH:./
   ```

Since the **cygstart** command was in one of the paths that existed in your PATH environment variable, you didn't need to precede it with any directory name such as ./ which implies *current directory*.

1.5.3 Developing in a Mac Platform

As we discussed in Section 1.5.2, all you need to know to execute the programs in Part I and display the results is how to display a BMP image within the directory that the image ends up in. This is true for a Mac computer too. A Mac has a built-in Unix terminal, or a downloadable "iterm," so it doesn't need something like Cygwin64. In other words, a Mac is a Unix computer at heart. The only minor discrepancies you might see from what I am explaining will surface if you use an IDE like Xcode in Mac. If you use Notepad++, everything should work exactly the same way I described. However, if you see yourself developing a lot of parallel programs, Xcode is great and it is a free download when you create a developer account on Apple.com. It is worth the effort. To display images, Mac has its own programs, so just double-click on the BMP images to display them. Mac will also have a corresponding directory for each terminal directory. So, find the directory where you are developing the application, and double click on the BMP images from the desktop.

1.5.4 Developing in a Unix Platform

If you have a GUI-based Unix box running Ubuntu, Fedora, or CentOS, they will all have a command line terminal. I am using the term "box" to mean a generic or a brand-name computer with an INTEL or AMD CPU. A Unix box will have either an *xterm*, or a strictly text-based terminal, like *bash*. Either one will work to compile and run the programs described here. You can, then, figure out where the directory is where you are running the program, and double-click on the BMP images to display them. Instead of drag-and-dropping into a program, if you simply double-click on them, you are asking the OS to run the *default* program to display them. You can change this program through system settings.

1.6 CRASH COURSE ON UNIX

Almost half of my students needed a *starter* course on Unix every year in the past five years of my GPU teaching. So, I will provide it in this section. If you are comfortable with Unix, you can skip this section. Only the key concepts and commands will be provided here which should be sufficient for you to follow everything in this book. A more thorough Unix education will come with practice and possibly a book dedicated strictly to Unix.

1.6.1 Unix Directory-Related Commands

Unix directory structure starts with your home directory, which has a special tilde character (~) to represent it. Any directory that *you* create is created somewhere under *your* "~/" home directory. For example, if you create a directory called **cuda** in your home directory, the Unix way to represent this directory is ~/cuda using the tilde notation. You should organize your files to be in a neat order, and create directories under directories to make them hierarchical. For example, the examples we have in this book could be placed under a **cuda** directory and each chapter's examples could be placed under sub-directories such as ch1, ch2, ch3, ... They would have directory names ~/cuda/ch1 etc.

Some commonly used directory creation/removal commands in Unix are:

```
ls               # show me the listing of the root directory
mkdir cuda       # make a directory named 'cuda' right here
cd cuda          # go into (change directory to) 'cuda'
ls               # list the contents of this directory
mkdir ch1        # make a sub-directory named 'ch1'
mkdir ch2        # make another sub-directory named 'ch2'
mkdir ch33       # make a third. Oops, I mis-typed. I meant 'ch3'.
rmdir ch33       # Too late. Let's remove the wrong directory 'ch33'
mkdir ch3        # Now, make the correct third sub-directory 'ch3'
mkdir ch4        # Oops, I meant to create a directory named ch5, not ch4
mv ch4 ch5       # move the directory to ch5 (effectively renames ch4)
ls -al           # List detailed contents with the -al switch
ls ..            # List contents of the above directory
pwd              # print-working-directory. Where am I in the hierarchy ?
cd ..            # Two special directories: . is 'here' .. is 'one above'
pwd              # Where am I again ? Did I go 'one above' with cd .. ?
ls -al           # detailed listing again. I should be at 'cuda'
cd               # go to my home directory
rm -r dirname    # removes a dir. and all of its subdirectories even if not empty
cat /proc/cpuinfo # get information about the CPU you have in your computer
```

The ls -al command enables you to see the sizes and permissions of a directory and the files contained in it (i.e., detailed listing) no matter which directory you are in. You will also see two directories Unix created for you automatically with special names. (meaning *this*) and .. (meaning *upper*) directory in relationship to where you are. So, for example, the command ./imflip ... is telling Unix to run imflip from *this* directory.

Using the pwd command to find out where you are, you will get a directory that doesn't start with a tilde, but rather looks something like /home/Tolga/cuda Why? because pwd reports where you are in relationship to the Unix root, which is / rather than your home directory /home/Tolga/ or the ~/ short notation. While the cd command will take you to your home directory, cd / command will take you to the Unix root, where you will see the directory named **home**. You can drill down into home/Tolga with commands cd home/Tolga and end up at your home directory, but clearly the short notation cd is much more convenient.

rmdir command removes a directory as long as it is empty. However, if it has files or other directories in it (i.e., subdirectories), you will get an error indicating that the directory is not empty and cannot be removed. If you want to remove a directory that has files in it, use the file deletion command rm with the switch "-r" that implies "recursive." What rm -r dirname means is: *remove every file from the directory* **dirname** *along with all of its subdirectories.* There is possibly no need to emphasize how dangerous this command is. Once you issue this command, the directory is *gone*, so are the entire contents inside it, not to mention all of the subdirectories. So, use this command with extreme caution.

mv command works for files and also directories. For example, mv dir1 dir2 "moves" a directory **dir1** into **dir2**. This effectively *renames* the directory **dir1** as **dir2** and the old directory **dir1** is gone. When you ls, you will only see the new directory **dir2**.

1.6.2 Unix File-Related Commands

Once the directories (aka *folders*) are created, you will need to create/delete files in them. These files are your programs, the input files that your program needs, and the files that your program generates. For example, to run the imflip.c serial program we saw in Section 1.4, which flips a picture, you need the program itself, you need to compile it, and when this program generates an output BMP picture, you need to be able to see that picture. You also need to bring (copy) the picture into this directory. There is also a Makefile that I created for you which helps the compilation. Here are some common Unix commands for file manipulation:

```
clear                 # clear the screen
ls                    # let's see the files. dogL.bmp is the dog picture
cat Makefile          # see the contents of the text file named Makefile
more Makefile         # great for displaying contents of a multi-page file
cat > mytest.c        # the quickest way to create a file mytest.c. End with ^D
make imflip           # run the entry in the Makefile to compile imflip.c
ls -al                # let's see the files sizes, etc. to investigate
ls -al imflip         # Show me the details of the executable file imflip
cp imflip if1         # make a copy of this executable file, named if1
man cp                # display a manual for the unix command ''cp''
imflip                # without command line arguments, we get a warning
imflip dogL.bmp dogH.bmp h # run imflip with correct parameters
cat Makefile | grep imflip # look for text ''imflip'' inside the Makefile
ls -al | grep imflip      # pipe the listing to grep to search ''imflip''
ls imf*               # list everyfile whose file names start with ''imf''
rm imf*.exe           # remove every file starting with imf and ending with .exe
diff f1 f2            # compare files f1 and f2. Display differences
touch imflip          # set the last-access date of imflip to ''now''
rm imflip             # or imflip.exe in Windows. This removes a file.
mv f1 f2              # move a file from an old name f1 to a new one f2.
mv f1 ../f2           # move file f1 to the above directory and rename as f2.
mv f1 ../             # move file f1 to the above directory. Keep the same name.
mv ../f1 f2           # move file f1 from the above dir. to here. rename as f2
history               # show the history of my commands
```

- # (hash) is the comment symbol and anything after it is ignored.

- clear command erases the clutter on your terminal screen.

- cat Makefile displays the contents of Makefile from the command line, without having to use another external program like Notepad++.

- more Makefile displays the contents of Makefile and also allows you to scroll through the pages one by one. This is great for multipage files.

- cat > filename is the fastest way to create a text file named filename. This makes Unix go into text-entry mode. Text-entry mode takes everything you are typing and sends it to the file that comes after the > you type (e.g., mytest.c). End the text-input mode by typing CTRL-D (the CTRL and D keys together, which is the EOT character, ASCII code 4, indicating *end of transmission*). This method for text entry is great if you do not even want to wait to use an editor like Notepad++. It is perfect for programs that are just a few lines, although nothing prevents you from typing an entire program using this method!

- | is the "pipe" command that channels (i.e., pipes) the output of one Unix command or a program into another. This allows the user to run two separate commands using only a single command line. The second command accepts the output of the first one as its input. Piping can be done more than once, but it is uncommon.

- cat Makefile | grep imflip pipes the output of the cat command into another command grep that looks and lists the lines containing the keyword imflip. grep is excellent for searching some text strings inside text files. The output of any Unix command could be piped into grep.

- ls -al | grep imflip pipes the output of the ls command into the grep imflip. This is effectively looking for the string imflip in the directory listing. This is very useful in determining file names that contain a certain string.

- make imflip finds   imflip : file1 file2 file3 ...   inside Makefile and *remakes* imflip if a file has been modified in the list.

- cp imflip if1 copies the executable file imflip that you just created under a different filename if1, so you do not lose it.

- man cp displays a help file for the cp command. This is great to display detailed information about any Unix command.

- ls -al can be used to show the permissions and file sizes of source files and input/output files. For example, it is perfect to check whether the sizes of the input and output BMP files dogL.bmp and dogH.bmp are identical. If they are not, this is an early indication of a bug!

- ls imf* lists every file whose names start with "imf." This is great for listing files that you know contain this "imf" prefix, like the ones we are creating in this book named imflip, imflipP, ... Start (*) is a wildcard that means "anything." Of course, you can get fancier with the * like : ls imf*12 that means *files starting with "imf" and ending with "12."* Another example is ls imf*12* that means *files starting with "imf" and having "12" in the middle of the file name.*

- diff file1 file2 displays the differences between two text files. This is great to determine if a file has changed. It can also be used for binary files.

- imflip or imflip dog... runs the program if ./ is in your $PATH. Otherwise, you have to use ./imflip dog...

- touch imflip updates the "last access time" of the file imflip.

- rm imflip deletes the imflip executable file.

- mv command, just like renaming directories, can also be used to rename files as well as truly *move* them. mv file1 file2 renames file1 as file2 and keeps it in the same directory. If you want to move files from one directory to another, precede the filename with the directory name and it will move to that directory. You can also move a file without renaming them. Most Unix commands allow such versatility. For example, the cp command can be used exactly the way mv is used to copy files from one directory to another.

- history lists the commands you used since you opened the terminal.

The Unix commands to compile our first serial program imflip.c and turn it into the executable imflip (or, imflip.exe in Windows) will produce an output that looks something like the listing below. Only the important commands that the user entered are shown on the left and the Unix output is shown on a right-indentation:

```
ls
        ImageStuff.c ImageStuff.h Makefile dogL.bmp imflip.c
cat Makefile
        imflip  :  imflip.c ImageStuff.c ImageStuff.h
                   g++ imflip.c ImageStuff.c -o imflip
make imflip
ls
        ImageStuff.c ImageStuff.h Makefile dogL.bmp imflip.c imflip
imflip
        Usage : imflip [input][output][v/h]
imflip dogL.bmp dogH.bmp h
          Input BMP File Name : dogL.bmp (3200x2400)
          Output BMP File Name : dogH.bmp (3200x2400)

          Total Execution time : 83.0775 ms (10.817 ns/pixel)
ls -al
          ...
          -rwxr-x   1     Tolga     23020054     Jul 18   15:01   dogL.bmp
          -rwxr-x   1     Tolga     23020054     Jul 18   15:08   dogH.bmp
          ...
rm imflip
history
```

In this listing, each file's *permissions* are shown as -rwxr-x, etc. Your output might be slightly different depending on the computer or organization you are running these commands at. The Unix command chmod is used to change these permissions to make them read-only, etc.

The Unix make tool allows us to automate routinely executed commands and makes it easy to compile a file, among other tasks. In our case, "make imflip" asks Unix to look inside the Makefile and execute the line "gcc imflip.c ImageStuff.c -o imflip" which will invoke the gcc compiler and will compile imflip.c and ImageStuff.c source files and will produce an executable file named imflip. In our Makefile, the first line is showing *file dependencies*: It instructs make to make the executable file imflip only if one of the listed source files, imflip.c, ImageStuff.c, or ImageStuff.h have changed. To force a compile, you can use the touch Unix command.

1.7 DEBUGGING YOUR PROGRAMS

Debugging code is something that you will have to do eventually. At some point, you will write code that is supposed to work but instead throws a segmentation fault or some other random error that you have never seen before. This process can be extremely frustrating and oftentimes is a result of a simple typo or logic error that is nearly impossible to find. Other bugs in your code may not even show up during runtime and you won't see the effects of it at first glance. These are the worst kinds of bugs because the compiler does not find them and they are not obvious during runtime. Some errors, like memory leaks, will not show up at runtime. A good practice during your code development is to regularly run debugger tools like gdb and valgrind to potentially determine where your segmentation fault happened. To run your codes in a debugger, you need to compile it with a

```
$ gdb imflip
GNU gdb (Ubuntu 7.7.1-0ubuntu5~14.04.2) 7.7.1
Copyright (C) 2014 Free Software Foundation, Inc.
License GPLv3+: GNU GPL version 3 or later <http://gnu.org/licenses/gpl.html>
This is free software: you are free to change and redistribute it.
There is NO WARRANTY, to the extent permitted by law.  Type "show copying"
and "show warranty" for details.
This GDB was configured as "x86_64-linux-gnu".
Type "show configuration" for configuration details.
For bug reporting instructions, please see:
<http://www.gnu.org/software/gdb/bugs/>.
Find the GDB manual and other documentation resources online at:
<http://www.gnu.org/software/gdb/documentation/>.
For help, type "help".
Type "apropos word" to search for commands related to "word"...
Reading symbols from imflip...done.
(gdb) set args dogL.bmp flipped.bmp V
(gdb) run
Starting program: imflip dogL.bmp flipped.bmp V


Program received signal SIGSEGV, Segmentation fault.
0x0000000000401172 in WriteBMP (img=0x603250, filename=0x7fffffffe865 "flipped.bmp")
at ImageStuff.c:73
73                              temp=img[x][y];
(gdb) where
#0  0x0000000000401172 in WriteBMP (img=0x603250, filename=0x7fffffffe865
"flipped.bmp") at ImageStuff.c:73
#1  0x0000000000400e5d in main (argc=4, argv=0x7fffffffe5f8) at imflip.c:98
```

FIGURE 1.4 Running gdb to catch a segmentation fault.

debug flag, typically "-g". This tells the compiler to include debug symbols, which includes things like line numbers, to tell you where your code went wrong. An example is shown below:

$ gcc imflip.c imageStuff.c -o imflip -g

1.7.1 gdb

For the sake of showing what happens when you mess up your code, I've inserted a memory free() into imflip.c before the code is done using the data. This will knowingly cause a segmentation error in the code as shown below:

$ gcc imflip.c imageStuff.c -o imflip -g

$ ./imflip dogL.bmp flipped.bmp V

Segmentation fault (core dumped)

Since imflip was compiled with debug symbols, gdb, the GNU debugger, can be run to try to figure out where the segmentation fault is happening. The output from gdb is given in Figure 1.4. gdb is first called by running

$ gdb ./imflip

Once gdb is running, the program arguments are set by the command:

set args dogL.bmp flipped.bmp V

TABLE 1.1   A list of common gdb commands and functionality.

Task	Command	Example Use
Start GDB	gdb	gdb ./imflip
Set the program arguments (once in gdb)	set args	set args input.bmp output.bmp H
Run the debugger	run	run
List commands	help	help
Add a breakpoint at a line	break	break 13
Break at a function	break	break FlipImageV
Display where error happened	where	where

After this, the program is run using the simple run command. gdb then proceeds to spit out a bunch of errors saying that your code is all messed up. The where command helps give a little more information as to where the code went wrong. Initially, gdb thought the error was in the WriteBMP() function within ImageStuff.c at line 73, but the where command narrows it down to line 98 in imflip.c. Further inspection in imflip.c code reveals that a free(data) command was called before writing data to a BMP image with the WriteBMP() function. This is just a simple example, but gdb can be expanded to use break points, watching specific variables, and a host of other options. A sample of common commands is listed in Table 1.1.

Most integrated development environments (IDEs) have a built-in debugging module that makes debugging quite easy to use. Typically the back-end is still gdb or some proprietary debugging engine. Regardless of whether you have an IDE to use, gdb is still available from the command line and contains just as much, if not more functionality compared to your IDE (depending on your IDE of choice).

1.7.2   Old School Debugging

This is possibly the correct term for the type of debugging that programmers used in the old days —1940s, 1950s, 1960s, 1970s— and continue to use to date. I see no reason why old school debugging should go away in the foreseeable future. After all, a "real" debugger that we use to debug our code —such as gdb— is nothing more than an automated implementation of old school debugging. I will provide a much more detailed description of the old school debugging concepts within the context of GPU programming in Section 7.9. All of what I will describe in Section 7.9 is applicable to the CPU world. So, you can either hang tight (continue reading this chapter) or peek at Section 7.9 now.

In every debugger the idea is to insert *breakpoints* into the code, in an attempt to print/display some sort of values related to the system status at that point. This status could be variable values or the status of a peripheral, you name it. The execution can be either halted at the breakpoint or continue, while printing multiple statuses along the way.

Lights: In the very early days when the Machine Code programmers were writing their programs bit-by-bit by flicking the associated switches, the breakpoints were possibly a single light to display a single bit's value; today, FPGA programmers use 8 LEDs to display the value of an 8-bit Verilog variable (Note: Verilog is a hardware description language). However, this requires an incredibly experienced programmer to infer the system status from just a few bits.

printf: In a regular C program, it is extremely common for the programmers to insert a bunch of printf() commands to display the value of one or more variables at certain locations of the code. These are nothing more than manual breakpoints. If you feel that the bug in your code is fairly easy to discover, there is no reason to go through the lengthy gdb process, as I described in Section 1.7.1. Stick a bunch of printf()'s inside the code and they will tell you what is going on. A printf() can display a lot of information about multiple variables, clearly being a much more powerful tool than the few lights.

assert: An `assert` statement does not do anything unless a condition — that you specified — is violated as opposed to printf(), which *always* prints something. For example, if your code had the following lines:

```
ImgPtr=malloc(...);
assert(ImgPtr != NULL);
```

In this case, you are simply trying to make sure that the pointer that is given to us is not NULL, which red flags a huge problem with memory allocation. While assert() would do nothing under normal circumstances, it would issue an error like the one shown below if the condition is violated:

Assertion violation: file mycode.c, line 36:   ImgPtr != NULL

Comment lines: Surprisingly enough, there is something easier than sticking a bunch of printf()'s in your code. Although C doesn't care about the "lines," it is fairly common that the C programmers write their code pretty similarly to a line-by-line fashion, much like Python. This is why Python received some criticism for making the line-by-line style the actual syntax of the language, rather than an option as in C. In the commenting-driven debugging, you simply comment out a line that you are suspicious of, recompile, re-execute to see if the problem went away, although the result is definitely no longer correct. This is perfect in situations where you are getting core dumps, etc. In the example below, your program will give you a Divide By 0 error if the user enters 0 for `speed`. You insert the printf() there to give you an idea about where it might be crashing, but an assert() is much better, because assert() would do nothing under normal circumstances avoiding the clutter on your screen during debugging.

```
scanf(&speed);
  printf("DEBUG: user entered speed=%d\n",speed);
  assert(speed != 0);
distance=100;           time=distance/speed;
```

Comments are super practical, because you can insert them in the middle of the code in case there are multiple C statements in your code as shown below:

```
scanf(&speed);
distance=100;        // time=distance/speed;
```

1.7.3 valgrind

Another tool to debug that can be extremely useful is a framework called valgrind. Once the code has been compiled with debug symbols on, valgrind is simple to run. It has a host of options to run with, similar to GDB, but the basic usage is quite easy. The output from valgrind on the same imflip code with a memory error is shown in Figure 1.5. It catches quite a few more errors and even locates the proper line of the error on line 96 in imflip.c where the improper free() command is located.

```
$ valgrind ./imflip dogL.bmp flipped.bmp V
==29048== Memcheck, a memory error detector
==29048== Copyright (C) 2002-2013, and GNU GPL'd, by Julian Seward et al.
==29048== Using Valgrind-3.10.0.SVN and LibVEX; rerun with -h for copyright info
==29048== Command: ./imflip dogL.bmp flipped.bmp V
==29048==

==29048== Invalid read of size 8
==29048==    at 0x401168: WriteBMP (ImageStuff.c:73)
==29048==    by 0x400E5C: main (imflip.c:98)
==29048==  Address 0x51fc2c0 is 0 bytes inside a block of size 19,200 free'd
==29048==    at 0x4C2BDEC: free (in /usr/lib/valgrind/vgpreload_memcheck-amd64-
linux.so)
==29048==    by 0x400E42: main (imflip.c:96)
==29048==
==29048==
==29048== More than 10000000 total errors detected.  I'm not reporting any more.
==29048== Final error counts will be inaccurate.  Go fix your program!
==29048== Rerun with --error-limit=no to disable this cutoff.  Note
==29048== that errors may occur in your program without prior warning from
==29048== Valgrind, because errors are no longer being displayed.
==29048==
```

FIGURE 1.5 Running valgrind to catch a memory access error.

valgrind also excels at finding memory errors that won't show up during run time. Typically memory leaks are harder to find with simple print statements or a debugger like gdb. For example, if imflip did not free any of the memory at the end, a memory leak would be present, and valgrind would pick up on this. valgrind also has a module called cachegrind that helps simulate how your code interacts with the CPU's cache memory system. cachegrind is called with the –tool=cachegrind command line option. Further options and documentation can be found at http://valgrind.org.

1.8 PERFORMANCE OF OUR FIRST SERIAL PROGRAM

Let's get a good grasp of the performance of our first serial program, imflip.c. Since many random events are always executing inside your operating system (OS), it is a good idea to run the same program many times to make sure that we are getting consistent results. So, let's run the imflip.c a few times through the command prompt. When we do that, we get results like 81.022 ms, 82.7132 ms, 81.9845 ms, ... We can call it roughly 82 ms. This is good enough. This corresponds to 10.724 ns/pixel, since there are 3200×2400 pixels in this dogL.bmp expanded dog image.

To be able to measure the performance of the program a little bit more accurately, I put a repetition for() loop in the code that executes the same code many times (e.g., 129) and divides the execution time by that same amount, 129. This will allow us to take 129 times longer than usual, thereby allowing an inaccurate UNIX system timer to be used to achieve much more accurate timing information. Most machines cannot provide hardware clock at better accuracy than 10 ms, or even worse. If a program is taking only, say, 50 ms to execute, you will get a highly inaccurate performance reading (with a 10-ms-accuracy clock) even if you repeat the measurement many times as described above. However, if you repeat the same program 129 times and measure the time at the very beginning and at the very end of 129 repetitions, and divide it by 129, your 10 ms becomes effectively a 10/129 ms accuracy, which is sufficient for our purposes. Notice that this must be an odd number, otherwise, the resulting dog picture won't be flipped!

1.8.1 Can We Estimate the Execution Time?

With this modification, the results in the 81–82 ms, shown above, were obtained for the imflip.c program that is horizontally flipping the dog picture. Since we are curious about what happens when we run the same program on the smaller dog.bmp dog picture, which is only a 901 KB bitmap image, let's run exactly the same program, except on the original 640×480 dog.bmp file. We get a run time of 3.2636 ms which is 10.624 ns/pixel. In other words, when the picture shrunk $25\times$, the run time almost perfectly reduced by that much. One weird thing though: we get *exactly* the same execution time every time. Although, this shows that we were able to calculate the execution time with an incredible accuracy, please hold off on celebrating this discovery, since we will encounter such complexities that it will rock our world!

Indeed, first weirdness showed up its face already. Can you answer this question: Although we are getting almost identical (per pixel) execution times, why is the execution time not changing for the smaller image? Identical up to the 4 decimal digits, whereas, the execution time for the bigger dog image is changing within 1–2%. Although this might look like a statistical anomaly, it is not! It has a very clear explanation. Just to prevent you from losing sleep, I will give you the answer and will not make you wait until the next chapter: When we processed the 22 MB image, dogL.bmp, as opposed to the original 901 KB version, dog.bmp, what changed? The answer is that during the processing of dogL.bmp, the CPU cannot keep the entire image in its last level L3 cache memory (L3$), which is 8 MB. This means that, to access the image, it was continuously emptying and refilling its L3$ during execution. Alternatively, when working on the 901 KB little dog image dogL.bmp, all it takes is one turn of processing to completely soak the data into the L3$, and the CPU owns that data during all 129 loops of execution. Note that I will be using the notation L3$, which is pronounced *"el-three-cash"* to denote L3$.

1.8.2 What Does the OS Do When Our Code Is Executing?

The reason for the higher amount of variability in the execution time for the big dog image is that there is a lot more uncertainty in accessing the memory than the inside of a core. Since imflip.c is a serial program, we really need a "1T" to execute our program. On the other hand, our CPU has luxurious resources such as 4C/8T. This means that once we start running our program which consists of an extremely active thread, the OS almost instantly realizes that giving us one completely dedicated CPU thread (or even a core) is in everybody's best interest, so our application can fully utilize this resource. All said and done, this is the OS's job: to allocate resources intelligently. Whether it is Windows or a Unix clone, all of today's OS code is very smart in understanding these patterns in program execution. If a program is passionately asking for a single thread, but nothing more, unless you are running many other programs, the best course of action for the OS is to give you a nearly VIP access to a single thread (even, possibly a complete core).

However, the story is completely different for the main memory, which is accessed by every active thread in the OS. Think of it as 1M! There is no 2M! So, the OS must share it among every thread. Memory is where all OS data is, where every thread's data is, and main memory is where all of your coconuts (oops, sorry, I meant, your image data elements) are. So, not only the OS has to figure out how your imflip.c can access the image data from the main memory, but even *itself*. Another important job of the OS is to ensure *fairness*. If a thread is starved of data, and another is feasting, the OS is not doing its job. It has to be fair to everyone, including itself. When you have so much more moving parts in *main memory access*, you can see why there is a lot more uncertainty in determining what the main memory access time should be. Alternatively, if we have almost a completely dedicated core when processing the small image, we are running the entire program inside the core itself, without needing to go to the main memory. And, we are not sharing that core with anyone. So, there is nearly zero uncertainty in determining the execution time. If these concepts are slightly blurry, do not worry about them, there will be an entire chapter dedicated to the CPU architecture. Here is the meaning of the C/T (cores/threads) notation:

> ➢ *The C/T (cores/threads) notation denotes:*
>   *e.g., 4C/8T means 4 cores, 8 threads,*
>   *4C means that the processor has 4 cores,*
>   *8T means that each core houses 2 threads,*
>
> ➢   *So, the 4C/8T processor can execute 8 threads simultaneously,*
>   *However each thread pair has to share internal core resources,*

1.8.3 How Do We Parallelize It?

There is so much detail to be aware of even when running the serial version of our code that instead of shoving the parallel version of the code into this tiny section, I would rather make it an entire chapter of its own. Indeed, the next few chapters will be completely dedicated to the parallel version of the code and a deep analysis of its performance implications. For now, let's get warmed up to the parallel version by thinking in terms of coconuts! Answer these questions:

In Analogy 1.1, what would happen if there were two tractors with two farmers in each one of the tractors? In this case, you have four threads executing ... Because there are two — physically separate — tractors, everything is comfortable inside the tractors (i.e., cores). However, now, instead of two, four people will be sharing the road from the tractors to the coconuts (i.e., multiple threads accessing the main memory). Keep going ... What if 8 farmers went to coconut harvesting? How about 16? In other words, even in the 8C/16T scenario, where you have 8 cores and 16 threads, you have 8 tractors to satisfy every farmer and two of them can share the coconut crackers, etc. But, what about the main memory access? The more farmers you have harvesting, the more they will start waiting for that road to be available to *get* the coconuts. In CPU terms, the *memory bandwidth* will sooner or later saturate. Indeed, even in the next chapter, I will show you a program where this happens. This means that before we move onto parallelizing our program, we have to do some thinking in terms of which resources our threads will access during execution.

1.8.4 Thinking About the Resources

Even if you had the answers to the questions above there is yet another question: Would the magic of parallelism work the same way on every possible resource scenario? In other words, is the concept of *parallelism* separate from the *resources* it is applied to? To exemplify, would the 2C/4T core scenario always give us the same performance improvement *regardless of the memory bandwidth* or, if the memory bandwidth is really bad, the added performance gain from the extra core would disappear? For now, just think about them, but the entire Part I of the book will be spent on answering these questions. So, don't stress over them at this point.

Ok, this is good enough brain warm up ... Let's write our first parallel program.

## CHAPTER 2

Developing Our First Parallel CPU Program

THIS chapter is dedicated to understanding our very first CPU parallel program, imflipP.c. Notice the 'P' at the end of the file name that indicates *parallel*. For the CPU parallel programs, the development platform makes no difference. In this chapter, I will slowly start introducing the concepts that are the backbone of a parallel program, and these concepts will be readily applicable to GPU programming when we start developing GPU programs in Part II. As you might have noticed, I never say *GPU Parallel Programming*, but, rather, *GPU Programming*. This is much like there is no reason to say *a car with wheels*; it suffices to say *a car*. In other words, there is no GPU serial programming, which would mean using one GPU thread out of the available 100,000s! So, GPU programming by definition implies GPU parallel programming.

2.1 OUR FIRST PARALLEL PROGRAM

It is time to write our first parallel program imflipP.c, which is the parallel version of our serial imflip.c program, introduced in Chapter 1.4. To parallelize imflip.c, we will simply have the main() function create multiple threads and let them do a portion of the work and exit. If, for example, in the simplest case, we are trying to run the two-threaded version of our program, the main() will create two threads, let them each do half the work, join the threads, and exit. In this scenario, main() is nothing but the *organizer* of events. It is not doing actual heavy-lifting.

To do what we just described, main() needs to be able to create, terminate, and organize threads and assign tasks to threads. The functions that allow it to perform such tasks are a part of the *Pthreads* library. *Pthreads* only work in a POSIX-compliant operating system. Ironically, Windows is not POSIX compliant! However, the Cygwin64 allows Pthreads code to run in Windows by performing some sort of API-by-API translation between POSIX and Windows. This is why everything we describe here will work in Windows, and hence the reason for me to use Cygwin64 in case you have a Windows PC. Here are a few functions that we will use from the Pthreads library:

1. pthread_create() allows you to create a thread.

2. pthread_join() allows you to join any given thread into the thread that originally created it. Think of the "join" process as "uncreating" threads, or like the top thread "swallowing" the thread if just created.

3. pthread_attr() allows you to initialize attributes for threads.

4. pthread_attr_setdetachstate() allows you to set attributes for the threads you just initialized.

2.1.1 The main() Function in imflipP.c

Our serial program imflip.c, shown in Code 1.1, read a few command line parameters and flipped an input image either vertically or horizontally, as specified by the user's command line entry. The same flip operation was repeated an odd number of times (e.g., 129) to improve the accuracy of the system time read by clock().

Code 2.1 and Code 2.2 show the same main() function in imflipP.c with one exception: Code 2.1 shows main() {..., which means the "first part" of main(), which is further emphasized by the ... at the end of this listing. This part is for command line parsing and other routine work. In Code 2.2, an opposite main() ...} notation is used along with the ... in the beginning of the listing, indicating the "second part" of the main() function, which is for launching threads and assigning tasks to threads.

To improve readability, I might repeat some of the code in both parts, such as the time-stamping with gettimeofday() and the image reading with our own function ReadBMP() that will be detailed very soon. This will allow the readers to clearly follow the beginning and connection points within the two separate parts. As you might have noticed already, whenever a function is entirely listed, the "func() {...}" notation is used. When a function and some surrounding code is listed, "... func() {...}" notation will be used, denoting "some common code ... followed by a complete listing of func()."

Here is the part of main() which parses command arguments, given in the argv[] array (a total of argc of them). It issues errors if the user enters an unaccepted number of arguments. It saves the flip direction that the user requested in a variable called Flip for further use. The global variable NumThreads is also determined based on user input and is used later in the functions that actually perform the flip.

```c
int main(int argc, char** argv)
{
    ...
    switch (argc){
        case 3: NumThreads=1;          Flip='V';                    break;
        case 4: NumThreads=1;          Flip=toupper(argv[3][0]); break;
        case 5: NumThreads=atoi(argv[4]); Flip=toupper(argv[3][0]); break;
        default:printf("Usage: imflipP input output [v/h] [threads]");
            printf("Example: imflipP infile.bmp out.bmp h 8\n\n");
            return 0;
    }
    if((Flip != 'V') && (Flip != 'H')) {
        printf("Invalid option '%c' ... Exiting...\n",Flip);
        exit(EXIT_FAILURE);
    }
    if((NumThreads<1) || (NumThreads>MAXTHREADS)){
        printf("Threads must be in [1..%u]... Exiting...\n",MAXTHREADS);
        exit(EXIT_FAILURE);
    }else{
    ...
```

2.1.2 Timing the Execution

When we have more than one thread executing, we want to be able to quantify the speed-up. We used the clock() function in our serial code, which was included in the time.h header file and was only millisecond-accurate.

The gettimeofday() function we will use in imflipP.c will get us down to μs-accuracy. gettimeofday() requires us to #include the sys/time.h header file and provides the time in two parts of a struct: one for seconds through the .tv_sec member and one for the micro-seconds through the .tv_usec member. Both of these members are int types and are combined to produce a double time value before being displayed.

An important note here is that the accuracy of the timing does not depend on the C function itself, but, rather, the *hardware*. If your computer's OS or hardware cannot provide a μs-accurate timestamp, gettimeofday() will provide only as accurate of a result as it can obtain from the OS (which itself gets its value from a hardware clock unit). For example, Cygwin64 does not achieve μs-accuracy even with the gettimeofday() function due to its reliance on the underlying Windows APIs.

```
#include <sys/time.h>
...
struct timeval      t;
double              StartTime, EndTime;
double              TimeElapsed;
    ...
    gettimeofday(&t, NULL);
    StartTime = (double)t.tv_sec*1000000.0 + ((double)t.tv_usec);
    // work is done here : thread creation, task/data assignment, join
    ...
    gettimeofday(&t, NULL);
    EndTime = (double)t.tv_sec*1000000.0 + ((double)t.tv_usec);
    TimeElapsed=(EndTime-StartTime)/1000.00;
    TimeElapsed/=(double)REPS;
    ...
    printf("\n\nTotal execution time: %9.4f ms ...",TimeElapsed,...
```

2.1.3 Split Code Listing for main() in imflipP.c

I am intentionally staying away from providing one long listing for the main() function in a single code fragment. This is because, as you can see from this first example, Code 2.1 and Code 2.2 provide listings for entirely different functionality: Code 2.1 provides a listing for flat-out "boring" functionality for getting command line arguments, parsing them, and warning the user. On the other hand, Code 2.2 is where the "cool action" of creating and joining threads happens. Most of the time, I will arrange my code to allow such partitioning and try to put a lot more emphasis on the important part of the code.

CODE 2.1: imflipP.c   ... main() {...

The first part of the main() function of the imflipP.c reads and parses command line options. Issues errors if needed. The BMP image is read into a memory array and the timer is started. This part determines whether the multithreaded code will even run.

```c
#define MAXTHREADS 128
...
int main(int argc, char** argv)
{
    char          Flip;
    int           a,i,ThErr;
    struct timeval t;
    double        StartTime, EndTime;
    double        TimeElapsed;

    switch (argc){
      case 3: NumThreads=1;        Flip='V';                break;
      case 4: NumThreads=1;        Flip=toupper(argv[3][0]); break;
      case 5: NumThreads=atoi(argv[4]); Flip=toupper(argv[3][0]); break;
      default:printf("Usage: imflipP input output [v/h] [threads]");
              printf("Example: imflipP infile.bmp out.bmp h 8\n\n");
              return 0;
    }
    if((Flip != 'V') && (Flip != 'H')) {
      printf("Invalid option '%c' ... Exiting...\n",Flip);
      exit(EXIT_FAILURE);
    }
    if((NumThreads<1) || (NumThreads>MAXTHREADS)){
      printf("Threads must be in [1..%u]... Exiting...\n",MAXTHREADS);
      exit(EXIT_FAILURE);
    }else{
      if(NumThreads != 1){
        printf("\nExecuting %u threads...\n",NumThreads);
        MTFlipFunc = (Flip=='V') ? MTFlipV:MTFlipH;
      }else{
        printf("\nExecuting the serial version ...\n");
        FlipFunc = (Flip=='V') ? FlipImageV:FlipImageH;
      }
    }
    TheImage = ReadBMP(argv[1]);

    gettimeofday(&t, NULL);
    StartTime = (double)t.tv_sec*1000000.0 + ((double)t.tv_usec);
    ...
}
```

CODE 2.2:   imflipP.c   ... main() ...}

The second part of the main() function in imflipP.c creates multiple threads and assigns tasks to them. Each thread executes its assigned task and returns. When every thread is done, main() joins (i.e., terminates) the threads and reports the elapsed time.

```c
#define REPS       129
...
int main(int argc, char** argv)
{
    ...
    gettimeofday(&t, NULL);
    StartTime = (double)t.tv_sec*1000000.0 + ((double)t.tv_usec);
    if(NumThreads >1){
        pthread_attr_init(&ThAttr);
        pthread_attr_setdetachstate(&ThAttr, PTHREAD_CREATE_JOINABLE);
        for(a=0; a<REPS; a++){
            for(i=0; i<NumThreads; i++){
                ThParam[i] = i;
                ThErr = pthread_create(&ThHandle[i], &ThAttr,
                                    MTFlipFunc, (void *)&ThParam[i]);
                if(ThErr != 0){
                    printf("Create Error %d. Exiting abruptly...\n",ThErr);
                    exit(EXIT_FAILURE);
                }
            }
            pthread_attr_destroy(&ThAttr);
            for(i=0; i<NumThreads; i++){ pthread_join(ThHandle[i], NULL); }
        }
    }else{
        for(a=0; a<REPS; a++){ (*FlipFunc)(TheImage); }
    }
    gettimeofday(&t, NULL);
    EndTime = (double)t.tv_sec*1000000.0 + ((double)t.tv_usec);
    TimeElapsed=(EndTime-StartTime)/1000.00;
    TimeElapsed/=(double)REPS;
    // merge with header and write to file
    WriteBMP(TheImage, argv[2]);
    // free() the allocated memory for the image
    for(i = 0; i < ip.Vpixels; i++) { free(TheImage[i]); }
    free(TheImage);
    printf("\n\nTotal execution time: %9.4f ms (%s flip)",TimeElapsed,
        Flip=='V'?"Vertical":"Horizontal");
    printf(" (%6.3f ns/pixel)\n",
        1000000*TimeElapsed/(double)(ip.Hpixels*ip.Vpixels));
    return (EXIT_SUCCESS);
}
```

2.1.4 Thread Initialization

Here is the part of the code that initializes the threads and runs the multithreaded code multiple times. To initialize threads, we tell the OS, through the APIs pthread_attr_init() and pthread_attr_setdetachstate() that we are getting ready to launch a bunch of threads that will be joined later ... The loop that repeats the same code 129 times simply "slows down" the time! Instead of measuring how long it takes to execute once, if you execute 129 times and divide the total time elapsed by 129, nothing changes, except you are a lot less susceptible to the inaccuracy of the Unix time measurement APIs.

```
#include <pthread.h>
...
#define REPS       129
#define MAXTHREADS 128
...
long           NumThreads;              // Total # of parallel threads
int            ThParam[MAXTHREADS];   // Thread parameters ...
pthread_t      ThHandle[MAXTHREADS];  // Thread handles
pthread_attr_t ThAttr;                  // Pthread attributes
...
   pthread_attr_init(&ThAttr);
   pthread_attr_setdetachstate(&ThAttr, PTHREAD_CREATE_JOINABLE);
   for(a=0; a<REPS; a++){
      ...
   }
```

2.1.5 Thread Creation

Here is where the good stuff happens : Look at the code below. Each thread is *created* by using the API function pthread_create() and starts *executing* the moment it is created. What is this thread going to do? This is what the third argument tells the thread to execute: MTFlipFunc. It is as if we called a function named MTFlipFunc() which starts executing *on its own*, i.e., *in parallel with us*. Our main() just created a child whose name is MTFlipFunc() and starts executing immediately in parallel. The question is that, if main() is creating 2, 4, 8 of these threads, how does each thread know who (s)he is? That's the fourth argument, which, after some pointer manipulation, boils down to ThParam[i].

```
   for(i=0; i<NumThreads; i++){
      ThParam[i] = i;
      ThErr = pthread_create(&ThHandle[i], &ThAttr,
                             MTFlipFunc, (void *)&ThParam[i]);
      if(ThErr != 0){
         printf("Create Error %d. Exiting abruptly...\n",ThErr);
         exit(EXIT_FAILURE);
      }
   }
```

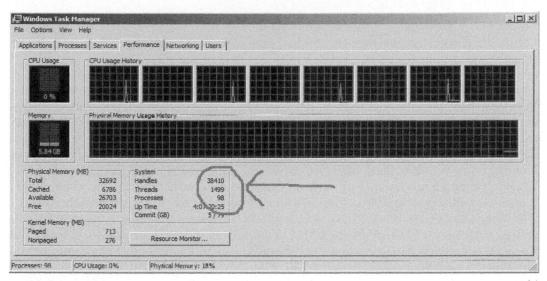

FIGURE 2.1 Windows Task Manager, showing 1499 threads, however, there is 0% CPU utilization.

The OS needs about the first and second arguments: The second argument, &ThAttr, is the same for all threads and contains the thread attributes. The first argument contains the "handles" of each thread and is very important to the OS, to be able to keep track of the threads. If the OS cannot create a thread for any reason, it will return a NULL (i.e., 0), and this is our clue to know that we can no longer create a thread. This is a show-stopper, so our program issues a runtime error and exits.

Here is the interesting question: If main() creates two threads, is our program a *dual-threaded* program? As we will see shortly, when main() creates two threads using pthread_create(), the best we can expect is a 2x program speed-up. What about the main() itself? It turns out main() itself is most definitely a thread too. So, there are **3 threads involved** in a program where main() created two child threads. The reason we only expect a 2x speed-up is the fact that, while main() is only doing *trivial work*, the other two threads are doing *heavy work*.

To quantify this: the main() function creates threads, assigns tasks to them, and joins them, which constitutes, say, 1% of the *activity*, while the other 99% of the activity is caused by the other two threads doing the actual heavy work (about 49.5% each). That being the case, the amount of time that the third thread takes, running the main() function, is negligible. Figure 2.1 shows my PC's Windows Task Manager, which indicates 1499 active threads. However, the *CPU load* is negligible (almost 0%). These 1499 are the threads that the Windows OS created to listen to network packets, keyboard strokes, other interrupts, etc. If, for example, the OS realizes that a network packet has arrived, it wakes up the responsible thread, immediately processes that packet in a very short period of time and the thread goes back to sleep, although still active. Remember: the CPU is drastically faster than the network packet.

2.1.6   Thread Launch/Execution

Figure 1.2 shows that although the OS has 1499 threads that are very *sleepy*, the threads that the main() function creates have a totally different personality: The moment the main() creates the two child threads, the number of threads is now 1501. However, these two new guys are adrenaline-driven crazy guys! They go crazy for 82 ms and during their execution Windows Task Manager will peak at 100% for two of the virtual CPUs, and these threads get swallowed by main() after pthread_join() after 82 ms. At that point, you are back down to 1499 threads. The main() function itself doesn't die until it reaches the very end line, where it reports the time and exits, at which point, we are down to 1498 threads. So, if you looked at the Windows Task Manager during the 129 repetitions of the code, where two threads are on an adrenaline rush — from thread launch to thread join, you would see the 2 of the 8 CPUs peak at 100%. The CPU in my computer has 4 cores, 8 threads (4C/8T). The Windows OS sees this CPU as "8 virtual CPUs," hence the reason to see 8 "CPUs" in the Task Manager. When you have a Mac or a Unix machine, the situation will be similar. To summarize what happens when our *2-threaded* code runs, remember the command line you typed to run your code:

imflipP dogL.bmp dogH.bmp H 2

This command line instructs the OS to *load* and *execute* the binary code imflipP. Execution involves creating the very first thread, assigning function main() to it, and passing arguments `argc` and `argv[]` to that thread. This sounds very similar to what we did to create our child threads.

When the OS finishes loading the executable binary imflipP, it passes the control to main() as if it called a function main() and passed the variables `argc` and `argv[]` array to it. The main() starts running ... Somewhere during the execution, main() asks the OS to create two more threads ...

```
...pthread_create(&ThHandle[0], ...,MTFlipFunc, (void *)&ThParam[0]);
...pthread_create(&ThHandle[1], ...,MTFlipFunc, (void *)&ThParam[1]);
```

and the OS sets up the memory and stack area and allocates these two super-active threads 2 of its available virtual CPUs. After successful creation of the threads, they must be *launched*. pthread_create() also implies launching a thread that has just been created. Launching a thread effectively corresponds to calling the following functions:

```
(*MTFlipFunc)(ThParam[0]);
(*MTFlipFunc)(ThParam[1]);
```

which will turn into either one of the horizontal or vertical flip functions, determined at runtime, based on user input. If the user chose 'H' as the flip option, the launch will be effectively equivalent to this:

```
...
MTFlipFunc=MTFlipH;
...
(*MTFlipH)(ThParam[0]);
(*MTFlipH)(ThParam[1]);
```

2.1.7 Thread Termination (Join)

After launch, a tornado will hit the CPU! They will use the two virtual CPUs super actively, and they eventually return() one by one, which allows the main() to execute the pthread_join() on each thread one by one:

```
pthread_join(ThHandle[0], NULL);
pthread_join(ThHandle[1], NULL);
```

After the first pthread_join(), we are down to 1500 threads. The first child thread got swallowed by main(). After the second pthread_join(), we are down to 1499 threads. The second child thread got gobbled up too. This stops the tornado! And, a few ms later, the main() reports and time and exits. As we will see in Code 2.5, imageStuff.c contains code to dynamically allocate the memory area to store the image that is read from the disk. malloc() function is used for dynamic (i.e., at run time) memory allocation. Before exiting main(), all of this memory area is unallocated using free() as shown below.

```
   ...
   // free() the allocated memory for the image
   for(i = 0; i < ip.Vpixels; i++) { free(TheImage[i]); }
   free(TheImage);
   printf("\n\nTotal execution time: %9.4f ms (%s flip)",TimeElapsed,
       Flip=='V'?"Vertical":"Horizontal");
   printf(" (%6.3f ns/pixel)\n",
       1000000*TimeElapsed/(double)(ip.Hpixels*ip.Vpixels));
   return (EXIT_SUCCESS);
}
```

When main() exits, the parent OS thread swallows the child thread running main(). These threads are an interesting life form, they are like some sort of bacteria that create and swallow each other!

2.1.8 Thread Task and Data Splitting

Ok, the operation of these bacterial lifeforms, called *threads*, is pretty clear now. What about *data*? The entire purpose of creating more than one thread was to execute things faster. This, by definition, means that the more threads we create, the more *task splitting* we have to do, and the more *data splitting* we have to do. To understand this, look at Analogy 2.1.

ANALOGY 2.1: *Task and Data Splitting in Multithreading.*

Cocotown is a major producer of coconuts, harvesting 1800 trees annually. Trees are numbered from 0 to 1799. Every year, the entire town helps coconut harvesting. Due to the increasing number of participants, the town devised the following strategy to speed-up harvesting:

Farmers who are willing to help show up at the harvesting place and get a one-page instruction manual. This page has a top and bottom part. The top part is the same for every farmer which says: *"crack the shell, peel the skin, ..., and do this for the following trees:"*

When 2 farmers show up, the bottom part says: *Process only trees numbered* [0...899] for the first farmer, and [900...1799] for the second farmer. But if 5 farmers showed up, their bottom part of the instructions would have the following numbers: [0..359] for the first farmer, [360..719] for the second farmer, ..., and [1440..1799] for the last farmer.

In Analogy 2.1, harvesting a portion of the coconut trees is **the task of each thread** and it is the same for every farmer, regardless of how many farmers show up. The farmers are **threads that are executing**. Each farmer must be given a unique ID to know *which part* of the trees he or she must harvest. This unique ID is analogous to a Thread ID, or, tid. The number of trees is 1800, which is **all of the data elements to process**. The most interesting thing is that the **task** (i.e., the top part of the instructions) can be separated from the **data** (i.e., the bottom part of the instructions). While the task is the same for every farmer, the data is completely different. So, in a sense, there is only a single task, applied to different data elements, determined by **tid**.

It is clear that the task can be completely predetermined at *compile-time*, which means, during the preparation of the instructions. However, it looks like the data portion must be determined at *runtime*, i.e., when everybody shows up and we know exactly how many farmers we have. The key question is whether the data portion can also be determined at compile-time. In other words, the mayor of the town can write only one set of instructions and make, say, 60 photocopies (i.e., the maximum number of farmers that is ever expected) and never have to prepare anything else when the farmers show up. If 2 farmers show up, the mayor hands out 2 instructions and assigns $tid=0$ and $tid=1$ to them. If 5 farmers show up, he hands out 5 instructions and assigns them $tid=0$, $tid=1$, $tid=2$, $tid=3$, $tid=4$.

More generally, the only thing that must be determined at runtime is the *tid* assignment, i.e., $tid = 0 \ ... \ tid = N-1$. Everything else is determined at compile time, including the parameterized task. Is this possible? It turns out, it is most definitely possible. In the end, for N farmers, we clearly know what the *data splitting* will look like: Each farmer will get $1800/N$ coconut trees to harvest, and farmer number *tid* will have to harvest trees in the range

$$\frac{1800}{N} \times tid \quad ... \quad \frac{1800}{N} \times (tid + 1) - 1 \tag{2.1}$$

To validate this, let us calculate the data split for $tid = [0...4]$ (5 farmers). They end up being $[360 \times tid \ ... \ 360 \times tid + 359]$ for a given *tid*. Therefore, for 5 farmers, the data split ends up being [0...359], [360...719], [720...1079], [1080...1439], and [1440...1799]. This is precisely what we wanted. This means that for an N-threaded program, such as flipping a picture horizontally, we really need to write a *single function* that will get assigned to a launched thread at runtime. All we need to do is to let the launched thread know what its *tid* is ... The thread, then, will be able to figure out exactly what portion of the data to process at run time using an equation similar to Equation 2.1.

One important note here is that none of the data elements have *dependencies*, i.e., they can be processed independently and in parallel. Therefore, we expect that when we launch N threads, the entire task (i.e., 1800 coconut trees) can be processed N times faster. In other words, if 1800 coconut trees took 1800 hours to harvest, when 5 farmers show up, we expect it to take 360 hours. As we will see shortly, this perfectness proves to be difficult to achieve. There is an inherent overhead in parallelizing a task. It is called the *parallelization overhead*. Because of this overhead may 5 farmers would take 400 hours to complete the job. The details will depend on the hardware and the way we wrote the function for each thread. We will be paying a lot of attention to this issue in the coming chapters.

2.2 WORKING WITH BITMAP (BMP) FILES

Now that we understand how the task and data splitting must happen in a multithreaded program, let us apply this knowledge to our first parallel program imflipP.c. Before we do this, we need to understand the format of a bitmap (BMP) image file and how to read/write these files.

2.2.1 BMP is a Non-Lossy/Uncompressed File Format

A Bitmap (BMP) file is an uncompressed image file. This means that by knowing the size of the image, you can easily determine the size of the file that the image is stored in. For example, a 24-bit-per-pixel BMP image occupies 3 bytes per pixel (i.e., one byte per R, G, and B). This format also needs 54 additional bytes to store "header" information. I will provide and explain the exact formula in Section 2.2.2, but, for now, let's focus on the concept of *compression*.

ANALOGY 2.2: *Data Compression.*

Cocotown record keeping division wanted to store the picture of the 1800 coconut trees they harvested at the beginning and end of year 2015. The office clerk stored the following information in a file named 1800trees.txt:

On January 1st of 2015, there were 1800 identical trees, arranged in a 40 wide and 45 long rectangle. I took the picture of a single tree and saved it in a file named OneTree.BMP. Take this picture, make a 40x45 tile out of it. I only noticed a single tree that was different at location (30,35) and stored its picture in another picture file DifferentTree.BMP. The other 1799 are identical.

On December 31, 2015, the trees looked different because they grew. I took the picture of one tree and saved it in GrownTree.BMP. Although they grew, 1798 of them were still identical on Dec. 31, 2015, while two of them were different. Make a 40x45 tile out of GrownTree.BMP and replace the two different trees, at locations (32,36) and (32,38) by the files Grown3236.BMP and Grown3238.BMP.

If you look at Analogy 2.2, the clerk was able to get all of the *information* that is necessary to draw an entire picture of the 40x45 tree farm by providing only a single tree picture (OneTree.BMP) and the picture of the one that looked a little different than the other 1798 (DifferentTree.BMP). Assume that each of these pictures occupies 1 KB of storage. Including the text file that the clerk provided, this information fits in roughly 3 KB on January 1, 2015. If we were to make a BMP file out of the entire 40x45 tree farm, we would need 1 KB for all 1800 trees, occupying 1800 KB. Repetitious (i.e., redundant) data allowed the clerk to substantially reduce the size of the file we need to deliver the same information.

This concept is called *data compression* and can be applied to any kind of data that has redundancies in it. This is why an *uncompressed* image format like BMP will be drastically larger in size in comparison to a *compressed* image format like JPEG (or JPG) that compresses the information before it stores it; the techniques used in compression require the knowledge of frequency domain analysis, however, the abstract idea is simple and is exactly what is conceptualized in Analogy 2.2.

A BMP file stores "raw" image pixels without compressing them; because compression is not performed, no additional processing is necessary before each pixel is stored in a BMP file. This contrasts with a JPEG file, which first applies a frequency-domain transformation like Cosine Transform. Another interesting artifact of a JPEG file is that only 90–99% of the actual image information might be there; this concept of *losing* part of the image information — though not noticeable to the eye — means that a JPEG file is a *lossy* image storage format, whereas no information is lost in a BMP file because each pixel is stored without any transformation. Considering that a 20 MB BMP file could be stored as a 1 MB JPG file if we could tolerate a 1% loss in image data, this trade-off is perfectly acceptable to almost any user. This is why almost every smartphone stores images in JPG format to avoid quickly filling your storage space.

2.2.2 BMP Image File Format

Although BMP files support grayscale and various color depths (e.g., 8-bit or 24-bit), I will only use the 24-bit RGB files in our programs. These files have a 54-byte header followed by the RGB colors of each pixel. Unlike JPEG files, BMP files are not compressed, so each pixel takes up 3 bytes and it is possible to determine the exact size of the BMP files according to this formula:

$$Hbytes = (Hpixels \times 3 + 3) \wedge (11...1100)_2$$

$$24b \text{ RGB BMP File Size} = 54 + Vpixels \times Hbytes$$

(2.2)

where $Vpixels$ and $Hpixels$ are the height and width of the image (e.g., $Vpixels = 480$ and $Hpixels = 640$ for the 640×480 dog.bmp image file). Per Equation 2.2, the dog.bmp occupies $54 + 3 \times 640 \times 480 = 921,654$ bytes whereas the 3200×2400 dogL.bmp image has a file size of $23,040,054$ bytes ($\approx$ 22MB).

The conversion from $Hpixels$ to $Hbytes$ should be as straightforward as

$$Hbytes = 3 \times Hpixels,$$

however, it must be rounded up to the next integer that is divisible by 4 to ensure that the BMP image size is a multiple of 4. This is achieved at the top line of Equation 2.2 by adding 3 and wiping out the LSB two bits of the resulting number (i.e., by ANDing the last 2 bits with 00).

Here are some example BMP size computations:

- A 24-bit 1024x1024 BMP file would need 3,145,782 bytes of storage ($54+1024\times1024\times3$).

- A 24-bit 321×127 BMP would need 122,482 bytes ($54 + (321 \times 3 + 1) \times 127$).

CODE 2.3:   ImageStuff.h

The header file ImageStuff.h contains the descriptions for the two BMP image manipulation files as well as `struct` definitions for our images.

```
struct ImgProp{
    int Hpixels;
    int Vpixels;
    unsigned char HeaderInfo[54];
    unsigned long int Hbytes;
};
struct Pixel{
    unsigned char R;
    unsigned char G;
    unsigned char B;
};

unsigned char** ReadBMP(char* );
void WriteBMP(unsigned char** , char*);

extern struct ImgProp  ip;
```

2.2.3   Header File ImageStuff.h

Since I will be using exactly the same image format through Part I, I put all of the BMP image manipulation files and the associated header file outside our actual code. The ImageStuff.h header file contains the headers of the functions and `struct` definitions related to images and needs to be included in all of our programs and is shown in Code 2.3. Instead of the ImageStuff.h file, you could use more professional grade image helper packages like ImageMagick. However, because ImageStuff.h is, in a sense, "open-source," I strongly suggest the readers to understand this file before starting to use another package like ImageMagick or OpenCV. This will allow you to get a good grasp of the low-level concepts related to images. We will switch over to other easier-to-use packages in Part II of the book.

In ImageStuff.h, a `struct` is defined for an image that contains the previously mentioned **Hpixels** and **Vpixels** of that image. The header of the image that is being processed is saved into the **HeaderInfo[54]** to be restored when writing back the image after the flip operation, etc. **Hbytes** contains the number of bytes each row occupies in memory, rounded up to the nearest integer that is divisible by 4. For example, if a BMP image has 640 horizontal pixels, $\text{Hbytes} = 3 \times 640 = 1920$. However, for an image that has, say, 201 horizontal pixels, $\text{Hbytes} = 3 \times 201 = 603 \rightarrow 604$. So, each row will take up 604 bytes and there will be one wasted byte on each row.

The ImageStuff.h file also contains the headers for the BMP image read and write functions ReadBMP() and WriteBMP(), that are provided in the ImageStuff.c file. The actual C variable **ip** contains the properties of our primary image, which is the dog picture in many of our examples. Since this variable is defined inside the actual program, imflipP.c in this chapter, it must be included within ImageStuff.h as an `extern struct`, so the ReadBMP() and WriteBMP() functions can reference them without a problem.

CODE 2.4: ImageStuff.c   WriteBMP() {...}

WriteBMP() writes a BMP image after the processing. Variable `img` contains a pointer to a struct that contains the image to be written.

```c
void WriteBMP(unsigned char** img, char* filename)
{
    FILE* f = fopen(filename, "wb");
    if(f == NULL){
        printf("\n\nFILE CREATION ERROR: %s\n\n",filename);
        exit(1);
    }
    unsigned long int x,y;
    char temp;
    //write header
    for(x=0; x<54; x++) { fputc(ip.HeaderInfo[x],f); }
    //write image data on byte at a time
    for(x=0; x<ip.Vpixels; x++){
        for(y=0; y<ip.Hbytes; y++){
            temp=img[x][y];
            fputc(temp,f);
        }
    }
    printf("\n Output BMP File name: %20s (%u x %u)",filename,ip.Hpixels,ip.Vpixels);
    fclose(f);
}
```

2.2.4 Image Manipulation Routines in ImageStuff.c

The file ImageStuff.c contains two functions, responsible for reading and writing BMP files. Encapsulating these two functions and the surrounding variable definitions, etc. into the ImageStuff.c and ImageStuff.h files allows us to worry about the details of these files only once here for the entire Part I of this book. Whether we are developing CPU or GPU programs, we will read the BMP images using these functions. Even when developing GPU programs, the image will be first read into the CPU area and later transferred into the GPU memory as will be detailed in Part II of the book.

Code 2.4 shows the WriteBMP() function that writes the processed BMP file back to disk. This function takes a `struct` pointer named `img` that contains the header for the image being written, as well as the output file name that is contained in a string named `filename`. The header of the output BMP file is 54 bytes and was saved during the BMP read operation.

The ReadBMP() function, shown in Code 2.5, allocates memory for the image one row at a time, using the `new` keyword. Each pixel is treated as a `struct` of three bytes, containing the RGB values of that pixel, during processing. However, when reading an image from disk, we can simply read each row `Hbytes` at a time and write it into a consecutive array of `Hbytes` unsigned char values and not worry about individual pixels.

CODE 2.5:   ImageStuff.c   ... ReadBMP() {...}

ReadBMP() reads a BMP image and allocates memory for it. Required image parameters, such as Hpixels and Vpixels, are extracted from the BMP file and written into the struct. Hbytes is computed using Equation 2.2.

```c
#include <stdlib.h>
#include <stdio.h>
#include <time.h>
#include "ImageStuff.h"

unsigned char** ReadBMP(char* filename)
{
   int i;
   FILE* f = fopen(filename, "rb");
   if(f == NULL){    printf("\n\n%s NOT FOUND\n\n",filename); exit(1);  }

   unsigned char HeaderInfo[54];
   fread(HeaderInfo, sizeof(unsigned char), 54, f); // read 54b header
   // extract image height and width from header
   int width = *(int*)&HeaderInfo[18];
   int height = *(int*)&HeaderInfo[22];
   //copy header for re-use
   for(i=0; i<54; i++){       ip.HeaderInfo[i]=HeaderInfo[i];       }

   ip.Vpixels = height;
   ip.Hpixels = width;
   int RowBytes = (width*3 + 3) & (~3);
   ip.Hbytes = RowBytes;
   printf("\n  Input BMP File name: %20s (%u x %u)",filename,ip.Hpixels,ip.Vpixels);
   unsigned char tmp;
   unsigned char **TheImage = (unsigned char **)malloc(height *
                                   sizeof(unsigned char*));
   for(i=0; i<height; i++) {
      TheImage[i] = (unsigned char *)malloc(RowBytes * sizeof(unsigned char));
   }
   for(i = 0; i < height; i++) {
      fread(TheImage[i], sizeof(unsigned char), RowBytes, f);
   }
   fclose(f);
   return TheImage; // remember to free() it in caller!
}
```

ReadBMP() extracts Hpixels and Vpixels values from the BMP header, i.e., the first 54 bytes of the BMP file, and calculates Hbytes from Equation 2.2. It dynamically allocates sufficient memory for the image using the malloc() function, which will be released using the free() function at the end of main(). The image is read from a user-specified file name that is passed onto ReadBMP() within the string filename. This BMP file header is saved in HeaderInfo[] to use when we need to write the processed file back to the disk.

With the ReadBMP() and WriteBMP() functions use the C library function fopen() with either the **"rb"** or **"wb"** options which mean read or write a binary file. If the file cannot be opened by the OS, the return value of fopen() is NULL, an error is issued to the user.

This would happen due to a wrong file name or an existing lock on the file. fopen() allocates a file handle and a read/write buffer area for the new file and returns it to the caller.

Depending on the fopen() parameters, it also places a lock on the file to prevent multiple programs from corrupting the file due to a simultaneous access. Each byte is read/written from/to the file one byte at a time (i.e., the C variable type unsigned char) by using this buffer. Function fclose() de-allocates this buffer and removes the lock (if any) from the file.

2.3   TASK EXECUTION BY THREADS

Now that we know how to time our CPU code and how to read/write BMP images in detail, let us flip an image using multiple threads. The responsibilities of each party in a multithreaded program are as follows:

- main() is responsible for creating the threads and assigning a unique tid to each one at runtime (e.g., ThParam[i] shown below).

- main() invokes a function for each thread (function pointer MTFlipFunc).

- main() must also pass other necessary values to the thread, if any (also passed in ThParam[i] below).

```
for(i=0; i<NumThreads; i++){
    ThParam[i] = i;
    ThErr = pthread_create(&ThHandle[i], &ThAttr, MTFlipFunc,
                           (void *)&ThParam[i]);
```

- main() is also responsible to let the OS know what type of thread it is creating (i.e., thread attributes, passed in &ThAttr). In the end, main() is nothing but another thread itself and it is *speaking on behalf of* the other child threads it will create in a moment.

- The OS is responsible for deciding whether a thread can be created. Threads are nothing but *resources* that must be managed by the OS. If a thread can be created, the OS is responsible for assigning a *handle* to that thread (ThHandle[i]). If not, OS returns NULL (ThErr).

- If the OS cannot create a thread, main() is also responsible for either exiting or some other action.

```
if(ThErr != 0){
    printf("\nThread Creation Error %d. Exiting abruptly...
        \n",ThErr);
    exit(EXIT_FAILURE);
}
```

- The responsibility of each thread is to receive its tid and perform its task MTFlipFunc() only on the data portion that it is required to process. We will spend multiple pages on this.

- The final responsibility of main() is to wait for the threads to be done, and join them. This instructs the OS to de-allocate the thread resources.

```
      pthread_attr_destroy(&ThAttr);
      for(i=0; i<NumThreads; i++){
        pthread_join(ThHandle[i], NULL);
      }
```

2.3.1 Launching a Thread

Let us look at the way function pointers are used to launch threads. The pthread_create() function is expecting a *function pointer* as its third parameter, which is MTFlipFunc. Where did this pointer come from? To be able to determine this, let us list all of the lines in imflipP.c that participate in "computing" the variable MTFlipFunc. They are listed in Code 2.6. Our aim is to give sufficient flexibility to main(), so, we can launch a thread with any function we want. Code 2.6 lists four different functions:

```
void FlipImageV(unsigned char** img)
void FlipImageH(unsigned char** img)
void *MTFlipV(void* tid)
void *MTFlipH(void* tid)
```

The first two functions are exactly what we had in Chapter 1.1. These are the *serial* functions that flip an image in the vertical or horizontal direction. We just introduced their multithreaded versions (the last two above), that will do exactly what the serial version did, except faster by using multiple threads (hopefully)! Note that the multithreaded versions will need the tid as we described before, whereas the serial versions don't ...

Now, our goal is to understand how we pass the *function pointer* and *data* to each launched thread. The serial versions of the function are slightly modified to eliminate the return value (i.e., void), so they are consistent with the multithreaded versions that also do not return a value. All four of these functions simply modify the image that is pointed to by the pointer TheImage. It turns out, we do not really have to *pass* the function pointer to the thread. Instead, we have to *call* the function that is pointed to by the function pointer. This process is called *thread launch*.

The way we pass the *data* and launch a thread differs based on whether we are launching the serial or multithreaded version of the function. I designed imflipP.c to be able to run the older serial versions of the code as well as the new multithreaded versions, based on the user command-line parameters. Since the input variables of the two families of functions are slightly different, it was easier to define two separate function pointers, FlipFunc and MTFlipFunc, that were responsible for launching the serial and multithreaded version of the functions. I maintained two function pointers shown below:

```
void (*FlipFunc)(unsigned char** img);   // Serial flip function ptr
void* (*MTFlipFunc)(void *arg);          // Multi-threaded flip func ptr
```

Let us clarify the difference between *creating* and *launching* a thread, both of which are implied in pthread_create(). *Creating* a thread involves a request/grant mechanism between the parent thread main() and the OS. If the OS says No, nothing else can happen. So, it is the OS that actually creates the thread and sets up a memory area, a handle, a virtual CPU, and a stack area for it and gives a nonzero *thread handle* to the parent thread, thereby granting the parent permission to *launch* (aka *run*) another thread in parallel.

CODE 2.6:   imflipP.c   Thread Function Pointers

Code that determines the function pointer that is passed onto the launched thread as a parameter. This is how the thread will know what to execute.

```c
...
void (*FlipFunc)(unsigned char** img);  // Serial flip function ptr
void* (*MTFlipFunc)(void *arg);         // Multi-threaded flip func ptr
...
void FlipImageV(unsigned char** img)
{
  ...
}

void FlipImageH(unsigned char** img)
{
  ...
}

void *MTFlipV(void* tid)
{
  ...
}

void *MTFlipH(void* tid)
{
  ...
}

int main(int argc, char** argv)
{
   char        Flip;
   ...
   ...   if(NumThreads != 1){ // multi-threaded version
            printf("\nExecuting the multi-threaded version..."...);
            MTFlipFunc = (Flip=='V') ? MTFlipV:MTFlipH;
         }else{ // serial version
            printf("\nExecuting the serial version ...\n");
            FlipFunc = (Flip=='V') ? FlipImageV:FlipImageH;
     ...
   if(NumThreads >1){ // multi-threaded version
     ...
       for(i=0; i<NumThreads; i++){
          ThParam[i] = i;
          ThErr=pthread_create( , , MTFlipFunc,(void *)&ThParam[i]);
          ...
   }else{ // if running the serial version
       ...
       (*FlipFunc)(TheImage);
       ...
```

Notice, although the parent now has the license to run, nothing is happening yet. Launching a thread is effectively a *parallel function call*. In other words, main() knows that another child thread is running after the launch, and can communicate with it if it needs to.

The main() function may never communicate (e.g., pass data back and forth) with its child thread(s), as exemplified in Code 2.2, since it doesn't need to. Child threads modify the required memory areas and return. The tasks assigned to child threads, in this specific case, leave only a single responsibility to main(): Wait until the child threads are done and terminate (join) them. Therefore, the only thing that main() cares about: it has the *handle* of the new thread, and it can determine when that thread has finished execution (i.e., returned) by using pthread_join(). So, effectively, pthread_join(x) means, *wait until thread with the handle number x is done*. When that thread is done, well, it means that it executed a `return` and finished its job. There is no reason to keep him around.

When the thread (with handle x) joins main(), the OS gets rid of all of the memory, virtual CPU, and stack areas it allocated to that thread, and this thread disappears. However, main() is still alive and well ... until it reaches the last code line and returns (last line in Code 2.2) ... When main() executes a return, the OS de-allocates all of the resources it allocated for main() (i.e., *the* imflipP program). The program has just completed its execution. You then get a prompt back in Unix, waiting for your next Unix command, since the execution of imflipP has just been completed.

2.3.2 Multithreaded Vertical Flip: MTFlipV()

Now that we know how a multithreaded program should work, it is time to look at the *task* that each thread executes in the multithreaded version of the program. This is like our Analogy 2.1, where one farmer had to harvest all 1800 trees (from 0 to 1799) if he or she was alone, whereas if two farmers came to harvest, they could split it into two segments [0...899] and [900...1799] and the split assigns a shrinking range of coconut tree ranges (data ranges) as the number of farmers increases. The magic formula was our Equation 2.1, which specified these split ranges based on only a single parameter called `tid`. Therefore, assigning the same *task* to every single thread (i.e., writing a function that each thread will execute at runtime) and assigning a unique `tid` to each thread at runtime proved to be perfectly sufficient to write a multithreaded program.

If we remember the serial version of our vertical flip code, displayed in Code 1.2, it went through each column one by one, and it swapped each column's pixel with its vertical mirror. As an example, in our 640×480 image named dog.bmp, row 0 (the very first row) had horizontal pixels [0][0...639] and its vertical mirror row 479 (the last row) had pixels [479][0...639]. So, to vertically flip the image, our serial function FlipImageV() had to swap each row's pixels one by one as follows. The $\longleftrightarrow$ symbol denotes a swap.

Row [0]:	[0][0]$\longleftrightarrow$[479][0] ,	[0][1]$\longleftrightarrow$[479][1]	...	[0][639]$\longleftrightarrow$[479][639]
Row [1]:	[1][0]$\longleftrightarrow$[478][0] ,	[1][1]$\longleftrightarrow$[478][1]	...	[1][639]$\longleftrightarrow$[478][639]
Row [2]:	[2][0]$\longleftrightarrow$[477][0] ,	[2][1]$\longleftrightarrow$[477][1]	...	[2][639]$\longleftrightarrow$[477][639]
Row [3]:	[3][0]$\longleftrightarrow$[476][0] ,	[3][1]$\longleftrightarrow$[476][1]	...	[3][639]$\longleftrightarrow$[476][639]
...	...	...	...	...
Row [239]:	[239][0]$\longleftrightarrow$[240][0] ,	[239][1]$\longleftrightarrow$[240][1]	...	[239][639]$\longleftrightarrow$[240][639]

Refreshing our memory with Code 1.2, FlipImageV() function that swaps pixels looked something like this. Note: the return value type is modified to be void to be consistent with the multithreaded versions of the same program. Otherwise, the rest of the code below looks exactly like Code 1.2.

```
void FlipImageV(unsigned char** img)
{
    struct Pixel pix; //temp swap pixel
    int row, col;

    //vertical flip
    for(col=0; col<ip.Hbytes; col+=3){
        row = 0;
        while(row<ip.Vpixels/2){
            pix.B = img[row][col];
            ...
            row++;
        }
    }
    return img;
}
```

The question now is: how to modify this FlipImageV() function to allow multithreading? The multithreaded version of the function, MTFlipV(), will receive one parameter named tid as we emphasized before. The image it will work on is a global variable TheImage, so it doesn't need to be passed as an additional input. Since our friend pthread_create() expects us to give it a *function pointer*, we will define MTFlipV() as follows:

```
void *MTFlipV(void* tid)
{
    ...
}
```

During the course of this book, we will be encountering other types of functions that are not so amenable to *being parallelized*. A function that doesn't easily parallelize is commonly referred to as a function that *doesn't thread well*. There should be no question in any of the readers' mind at this point that, if a function doesn't thread well, it is expected to be *not GPU-friendly*. Here, in this section, I am also making the point that such a function is also possibly *not CPU multithreading friendly*.

So, what do we do when a task is "born to be serial"? You clearly do not run this task on a GPU. You keep it on the CPU ... keep it serial ... run it fast. Most modern CPUs, such as the i7-5960x [11] I mentioned in Section 1.1, have a feature called *Turbo Boost* that allows the CPU to achieve very high performance on a single thread when running a serial (single-threaded) code. They achieve this by clocking one of the cores at, say, 4 GHz, while other cores are at, say, 3 GHz, thereby significantly boosting the performance of single-threaded code. This allows the CPU to achieve a good performance for both modern and *old-fashioned serial* code ...

CODE 2.7: imflipP.c ... MTFlipV() {...}

Multithreaded version of the FlipImageV() in Code 1.2 that expects a `tid` to be provided. The only difference between this and the serial version is the portion of the data area it processes, rather than the entire data.

```c
...
long                NumThreads;   // Total # threads working in parallel
unsigned char**     TheImage;     // This is the main image
struct ImgProp      ip;
...
void *MTFlipV(void* tid)
{
    struct Pixel pix; //temp swap pixel
    int row, col;

    long ts = *((int *) tid);      // My thread ID is stored here
    ts *= ip.Hbytes/NumThreads;    // start index
    long te = ts+ip.Hbytes/NumThreads-1;  // end index

    for(col=ts; col<=te; col+=3)
    {
        row=0;
        while(row<ip.Vpixels/2)
        {
            pix.B = TheImage[row][col];
            pix.G = TheImage[row][col+1];
            pix.R = TheImage[row][col+2];

            TheImage[row][col] = TheImage[ip.Vpixels-(row+1)][col];
            TheImage[row][col+1] = TheImage[ip.Vpixels-(row+1)][col+1];
            TheImage[row][col+2] = TheImage[ip.Vpixels-(row+1)][col+2];

            TheImage[ip.Vpixels-(row+1)][col] = pix.B;
            TheImage[ip.Vpixels-(row+1)][col+1] = pix.G;
            TheImage[ip.Vpixels-(row+1)][col+2] = pix.R;

            row++;
        }
    }
    pthread_exit(NULL);
}
```

The entire code listing for MTFlipV() is shown in Code 2.7. Comparing this to the serial version of the function, shown in Code 1.2, there aren't really a lot of differences other than the concept of `tid`, which acts as the *data partitioning agent*. Please note that this code is an overly simple multithreaded code. Normally, what each thread does completely depends on the logic of the programmer. For our purposes, though, this simple example is perfect to demonstrate the basic ideas. Additionally, the FlipImageV() function is a well-mannered function that is very amenable to multithreading.

2.3.3 Comparing FlipImageV() and MTFlipV()

Here are the major differences between the serial version of the vertical-flipper function FlipImageV() and its parallel version MTFlipV():

- FlipImageV() is defined as a *function*, whereas MTFlipV() is defined as a *function pointer*. This is to make it easy for us to use this pointer in launching the thread using pthread_create().

```
void (*FlipFunc)(unsigned char** img);  // Serial flip func ptr
void* (*MTFlipFunc)(void *arg);          // Multi-th flip func ptr
...
void FlipImageV(unsigned char** img)
{
   ...
}

void *MTFlipV(void* tid)
{
   ...
}
```

- FlipImageV() is designed to process the entire image, while its parallel counterpart MTFlipV() is designed to process only a portion of the image defined by an equation similar to Equation 2.1. Therefore, MTFlipV() needs a variable tid passed to it to know who he is. This is done when launching the thread using pthread_create().

- Besides the option of using the MTFlipFunc function pointer in launching threads using pthread_create(), nothing prevents us from simply *calling* the function ourselves by using the MTFlipFunc function pointer (and its serial version FlipFunc). To *call* the functions that these pointers are pointing to, the following notation has to be used:

```
FlipFunc = FlipImageV;
MTFlipFunc = MTFlipV;
...
(*FlipFunc)(TheImage);                  // call the serial version
(*MTFlipFunc)(void *(&ThParam[0]));     // call the multi-threaded version
```

- Each image row occupies ip.Hbytes bytes. For example, for the 640×480 image dog.bmp, ip.Hbytes= 1920 bytes according to Equation 2.2. The serial function FlipImageV() clearly has to loop through every byte in the range [0...1919]. However, the multithreaded version MTFlipV() partitions these horizontal 1920 bytes based on tid. If 4 threads are launched, the byte (and pixel) range that has to be processed for each thread is:

$tid = 0$:	Pixels [0...159]	Hbytes [0...477]
$tid = 1$:	Pixels [160...319]	Hbytes [480...959]
$tid = 2$:	Pixels [320...479]	Hbytes [960...1439]
$tid = 3$:	Pixels [480...639]	Hbytes [1440...1919]

- Multi-threaded function's first task is to calculate which data range *it* has to process. If every thread does this, all 4 of these pixel ranges shown above can be processed *in parallel*. Here is how each thread calculates its own range:

```
void *MTFlipV(void* tid)
{
    struct Pixel pix; //temp swap pixel
    int row, col;

    long ts = *((int *) tid);       // My thread ID is stored here
    ts *= ip.Hbytes/NumThreads;     // start index
    long te = ts+ip.Hbytes/NumThreads-1;   // end index

    for(col=ts; col<=te; col+=3)
    {
        row=0;
        ...
```

The thread, as its very first task, is calculating its `ts` and `te` values (*thread start* and *thread end*). These are the `Hbytes` ranges, similar to the ones shown above and the split is based on Equation 2.1. Since each pixel occupies 3 bytes (one byte for each of the RGB colors), the function is adding 3 to the `col` variable in the `for` loop. The FlipImageV() function doesn't have to do such a computation since it is expected to process *everything*, i.e., `Hbytes` range 0...1919.

- The image to process is passed via a local variable `img` in the serial FlipImageV() to be compatible with the version introduced in Chapter 1.1, whereas a global variable (`TheImage`) is used in MTFlipV() for reasons that will be clear in the coming chapters.

- The multithreaded function executes pthread exit() to let main() know that it is done. This is when the pthread_join() function *advances* to the next line for the thread that finished.

As an interesting note, if we launched only a single thread using pthread_create(), we are technically running a multithreaded program where the `tid` range is `tid=` $[0...0]$. This thread would still calculate its data range just to find that it has to process the entire range anyway. In the imflipP.c program, the FlipImageV() function is referred to as a *serial* version, whereas launching the multithreaded version with 1 thread is allowed which is referred to as the *1-threaded* version.

By comparing serial Code 1.2 and its parallel version Code 2.7, it is easy to see that, as long as the function is written carefully at the beginning, it is easy to parallelize it with minor modifications. This will be a very useful concept to remember when we are writing GPU versions of certain serial CPU code. Since GPU code by definition means parallel code, this concept might allow us to *port* CPU code to the GPU world with minimal effort. This will prove to be the case sometimes. Not always!

CODE 2.8:   imflipP.c   ... MTFlipH() {...}

Multithreaded version of the FlipImageH() function in Code 1.3.

```
...
long            NumThreads;   // Total # threads working in parallel
unsigned char** TheImage;     // This is the main image
struct ImgProp  ip;
...
void *MTFlipH(void* tid)
{
    struct Pixel pix; //temp swap pixel
    int row, col;

    long ts = *((int *) tid);      // My thread ID is stored here
    ts *= ip.Vpixels/NumThreads;   // start index
    long te = ts+ip.Vpixels/NumThreads-1;  // end index

    for(row=ts; row<=te; row++)
    {
        col=0;
        while(col<ip.Hpixels*3/2)
        {
            pix.B = TheImage[row][col];
            pix.G = TheImage[row][col+1];
            pix.R = TheImage[row][col+2];

            TheImage[row][col] = TheImage[row][ip.Hpixels*3-(col+3)];
            TheImage[row][col+1] = TheImage[row][ip.Hpixels*3-(col+2)];
            TheImage[row][col+2] = TheImage[row][ip.Hpixels*3-(col+1)];

            TheImage[row][ip.Hpixels*3-(col+3)] = pix.B;
            TheImage[row][ip.Hpixels*3-(col+2)] = pix.G;
            TheImage[row][ip.Hpixels*3-(col+1)] = pix.R;

            col+=3;
        }
    }
    pthread_exit(NULL);
}
```

2.3.4 Multithreaded Horizontal Flip: MTFlipH()

The serial function FlipImageH(), shown in Code 1.3, is parallelized and its multithreaded version, MTFlipH() is shown in Code 2.8. Very much like its vertical version, the horizontal version of the multithreaded code looks at `tid` to determine which data partition it has to process. For an example 640×480 image (480 rows) using 4 threads, these pixels are flipped:

tid=0 : Rows [0...119]		tid=1 : Rows [120...239]
tid=2 : Rows [240...359]		tid=3 : Rows [360...479]

TABLE 2.1 Serial and multithreaded execution time of imflipP.c, both for vertical flip and horizontal flip, on an i7-960 (4C/8T) CPU.

#Threads	Command line	Run time (ms)
Serial	imflipP dogL.bmp dogV.bmp v	131
2	imflipP dogL.bmp dogV2.bmp v 2	70
3	imflipP dogL.bmp dogV3.bmp v 3	46
4	imflipP dogL.bmp dogV4.bmp v 4	67
5	imflipP dogL.bmp dogV5.bmp v 5	55
6	imflipP dogL.bmp dogV6.bmp v 6	51
8	imflipP dogL.bmp dogV8.bmp v 8	52
9	imflipP dogL.bmp dogV9.bmp v 9	47
10	imflipP dogL.bmp dogV10.bmp v 10	51
12	imflipP dogL.bmp dogV10.bmp v 12	44
Serial	imflipP dogL.bmp dogH.bmp h	81
2	imflipP dogL.bmp dogH2.bmp h 2	41
3	imflipP dogL.bmp dogH3.bmp h 3	28
4	imflipP dogL.bmp dogH4.bmp h 4	41
5	imflipP dogL.bmp dogH5.bmp h 5	33
6	imflipP dogL.bmp dogH6.bmp h 6	28
8	imflipP dogL.bmp dogH8.bmp h 8	32
9	imflipP dogL.bmp dogH9.bmp h 9	30
10	imflipP dogL.bmp dogH10.bmp h 10	33
12	imflipP dogL.bmp dogH7.bmp h 12	29

For each row that a thread is responsible for, each pixel's 3-byte RGB values are swapped with its horizontal mirror. This swap starts at col= $[0...2]$ which holds the RGB values of pixel 0, and continues until the last RGB (3 byte) value has been swapped. For a 640×480 image, since Hbytes= 1920, and there is no wasted byte, the last pixel (i.e., pixel 639) is at col= $[1917...1919]$.

2.4 TESTING/TIMING THE MULTITHREADED CODE

Now that we know how the imflipP program works in detail, it is time to test it. The program command line syntax is determined via the parsing portion of main() as shown in Code 2.1. To run imflipP, the general command line syntax is:

imflipP InputfileName OutputfileName [v/h/V/H] [1-128]

where InputFileName and OutputFileName are the BMP files to read and write, respectively. The optional command line argument [**v/h/V/H**] is to specify the flip direction ('V' is the default). The next optional argument is the number of threads and can be specified between 1 and MAXTHREADS (128), with a default value of 1 (serial).

Table 2.1 shows the run times of the same program by using 1 through 10 threads on an Intel i7-960 CPU that has 4C/8T (4 cores, 8 threads). Results all the way to 10 threads is reported, not that we expect an improvement beyond 8 threads, but as a *sanity check*. These kinds of checks are useful to quickly discover a potentially hidden bug. The functionality is also confirmed by looking at the picture, and checking the file size, and running a comparison program that checks two binary files, the Unix diff command.

So, what do these results tell us ? First of all, in both the vertical and horizontal flip case, it is clear that using more than a single thread helps. So, our efforts to parallelize the program weren't for nothing. However, the troubling news is that, beyond 3 threads, there seems to be no performance improvement at all, in both the vertical and horizontal case. For ≥ 4 threads, you can simply regard the data as *noise*!

What Table 2.1 clearly shows is that *multithreading helps up to 3 threads*. Of course, this is not a generalized statement. This statement strictly applies to my i7-960 test CPU (4C/8T) and the code I have shown in Code 2.7 and Code 2.8, which are the heart of the imflipP.c code. By this time, you should have a thousand questions in your mind. Here are some of them:

- Would the results be different with a less powerful CPU like a 2C/2T?

- How about a more powerful CPU, such as a 6C/12T?

- In Table 2.1, considering the fact that we tested it on a 4C/8T CPU, shouldn't we have gotten better results up to 8 threads? Or, at least 6 or something? Why does the performance collapse beyond 3 threads?

- What if we processed a smaller 640×480 image, such as dog.bmp, instead of the giant 3200×2400 image dog.bmp? Would the *inflection point* for performance be at a different thread count?

- Or, for a smaller image, would there even be an inflection point?

- Why is the horizontal flip faster, considering that vertical is also processing the same number of pixels; 3200×2400?

- ...

The list goes on ... Don't lose sleep over Table 2.1. For this chapter, we have achieved our goal. We know how to write multithreaded programs now, and we are getting *some* speedup. This is more than enough so early in our parallel programming journey. I can guarantee that you will ask another 1000 questions to yourself about why this program isn't as fast as we hoped. I can also guarantee that you will **NOT** ask some of the key questions that actually contribute to this lackluster performance. The answers to these questions deserves an entire chapter, and this is what I will do. In Chapter 3, I will answer all of the above questions and more of the ones you are not asking. For now, I invite you to think about what you might **NOT** be asking ...

## CHAPTER 3

Improving Our First Parallel CPU Program

W E parallelized our first serial program imflip.c and developed its parallel version imflipP.c in Chapter 2. The parallel version achieved a reasonable speed-up using pthreads, as shown in Table 2.1. Using multiple threads reduced the execution time from 131 ms (serial version) down to 70 ms, and 46 ms when we launched two and three threads, respectively, on an i7-960 CPU with 4C/8T. Introducing more threads (i.e., ≥ 4) didn't help. In this chapter, we want to understand the factors that contributed to the performance numbers that were reported in Table 2.1. We might not be able to improve them, but we have to be able to explain why we cannot improve them. We do not want to achieve good performance by *luck*!

3.1 EFFECT OF THE "PROGRAMMER" ON PERFORMANCE

Understanding the *hardware* and the *compiler* helps a programmer write good code. Over many years, CPU architects and compiler designers kept improving their CPU architecture and the optimization capabilities of compilers. A lot of these efforts helped ease the burden on the software developer, so, he or she can write code without worrying too much about low-level hardware details. However, as we will see in this chapter, understanding the underlying hardware and utilizing the hardware efficiently might allow a programmer to develop 10x higher performance code in some cases.

Not only is this statement true for CPUs, but also the potential GPU performance improvement is even more emphasized when the hardware is utilized efficiently, since a lot of the impressive GPU performance comes from shifting the performance-improvement responsibility from the *hardware* to the *programmer*. This chapter explains the interaction among all parties that contribute to the performance of a program. They are the *programmer*, *compiler*, *OS*, and *hardware* (and, to a certain degree, the *user*).

- **Programmer** is the ultimate intelligence and should understand the capabilities of the other pieces. No software or hardware can match what the programmer can do, since the programmer brings in the most valuable asset to the game: *the logic*. Good programming logic requires a complete understanding of all pieces of the puzzle.

- **Compiler** is a large piece of code that is packed with *routine* functionality for two things: (1) compilation, and (2) optimization. (1) is the compiler's *job* and (2) is the additional work the compiler has to do at *compile time* to potentially optimize the inefficient code that the programmer wrote. So, the compiler is the "organizer" at *compile time*. At compile time, time is *frozen*, meaning that the compiler could contemplate many alternative scenarios that can happen at *run time* and produce the best code for run time. When we run the program, the clock starts ticking. One thing the compiler cannot know is the *data*, which could completely change the flow of the program. The data can only be known at runtime, when the OS and CPU are in action.

- The Operating System (**OS**) is the software that is the "boss" or the "manager" of the hardware at *run time*. Its job is to allocate and map the hardware resources efficiently at *run time*. Hardware resources include the virtual CPUs (i.e., threads), memory, hard disk, flash drives (via Universal Serial Bus [USB] ports), network cards, keyboard, monitor, GPU (to a certain degree), and more. A good OS knows its resources and how to map them very well. Why is this important? Because the resources themselves (e.g., CPU) have no idea what to do. They simply follow orders. The OS is the general and the threads are the soldiers.

- **Hardware** is the CPU+memory+peripherals. OS takes the binary code that the compiler produced and assigns them to virtual cores at run time. Virtual cores execute them as fast as possible at *run time*. OS also facilitates the data movement between the CPU and the memory, disk, keyboard, network card, etc.

- **User** is the final piece of the puzzle: Understanding the user is also important in writing good code. The user of a program isn't a programmer, yet the programmer has to appeal to the user and has to *communicate* with him or her. This is not an easy task!

In this book, the major focus will be on *hardware*, particularly the CPU and memory (and, later in Part II, GPU and memory). Understanding the hardware holds the key to developing high-performance code, whether for CPU or GPU. In this chapter, we will discover the truth about whether it is possible to speed-up our first parallel program, imflipP.c. If we can, *how?* The only problem is: we don't know which part of the hardware we can use more efficiently for performance improvement. So, we will look at everything.

3.2 EFFECT OF THE "CPU" ON PERFORMANCE

In Section 2.3.3, I explained the sequence of events that takes place when we launch multithreaded code. In Section 2.4, I also listed numerous questions you might be asking yourself to explain Table 2.1. Let us answer the very first and the most obvious family of questions:

- How would the results differ on different CPUs?

- Is it the *speed* of the CPU, or the *number of cores, number of threads*?

- Anything else related to the CPU? Like *cache memory?*

Perhaps, the most *fun* way to answer this question is go ahead and run the same program on many different CPUs. Once the results are in place, we can try to make sense out of them. The more variety of CPUs we test this code on, the better idea we might get. I will go ahead and run this program on 6 different CPUs shown in Table 3.1.

Table 3.1 lists important CPU parameters, such as the number of cores and threads (C/T), per-core L1\$ and L2\$ cache memory sizes, denoted as L1\$/C and L2\$/C, and the shared L3\$ cache memory size (shared by all 4, 6, or 8 cores). Table 3.1 also lists the

TABLE 3.1 Different CPUs used in testing the imflipP.c program.

Feature	CPU1	CPU2	CPU3	CPU4	CPU5	CPU6
Name	i5-4200M	i7-960	i7-4770K	i7-3820	i7-5930K	E5-2650
C/T	2C/4T	4C/8T	4C/8T	4C/8T	6C/12T	8C/16T
Speed:GHz	2.5-3.1	3.2-3.46	3.5-3.9	3.6-3.8	3.5-3.7	2.0-2.8
L1$/C	64KB	64KB	64KB	64KB	64KB	64KB
L2$/C	256KB	256KB	256KB	256KB	256KB	256KB
shared L3$	3MB	8MB	8MB	10MB	15MB	20MB
Memory	8GB	12GB	32GB	32GB	64GB	16GB
BW:GB/s	25.6	25.6	25.6	51.2	68	51.2

amount of memory and *memory bandwidth* (BW) in Giga-Bytes-per-second (GB/s) for each computer that a given column indicates. In this section, we focus on the CPU's role on performance, but the role of the memory in determining the performance will be explained thoroughly in this book. We will also look at the operation of memory in this chapter. For now, just in case the performance numbers had anything to do with the memory instead of the CPU, they are also listed in Table 3.1.

In this section, we have no intention to make an intelligent assessment on how the performance results could differ among these 6 CPUs. We simply want to watch the CPU horse race and have fun! As we see the numbers, we will develop theories about what could be the most responsible party in determining the overall performance of our program. In other words, we are looking at things from long distance at this point. We will dive deep into details later, but the experimental data we gather first will help us develop methods to improve program performance later.

3.2.1   In-Order versus Out-Of-Order Cores

Aside from *how many* cores a CPU has, there is another consideration that relates to the cores; almost every CPU manufacturer started manufacturing **in order** cores and upgraded their design to **out of order** in their more advanced family of offerings. Abbreviations inO and OoO will be used going forward. For example, MIPS R2000 was an inO CPU, while the more advanced R10000 was OoO. Similarly, Intel 8086, 80286, 80386, and the newer Atom CPUs are inO, whereas Core i3, i5, and i7, as well as every Xeon is OoO.

The difference between inO andOoO is the way the CPU is capable of executing a given set of instructions; while an inO CPU can only execute the instructions in precisely the order that is listed in the binary code, an OoO CPU executes them in the order of *operand availability*. In other words, an OoO CPU can find a lot more work to do in the later list of instructions, whereas an inO CPU simply sits idle if the next instruction in the order of all given instructions does not have data available, perhaps because the memory controller did not yet read the necessary data from the memory.

This allows an OoO CPU to execute instructions a lot faster due to the ability to avoid getting stuck when the next instruction cannot be immediately executed until the operands are available. However, this luxury comes at a price: an inO CPU takes up a lot less chip area, therefore allowing the manufacturer to fit a lot more inO cores in the same integrated circuit chip. Because of this very reason, each inO core might actually be clocked a little faster, since they are *simpler*. Time for an analogy ...

ANALOGY 3.1: *In order versus Out of Order Execution.*

Cocotown had a competition in which two teams of farmer families had to make coconut pudding. These were the instructions provided to them: (1) crack the coconut with the automated cracker machine, (2) grind the coconuts that come out of the cracker machine using the grinder machine, (3) boil milk, and (4) put cocoa in milk and boil more, (5) put the ground coconuts in the cocoa-mixed milk and boil more.

Each step took 10 minutes. Team 1 finished their pudding in 50 minutes, while Team 2 shocked everyone by finishing in 30 minutes. Their secret became obvious after the competition: they started cracking the coconut (step 1) and boiling the milk (step 3) at the same time. These two tasks did not depend on each other and could be started at the same time. Within 10 minutes, both of them were done and the coconuts could be placed in the grinder (step 2), while, in parallel, cocoa is mixed with the milk and boiled (step 4). So, in 20 minutes, they were done with steps 1–4.

Unfortunately, step 5 had to wait for steps 1–4 to be done, making their total execution time 30 minutes. So, the secret of Team 2 was to execute the tasks *out of order*, rather than *in the order* they were specified. In other words, they could start executing any step if it did not depend on the results of a previous step.

Analogy 3.1 emphasizes the performance advantage of OoO execution; an OoO core can execute independent *dependence-chains* (i.e., chains of CPU instructions that do not have dependent results) in parallel, without having to wait for the next instruction to finish, achieving a healthy speed-up. But, there are other trade-offs when a CPU is designed using one of the two paradigms. One wonders which one is a better design idea: (1) more cores that are inO, or (2) fewer cores that are OoO? What if we took the idea to the extreme and placed, say, 60 cores in a CPU that are all inO? Would this work faster than a CPU that has 8 OoO cores? The answer is not as easy as just picking one of them.

Here are the facts related to inO versus OoO CPUs:

- Since both design ideas are valid, there is a real inO CPU design like this, called *Xeon Phi*, manufactured by Intel. One model, Xeon Phi 5110P, has 60 inO cores and 4 threads in each core, making it capable of executing 240 threads. It is considered a Many Integrated Core (MIC) rather than a CPU; each core works at a very low speed like 1 GHz, but it gets its computational advantage from the sheer number of cores and threads. Since inO cores consume much less power, a 60C/240T Xeon Phi power consumption is only slightly higher than a comparable 6C/12T Core i7 CPU. I will be providing execution times on Xeon 5110P shortly.

- An inO CPU would only benefit a restricted set of applications; not every application can take advantage of so many cores or threads. In most applications, we get diminishing returns beyond a certain number of cores or threads. Generally, image and signal processing applications are perfect for inO CPUs or MICs. Scientific high performance processing applications are also typically a good candidate for inO CPUs.

- Another advantage of inO cores is their low power consumption. Since each core is much simpler, it does not consume as much power as a comparable OoO core. This is why most of today's netbooks incorporate Intel Atom CPUs, which have inO cores. An Atom CPU consumes only 2–10 Watts. The Xeon Phi MIC is basically 60 Atom cores, with 4 threads/core, stuffed into a chip.

- If having so many cores and threads can benefit even just a small set of applications, why not take this idea even farther and put thousands of cores in a compute unit that can execute even more than 4 threads per core? It turns out, even this idea is valid. Such a processor, which could execute something like hundreds of thousands of threads in thousands of cores, is called a *GPU*, which is what this book is all about!

3.2.2 Thin versus Thick Threads

When a multithreaded program is being executed, such as imflipP.c, a core can be assigned more than one thread to execute at run time. For example, in a 4C/8T CPU, what is the difference between having two threads in two separate cores versus both threads in the same core? The answer is: when two threads are *sharing* a core, they are sharing all of the core resources, such as the cache memory, integer, and floating point units.

In the event that two threads that need a large amount of cache memory are assigned to the same core, they will both be wasting time evicting data in and out of the cache memory, thereby not benefiting from multithreading. Assume that one thread needed a huge amount of cache memory access and another one needed only the integer unit and no near-zero cache memory access. These two threads would be good candidates to place in the same core during execution because they do not need the same resources during execution.

On the other hand, if a thread was designed by the programmer to require minimal core resources, it could benefit significantly from multithreading. These kinds of threads are called *thin threads*, whereas the ones that require excessive amounts of core resources are called *thick threads*. It is the responsibility of the programmer to design the threads carefully to avoid significant core resource usage; if every thread is thick, the benefit from the additional threads will be minimal. This is why OS designers, such as Microsoft, design their threads to be think to avoid interfering with the performance of the multithreaded applications. In the end, the OS is the ultimate multithreaded application.

3.3 PERFORMANCE OF IMFLIPP

Table 3.2 lists the execution times (in ms) of imflipP.c for the CPUs listed in Table 3.1. The total number of threads is only reported for values that make a difference. CPU2 results are a repeat of Table 2.1. The pattern for every CPU seems to be very similar: the performance gets better up to a certain number of threads and hits a brick wall! Launching more threads doesn't help beyond that inflection point and this point depends on the CPU.

Table 3.2 raises quite a few questions, such as:

- What does it mean to launch 9 threads on a 4C/8T CPU that is known to be able to execute a maximum of 8 threads (.../8T)?

- Maybe the right way to phrase this question is: is there a difference between *launching* and *executing* a thread?

- When we design a program to be "8-threaded," what are we assuming about *runtime*? Are we assuming that all 8 threads are *executing*?

- Remember from Section 2.1.5: there were 1499 threads launched on the computer, yet the CPU utilization was 0%. So, not every thread is *executing* in parallel. Otherwise, CPU utilization would be hitting the roof. If a thread is not *executing*, what is it doing? Who is managing these threads at runtime?

TABLE 3.2 imflipP.c execution times (ms) for the CPUs listed in Table 3.1.

#Threads		CPU1	CPU2	CPU3	CPU4	CPU5	CPU6
Serial	V	109	131	159	117	181	185
2	V	93	70	50	58	104	95
3	V	78	46	33	43	75	64
4	V	78	67	49	59	54	49
5	V	93	55	40	52	35	57
6	V	78	51	35	55	35	48
8	V	78	52	37	53	26	37
9	V		47	34	52	25	49
10	V			40		23	45
12	V			35		28	38
Serial	H	62	81	50	60	66	73
2	H	31	41	25	36	57	38
3	H	46	28	16	29	39	25
4	H	46	41	25	41	23	19
5	H		33	20	34	13	28
6	H		28	18	31	17	24
8	H		32	20	23	13	18
9	H		30	19	21	12	24
10	H			20		11	22
12	H			18		14	19

- Probably, the answer for why ≥ 4 threads is not helping our performance in Table 3.2 is hidden in these questions.

- Yet another question is whether the thick versus thin threads could change this answer.

- Another obvious question is: all of the CPUs in Table 3.1 are OoO.

3.4 EFFECT OF THE "OS" ON PERFORMANCE

There are numerous other questions we can raise which all ask the same thing: *what happens to a thread at runtime?* In other words, we know that the OS is responsible for *managing* the creation/joining of the virtual CPUs (threads), but it is time to understand the details. To go back to our list of *who is responsible for what*:

- **Programmer** determines *what a thread does* by writing a function for each thread. This determination is made **at compile time**, when no runtime information is available. The function is written in a *language* that is much higher level than the *machine code*: the only language the CPU understands. In the old days, programmers wrote machine code, which made program development possibly 100x more difficult. Now we have high level languages and compilers, so we can shift the burden to the compiler in a big way. The final product of the programmer is a *program*, which is a set of tasks to execute in a given order, packed with what-if scenarios. The purpose of the what-if scenarios is to respond well to **runtime** events.

- **Compiler** compiles the thread creation routines to machine code (CPU language), **at compile time**. The final product of the compiler is an *executable instruction list*, or the *binary executable*. Note that the compiler has minimal information (or idea) about what is going to happen **at runtime** when it is compiling from the programming language to machine code.

- The **OS** is responsible for **the runtime**. Why do we need such an intermediary? It is because a multitude of different things can happen when executing the binary that the compiler produced. BAD THINGS could happen such as: (1) the disk could be full, (2) memory could be full, (3) the user could enter a response that causes the program to crash, (4) the program could request an extreme number of threads, for which there are no available thread handles. Alternatively, even if nothing goes wrong, somebody has to be responsible for RESOURCE EFFICIENCY, i.e., running programs efficiently by taking care things like (1) who gets which virtual CPU, (2) when a program asks to create memory, should it get it, if so, what is the pointer, (3) if a program wants to create a child thread, do we have enough resources to create it? If so, what thread handle to give, (4) accessing disk resources, (5) network resources, (6) any other resource you can imagine. Resources are managed **at runtime** and there is no way to know them precisely at compile time.

- **Hardware** executes the machine code. The machine code for the CPU to execute is assigned to the CPU **at runtime** by the OS. Similarly, the memory is read and transferred mostly by the peripherals that are under the OS's control (e.g., Direct Memory Access – DMA controller).

- **User** enjoys the program which produces excellent results if the program was written well and everything goes right at runtime.

3.4.1 Thread Creation

The OS knows what resources it has, since most of them are statically determined once the computer turns on. The number of virtual CPUs is one of them, and the one we are most interested in understanding. If the OS determines that it is running on a processor with 8 virtual CPUs (as we would find on a 4C/8T machine), it assigns these virtual CPUs names, such as vCPU0, vCPU1, vCPU2, vCPU3, ..., vCPU7. So, in this case, the OS has 8 virtual CPU resources and is responsible for managing them.

When the program launches a thread using pthread_create(), the OS assigns a thread handle to that thread; say, 1763. So, the program sees `ThHandle[1]`=1763 at runtime. The program is interpreting this as "tid=1 was assigned the handle `ThHandle[1]`=1763." The program only cares about `tid`=1 and the OS only cares about 1763 inside its handle list. Although, the program must save this handle (1763), since this handle is its only way to tell the OS which thread it is talking about ... `tid`=1 or `ThHandle[]` are nothing more than program variables and have no significance to the internal workings of the OS.

3.4.2 Thread Launch and Execution

When the OS gives the parent thread `ThHandle[1]`= 1763 at runtime, the parent thread understands that it has the authorization to execute some function using this child thread. It launches the code with the function name that is built into pthread_create(). What this is telling the OS is that, in addition to *creating* the thread, now the parent wants to *launch* this thread. While creating a thread required a thread handle, launching a thread requires a virtual CPU to be assigned (i.e., find somebody to do the work). In other words, the parent thread is saying: *find a virtual CPU, and run this code on it.*

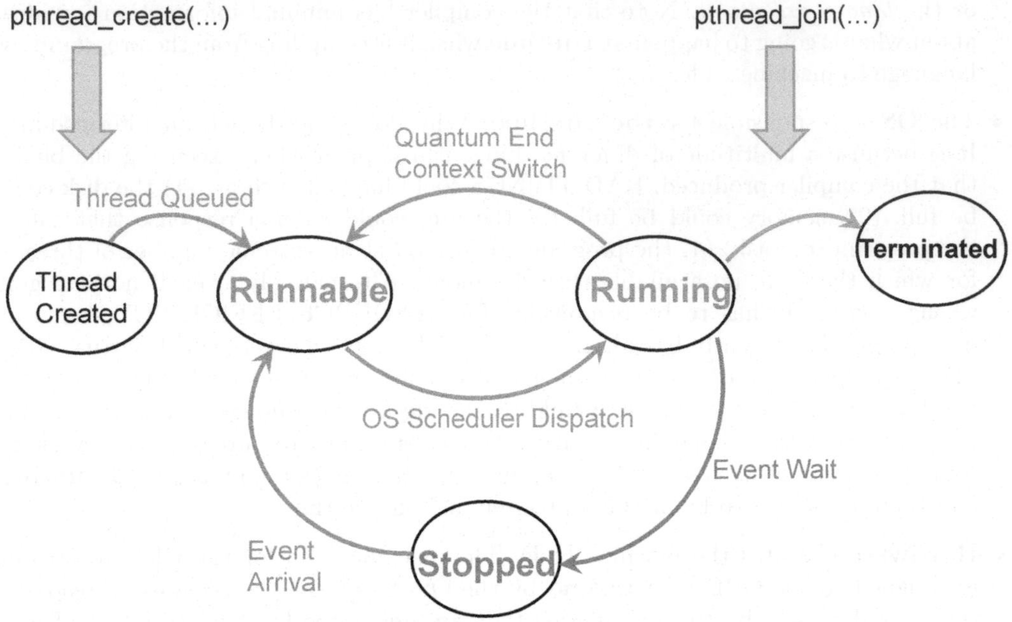

FIGURE 3.1 The life cycle of a thread. From the creation to its termination, a thread is cycled through many different statuses, assigned by the OS.

The OS, then, tries to find a virtual CPU resource that is available to execute this code. The parent thread does not care which virtual CPU the OS chooses, since this is a *resource management* issue that the OS is responsible for. The OS maps the thread handle it just assigned to an available virtual CPU at runtime (e.g., handle 1763 → vCPU4), assuming that virtual CPU 4 (vCPU4) is available right at the time that the pthread_create() is called.

3.4.3 Thread Status

Remember from Section 2.1.5 that there were 1499 threads launched on the computer, yet the CPU utilization was 0%. So, not every thread is *executing* in parallel. If a thread is not *executing*, then what is it doing? In summary, if a CPU has 8 virtual CPUs (as in a 4C/8T processor), no more than 8 threads could have the **running** status. This is the status that a thread possesses when it is executing; rather than being a *thought* by the OS when it was **runnable**, this thread is now on the CPU, executing (i.e., **running**). Aside from **running**, a thread could have the status of **runnable, stopped**, or in case it is done with its job, **terminated**, as depicted in Figure 3.1.

When our application calls pthread_create() to launch a thread, the OS immediately determines one thing: *do I have a sufficient amount of resources to assign a handle and create this thread?* If the answer is "Yes," the thread handle is assigned to that thread, and all of the necessary memory, stack areas are created. Right at that time, the status of the thread is recorded within that handle as **runnable**. This means that the thread *can* run, but it *is not running* yet. It basically goes into a queue of *runnable* threads and waits for its turn to come. At some point in time, the OS decides to start executing this thread.

For that, two things must happen: (1) a virtual CPU (vCPU) is found for that thread to execute on, (2) the status of the thread now changes to **running**.

The **Runnable⟹Running** status change is handled by a part of the OS called *the dispatcher*. The OS treats each one of the available CPU threads as a virtual CPU (vCPU); so, for example, a 8C/16T CPU has 16 vCPUs. The thread that was waiting in the queue starts running on a vCPU. A sophisticated OS will pay attention to where to place the threads to optimize the performance. This placement, called the *core affinity* can actually be manually modified by the user to override a potentially suboptimal placement by the OS.

The OS allows each thread to run for a certain period of time (called *quantum*) before it switches to another thread that has been waiting in the **Runnable** status. This is necessary to avoid *starvation*, i.e., a thread being stuck forever in the **Runnable** status. When a thread is switched from **Running⟹Runnable**, all of its register information – and more – has to be saved in an area; this information is called the *context* of a thread. Similarly, the **Running⟹Runnable** status change is called a *context switch*. A context switch takes a certain amount of time to complete and has performance implications, although it is an unavoidable reality.

During execution (in the **Running** status), a thread might call a function, say scanf(), to read a keyboard input. Reading a keyboard is much slower than any other CPU operation; so, there is no reason why the OS should make our thread keep in the **Running** status while waiting for the keyboard input, which would starve other threads of core time. In this case, the OS cannot switch this thread to **Runnable** either, since the **Runnable** queue is dedicated to the threads that can be immediately switched over to the **Running** status when the time is right. A thread that is waiting for a keyboard input could wait for an indefinite amount of time; it could happen immediately or it could happen within 10 minutes, in case the user has left to get coffee! So, there is another status to distinguish this specific status; it is called **Stopped**.

A thread undergoes a **Running⟹Stopped** status switch when it requests a resource that is not going to be available for a period of time, or it has to wait for an event to happen for an indeterminate amount of time. When the requested resource (or data) becomes available, the thread undergoes a **Stopped⟹Runnable** status switch and is placed in the queue of **Runnable** threads that are waiting for their time to start executing again. It would make no sense for the OS to switch this thread to **Running** either, as this would mean a chaotic unscheduling of another peacefully executing thread, i.e., kicking it out of the core! So, to do things in a calm and orderly fashion, the OS places the **Stopped** thread back in the **Runnable** queue and decides when to allow it to execute again later. It might, however, assign a different *priority* to threads that should be dispatched ahead of others for whatever reason.

Finally, when a thread completes execution, upon calling, say, the pthread_join() function, the OS makes the **Running⟹Terminated** status switch and the thread is permanently out of the **Runnable** queue. Once its memory areas etc. are cleaned up, the handle for that thread is destroyed and it is available later for another pthread_create().

3.4.4 Mapping Software Threads to Hardware Threads

Section 3.4.3 answers our question about the 1499 threads in Figure 2.1: we know that out of the 1499 threads that we see in Figure 2.1, at least 1491 threads must be either **Runnable** or **Stopped** on a 4C/8T CPU, since no more than 8 threads could be in the **Running** status. Think of 1499 as the number of tasks to do, but there are only 8 people to do it! The OS simply does not have physical resources to "do" (i.e., execute) more than 8 things at any given point in time. It picks one of the 1499 tasks, and assigns one of the people to do it. If another task becomes more urgent for that one person (for example, if a network packet arrives requiring immediate attention), the OS switches that person to doing that more urgent task and suspends what he or she was currently doing.

We are curious about how these status switches affect our application's performance. In the case of 1499 threads in Figure 2.1, it is highly likely that something like 1495 threads are **Stopped** or **Runnable**, waiting for you to hit some key on the keyboard or a network packet to arrive, and only four threads are **Running**, probably your multithreaded application code. Here is an analogy:

ANALOGY 3.2: *Thread Status.*

You look through your window and see 1499 pieces of paper on the curb with written tasks on them. You also see 8 people outside, all sitting on their chair until the manager gives them a task to execute. At some point, the manager tells person #1 to grab paper #1256. Then, person #1 starts doing whatever is written on paper #1256. All of a sudden, the manager tells person #1 to bring that paper #1256 back, stop executing task #1256 and grab paper #867 and start doing what is written on paper #867 ...

Since the execution of task #1256 is not yet complete, all of the notes that person #1 took during the execution of task #1256 must be written somewhere, on the manager's notebook, for person #1 to remember it later. Indeed, this task might not even be completed by the same person. If the task on paper #867 is finished, it could be crumpled up and thrown into the waste basket. That task is complete.

In Analogy 3.2, *sitting on the chair* corresponds to the thread status **Runnable** and *doing what is written on the paper* corresponds to the thread status **Running** and the *people* are the *virtual CPUs*. The *manager*, who is allowed to switch the status of people, is the OS, whereas his or her notebook is where the *thread contexts* are saved to be used later during a context switch. Crumpling up a paper (task) is equivalent to switching it to the **Terminated** status.

The number of the launched threads could fluctuate from 1499 to 1505 down to 1365, etc., but the number of available virtual CPUs cannot change (e.g., 8 in this example), since they are a "physical" entity. A good way to define the 1499 quantity is **software threads**, i.e., the threads that the OS creates. The available number of physical threads (virtual CPUs) is the **hardware threads**, i.e., the maximum number of threads that the CPU manufacturer designed the CPU to be able to execute. It is a little bit confusing that both of them are called "threads," since the software threads are nothing but a data structure containing information about the task that the thread will perform as well as the thread handle, memory areas, etc., whereas the hardware threads are the physical hardware component of the CPU that is executing machine code (i.e., the compiled version of the task). The job of the OS is to find an available hardware thread for each one of the software threads it is managing. The OS is responsible for managing the virtual CPUs, which are hardware resources, much like the available memory.

3.4.5 Program Performance versus Launched Pthreads

The maximum number of *software threads* is only limited by the internal OS parameters, whereas the number of *hardware threads* is set in stone at the time the CPU is designed. When you launch a program that executes two heavily active threads, the OS will do its best to bring them into the **Running** status as soon as possible. Possibly one more thread, belonging to the OS's thread scheduler is very active, making the highly active threads 3.

So, how does this help in explaining the results in Table 3.2? Although the exact answer depends on the model of the CPU, there are some highly distinguishable patterns that can be explained with what we just learned. Let us pick one CPU2 as an example. While CPU2 should be able to execute 8 threads in parallel (it is a 4C/8T), the performance falls off a cliff beyond 3 launched threads. Why? Let's try to guess this by fact-checking:

- Remember our Analogy 1.1, where two farmers were sharing a tractor. By timing the tasks perfectly, together, they could get 2x more work done. This is the hope behind 4C/8T getting an 8T performance, otherwise, you really have only 4 physical cores (i.e., tractors).

- If this best case scenario happened here in our Code 2.1 and Code 2.2, we should expect the performance improvement to continue to 8 threads, or at least, something like 6 or 7. This is not what we see in Table 3.2!

- So, what if one of the tasks required one of the farmers to use the hammer and other resources in the tractor in a chaotic way? The other wouldn't be able to do anything useful since they would be continuously bumping into each other and keep falling, etc; the performance wouldn't even be close to 2x (i.e., $1+1=2$)! It would be more like 0.9x! As far as efficiency is concerned, $1+1=0.9$ sounds pretty bad! In other words, if both threads are "thick threads," they are not meant to work simultaneously with another thread inside the same core ... I mean, *efficiently* ... This must be somehow the case in Code 2.1 and Code 2.2, since we are not getting anything out of the dual threads inside each core ...

- What about memory? We will see an entire architectural organization of the cores and memory in Chapter 4. But, for now, it suffices to say that, no matter how many cores/threads you have in a CPU, you only have a single *main memory* for all of the threads to share. So, if one thread was a memory-unfriendly thread, it would mess up everybody's memory accesses. This is another possibility in explaining why the performance hits a brick wall at ≥ 4 threads.

- Let's say that we explained the problems with why we are not able to use the double-threads in each core (called hyper-threading by Intel), but why does the performance stop improving at 3 threads, not 4? The performance from 3 to 4 threads is lower, which is counterintuitive. Are these threads not even able to use all of the cores? A similar pattern is visible in almost every CPU, although the exact thread count depends on the maximum available threads and varies from CPU to CPU.

3.5 IMPROVING IMFLIPP

Instead of answering all of these questions, let's see if we can improve the program without knowing all of these answers, but simply guessing what the problem could be. After all, we have enough intuition to be able to make educated guesses. In this section, we will analyze the code and will try to identify the parts of the code that could be causing inefficiencies and will suggest a fix. After implementing this fix, we will see how it performed and will explain why it worked (or didn't).

Where is the best place to start? If you want to improve a computer program's performance, the best place to start is the innermost loops. Let's start with the MTFlipH() function shown in Code 2.8. This function is taking a pixel value and moving it to another memory area one byte at a time. The MTFlipV() function shown in Code 2.7 is very similar. For each pixel, both functions move R, G, and B values one byte at a time. What is wrong with this picture? A lot! When we go through the details of the CPU and memory architecture in Chapter 4, you will be amazed with how horribly inefficient Code 2.7 and Code 2.8 are. But, for now, we just want to find obvious fixes and apply them and observe the improvements quantitatively. We will not comment on them until we learn more about the memory/core architecture in Chapter 4.

3.5.1 Analyzing Memory Access Patterns in MTFlipH()

The MTFlipH() function is clearly a "memory-intensive" function. For each pixel, there is really no "computation" done. Instead, one byte is moved from one memory location to another. When I say "computation," I mean something like making each pixel value darker by reducing the RGB values, or turning the image into a B&W image by recalculating a new value for each pixel, etc. None of these computations are being performed here. The innermost loop of MTFlipH() looks something like this:

```
...
for(row=ts; row<=te; row++) {
   col=0;
   while(col<ip.Hpixels*3/2){
      // example: Swap pixel[42][0] , pixel[42][3199]
      pix.B = TheImage[row][col];
      pix.G = TheImage[row][col+1];
      pix.R = TheImage[row][col+2];
      TheImage[row][col] = TheImage[row][ip.Hpixels*3-(col+3)];
      ...
```

So, to improve this program, we should strictly analyze the memory access patterns. Figure 3.2 shows the memory access patterns of the MTFlipH() function during the processing of the 22 MB image dogL.bmp. This dog picture consists of 2400 rows and 3200 columns. For example, when flipping Row 42 horizontally (no specific reason for choosing this number), here is the swap pattern for pixels (also shown in Figure 3.2):

[42][0]⟷[42][3199], [42][1]⟷[42][3198] ... [42][1598]⟷[42][1601], [42][1599]⟷[42][1600]

3.5.2 Multithreaded Memory Access of MTFlipH()

Not only the byte-by-byte memory access of MTFlipH() sounds bad, but also remember that this function is running in a multithreaded environment. First, if we launched only a single thread, let's look at what a single thread's memory access patterns look like: This single thread would want to flip all 2400 rows by itself, starting at row 0, and continuing with row 1, 2, ... 2399. During this loop, when it is flipping row 42 as an example again, which "bytes" does MTHFlip() really swap? Let's take the very first pixel swap as an example. It involves the following operations:

Swap pixel[42][0] with pixel[42][3199], which corresponds to

Swap bytes[0..2] of row 42 with bytes [9597..9599] of row 42, consecutively.

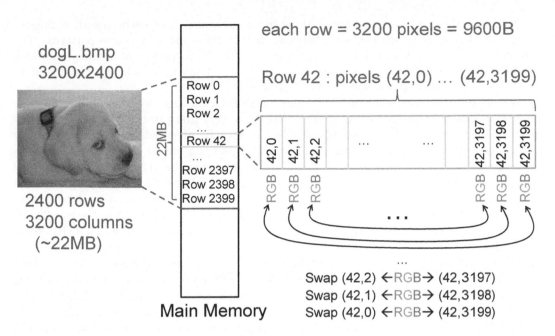

each row = 3200 pixels = 9600B

dogL.bmp
3200x2400

2400 rows
3200 columns
(~22MB)

Row 42 : pixels (42,0) ... (42,3199)

Swap (42,2) ←RGB→ (42,3197)
Swap (42,1) ←RGB→ (42,3198)
Swap (42,0) ←RGB→ (42,3199)

Main Memory

FIGURE 3.2 Memory access patterns of MTFlipH() in Code 2.8. A total of 3200 pixels' RGB values (9600 Bytes) are flipped for each row.

In Figure 3.2, notice that each pixel corresponds to 3 consecutive bytes holding that pixel's RGB values. During just this one pixel swap, the function MTFlipH() requests 6 memory accesses, 3 to read the bytes [0..2] and 3 to write them into the flipped pixel location held at bytes [9597..9599]. This means that, to merely flip one row, our MTFlipH() function requests $3200 \times 6 = 19,200$ memory accesses, with mixed read and writes. Now, let's see what happens when, say, 4 threads are launched. Each thread is trying to finish the following tasks, consisting of flipping 600 rows.

```
tid=0 :  Flip Row[0]    , Flip Row[1]      ...  Flip Row [598] , Flip Row [599]
tid=1 :  Flip Row[600]  , Flip Row[601]    ...  Flip Row [1198] , Flip Row [1199]
tid=2 :  Flip Row[1200] , Flip Row[1201]   ...  Flip Row [1798] , Flip Row [1799]
tid=3 :  Flip Row[1800] , Flip Row[1801]   ...  Flip Row [2398] , Flip Row [2399]
```

Notice that each one of the 4 threads is requesting as frequent memory accesses as the single thread does. If each thread was designed improperly, causing chaotic memory access requests, 4 of them together will have 4x the mess! Let's look at the very early part of the execution when main() launches all 4 threads and assigns them the MTHFlip() function to execute. If we assume that all 4 threads started executing at exactly the same time, this is what all 4 threads are trying to do *simultaneously* for the first few bytes:

```
tid=0 : Flip Row[0]    : mem(00000000..00000002)⟷mem(00009597..00009599) , ...
tid=1 : Flip Row[600]  : mem(05760000..05760002)⟷mem(05769597..05769599) , ...
tid=2 : Flip Row[1200] : mem(11520000..11520002)⟷mem(11529597..11529599) , ...
tid=3 : Flip Row[1800] : mem(17280000..17280002)⟷mem(17289597..17289599) , ...
```

Although there will be slight variations in the progression of each thread, it doesn't change the story. What do you see when you look at these memory access patterns? The very first thread, `tid`=0 is trying to read the pixel [0][0], whose value is at memory addresses mem(00000000..00000002). This is the very beginning of `tid`= 0's task that requires it to swap the entire row 0 first before it moves on to row 1.

While `tid`=0 is waiting for its 3 bytes to come in from the memory, at precisely the same time, `tid`=1 is trying to read pixel[600][0] that is the very first pixel of row 600, located at memory addresses mem(05760000..05760002), i.e., 5.5 MB (Megabytes) away from the very first request. Hold on, `tid`=2 is not standing still. It is trying to do its own job that starts by swapping the entire row 1200. The first pixel to be read is pixel[1200][0], located at the 3 consecutive bytes with memory addresses mem(11520000..11520002), i.e., 11 MB away from the 3 bytes that `tid`=0 is trying to read. Similarly, `tid`=3 is trying to read the 3 bytes that are 16.5 MB away from the very first 3 bytes ... Remember that the total image was 22 MB and the processing of it was divided into 4 threads, each responsible for a 5.5 MB chunk (i.e., 600 rows).

When we learn the detailed inner-workings of a DRAM (Dynamic Random Access Memory) in Chapter 4, we will understand why this kind of a memory access pattern is nothing but a disaster, but, for now, we will find a very simple fix for this problem. For the folks who are craving to get into the GPU world, let me make a comment here that the DRAM in the CPU and the GPU are almost identical operationally. So, anything we learn here will be readily applicable to the GPU memory with some exceptions resulting from the massive parallelism of the GPUs. An identical "disaster memory access" example will be provided for the GPUs and you will be able to immediately guess what the problem is by relying on what you learned in the CPU world.

3.5.3 DRAM Access Rules of Thumb

While it will take a good portion of Chapter 4 to understand why these kinds of *choppy* memory accesses are bad for DRAM performance, fixing the problem is surprisingly simple and intuitive. All you need to know is the rules of thumb in Table 3.3 that are a good guide to achieving good DRAM performance. Let's look at this table and make sense out of it. These rules are based on the architecture of the DRAM that is designed to allow the data to be shared by every CPU core, and they all say the same thing one way or another:

> ➤ *When accessing DRAM, access big consecutive chunks like*
> *1 KB, 4 KB, etc., rather than small onesy-twosy bytes ...*

While this is an excellent guide to improving the performance of our first parallel program imflipP.c, let's first check to see if we were obeying these rules in the first place. Here is the summary of the MTFlipH() function's memory access patterns (Code 2.8):

- **Granularity** rule is clearly violated, since we are trying to access one byte at a time.

- **Locality** rule wouldn't be violated if there was only a single thread. However, multiple simultaneous (and distant) accesses by different threads cause violations.

- **L1, L2, L3** caching do not help us at all since there isn't a good "data reuse" scenario. This is because we never need any data element more than once.

With almost every rule violated, it is no wonder that the performance of imflipP.c is miserable. Unless we obey the access rules of DRAM, we are just creating massive inefficient memory access patterns that cripple the overall performance.

TABLE 3.3 Rules of thumb for achieving good DRAM performance.

Rule	Ideal Values	Description
Granularity	8 B...64 B	The size of each read/write. "Too small" values are extremely inefficient (e.g., one byte at a time)
Locality	1 KB ... 4 KB	If consecutive accesses are too far from each other, they force the row buffer to be flushed (i.e., they trigger a new DRAM row-read)
L1, L2 Caching	64 KB ... 256 KB	If the total number of bytes read/written repeatedly by a single thread are confined into a small region like this, the data can be L1- or L2-cached, thereby dramatically improving re-access speed by that thread
L3 Caching	8 MB ... 20 MB	If the total number of bytes read/written repeatedly by all threads are confined into a small region like this, the data can be L3-cached, thereby dramatically improving re-access speed by every core

3.6 IMFLIPPM: OBEYING DRAM RULES OF THUMB

It is time to improve imflipP.c by making it obey the rules of thumb in Table 3.3. The improved program is imflipPM.c ("M" for "memory-friendly").

3.6.1 Chaotic Memory Access Patterns of imflipP

Again, let's start by analyzing MTFlipH(), one of the un-memory-friendly functions of imflipP.c. When we are reading bytes and replacing with other bytes, we are accessing DRAM for every single byte of the pixel individually as follows:

```
for(row=ts; row<=te; row++) {
    col=0;
    while(col<ip.Hpixels*3/2){
        pix.B = TheImage[row][col];
        pix.G = TheImage[row][col+1];
        pix.R = TheImage[row][col+2];
        TheImage[row][col]   = TheImage[row][ip.Hpixels*3-(col+3)];
        TheImage[row][col+1] = TheImage[row][ip.Hpixels*3-(col+2)];
        TheImage[row][col+2] = TheImage[row][ip.Hpixels*3-(col+1)];
        TheImage[row][ip.Hpixels*3-(col+3)] = pix.B;
        TheImage[row][ip.Hpixels*3-(col+2)] = pix.G;
        TheImage[row][ip.Hpixels*3-(col+1)] = pix.R;
        col+=3;
    }
    ...
```

The key observation is that, since the dog picture is in the main memory (i.e., DRAM), every single pixel-read triggers an access to DRAM. According to Table 3.3, we know that DRAM doesn't like to be bothered frequently.

3.6.2 Improving Memory Access Patterns of imflipP

What if we read the entire row of the image (all 3200 pixels, totaling 9600 bytes) into a temporary area (somewhere other than DRAM) and then processed it in that confined area without ever bothering the DRAM during the processing of that row? We will call this area a *Buffer*. Since the buffer is small enough, it will be cached inside L1$ and will allow us to take advantage of L1 caching. Well, at least we are now using the cache memory and obeying the cache-friendliness rules in Table 3.3.

```
unsigned char Buffer[16384]; // This is the buffer to use to get the entire row
...
for(row=ts; row<=te; row++) {
   // bulk copy from DRAM to cache
   memcpy((void *) Buffer, (void *) TheImage[row], (size_t) ip.Hbytes);
   col=0;
   while(col<ip.Hpixels*3/2){
      pix.B = Buffer[col];
      pix.G = Buffer[col+1];
      pix.R = Buffer[col+2];
       Buffer[col]   = Buffer[ip.Hpixels*3-(col+3)];
       Buffer[col+1] = Buffer[ip.Hpixels*3-(col+2)];
       Buffer[col+2] = Buffer[ip.Hpixels*3-(col+1)];
      Buffer[ip.Hpixels*3-(col+3)] = pix.B;
      Buffer[ip.Hpixels*3-(col+2)] = pix.G;
      Buffer[ip.Hpixels*3-(col+1)] = pix.R;
      col+=3;
   }
   // bulk copy back from cache to DRAM
   memcpy((void *) TheImage[row], (void *) Buffer, (size_t) ip.Hbytes);
   ...
```

When we transfer the 9600 B from the main memory into the `Buffer`, we are relying on the efficiency of the memcpy() function, which is provided as part of the standard C library. During the execution of memcpy(), 9600 bytes are transferred from the main memory into the memory area that we name `Buffer`. This access is super efficient, since it only involves a single continuous memory transfer that obeys every rule in Table 3.3.

Let's not kid ourselves: `Buffer` is also in the main memory; however, there is a huge difference in the way we will use these 9600 bytes. Since we will access them continuously, they will be cached and will no longer bother DRAM. This is what will allow the accesses to the `Buffer` memory area to be significantly more efficient, thereby obeying most of the rules in Table 3.3. Let us now re-engineer the code to use the `Buffer`.

CODE 3.1: imflipPM.c   MTFlipHM() {...}

Memory-friendly version of MTFlipH() (Code 2.8) that obeys the rules in Table 3.3.

```c
void *MTFlipHM(void* tid)
{
   struct Pixel pix;  //temp swap pixel
   int row, col;
   unsigned char Buffer[16384];  // This is the buffer to use to get the entire row

   long ts = *((int *) tid);     // My thread ID is stored here
   ts *= ip.Vpixels/NumThreads;  // start index
   long te = ts+ip.Vpixels/NumThreads-1;  // end index

   for(row=ts; row<=te; row++){
      // bulk copy from DRAM to cache
      memcpy((void *) Buffer, (void *) TheImage[row], (size_t) ip.Hbytes);
      col=0;
      while(col<ip.Hpixels*3/2){
         pix.B = Buffer[col];
         pix.G = Buffer[col+1];
         pix.R = Buffer[col+2];
          Buffer[col]   = Buffer[ip.Hpixels*3-(col+3)];
          Buffer[col+1] = Buffer[ip.Hpixels*3-(col+2)];
          Buffer[col+2] = Buffer[ip.Hpixels*3-(col+1)];
         Buffer[ip.Hpixels*3-(col+3)] = pix.B;
         Buffer[ip.Hpixels*3-(col+2)] = pix.G;
         Buffer[ip.Hpixels*3-(col+1)] = pix.R;
         col+=3;
      }
      // bulk copy back from cache to DRAM
      memcpy((void *) TheImage[row], (void *) Buffer, (size_t) ip.Hbytes);
   }
   pthread_exit(NULL);
}
```

3.6.3   MTFlipHM(): The Memory Friendly MTFlipH()

The memory-friendly version of the MTFlipH() function in Code 2.8 is the MTFlipHM() function shown in Code 3.1. They are virtually identical with one distinctive difference: To flip each row's pixels, MTFlipHM() accesses the DRAM only once to read a large chunk of data using the memcpy() function (entire row of the image, e.g., 9600 B for dogL.bmp). A 16 KB buffer memory array is defined as a local array variable and the entire contents of the row are copied into the buffer before even getting into the innermost loop that starts swapping the pixels. We could have very well defined a 9600 B buffer since this is all we need for our image, but a larger buffer allows scalability to larger images.

Although the innermost loops are identical inside both functions, notice that only the Buffer[] array is accessed inside the while loop of MTFlipHM(). We know that the OS assigns a stack area for all local variables and we will get into great details of this in Chapter 5. But, for now, the most important observation is the definition of a localized storage area in the 16 KB range that will allow MTFlipHM() to be compliant with the L1 caching rule of thumb, shown in Table 3.3.

Here are some lines from Code 3.1, highlighting the buffering in MTFlipHM(). Note that the global array `TheImage[]` is in DRAM, since it was read into DRAM by the ReadBMP() function (see Code 2.5). This is the variable that should obey strict DRAM rules in Table 3.3. I guess we cannot do better than accessing it once to read 9600 B of data and copying this data into our local memory area. This makes it 100% DRAM-friendly.

```
unsigned char Buffer[16384];  // This is the buffer to use to get the entire row
...
for(...){
   // bulk copy from DRAM to cache
   memcpy((void *) Buffer, (void *) TheImage[row], (size_t) ip.Hbytes);
   ...
   while(...){
      ...  =Buffer[...]
      ...  =Buffer[...]
      ...  =Buffer[...]
      Buffer[...]=Buffer[...]
      ...
      Buffer[...]=...
      Buffer[...]=...
      ...
   }
   // bulk copy back from cache to DRAM
   memcpy((void *) TheImage[row], (void *) Buffer, (size_t) ip.Hbytes);
   ...
```

The big question is: Why is the local variable `Buffer[]` fair game? We modified the innermost loop and made it access the `Buffer[]` array as terribly as we were accessing `TheImage[]` before. What is so different with the `Buffer[]` array? Also, another nagging question is the claim that the contents of the `Buffer[]` array will be "cached." Where did that come from? There is no indication in the code that says "put these 9600 bytes into the cache." How are we so sure that it does go into cache? The answer is actually surprisingly simple and has everything to do with the design of the CPU architecture.

A CPU caching *algorithm* predicts which values inside DRAM (the "bad area") should be temporarily brought into the cache (the "good area"). These guesses do not have to be 100% accurate, since if the guess is bad, it could always be corrected later. The result is an efficiency penalty rather than a crash or something. Bringing "recently used DRAM contents" into the cache memory is called **caching**. The CPU could get lazy and bring *everything* into the cache, but this is not possible since there are only small amounts of cache memory available. In the i7 family processors, L1 cache is 32 KB for data elements and L2 cache is 256 KB. L1 is faster to access than L2. Caching helps for three major reasons:

- **Access Patterns:** Cache memory is SRAM (static random access memory), not DRAM like the main memory. The rules governing SRAM access patterns are a lot less strict as compared to the DRAM efficiency rules listed in Table 3.3.
- **Speed:** Since SRAM is much faster than DRAM, accessing cache is substantially faster once something is cached.
- **Isolation:** Each core has its own cache memory (L1$ and L2$). So, if each thread was accessing up to a 256 KB of data frequently, this data would be very efficiently cached in that core's cache and would not bother the DRAM.

CODE 3.2:   imflipPM.c   MTFlipVM() {...}

Memory-friendly version of MTFlipV() (Code 2.7) that obeys the rules in Table 3.3.

```
void *MTFlipVM(void* tid)
{
   struct Pixel pix;          //temp swap pixel
   int row, row2, col;        // need another index pointer ...
   unsigned char Buffer[16384]; // This is the buffer to get the first row
   unsigned char Buffer2[16384]; // This is the buffer to get the second row

   long ts = *((int *) tid);            // My thread ID is stored here
   ts *= ip.Vpixels/NumThreads/2;       // start index
   long te = ts+(ip.Vpixels/NumThreads/2)-1;  // end index

   for(row=ts; row<=te; row++){
      memcpy((void *) Buffer, (void *) TheImage[row], (size_t) ip.Hbytes);
      row2=ip.Vpixels-(row+1);
      memcpy((void *) Buffer2, (void *) TheImage[row2], (size_t) ip.Hbytes);
      // swap row with row2
      memcpy((void *) TheImage[row], (void *) Buffer2, (size_t) ip.Hbytes);
      memcpy((void *) TheImage[row2], (void *) Buffer, (size_t) ip.Hbytes);
   }
   pthread_exit(NULL);
}
```

We will get into the details of how the CPU cores and CPU main memory work together in Chapter 4. However, we have learned enough so far about the concept of *buffering* to improve our code. Note that caching is extremely important for both CPUs and even more so in GPUs. So, understanding the buffering concept that causes data to be cached is extremely important. There is no way to tell the CPU to cache something explicitly, although some theoretical research has investigated this topic. It is done completely automatically by the CPU. However, the programmer can influence the caching dramatically by the memory access patterns of the code. We experienced first-hand what happens when the memory access patterns are chaotic like the ones shown in Code 2.7 and Code 2.8. The CPU caching algorithms simply cannot correct for these chaotic patterns, since their simplistic caching/ eviction algorithms throw in the towel. The compiler cannot correct for these either, since it literally requires the compiler to read the programmer's mind in many cases! The only thing that can help the performance is the *logic* of the programmer.

3.6.4 MTFlipVM(): The Memory Friendly MTFlipV()

Now, let's take a look at the redesigned MTFlipVM() function in Code 3.2. We see some major differences between this code and its inefficient version, the MTFlipV() function, in Code 2.7. Here are the differences between MTFlipVM() and MTFlipV():

- There are two buffers used in the improved version: 16 KB each.
- In the outermost loop, the first buffer is used to read an entire *start image row*, where the second buffer is used to read an entire *end image row*. They are swapped.
- The innermost loop is eliminated and replaced with bulk memory transfers using the buffers, although the outermost loop is identical.

TABLE 3.4 imflipPM.c execution times (ms) for the CPUs listed in Table 3.1.

#Threads		CPU1	CPU2	CPU3	CPU4	CPU5	CPU6
Serial	W	4.116	5.49	3.35	4.11	5.24	3.87
2	W	3.3861	3.32	2.76	2.43	3.51	2.41
3	W	3.0233	2.90	2.66	1.96	2.78	2.52
4	W	3.1442	3.48	2.81	2.21	1.57	1.95
5	W	3.1442	3.27	2.71	2.17	1.47	2.07
6	W		3.05	2.73	2.04	1.69	2.00
8	W		3.02	2.75	2.03	1.45	2.09
9	W			2.74		1.45	2.26
10	W			2.74	1.98	1.45	1.93
12	W			2.75		1.33	1.91
Serial	I	35.8	49.4	29.0	34.6	45.3	42.6
2	I	23.7	25.2	14.7	17.6	34.5	21.4
3	I	21.2	17.4	9.8	12.3	19.5	14.3
4	I	22.7	20.1	14.6	17.6	12.5	10.9
5	I	22.3	17.1	11.8	14.3	8.8	15.8
6	I	21.8	15.8	10.5	11.8	10.5	13.2
8	I		18.4	10.4	12.1	8.3	10.0
9	I			9.8		7.5	13.5
10	I		16.6	9.5	11.6	6.9	12.3
12	I			9.2		8.6	11.2

3.7 PERFORMANCE OF IMFLIPPM.C

To run the improved program imflipPM.c, the following command line is used:

imflipPM InputfileName OutputfileName [v/V/h/H/w/W/i/I] [1-128]

where the newly added command line options **W** and **I** are to allow using the memory-friendly MTFlipVM() and MTFlipHM() functions, respectively. Upper or lower case does not matter, hence the option listing **W/w** and **I/i**. The older **V** and **H** options still work and allow access to the memory-unfriendly functions MTFlipV() and MTFlipH(), respectively. This helps us to compare the two families of functions by running only a single program.

Execution times of the improved program imflipPM.c are shown in Table 3.4. When we compare these results to their "memory unfriendly" counterpart imflipP.c (listed in Table 3.2), we see major improvement in performance all across the board. This shouldn't be surprising to the readers, since I wouldn't drag you through an entire chapter to show marginal improvements! This wouldn't make happy readers!

In addition to being substantial, the improvements also differ substantially based on whether it is for the vertical or horizontal flip. Therefore, instead of making generic comments, let's dig deep into the results by picking an example CPU and listing the memory-friendly and memory-unfriendly results side by side. Since almost every CPU is showing identical improvement patterns, there is no harm in commenting on a representative CPU. The best pick is CPU5 due to the richness of the results and the possibility of extending the analysis beyond just a few cores.

3.7.1 Comparing Performances of imflipP.c and imflipPM.c

Table 3.5 lists the imflipP.c and imflipPM.c results only for CPU5. A new column is added to compare the improvement in speed (i.e., "Speedup") between the memory-friendly functions

TABLE 3.5 Comparing imflipP.c execution times (H, V type flips in Table 3.2) to imflipPM.c execution times (I, W type flips in Table 3.4).

#Threads	CPU5 V	CPU5 W	Speedup V → W	CPU5 H	CPU5 I	Speedup H → I
Serial	181	5.24	34×	66	45.3	1.5×
2	104	3.51	30×	57	34.5	1.7×
3	75	2.78	27×	39	19.5	2×
4	54	1.57	34×	23	12.5	1.8×
5	35	1.47	24×	13	8.8	1.5×
6	35	1.69	20×	17	10.5	1.6×
8	26	1.45	18×	13	8.3	1.6×
9	25	1.45	17×	12	7.5	1.6×
10	23	1.45	16×	11	6.9	1.6×
12	28	1.33	21×	14	8.6	1.6×

MTFlipVM() and MTFlipHM() and the unfriendly ones MTFlipV() and MTFlipH(). It is hard to make a generic comment such as "major improvement all across the board," since this is not really what we are seeing here. The improvements in the horizontal-family and vertical-family functions are so different that we need to comment on them separately.

3.7.2 Speed Improvement: MTFlipV() versus MTFlipVM()

First, let's look at the vertical flip function MTFlipVM(). There are some important observations when going from the MTFlipV() function ("V" column) in Table 3.5 to the MTFlipVM() function ("W" column):

- Speed improvement changes when we launch a different number of threads

- Launching more threads reduces the speedup gap (34× down to 16×)

- Speedup does continue even for a number of threads that the CPU cannot physically support (e.g., 9, 10).

3.7.3 Speed Improvement: MTFlipH() versus MTFlipHM()

Next, let's look at the horizontal flip function MTFlipHM(). Here are the observations when going from the MTFlipH() function ("H" column) in Table 3.5 to the MTFlipHM() function ("I" column):

- Speed improvement changes much less as compared to the vertical family.

- Launching more threads changes the speedup a little bit, but the exact trend is hard to quantify.

- It is almost possible to quantify the speedup as a "fixed 1.6," with some minor fluctuations.

3.7.4 Understanding the Speedup: MTFlipH() versus MTFlipHM()

Table 3.5 will take a little bit of time to digest. We will need to go through an entire Chapter 4 to appreciate what is going on. However, we can make some guesses in this chapter. To be able to make educated guesses, let's look at the facts. First, let's explain why there would be a difference in the vertical versus horizontal family of flips, although, in the end, both of the functions are flipping exactly the same number of pixels:

Comparing the MTFlipH() in Code 2.8 to its memory-friendly version MTFlipHM() in Code 3.1, we see that the only difference is the local buffering, and the rest of the code is identical. In other words, if there is any speedup between these two functions, it is strictly due to buffering. So, it is fair to say that

---

➢ *Local buffering allowed us to utilize cache memory,*
  which resulted in a 1.6× speedup.
➢ *This number fluctuates minimally with more threads.*

On the other hand, comparing the MTFlipV() in Code 2.7 to its memory-friendly version MTFlipVM() in Code 3.2, we see that we turned the function from a core-intensive function to a memory-intensive function. While MTFlipV() is picking at the data one byte at a time, and keeping the core's internal resources completely busy, MTFlipVM() uses the memcpy() bulk memory copy function and does everything through the bulk memory transfer, possibly completely bypassing the core involvement. The magical memcpy() function is extremely efficient to copy something from DRAM when you are grabbing a big chunk of data, like we are here. This is also consistent with our DRAM efficiency rules in Table 3.3.

If this is all true, why is the speedup somehow saturating? In other words, why are we getting a lower speedup when we launch more threads? It looks like the program execution time is not going below a $\approx 1.5\times$ speedup, no matter what the number of threads is. This can actually be explained intuitively as follows:

➢ *When a program is highly memory-intensive, its performance*
  will be strictly determined by the memory bandwidth.
➢ *We seemed to have saturated the memory bandwidth at ≈ 4 threads.*

3.8 PROCESS MEMORY MAP

What happens when you launch the following program at the command prompt?

imflipPM dogL.bmp Output.bmp V 4

First of all, we are asking the executable program imflipPM (or, imflipPM.exe in Windows) to be *launched*. To launch this program (i.e., start executing), the OS creates a *process* with a *Process ID* assigned to it. When this program is executing, it will need three different memory areas:

- A *stack* area to store the function call return addresses and arguments that are passed/returned onto/from the function calls. This area grows from top to bottom (from high address to the low addresses), since this is how each microprocessor uses a stack.

- A *heap* area to store the dynamically allocated memory contents using the malloc() function. This memory area grows in the opposite direction of the stack to allow the OS to use every possible byte of memory without bumping into the stack contents.

- A *code* area to store the program code and the constants that were declared within the program. This is the code area and is not modified. The constants within the program are stored here, since they also are not modified.

The memory map of the process that the OS created will look like Figure 3.3: First, since the program is launched with only a single thread that is running main(), the memory map looks like Figure 3.3 (left). As the four pthreads are launched using the pthread_create(), the memory map will look like Figure 3.3 (right). The stack of each thread is saved even if the OS decides to switch out of that thread to allow another one to run (i.e., *context switching*). The context of the thread is saved in the same memory area. Furthermore, the code is on the bottom memory area and the shared heap among all threads is just above the code. This is all the threads need to resume their operation when they are scheduled to run again, following a context switch.

The size of the stack and the heap are not known to the OS while launching imflipPM.c the first time. There are default settings that can be modified. Unix and Mac OS allow specifying them at the commands prompt using switches, while Windows allows changing them through the right click, modify application properties. Since the programmer is the one who has the best idea about how much stack, heap a program needs, generous stack, heap areas should be allocated to applications to avoid a core dump in case of a scenario when an invalid memory address access occurs due to these memory areas clashing with each other.

Let's look at our favorite Figure 2.1 again; it shows 1499 launched threads and 98 processes. What this means is that many of the processes that OS launched internally are multithreaded, very possibly all of them, resembling a memory map shown in Figure 3.3;

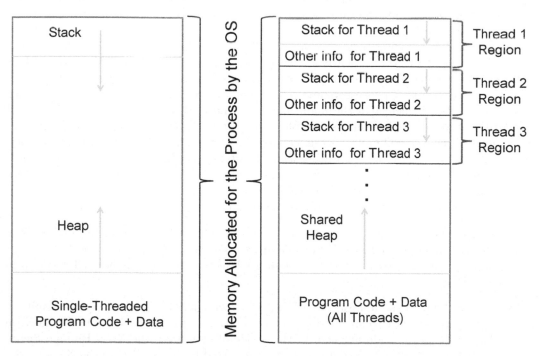

FIGURE 3.3 The memory map of a process when only a single thread is running within the process (left) or multiple threads are running in it (right).

each process seems to have launched an average of 15 threads, all of which must have a very low activity ratio. We saw what happens when even 5,6 threads are super active for a short period of time; in Figure 2.1, if all 1499 threads had a high activity ratio like the threads that we wrote so far, your CPU would possibly choke and you would not even be able to move your mouse on your computer.

There is another thing to keep in mind when it comes to the 1499 threads: the OS writers must design their threads to be as thin as possible to avoid the OS from interfering with application performance. In other words, if any of the OS threads are creating a lot of disturbance when changing their status from **Runnable** to **Running**, they will over-tax some of the core resources and will not allow their hyper-thread to work efficiently when one of your threads is scheduled alongside an OS thread. Of course, not every task can be performed so thinly, and what I just described about the OS has its limits. The other side of the coin is that the *application designed* should pay attention to making the application threads thin. We will only go into the details of this a little bit, since this is not a CPU parallel programming book, but rather, a GPU one. When I get a chance to make a comment about how a CPU thread could have been made a little thinner, I will make a point throughout the book though.

3.9 INTEL MIC ARCHITECTURE: XEON PHI

One interesting parallel computing platform that is close to GPUs is Many Integrated Core (MIC), an architecture introduced by Intel to compete with Nvidia and AMD GPU architectures. Xeon Phi model-named MIC architectures incorporate quite a few x86-compatible inO cores that are capable of running more than the two-threads-per-core that is standard in Intel's Core i7 OoO architectures. For example, the Xeon Phi 5110P unit that I will provide test results for incorporates 60 cores and 4 threads/core. Therefore, it is capable of executing 240 simultaneous threads.

As opposed to the cores inside a Core i7 CPU that work close to 4 GHz, each Xeon Phi core only works at 1.053 GHz, nearly 4× slower. To overcome this disadvantage, Xeon Phi is architected with a 30 MB cache memory, it has 16 memory channels, rather than the 4 that is in the modern core i7 processors and introduces a 320 GBps memory bandwidth, which is around 5 − 10× higher than the main memory bandwidth of a Core i7. Additionally, it has a 512-bit vector engine that is capable of performing 8 double-precision floating point operations each cycle. Therefore, it is capable of very high TFLOP (tera-floating point operating) processing capability. Rather than classifying the Xeon Phi as a CPU, it is more appropriate to classify it as a *throughput engine*, which is designed to process a lot of data (particularly scientific data) at a very high rate.

The Xeon Phi unit can be used in one of two ways:

- It could be used "almost" as a GPU with the OpenCL language. In this mode of operation, the Xeon Phi will be treated as a *device* that is outside the CPU; a device is connected to the CPU through an I/O bus, which will be PCI Express in our case.

- It could be used as "almost" a CPU with its own compiler, icc. After you compile it, you remote-connect to mic0 (which is a connection to the lightweight OS within Xeon Phi) and run the code in mic0. In this mode of operation, the Xeon Phi is a device again with its own OS, so data must be transferred from the CPU to the Xeon's work area. This transfer is done by using Unix commands that we use; scp (secure copy) is to transfer the data from the host into the Xeon Phi.

TABLE 3.6 Comparing imflipP.c execution times (H, V type flips in Table 3.2) to imflipPM.c execution times (I, W type flips in Table 3.4) for Xeon Phi 5110P.

Xeon Phi			Speedup			Speedup
#Threads	V	W	V → W	H	I	H → I
Serial	673	60.9	11×	358	150	2.4×
2	330	30.8	10.7×	179	75	2.4×
4	183	16.4	11.1×	90	38	2.35×
8	110	11.1	9.9×	52	22	2.35×
16	54	11.9	4.6×	27	15	1.8×
32	38	16.1	2.4×	22	18	1.18×
64	39	29.0	1.3×	28.6	29.4	0.98×
128	68	56.7	1.2×	48	53	0.91×
256	133	114	1.15×	90	130	0.69×
512	224	234	0.95×	205	234	0.87×

Here is the command line to compile imflipPM.c to execute on Xeon Phi to get the performance numbers in Table 3.6:

```
$ icc -mmic -pthread imflipPM.c ImageStuff.c -o imflipPM
$ scp imflipPM dogL.bmp mic0:~
$ ssh mic0
$ ./imflipPM dogL.bmp flipped.bmp H 60
```

Executing the multi-threaded version with 60 threads ...

Output BMP File name: flipped.bmp (3200 x 2400)

Total execution time: 27.4374 ms. (0.4573 ms per thread).

Flip Type = 'Horizontal' (H) (3.573 ns/pixel)

Performance results for Xeon Phi 5110P, executing the imflipPM.c program are shown in Table 3.6. While there is a healthy improvement from multiple threads up to 16 or 32 threads, the performance improvement limit is reached at 32 threads. Sixty-four threads provides no additional performance improvement. The primary reason for this is that the threads in our imflipPM.c program are so thick that they cannot take advantage of the multiple threads in each core.

3.10 WHAT ABOUT THE GPU?

Now that we understand how the CPU parallel programming story goes, what about the GPU? I promised that everything we learned in the CPU world would be applicable to the GPU world. Now, imagine that you had a CPU that allowed you to run 1000 or more cores/threads. That's the over-simplified story of the GPU. However, as you see, it is not easy to keep increasing the number of threads, since your performance eventually stops improving beyond a certain point. So, a GPU is not simply like a CPU with thousands of cores. Major architectural improvements had to be made inside the GPU to eliminate all of the core and memory bottlenecks we just discussed in this chapter. And, even after that, a lot of responsibility falls on the GPU programmer to make sure that his or her program is not encountering these bottlenecks.

The entire Part I of this book is dedicated to understanding how to "think parallel," in fact, this is not even enough. You have to start thinking "massively parallel." When we had 2, 4, 8 threads to execute in the examples shown before, it was somehow easy to adjust the sequence to make every thread do useful work. However, in the GPU world, you will be dealing with thousands of threads. Teaching how to think in such an absurdly parallel world should start by learning how to sequence two threads first! This is the reason why the CPU environment was perfect to warm up to parallelism, and is the philosophy of this book. By the time you finish Part I of this book, you will not only learn CPU parallelism, but will be totally ready to take on the massive parallelism that the GPUs bring you in Part II.

If you are still not convinced, let me mention this to you: GPUs actually support hundreds of thousands of threads, not just thousands! Convinced yet? A corporation like IBM with hundreds of thousands of employees can run as well as a corporation with 1 or 2 employees, and, yet, IBM is able to harvest the manpower of all of its employees. But, it takes extreme discipline and a systematic approach. This is what the GPU programming is all about. If you cannot wait until we reach Part II to learn GPU programming, you can peak at it now; but, unless you understand the concepts introduced in Part I, something will always be missing.

GPUs are here to give us shear computational power. A GPU program that works 10× faster than a comparable CPU program is better than one that works only 5× faster. If somebody could come in and rewrite the same GPU program to be 20× faster, that person is the king (or queen). There is no point in writing a GPU program unless you are targeting *speed*. There are three things that matter in a GPU program: *speed, speed,* and *speed*! So, the goal of this book is to get you to be a GPU programmer that writes super-fast GPU code. This doesn't happen unless we systematically learn every important concept, so, in the middle of a program, when we encounter some weird bottleneck, we can explain it and remove the bottleneck. Otherwise, if you are going to write slow GPU code, you might as well spend your time learning much better CPU multithreading techniques, since there is no point in using a GPU unless your code gets every little bit of extra speed out of it. This is the reason why we will time our code throughout the entire book and will find ways to make our GPU code faster.

3.11 CHAPTER SUMMARY

We now have an idea about how to write efficient multithreaded code. Where do we go from here? First, we need to be able to quantify everything we discussed in this chapter: What is memory bandwidth? How do the cores really operate? How do the cores get data from the memory? How do the threads share the data? Without answering these, we will only be *guessing* why we got any speedup.

It is time to quantify everything and get a 100% precise understanding of the architecture. This is what we will do in Chapter 4. Again, everything we learn will be readily applicable to the GPU world, although there will be distinct differences, which I will point out as they come up.

## CHAPTER 4

Understanding the Cores and Memory

Hen we say *hardware architecture*, what are we talking about? The answer is: *everything physical.* What does "everything physical" include? CPU, memory, I/O controller, PCI express bus, DMA controller, hard disk controller, hard disk(s), SSDs, CPU chipsets, USB ports, network cards, CD-ROM controller, DVDRW, I can go on for a while ... The big question is: which one of these do we *care about* as a programmer?

The answer is: CPU and memory, and more specifically, cores inside the CPU and memory. Especially if you are writing high performance programs (like the ones in this book), > 99% of your program performance will be determined by these two things. Since the purpose of this book is high-performance programming, let's dedicate this chapter to understanding the cores and memory.

4.1 ONCE UPON A TIME ... INTEL ...

Once upon a time, in the early 1970s, a tiny Silicon Valley company named INTEL, employing about 150 people at the time, designed a programmable chip in the beginning days of the concept of a "microprocessor," or "CPU," a digital device that can execute a *program* stored in *memory*. Every CPU had the capability to *address* a certain amount of memory, primarily determined at *design time* by the CPU designers, based on technological and business-related constraints.

The *program* (aka, *CPU instructions*), and the *data* (that was fed into the program) had to be stored somewhere. INTEL designed the 4004 processor to have 12 address bits, capable of *addressing* 4096 Bytes (i.e., 2^{12}). Each piece of information was stored in 8-bit units (a Byte). So, all together, 4004 could execute a program that was, say, 1 KB and could work on data that was in the remaining 3 KB. Or, maybe, some customers only needed to store 1 KB program and 1 KB data: so, they would have to buy the memory chips to attach to the 4004 somewhere else.

Although INTEL designed support chips, such as an I/O controller and memory controller to allow their customers to interface the 4004 CPU to the memory chips they bought somewhere else, they did not particularly focus on making memory chips themselves. This might look a little counterintuitive at first, but, from a business standpoint, it makes a lot of sense. At the time of the release of the 4004, one needed at least 6, 7 other types of interface chips to properly interface other important devices to the 4004 CPU. Out of those 6, 7 different chips, one was very special: the *memory* chips.

FIGURE 4.1 Inside a computer containing an i7-5930K CPU [10] (CPU5 in Table 3.1), and 64 GB of DDR4 memory. This PC has a GTX Titan Z GPU that will be used to test a lot of the programs in Part II.

4.2 CPU AND MEMORY MANUFACTURERS

Much like 30–40 years ago, INTEL still doesn't make the memory chips as of the year 2017, at the time of the projected publication of this book. The players in the memory manufacturing world are Kingston Technology, Samsung, and Crucial Technology, to name a few. So, the trend in the 4004 days four decades ago never changed. Memory manufacturers are different from CPU manufacturers, although a lot of the CPU manufacturers make their own support chips (chipsets). Figure 4.1 shows the inside of the PC containing the CPU5 in our Table 3.1. This CPU is an i7-5930K [10], made by INTEL. However, the memory chips in this computer (top left of Figure 4.1) are made by a completely different manufacturer: Kingston Technology Corp. No big surprise for the manufacturer of the GPU: Nvidia Corp! The SSD (solid state disk) is manufactured by Crucial Technology. Power supply is Corsair. Ironically, the CPU liquid cooler in Figure 4.1 is not made by INTEL. It is made by the same company that the chassis and power supply are made by (Corsair).

Support chips (which were later called *chipsets*) are made out of logic gates, AND, OR, XOR gates, etc. Their primary building blocks are metal oxide semiconductor (MOS) transistors. For example, the X99 chipset in Figure 4.1 was made by INTEL (not marked in the figure). This chip also controls the PCI Express bus that is connecting to the GPU to the CPU (marked in Figure 4.1), as well as the SATA3 bus that is connecting to the SSD.

4.3 DYNAMIC (DRAM) VERSUS STATIC (SRAM) MEMORY

While CPU manufacturers have all the interest in the world to design and manufacture their chipsets, why are the memory chips so special? Why do the CPU manufacturers have no interest in making memory chips? To answer this, first let's look at different types of memory.

4.3.1 Static Random Access Memory (SRAM)

This type of memory is still made with MOS transistors, something that is already the building block of the CPUs and their chipsets. It is easy for the CPU manufacturers to incorporate this type of memory into their CPU design, since it is made out of the same material. About 10 years after the introduction of the first CPUs, CPU designers introduced a type of SRAM that could be built right into the CPU. They called it *cache memory* that was able to buffer a very small portion of the main memory. Since most computer programs require access to very small portions of data in a repeated fashion, the effective speedup by the introduction of the cache memory was significant, although only very small amounts of this type of memory were possible to incorporate into the CPU.

4.3.2 Dynamic Random Access Memory (DRAM)

This type of memory was the only option to manufacture huge amounts of memory. For example, as of today, only 8–30 MB of cache memory can be built into the CPU, while a computer with 32 GB of main memory (DRAM) is mainstream (a stunning 1000× difference). To be able to build DRAMs, a completely different technology has to be employed: While the building block of chipsets and CPUs is primarily MOS transistors, the building block of DRAM is extremely small capacitors that store charge. This charge, with proper interface circuitry, is interpreted as *data*. There are a few things that are very different with DRAM:

- Since the charge is stored in extremely small capacitors, it drains (i.e., leaks) after a certain amount of time (something like 50 ms).

- Due to this leakage, data has to be read and put back on the DRAM (i.e., refreshing).

- It makes no sense to allow byte-at-a-time access to data, considering the disadvantages of refreshing. So, the data is accessed in big rows at a time (e.g., 4 KB at a time).

- There is a long delay in getting a row of data, although, once read, access to that row is extremely fast.

- In addition to being *row accessible*, DRAMs have all sorts of other delays, such as row-to-row delay, etc. Each one of these parameters is specified by the memory interface standards that are defined by a consortium of companies.

4.3.3 DRAM Interface Standards

How can CPU manufacturers control the compatibility of the memory chips that some other companies are making? The answer is: *memory interface standards*. From the first days of the introduction of the memory chips decades ago, there was a need to define standards for memory chips. SDRAM, DDR (double data rate), DDR2, DDR3, and finally, in 2015 the

DDR4 standard (contained inside the PC in Figure 4.1). These standards are determined by a large consortium of chip manufacturers that define such standards and precise timing of the memory chips. If the memory manufacturers design their memory to be 100% compliant with these standards, and INTEL designs their CPU to be 100% compliant, there is no need for both of these chips to be manufactured by the same company.

In the past four decades, the main memory was always made out of DRAM. A new DRAM standard was released every 2–3 years to take advantage of the exciting developments in the DRAM manufacturing technology. As the CPUs improved, so did the DRAM. However, not only the improvements in CPU and DRAM technology followed different patterns, but also *improvement* meant something different for CPU and DRAM:

- For CPU designs, *improvement* meant more work done per second.

- For DRAM designs, *improvement* meant more data read per second (bandwidth) as well as more storage (capacity).

CPU manufacturers improved their CPUs using better architectural designs and by taking advantage of more MOS transistors that became available at each shrinking technology node (130 nm, 90 nm, ... and 14 nm as of 2016). On the other hand, DRAM manufacturers improved their memories by the ability to continuously pack more capacitors into the same area, thereby resulting in more storage. Additionally, they were able to continuously improve their bandwidths by the newer standards that became available (e.g., DDR3, DDR4).

4.3.4 Influence of DRAM on our Program Performance

The most important question for us as programmers is this: how does this separation of the CPU versus DRAM manufacturing technology influence the performance of our programs? The characteristics of the SRAM versus DRAM, listed in Section 4.3, will stay relatively the same in the foreseeable future. While the CPU manufacturers will always try to increase the amount of cache memory inside their CPU by using more SRAM-based cache memory, DRAM manufacturers will always try to increase the bandwidth and storage area of the DRAMs. However, they will be less concerned about access speeds to small amounts of memory, since this should be remedied by the increasing amounts of cache memory inside the CPUs. The access speeds of DRAM are as follows:

- DRAM is accessed one row at a time, each row being the smallest accessible amount of DRAM memory. A row is approximately 2–8 KB in modern DRAMs.

- To access a row, a certain amount of time (latency) is required. However, once the row is accessed (brought into the row buffer internally by the DRAM), this row is practically free to access going forward.

- The latency in accessing DRAM is about 200–400 cycles for a CPU, while accessing subsequent elements in the same row is just a few cycles.

Considering all of these DRAM facts, the clear message to a programmer is this:

➤ *Write your programs in such a way that:*
  i)   *Accessing data from distant memory locations should be in large chunks, since we know that this data will be in DRAM.*
  ii) *Accessing small amounts should be highly localized and repetitive, since we know that this data will be in SRAM (cache).*
  iii) *Pay extreme attention to multiple threads, since there is only one memory, but multiple threads: simultaneous threads might cause bad access DRAM patterns although individually they look okay.*

4.3.5 Influence of SRAM (Cache) on our Program Performance

Since there is only one main memory (DRAM), as long as we obey the rules set forth in Table 3.3 and pay attention to the comments in Section 4.3.4, we should have fairly good DRAM performance. But, cache memory, made out of SRAM, is a little different. First of all, there are multiple types of cache memory, making their design *hierarchical*. For a modern CPU like the one shown in the PC in Figure 4.1, there are three different types of cache memory, built into the CPU:

- L1\$ is 32 KB for data (L1 data cache, or L1D\$), and 32 KB for instructions (L1 instruction cache, or L1I\$). Total L1\$ is 64 KB. Access to L1\$ is extremely fast (4 cycle load-to-use latency). Each core has its own L1\$.

- L2\$ is 256 KB. There is no separation between data or instructions. Access to L2\$ is fairly fast (11–12 cycles). Each core has its own L2\$.

- L3\$ is 15 MB. Access to L3\$ is faster than DRAM, but much slower than L2\$ (≈ 22 cycles). L3\$ is shared by all of the cores (6 in this CPU).

- The data that is brought into and evicted out of L1\$, L2\$, and L3\$ is purely controlled by the CPU and is completely out of the programmer's control. However, by keeping the data operations confined into small loops, the programmer can have a significant impact on the effectiveness of the cache memory operation.

- Alternatively, by disobeying the cache efficiency rules, the programmer could nearly render the cache memory useless.

- The best way for the programmer to take advantage of caching is to know the exact sizes of each cache hierarchy and design the programs to stay within these ranges.

Considering all of these SRAM facts, the clear message to a programmer is this:

➤ *To take advantage of caching, write your programs, so that:*
  i)   Each thread accesses 32 KB data regions repetitively
  ii) Try to confine broader accesses to 256 KB if possible,
  iii) When considering all of the launched threads:
      try to confine their cumulative data access within L3\$ (e.g., 15 MB)
  iv) If you must exceed this, make sure that there is
      heavy usage of L3\$ before exceeding this region

4.4   IMAGE ROTATION PROGRAM: IMROTATE.C

We covered a lot of information regarding the cache memory (SRAM) and main memory (DRAM). It is time to put this information to good use. In this section, I will introduce a program, imrotate.c, that rotates an image by a specified amount of degrees. This program will put a lot of pressure on the core and memory components of the CPU, thereby allowing us to understand their influence on the overall program performance. We will comment on the performance bottlenecks caused by over-stressing either the cores or the memory and will improve our program. The improved program, imrotateMC.c, the memory-friendly ("M") and core-friendly ("C") version of imrotate.c, will implement the improvements in an incremental fashion via multiple different steps, and each step will be explained in detail.

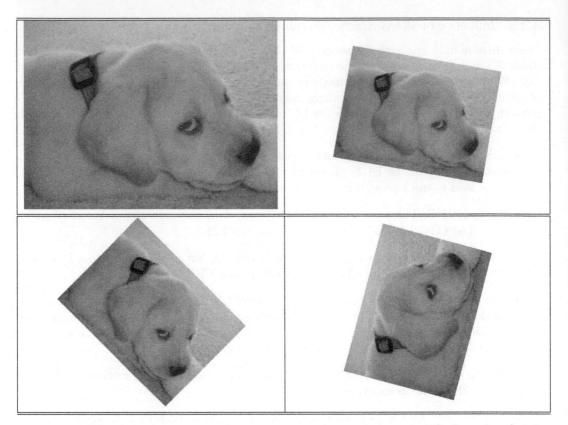

FIGURE 4.2 The imrotate.c program rotates a picture by a specified angle. Original dog (top left), rotated +10° (top right), +45° (bottom left), and −75° (bottom right) clockwise. Scaling is done to avoid cropping of the original image area.

4.4.1 Description of the imrotate.c

The purpose of imrotate.c is to create a program that is both memory-heavy and core-heavy. imrotate.c will take an image as shown in Figure 4.2 (top left) and will rotate it clockwise by a specified amount (in degrees). Example outputs of the program are shown in Figure 4.2 for a +10° rotation (top right), a +45° rotation (bottom left), and a −75° rotation (bottom right). These angels are all specified clockwise, thereby making the last −75° rotation effectively a counterclockwise +75° rotation. To run the program, the following command line is used:

    imrotate InputfileName OutputfileName [degrees] [1-128]

where **degrees** specifies the clockwise rotation amount and the next parameter [1–128] is the number of threads to launch, similar to the previous programs.

4.4.2 imrotate.c: Parametric Restrictions and Simplifications

Some simplifications had to be made in the program to avoid unnecessary complications and diversion of our attention to unrelated issues. These are

- For rectangular images where the width and height are not equal, part of the rotated image will end up outside the original image area.

- To avoid this cropping, the resulting image is scaled, so that the resulting image always fits in its original size.

- This scaling naturally implies empty areas in the resulting image (black pixels with RGB=000 values are used to fill the blank pixels).

- The scaling is not automatic, i.e., the same exact amount of scaling is applied at the beginning, thereby leaving more blank area for certain rotation amounts, as is clearly evidenced in Figure 4.2.

4.4.3 imrotate.c: Theory of Operation

Rotation of each pixel with the original coordinates (x, y) is achieved by multiplying these coordinates by a *rotation matrix* that yields the rotated coordinates $(x\prime, y\prime)$ as follows:

$$\begin{bmatrix} x\prime \\ y\prime \end{bmatrix} = \begin{bmatrix} \cos\theta & \sin\theta \\ -\sin\theta & \cos\theta \end{bmatrix} \times \begin{bmatrix} x \\ y \end{bmatrix} \tag{4.1}$$

where the θ is the rotation angle (specified in radians, θ_{rad}), with a simple conversion to user-specified degrees as:

$$\theta = \theta_{rad} = \frac{2\pi}{360} \times \theta_{deg} \tag{4.2}$$

When a pixel's destination coordinates $(x\prime, y\prime)$ are determined, all three color components RGB of that pixels are moved to that same $(x\prime, y\prime)$ location. Scaling (more accurately, *prescaling*) of the image is done by the following formula:

$$d = \sqrt{w^2 + h^2} \implies \begin{cases} \text{ScaleFactor} = \frac{w}{d}, & h > w \\ \text{ScaleFactor} = \frac{h}{d}, & w \leq h \end{cases} \tag{4.3}$$

where the width and height values are the previously introduced Hpixels and Vpixels attributes, respectively. The scale factor is determined to avoid cropping in case either one of these values is greater based on Equation 4.3. An important note about this program is that this rotation cannot be easily implemented without using additional image memory. For that purpose, an additional image function called CreateBlankBMP() (shown in Code 4.1) is introduced inside ImageStuff.c and its header is placed in ImageStuff.h.

CODE 4.1: ImageStuff.c CreateBlankBMP() {...}

The CreateBlankBMP() function creates an image filled with zeroes (blank pixels).

```
unsigned char** CreateBlankBMP()
{
int i,j;

   unsigned char** img=(unsigned char **)malloc(ip.Vpixels*sizeof(unsigned char*));
   for(i=0; i<ip.Vpixels; i++){
      img[i] = (unsigned char *)malloc(ip.Hbytes * sizeof(unsigned char));
      memset((void *)img[i],0,(size_t)ip.Hbytes); // zero out every pixel
   }
   return img;
}
```

CODE 4.2:   imrotate.c   ... main() {...

First part of the main() function in imrotate.c converts user supplied rotation in degrees to radians and calls the Rotate() function to rotate the image.

```c
#include <pthread.h>
#include <stdint.h>
#include <ctype.h>
#include <stdlib.h>
#include <stdio.h>
#include <string.h>
#include <math.h>
#include <sys/time.h>
#include "ImageStuff.h"
#define REPS        1
#define MAXTHREADS 128
long            NumThreads;           // Total number of threads
int             ThParam[MAXTHREADS];  // Thread parameters ...
double          RotAngle;             // rotation angle
pthread_t       ThHandle[MAXTHREADS]; // Thread handles
pthread_attr_t  ThAttr;               // Pthread attributes
void* (*RotateFunc)(void *arg);       // Func. ptr to rotate img
unsigned char** TheImage;             // This is the main image
unsigned char** CopyImage;            // This is the copy image
struct ImgProp  ip;
...
int main(int argc, char** argv)
{
    int             RotDegrees, a, i, ThErr;
    struct timeval  t;
    double          StartTime, EndTime, TimeElapsed;

    switch (argc){
        case 3 : NumThreads=1;              RotDegrees=45;              break;
        case 4 : NumThreads=1;              RotDegrees=atoi(argv[3]);   break;
        case 5 : NumThreads=atoi(argv[4]);  RotDegrees = atoi(argv[3]); break;
        default: printf("\n\nUsage: imrotate inputBMP outputBMP [RotAngle] [1-128]");
                 printf("\n\nExample: imrotate infilename.bmp outname.bmp 45 8\n\n");
                 printf("\n\nNothing executed ... Exiting ...\n\n");
                 exit(EXIT_FAILURE);
    }
    if((NumThreads<1) || (NumThreads>MAXTHREADS)){
        printf("\nNumber of threads must be between 1 and %u... \n",MAXTHREADS);
        printf("\n'1' means Pthreads version with a single thread\n");
        printf("\n\nNothing executed ... Exiting ...\n\n");    exit(EXIT_FAILURE);
    }
    if((RotDegrees<-360) || (RotDegrees>360)){
        printf("\nRotation angle of %d degrees is invalid ...\n",RotDegrees);
        printf("\nPlease enter an angle between -360 and +360 degrees ...\n");
        printf("\n\nNothing executed ... Exiting ...\n\n");    exit(EXIT_FAILURE);
    }
    ...
```

CODE 4.3: imrotate.c main() ...}

Second part of the main() function in imrotate.c creates a blank image and launches multiple threads to rotate TheImage[] and place it in CopyImage[].

```c
...
if((RotDegrees<-360) || (RotDegrees>360)){
   ...
}
printf("\nExecuting the Pthreads version with %u threads ...\n",NumThreads);
RotAngle=2*3.141592/360.000*(double) RotDegrees; // Convert the angle to radians
printf("\nRotating %d deg (%5.4f rad) ...\n",RotDegrees,RotAngle);
RotateFunc=Rotate;

TheImage = ReadBMP(argv[1]);
CopyImage = CreateBlankBMP();
gettimeofday(&t, NULL);
StartTime = (double)t.tv_sec*1000000.0 + ((double)t.tv_usec);
pthread_attr_init(&ThAttr);
pthread_attr_setdetachstate(&ThAttr, PTHREAD_CREATE_JOINABLE);
for(a=0; a<REPS; a++){
   for(i=0; i<NumThreads; i++){
      ThParam[i] = i;
      ThErr = pthread_create(&ThHandle[i], &ThAttr, RotateFunc,
                       (void *)&ThParam[i]);
      if(ThErr != 0){
         printf("\nThread Creation Error %d. Exiting abruptly... \n",ThErr);
         exit(EXIT_FAILURE);
      }
   }
   pthread_attr_destroy(&ThAttr);
   for(i=0; i<NumThreads; i++){ pthread_join(ThHandle[i], NULL); }
}
gettimeofday(&t, NULL);
EndTime = (double)t.tv_sec*1000000.0 + ((double)t.tv_usec);
TimeElapsed=(EndTime-StartTime)/1000.00;
TimeElapsed/=(double)REPS;
//merge with header and write to file
WriteBMP(CopyImage, argv[2]);

// free() the allocated area for the images
for(i = 0; i < ip.Vpixels; i++) { free(TheImage[i]); free(CopyImage[i]); }
free(TheImage); free(CopyImage);
printf("\n\nTotal execution time: %9.4f ms. ",TimeElapsed);
if(NumThreads>1) printf("(%9.4f ms per thread). ",
                        TimeElapsed/(double)NumThreads);
printf("\n (%6.3f ns/pixel)\n",
        1000000*TimeElapsed/(double)(ip.Hpixels*ip.Vpixels));
return (EXIT_SUCCESS);
}
```

CODE 4.4:   imrotate.c   Rotate() {...}

The Rotate() function takes each pixel from the TheImage[] array, scales it, and applies the rotation matrix in Equation 4.1 and writes the new pixel into the CopyImage[] array.

```c
void *Rotate(void* tid)
{
    long      tn;
    int       row,col,h,v,c, NewRow,NewCol;
    double    X, Y, newX, newY, ScaleFactor, Diagonal, H, V;
    struct    Pixel pix;

    tn = *((int *) tid);
    tn *= ip.Vpixels/NumThreads;

    for(row=tn; row<tn+ip.Vpixels/NumThreads; row++)
    {
        col=0;
        while(col<ip.Hpixels*3){
            // transpose image coordinates to Cartesian coordinates
            c=col/3;      h=ip.Hpixels/2; v=ip.Vpixels/2; // integer div
            X=(double)c-(double)h;
            Y=(double)v-(double)row;

            // image rotation matrix
            newX=cos(RotAngle)*X-sin(RotAngle)*Y;
            newY=sin(RotAngle)*X+cos(RotAngle)*Y;

            // Scale to fit everything in the image box
            H=(double)ip.Hpixels;
            V=(double)ip.Vpixels;
            Diagonal=sqrt(H*H+V*V);
            ScaleFactor=(ip.Hpixels>ip.Vpixels) ? V/Diagonal : H/Diagonal;
            newX=newX*ScaleFactor;
            newY=newY*ScaleFactor;

            // convert back from Cartesian to image coordinates
            NewCol=((int) newX+h);
            NewRow=v-(int)newY;
            if((NewCol>=0) && (NewRow>=0) && (NewCol<ip.Hpixels)
                                && (NewRow<ip.Vpixels)){
                NewCol*=3;
                CopyImage[NewRow][NewCol] = TheImage[row][col];
                CopyImage[NewRow][NewCol+1] = TheImage[row][col+1];
                CopyImage[NewRow][NewCol+2] = TheImage[row][col+2];
            }
            col+=3;
        }
    }
    pthread_exit(NULL);
}
```

TABLE 4.1 imrotate.c execution times for the CPUs in Table 3.1 ($+45°$ rotation).

# HW Threads	2C/4T i5-4200M CPU1	4C/8T i7-960 CPU2	4C/8T i7-4770K CPU3	4C/8T i7-3820 CPU4	6C/12T i7-5930K CPU5	8C/16T E5-2650 CPU6
# SW Threads						
1	951	1365	782	1090	1027	845
2	530	696	389	546	548	423
3	514	462	261	368	365	282
4	499	399	253	322	272	227
5	499	422	216	295	231	248
6		387	283	338	214	213
8		374	237	297	188	163
9			237		177	199
10		341	228	285	163	201
12			217		158	171

4.5 PERFORMANCE OF IMROTATE

Execution times of imrotate.c are shown in Table 4.1 for the same CPUs tabulated in Table 3.1.

4.5.1 Qualitative Analysis of Threading Efficiency

These results were obtained by rotating the image $+45°$ clockwise. Aside from being much slower than the corresponding horizontal/vertical flip programs, here are some observations from Table 4.1:

- If we look at the CPU5 performance (the CPU inside the personal computer shown in Figure 4.1), the performance improvement continues steadily, although the *threading efficiency* goes down significantly with the increasing number of hardware threads (noted as HW threads in Table 4.1).

- This is the first program that we are seeing where a 6C/12T CPU (CPU5) seems to be taking advantage of all of the 12 hardware threads, although the threading efficiency plummets for high software thread counts (noted as SW threads, i.e., software threads, in Table 4.1).

- Although certain CPUs have a relative advantage initially (e.g., CPU3 @788 ms) they lose this advantage with the increased number of SW threads, when compared against CPUs with higher HW threads (e.g., CPU5 is faster with 8 software threads @188 ms vs. 237 ms for CPU3).

- For now, ignore CPU6, since the Amazon environment is a multiuser environment and the explanation is the performance is not as straightforward.

4.5.2 Quantitative Analysis: Defining Threading Efficiency

It would be useful to define a metric called multithreading efficiency (or, in short, *threading efficiency*) that quantifies how additional software threads are improving program performance *relatively*. If we take CPU5 in Table 4.1 as an example and take the single-thread

TABLE 4.2 imrotate.c threading efficiency (η) and parallelization overhead $(1-\eta)$ for CPU3, CPU5. The last column reports the speedup achieved by using CPU5 that has more cores/threads, although there is no speedup up to 6 launched SW threads.

# SW Thr	CPU3: i7-4770K 4C/8T			CPU5:i7-5930K 6C/12T			Speedup CPU5→CPU3
	Time	η	$1-\eta$	Time	η	$1-\eta$	
1	782	100%	0%	1027	100%	0%	0.76×
2	389	100%	0%	548	94%	6%	0.71×
3	261	100%	0%	365	94%	6%	0.72×
4	253	77%	23%	272	95%	5%	0.93×
5	216	72%	28%	231	89%	11%	0.94×
6	283	46%	54%	214	80%	20%	1.32×
8	237	42%	58%	188	68%	32%	1.26×
9	237	41%	69%	177	65%	35%	1.34×
10	228	34%	66%	163	63%	37%	1.4×
12	217	30%	70%	158	54%	46%	1.37×

execution time of 1027 ms as our baseline (i.e., 100% efficiency), then, when we launch two threads, we ideally expect half of that execution time ($\frac{1027}{2} = 513.5$ ms). However, we see 548 ms. Not bad, but only $\approx 94\%$ of the performance we hoped for.

In other words, launching the additional thread improved the execution time, but hurt the *efficiency* of the CPU. Quantifying this efficiency metric (η) is fairly straightforward as shown in Equation 4.4. One corollary of Equation 4.4 is that parallelization has an *overhead* that can be defined as shown in Equation 4.5.

$$\eta = \text{Threading Efficiency} = \frac{(\text{Single-Thread Execution Time})}{(\text{N-Thread Execution Time}) \times \text{N}} \qquad (4.4)$$

$$\text{Parallelization Overhead} = 1 - \text{Threading Efficiency} = 1 - \eta \qquad (4.5)$$

In our case, launching 2 threads in CPU5 cost a 6% overhead (i.e., $1-0.94 = 0.06$) as the cost of parallelization. The threading efficiency of imrotate.c for CPU3 and CPU5 is shown in Table 4.2. CPU3 has 4 cores, 8 hardware threads, and its peak main DRAM memory bandwidth is 25.6 GB/s (Gigabytes per second) as per Table 3.1. On the other hand, CPU5 has 6 cores, 12 hardware threads, and its peak DRAM main memory bandwidth is 68 GB/s. The increase in the memory bandwidth should allow more hardware threads to bring data into the cores faster, thereby avoiding a memory bandwidth saturation when a higher number of launched software threads demand data from the DRAM main memory simultaneously. An observation of Table 4.2 shows exactly that.

In Table 4.2, while the decrease in threading efficiency (η) is evident for both CPUs as we launch more software threads, the rate of fall-off is much less for CPU5. If we consider 89% "highly efficient," then we observe from Table 4.2 that CPU3 is highly efficient for 3 software threads, while CPU5 is highly efficient for 5 software threads. Alternatively, if we define 67% as "highly inefficient" (i.e., wasting one out of three threads), then CPU3 becomes highly inefficient for > 5 threads, while CPU5 is highly inefficient for > 8 threads.

To provide a direct performance comparison between CPU3 and CPU5, the last column is added to Table 4.2. This column shows the "speedup" when running the same program on CPU5 (what is supposed to be a higher performance CPU) as compared to CPU3. Up to 5 launched software threads (i.e., before CPU3 starts becoming highly inefficient), we see that CPU3 beats CPU5. However, beyond 5 launched threads, CPU5 beats CPU3 for any number of threads. The reason has something to do with both the cores and memory as we will see shortly in the next section. Our imrotate.c program is by nature *designed* inefficiently, so, it is not taking advantage of the advanced architectural improvements that are built into CPU5.

We will get into the architectural details shortly, but, for now, it suffices to comment that just because a CPU is a later generation doesn't mean that it will work faster for every program. Newer generations CPUs' architectural improvements are typically geared toward programs that are written well. A program that causes choppy memory and core access patterns, like imrotate.c, will not benefit from the beautiful architectural improvements of newer generation CPUs. INTEL's message to the programmers is clear:

> ➢ *Newer generation CPUs and memories are always designed to work more efficiently when rules, like the ones in Table 3.3, are obeyed.*
> ➢ *The influence of these rules will keep increasing in future generations.*
> ➢ *CPU says: If you're gonna be a bad program designer, I'll be a bad CPU!*

4.6 THE ARCHITECTURE OF THE COMPUTER

In this section, we will understand what is going on inside the CPU and memory in detail. Based on this understanding, we will improve our results of the imrotate.c program. As mentioned previously, the function named Rotate() in Code 4.4 is what matters to this program's performance. So, we will only improve this function. To be able to do this, let's understand what happens, at runtime, when this function is being executed.

4.6.1 The Cores, L1$ and L2$

First, let's look at the structure of a CPU core. I will pick i7-5930K as an example (CPU5 in Table 4.2). The internal structure of an i7-5930K core is shown in Figure 4.3. Each core has a 64 KB L1 and 256 KB L2 cache memory. Let's see what these cache memories do:

- L1$ is broken into a 32 KB instruction cache (L1I$) and a 32 KB data cache (L1D$). Both of these cache memories are the fastest cache memories, as compared to L2$ and L3$. It takes the processor only a few cycles (e.g., 4 cycles) to access them.

- L1I$ stores the most recently used CPU instructions to avoid continuous fetching of something that has already been fetched; storing a replica of an instruction is called *caching* it.

- L1D$ caches a copy of the data elements, for example the pixels that are read from the memory.

- The L2$ is used to cache either instructions or data. It is the second fastest cache memory. The processor accessed L2$ in something like 11 cycles. This is why whatever is cached in L1$ always goes through L2$ first.

- Data or instructions that are brought from L3$ first land in the L2$ and then go into L1$. If the cache memory controller decides that it no longer needs some data, it *evicts* it (i.e., it un-caches it). However, since L2$ is bigger than L1$, a copy of the evicted data/instruction is probably still in the L2$ and can be brought back.

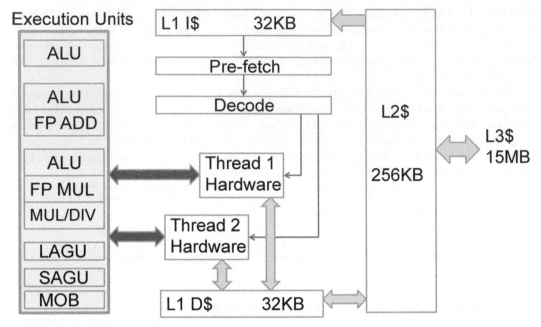

FIGURE 4.3 The architecture of one core of the i7-5930K CPU (the PC in Figure 4.1). This core is capable of executing two threads (hyper-threading, as defined by Intel). These two threads share most of the core resources, but have their own register files.

4.6.2 Internal Core Resources

An important observation in Figure 4.3 is that, although this core is capable of executing two threads, these threads share 90% of the core resources, and have a very minimal amount of dedicated hardware to themselves. Their dedicated hardware primarily consists of separate register files. Each thread (denoted as Thread 1 and Thread 2 in Figure 4.3) needs dedicated registers to save its results, since each thread executes different instructions (e.g., a different function) and produces results that might be completely unrelated. For example, while one thread is executing the Rotate() function of our code, the other thread could be executing a function that is a part of the OS, and has nothing to do with our Rotate() function. An example could be when the OS is processing a network packet that has just arrived while we are running our Rotate() function. Figure 4.3 shows how efficient the core architecture is. The two threads share the following components inside the core:

- The L2$ is shared by both threads to receive the instructions from the main memory via L3$ (I will explain L3$ shortly).

- The L1 I$ is shared by both threads to receive the instructions from L2$ that belongs to either thread.

- The L1 D$ is shared by both threads to receive/send data from/to either thread.

- Execution units fall into two categories: ALUs (arithmetic logic units) are responsible for integer operations, and logic operations such as OR, AND, XOR, etc. FPUs (floating point units) are responsible for floating point (FP) operations such as FP ADD and FP MUL (multiplication). Division (whether integer or FP) is significantly more sophisticated than addition and multiplication, so there is a separate unit for division. All of these *execution units* are shared by both threads.

- In each generation, more sophisticated computational units are available as shared execution units. However, multiple units are incorporated for common operations that might be executed by both threads, such as ALUs. Also, Figure 4.3 is overly simplified and the exact details of each generation might change. However, the ALU-FPU functionality separation has never changed in the past 3–4 decades of CPU designs.

- The addresses that both threads generate must be calculated to write the data from both threads back into the memory. For address computations, load and store address generation units (LAGU and SAGU) are shared by both threads, as well as a unit that properly orders the destination memory addresses (MOB).

- Instructions are prefetched and decoded only once and routed to the owner thread. Therefore, prefetcher and decoder are shared by both threads.

- Acronyms are:

    ALU=Arithmetic Logic Unit

    FPU=Floating Point Unit

    FPMUL=Floating Point Dedicated Multiplier

    FPADD = FP Adder

    MUL/DIV=Dedicated Multiplier/Divider

    LAGU=Load Address Generation Unit

    SAGU=Store Address Generation Unit

    MOB=Memory Order Buffer

The most important message from Figure 4.3 is as follows:

➤ *Our program performance will suffer if both threads inside a core are requesting exactly the same shared core resources.*

➤ *For example, if both threads require heavy floating point operations, they will pressure the FP resources.*

➤ *On the data side, if both threads are extremely memory intensive, they will pressure L1 D$ or L2$, and eventually L3$ and main memory.*

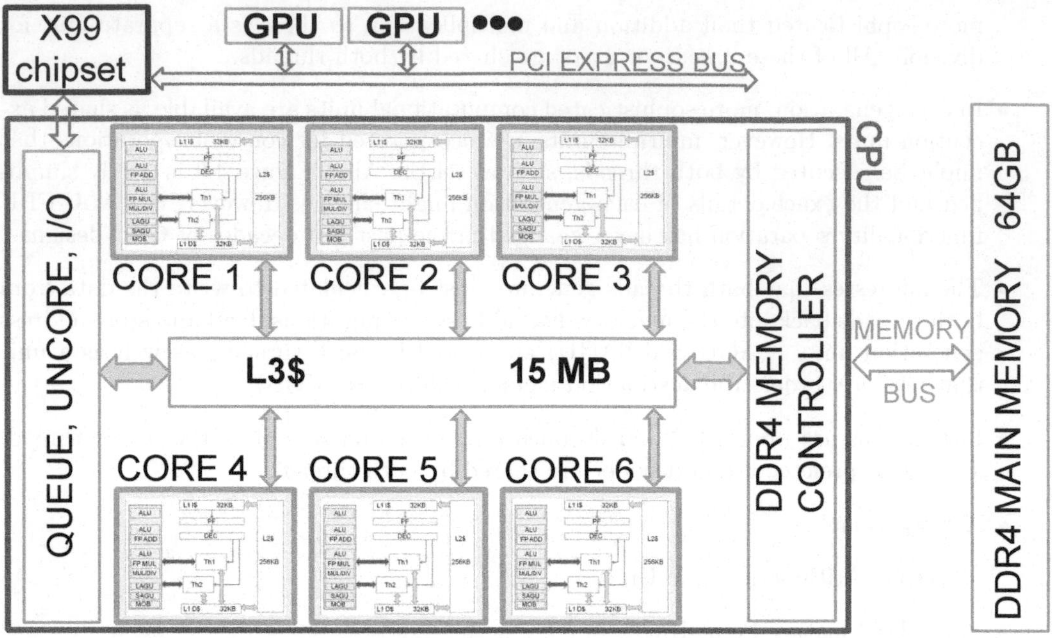

FIGURE 4.4 Architecture of the i7-5930K CPU (6C/12T). This CPU connects to the GPUs through an external PCI express bus and memory through the memory bus.

4.6.3 The Shared L3 Cache Memory (L3$)

The internal architecture of the i7-5930K CPU is shown in Figure 4.4. This CPU contains 6 cores, each capable of executing 2 threads as described in Section 4.6.2. While these "twin" threads in each core share their internal resources peacefully (yeah right!), all 12 threads share the 15 MB L3$. The impact of the L3$ is so high that INTEL dedicated 20% of the CPU die area to L3$. Although this gives around 2.5 MB to each core on average, cores that demand more L3$ could tap into a much larger portion of the L3$. The L3$ cache design inside i7-5930K is termed *Smart Cache* by INTEL and is completely managed by the CPU.

Although the programmer (i.e., the program) has no input in the way the cache memory manages its data, it has almost complete control of the efficiency of this cache by the *data demand patterns* of his or her programs. These data demand patterns by threads cause DRAM access patterns as I emphasized in the development of Code 3.1 and Code 3.2.

4.6.4 The Memory Controller

There is no direct access path from the cores to the main memory. Anything that is transferred from/to DRAM to the cores must first go through the L3$. The operational timing of the DRAM memory is so sophisticated that (regardless of whether it is DDR2, DDR3, or DDR4), the i7-5930K CPU architecture dedicates 18% of its chip area to a unit called the *memory controller*. The memory controller buffers and aggregates data coming from L3$ to somehow eliminate the inefficiencies resulting from the *block-transfer* nature of DRAM. Additionally, the memory controller is responsible for converting the data streams between the L3$ and DRAM into the proper format that the destination needs (whether L3$ or DRAM). For example, while DRAM works with *rows* of data, L3$ is an associative cache memory that takes data one *line* at a time.

4.6.5 The Main Memory

The 64GB DDR4 DRAM memory that the PC in Figure 4.1 has is capable of providing a peak 68 GB/s of data throughput. Such a high throughput will only be achieved if the program is reading massive consecutive blocks of data. In all of the programs we saw so far, how close did we get to this? Let's search for it. The fastest program we wrote so far was imflipPM.c that had the MTFlipVM() function as its innermost loop, shown in Code 3.2. The execution results are shown in Table 3.4. CPU5 was able to flip the image in 1.33 ms. If we analyze the innermost loop, in summary, the MTFlipVM() function is reading the entire image (i.e., reading 22 MB) and writing the entire vertically flipped image back (another 22 MB). So, all said and done, this is a 44MB data transfer. Using 12 threads, imflipPM.c was able to complete the entire operation in 1.33 ms.

If we moved 44 MB in 1.33 ms, what kind of a data rate does this translate to?

$$\text{Bandwidth Utilization} = \frac{\text{Data Moved}}{\text{Time Required}} \implies \frac{44\,\text{MB}}{1.33\,\text{ms}} \approx 33\,\text{GB/s} \quad (\text{CPU5}) \qquad (4.6)$$

Although this is only 49% of the actual CPU5's peak memory bandwidth of 68 GBps, it is far closer to the peak than any other program we have seen. This computation also clarifies the fact that increasing the number of threads is not helping nearly as much as it did with other programs, since we are getting fairly close to saturating the *Memory Bus*, as depicted in Figure 4.4. We know that we cannot ever exceed the 68 GB/s, since the memory controller (and the main memory) simply cannot transfer data that fast.

At this point, the next natural question to ask ourselves is the same comparison for another CPU. Why not CPU3 again? Since it is the same exact function MTFlipVM() that is running on both CPU3 and CPU5, we can expect CPU3 to be exposed to a similar memory bandwidth saturation when running MTFlipVM(). We see from Table 3.4 that CPU3 is not achieving an execution speed better than 2.66 ms, corresponding to a memory bandwidth of 16.5 GB/s. We know from Table 3.1 that CPU3 has a peak memory bandwidth of 25.6 GB/s. So CPU3 actually hit 65% of its peak bandwidth with the same program. The difference between CPU3 and CPU5 is that CPU3 is in a PC that has DDR3 memory: DDR3 is friendlier to smaller data chunk transfers, whereas DDR4 memory in CPU5's PC is designed to transfer much larger chunks efficiently and provides much higher bandwidth if one does so. This begs the question as to whether the CPU5 would do better if we increased the size of the buffer and made each thread responsible for processing much larger chunks. The answer is YES. I will provide an example in Chapter 5. The idea here is that you do not compare a Yugo and Porsche by making both of them go 30 mph! You are simply insulting the Porsche! The difference will not be seen until you go to 80 mph!

Code 3.2 barely reaches half of the peak bandwidth of CPU5. Reaching the full bandwidth will take a lot more engineering. First of all, there is a perfect "row size" for each DRAM (I called it the "chunk size" earlier, that corresponds to some physical feature of the DRAM). Although Code 3.2 is doing a good job by transferring large chunks at a time, we are not sure if the size of these chunks perfectly matches the optimum chunk size of the DRAM. Additionally, since multiple threads are requesting data simultaneously, they could be disrupting a good pattern that fewer threads would otherwise generate. It is very easy to observe this from Table 3.4, where a higher number of threads sometimes results in a lower performance. Figuring out the optimum size is left to the user as an exercise, however, when we get to the GPU coding, we will be analyzing GPU DRAM details a lot more closely. CPU DRAM and GPU DRAM are almost identical in operation with some differences related to supplying data in parallel to a lot more threads that are in the GPU.

4.6.6 Queue, Uncore, and I/O

Using Intel's fancy term, cores are connected to the external world through the part of the CPU called *uncore*. Uncore includes parts of the CPU that are not inside the cores, but must be included inside the CPU to achieve high performance. Memory controller and L3$ cache controller are two of the most important components of the uncore functionality. PCI Express controller is inside the X99 chipset that is designed for this CPU, as shown in Figure 4.4. Although the PCI Express bus (abbreviated PCIe) is managed by the chipset, there is a part of the CPU that interfaces to the chipset.

This part of the CPU takes up about 22% of the die area inside the CPU and is responsible for queuing and efficiently transferring the data between the PCIe bus and the L3$. When, for example, we are transferring data between the CPU and GPU, this part of the CPU gets heavily used: The X99 chipset is responsible for communicating the PCIe data between the GPUs and itself. Also, it is responsible for communicating the same data between itself and the CPU through the "I/O" portion of the CPU. We can have a hard disk controller, a network card, or a few GPUs hooked up to the I/O.

So, when I am describing the data transfers later in this book, I will pay attention to the memory, cores, and I/O. The programs we are developing are going to be *core-intensive*, *memory-intensive*, or *I/O-intensive*. By looking at Figure 4.4, you can see what they mean. A *core-intensive* program will heavily use the core resources, shown in Figure 4.3, while a *memory-intensive* program will use the memory controller heavily, on the right side of Figure 4.4. Alternatively, an *I/O-intensive* program will use the I/O controller of the CPU, shown on the left side of Figure 4.4.

These portions of the CPU can do their work in parallel, so there is nothing wrong with a program being core, memory, and I/O intensive all at the same time, which is using every part of the CPU. The only exception is that using all of these resources heavily might slow them down, as compared to using only one of them heavily; for example, the data you are transferring to the GPU from the main memory is going through the L3$, creating a bottleneck determined by L3$. Part II will describe this in detail.

4.7   IMROTATEMC: MAKING IMROTATE MORE EFFICIENT

Let's look at the execution times of the imrotate.c program in Table 4.1. The part of this program that is solely responsible for this performance is the innermost function named Rotate(). In this section, we will look at this function and will improve its performance. We will modify main() to allow us to run a different version of the Rotate() function named Rotate2(). As we keep improving this function that has direct influence on the performance of our program, we will name them Rotate3(), Rotate4(), Rotate5(), Rotate6(), and Rotate7(). To achieve this, a variable named **RotateFunc** was defined in our imflip.c program as follows:

```
...
void* (*RotateFunc)(void *arg);   // Func. ptr to rotate the image (multi-threaded)
...
void *Rotate(void* tid)
{
    ...
}
...
int main(int argc, char** argv)
{
    ...
    RotateFunc=Rotate;
    ...
}
```

To keep improving this function, we will design different versions of this function and will allow the user to select the desired function through the command line. To run the imrotateMC.c program, the following command line is used:

imrotateMC InputfileName OutputfileName [degrees] [threads] [func]

where **degrees** specifies the clockwise rotation, [threads] specifies the number of threads to launch as before, and the newly added [func] parameter (1–7) specifies which function to run (i.e., 1 is to run Rotate(), 2 to run Rotate2(), etc.) The improved functions will be consistently named Rotate2(), Rotate3(), etc. and the appropriate function pointer will be assigned to the **RotateFunc** based on the command line argument selection as shown in Code 4.5. The name of this new program constraints "MC" for "memory and core friendly."

CODE 4.5: imrotateMC.c   main() {...

The main() function in imrotateMC.c allows the user to specify which function to run, from Rotate() to Rotate7(). Parts of the code that are identical to imrotate.c, listed in Code 4.2 and 4.3, are not repeated.

```c
int main(int argc, char** argv)
{
    int             RotDegrees, Function;
    int             a,i,ThErr;
    struct timeval  t;
    double          StartTime, EndTime;
    double          TimeElapsed;
    char            FuncName[50];

    switch (argc){
        case 3 : NumThreads=1;    RotDegrees=45;   Function=1;              break;
        case 4 : NumThreads=1;    RotDegrees=at... Function=1;              break;
        case 5 : NumThreads=at... RotDegrees=at... Function=1;              break;
        case 6 : NumThreads=at... RotDegrees=at... Function=atoi(argv[5]); break;
        default: printf("\nUsage: %s inputBMP outBMP [RotAngle] [1-128] [1-7]...");
                printf("Example: %s infilename.bmp outname.bmp 125 4 3\n\n",argv[0]);
                printf("Nothing executed ... Exiting ...\n\n");
                exit(EXIT_FAILURE);
    }
    if((NumThreads<1) || (NumThreads>MAXTHREADS)){
        ...
    }
    if((RotDegrees<-360) || (RotDegrees>360)){
        ...
    }
    switch(Function){
        case 1: strcpy(FuncName,"Rotate()");    RotateFunc=Rotate;  break;
        case 2: strcpy(FuncName,"Rotate2()");   RotateFunc=Rotate2; break;
        case 3: strcpy(FuncName,"Rotate3()");   RotateFunc=Rotate3; break;
        case 4: strcpy(FuncName,"Rotate4()");   RotateFunc=Rotate4; break;
        case 5: strcpy(FuncName,"Rotate5()");   RotateFunc=Rotate5; break;
        case 6: strcpy(FuncName,"Rotate6()");   RotateFunc=Rotate6; break;
        case 7: strcpy(FuncName,"Rotate7()");   RotateFunc=Rotate7; break;
        //   case 8: strcpy(FuncName,"Rotate8()"); RotateFunc=Rotate8; break;
        //   case 9: strcpy(FuncName,"Rotate9()"); RotateFunc=Rotate9; break;
        default: printf("Wrong function %d ... \n",Function);
            printf("\n\nNothing executed ... Exiting ...\n\n");
            exit(EXIT_FAILURE);
    }
    printf("\nLaunching %d Pthreads using function: %s\n",NumThreads,FuncName);
    RotAngle=2*3.141592/360.000*(double) RotDegrees; // Convert the angle to radians
    printf("\nRotating image by %d degrees ...\n",RotDegrees);
    TheImage = ReadBMP(argv[1]);
    ...
}
```

CODE 4.6: imrotateMC.c Rotate2() {...}

In the Rotate2() function, the four lines that calculate H, V, Diagonal, and ScaleFactor are moved outside the two for loops, since they only need to be calculated once.

```
void *Rotate2(void* tid)
{
    int      row,col,h,v,c, NewRow,NewCol;
    double   X, Y, newX, newY, ScaleFactor, Diagonal, H, V;
    ...
    H=(double)ip.Hpixels;        //MOVE UP HERE
    V=(double)ip.Vpixels;        //MOVE UP HERE
    Diagonal=sqrt(H*H+V*V);      //MOVE UP HERE
    ScaleFactor=(ip.Hpixels>ip.Vpixels) ? V/Diagonal : H/Diagonal;   //MOVE UP HERE
    for(row=tn; row<tn+ip.Vpixels/NumThreads; row++){
        col=0;
        while(col<ip.Hpixels*3){
            ...
        newY=sin(RotAngle)*X+cos(RotAngle)*Y;
        //       MOVE THESE 4 INSTRUCTIONS OUTSIDE BOTH LOOPS
        //       H=(double)ip.Hpixels;          V=(double)ip.Vpixels;
        //       Diagonal=sqrt(H*H+V*V);
        //       ScaleFactor=(ip.Hpixels>ip.Vpixels) ? V/Diagonal : H/Diagonal;
        newX=newX*ScaleFactor;
            ...
}
```

4.7.1 Rotate2(): How Bad is Square Root and FP Division?

Now, let's look at the Rotate() function in Code 4.4. It calculates the new scaled X, Y coordinates and saves them in variables newX and newY. For this, it first has to calculate the ScaleFactor from Equation 4.3, and d (Diagonal). This calculation involves the code lines:

```
H=(double)ip.Hpixels;
V=(double)ip.Vpixels;
Diagonal=sqrt(H*H+V*V);
ScaleFactor=(ip.Hpixels>ip.Vpixels) ? V/Diagonal : H/Diagonal;
```

Moving these lines outside both loops will not change the functionality at all, since they really only need to be calculated once. We are particularly interested in understanding what parts of the CPU core in Figure 4.3 these computations use and how much speedup we will get from this move. Revised Rotate2() function is shown in Code 4.6. Identical lines compared to the original function Rotate() are not repeated and denoted as "..." to improve readability.

An entire list of execution times for every version of this function will be provided later in this chapter in Table 4.3. For now, let's quickly compare the single-threaded performance of Rotate2() versus Rotate() on CPU5. To run the single-threaded version of Rotate2(), type:

imrotateMC dogL.bmp d.bmp 45 1 2

where 45, 1, and 2 are rotation, number of threads (single), and function ID (Rotate2()). We reduced the single-threaded run time from 1027 ms down to 498 ms, a 2.06× improvement.

Now, let's analyze the instructions on the 4 lines that we moved to see why we had a 2.06× improvement. First two computations are integer-to-double precision floating point (FP) cast operations when we are calculating H and V. They are simple enough, but would use an FPU resource (shown in Figure 4.3). The next line in calculating `Diagonal` is absolutely a resource hog, since Square Root is very compute-intensive. This harmless looking line requires two FP multiplications (FP-MUL) to compute $H \times H$ and $V \times V$ and one FP-ADD to compute their sum. After this, the super-expensive Square Root operation is performed. As if sqrt is not enough torture for the core, we see a floating point division next, that is as bad as the square root, followed by an integer comparison! So, when the CPU core hits the instructions that computes these 4 lines, core resources are chewed up and spit out! When we move them outside both loops, it is no wonder that we get a 2.06× speedup.

4.7.2   Rotate3() and Rotate4(): How Bad Is sin() and cos()?

Why stop here? What else can we precompute? Look at these lines in the code:

```
newX=cos(RotAngle)*X-sin(RotAngle)*Y;
newY=sin(RotAngle)*X+cos(RotAngle)*Y;
```

We are simply forcing the CPU to compute the sin() and cos() for *every pixel*! There is no need for that, Rotate3() function defines a precomputed variable called CRA (meaning precomputed cosine of RotAngle) and uses it in the innermost loop whenever it needs to use cos(RotAngle). The revised Rotate3() function is shown in Code 4.7 and it is a single-threaded run time on CPU5 reduced from 498 ms to 376 ms, a 1.32× improvement.

Function Rotate4() (shown in Code 4.8) does the same thing by precomputing sin(RotAngle) as SRA. Rotate4() single-threaded run time is reduced from 376 ms to 235 ms, another 1.6× improvement. The simplified code lines in the Rotate4() function in Code 4.8 are

```
newX=CRA*X-SRA*Y;
newY=SRA*X+CRA*Y;
```

When we compare these two lines to the two lines above, the summary is as follows:

- Rotate3() needs the calculation of sin(), cos().

- Rotate3() performs 4 double-precision FP multiplications.

- Rotate3() performs 2 double-precision FP addition/subtractions.

- Rotate4() needs all of the above, except the sin(), cos().

CODE 4.7:   imrotateMC.c   Rotate3() {...}

Rotate3() function precomputes cos(RotAngle) as CRA outside both loops.

```
void *Rotate3(void* tid)
{
    int NewRow,NewCol;
    double X, Y, newX, newY, ScaleFactor;
    double Diagonal, H, V;
    double CRA;
    ...
    H=(double)ip.Hpixels;
    V=(double)ip.Vpixels;
    Diagonal=sqrt(H*H+V*V);
    ScaleFactor=(ip.Hpixels>ip.Vpixels) ? V/Diagonal : H/Diagonal;
    CRA=cos(RotAngle); /// MOVE UP HERE
    for(row=tn; row<tn+ip.Vpixels/NumThreads; row++){
        col=0;
        while(col<ip.Hpixels*3){
            ...
            newX=CRA*X-sin(RotAngle)*Y;    // USE PRE-COMPUTED CRA
            newY=sin(RotAngle)*X+CRA*Y;    // USE PRE-COMPUTED CRA
            //      newX=cos(RotAngle)*X-sin(RotAngle)*Y; // CHANGE
            //      newY=sin(RotAngle)*X+cos(RotAngle)*Y; // CHANGE
            newX=newX*ScaleFactor;
            ...
```

CODE 4.8:   imrotateMC.c   Rotate4() {...}

Rotate4() function precomputes sin(RotAngle) as SRA outside both loops.

```
void *Rotate4(void* tid)
{
    ...
    double CRA, SRA;
    ...
    ScaleFactor=(ip.Hpixels>ip.Vpixels) ? V/Diagonal : H/Diagonal;
    CRA=cos(RotAngle);
    SRA=sin(RotAngle); /// MOVE UP HERE
    for(row=tn; row<tn+ip.Vpixels/NumThreads; row++){
        col=0;
        while(col<ip.Hpixels*3){
            ...
            newX=CRA*X-SRA*Y;  // USE PRE-COMPUTED SRA, CRA
            newY=SRA*X+CRA*Y;  // USE PRE-COMPUTED SRA, CRA
            //      newX=cos(RotAngle)*X-sin(RotAngle)*Y; // CHANGE
            //      newY=sin(RotAngle)*X+cos(RotAngle)*Y; // CHANGE
    ...
```

4.7.3 Rotate5(): How Bad Is Integer Division/Multiplication?

We have almost completely exhausted the FP operations that we can precompute. It is time to look at integer operations now. We know that integer divisions are substantially slower than integer multiplications. Since every CPU that is used in testing is a 64-bit CPU, there isn't too much of a performance difference between 32-bit and 64-bit multiplications. However, divisions might be different depending on the CPU. It is time to put all of these ideas to test. The revised Rotate5() function is shown in Code 4.9.

For the Rotate5() function, our target is the following lines of code, containing integer computations:

```
for(row=tn; row<tn+ip.Vpixels/NumThreads; row++){
    col=0;
    while(col<ip.Hpixels*3){    // USE THE PRE-COMPUTED hp3 VALUE
        // transpose image coordinates to Cartesian coordinates
        c=col/3;    h=ip.Hpixels/2; v=ip.Vpixels/2; // integer div
```

We notice that we are calculating a value called `ip.Hpixels*3` just to turn back around and divide it by 3 a few lines below. Knowing that integer divisions are expensive, why not do this in a way where we can eliminate the integer division altogether? To do this, we observe that the variable `c` is doing nothing but mirror the same value as `col/3`. Since we are starting the variable `col` at 0 before we get into the `while()` loop, why not start the `c` variable at 0 also? Since we are incrementing the value of the `col` variable by 3 at the end of the `while` loop, we can simply increment the `c` variable by one. This will create two variables that completely track each other, with the relationship $col = 3 \times c$ without ever having to use an integer division.

The implementation of the Rotate5() function that implements this idea is shown in Code 4.9 and simply trades an integer division $c = col/3$ for an integer increment operation $c + +$. Additionally, to find the half-way point in the picture, `ip.Hpixels` and `ip.Vpixels` must be divided by 2 as shown above. Since this can also be precomputed, it is moved outside both loops in the implementation of the Rotate5(). All said and done, Rotate5() run time is reduced from 235 ms to 210 ms, an improvement of 1.12×. Not bad ...

4.7.4 Rotate6(): Consolidating Computations

To design the Rotate6(), we will target these lines

```
X=(double)c-(double)h;        Y=(double)v-(double)row;
// pixel rotation matrix
newX=CRA*X-SRA*Y;             newY=SRA*X+CRA*Y;
newX=newX*ScaleFactor;        newY=newY*ScaleFactor;
```

After all of our modifications, these lines ended up in this order. So, the natural question to ask ourselves is: can we consolidate these computations of `newX` and `newY` and can we precompute variables `X` or `Y`? An observation of the loop shows that, although the variable `X` is stuck inside the innermost loop, we can move the computation of variable `Y` outside the innermost loop, although we *cannot* move it outside *both* loops. But, considering that this will save us the repetitive computation of `Y` many times (3200 times for a 3200 row image!), the savings could be worth the effort. The revised Rotate6() is shown in Code 4.10 and implements these ideas by using two additional variables named `SRAYS` and `CRAYS`. Rotate6() runtime improved from 210 ms to 185 ms (1.14× better). Not bad ...

CODE 4.9:   imrotateMC.c   Rotate5() {...}

In Rotate5(), integer division, multiplications are taken outside both loops.

```c
void *Rotate5(void* tid)
{
  int hp3;
  double CRA,SRA;
  ...
  CRA=cos(RotAngle);    SRA=sin(RotAngle);
  h=ip.Hpixels/2;       v=ip.Vpixels/2;    // MOVE IT OUTSIDE BOTH LOOPS
  hp3=ip.Hpixels*3;                        // PRE-COMPUTE ip.Hpixels*3
  for(row=tn; row<tn+ip.Vpixels/NumThreads; row++){
    col=0;
    c=0;            // HAD TO DEFINE THIS
    while(col<hp3){    //INSTEAD OF col<ip.Hpixels*3
      // c=col/3; h=ip.Hpixels/2; v=ip.Vpixels/2; // MOVE OUT OF THE LOOP
      X=(double)c-(double)h;
      Y=(double)v-(double)row;
      // pixel rotation matrix
      newX=CRA*X-SRA*Y;                newY=SRA*X+CRA*Y;
      newX=newX*ScaleFactor;          newY=newY*ScaleFactor;
      ...
      col+=3;           c++;
      ...
```

CODE 4.10:   imrotateMC.c   Rotate6() {...}

In Rotate6(), FP operations are consolidated to reduce the FP operation count.

```c
void *Rotate6(void* tid)
{
  double CRA,SRA, CRAS, SRAS, SRAYS, CRAYS;
  ...
  CRA=cos(RotAngle);    SRA=sin(RotAngle);
  CRAS=ScaleFactor*CRA; SRAS=ScaleFactor*SRA;  // PRECOMPUTE ScaleFactor*SRA, CRA
  ...
  for(row=tn; row<tn+ip.Vpixels/NumThreads; row++){
    col=0;      c=0;
    Y=(double)v-(double)row;              // MOVE UP HERE
    SRAYS=SRAS*Y;          CRAYS=CRAS*Y;  // NEW PRE-COMPUTATIONS
    while(col<hp3){
      X=(double)c-(double)h;
      // Y=(double)v-(double)row;         // MOVE THIS OUT
      // pixel rotation matrix
      newX=CRAS*X-SRAYS;    // CALC NewX with pre-computed values
      newY=SRAS*X+CRAYS;    // CALC NewY with pre-computed values
      ...
```

CODE 4.11:   imrotateMC.c   Rotate7() {...}

In Rotate7(), every computation is expanded to avoid any redundant computation.

```c
void *Rotate7(void* tid)
{
   long tn;
   int row, col, h, v, c, hp3, NewRow, NewCol;
   double cc, ss, k1, k2, X, Y, newX, newY, ScaleFactor;
   double Diagonal, H, V, CRA,SRA, CRAS, SRAS, SRAYS, CRAYS;
   struct Pixel pix;

   tn = *((int *) tid);     tn *= ip.Vpixels/NumThreads;
   H=(double)ip.Hpixels;    V=(double)ip.Vpixels;    Diagonal=sqrt(H*H+V*V);
   ScaleFactor=(ip.Hpixels>ip.Vpixels) ? V/Diagonal : H/Diagonal;
   CRA=cos(RotAngle);       CRAS=ScaleFactor*CRA;
   SRA=sin(RotAngle);       SRAS=ScaleFactor*SRA;
   h=ip.Hpixels/2;          v=ip.Vpixels/2;          hp3=ip.Hpixels*3;
   for(row=tn; row<tn+ip.Vpixels/NumThreads; row++){
      col=0;                       cc=0.00;         ss=0.00;
      Y=(double)v-(double)row;     SRAYS=SRAS*Y;    CRAYS=CRAS*Y;
      k1=CRAS*(double)h + SRAYS;  k2=SRAS*(double)h - CRAYS;
      while(col<hp3){
         newX=cc-k1;                newY=ss-k2;
         NewCol=((int) newX+h);     NewRow=v-(int)newY;
         if((NewCol>=0) && (NewRow>=0) && (NewCol<ip.Hpixels) &&
            (NewRow<ip.Vpixels)){
           NewCol*=3;
           CopyImage[NewRow][NewCol] = TheImage[row][col];
           CopyImage[NewRow][NewCol+1] = TheImage[row][col+1];
           CopyImage[NewRow][NewCol+2] = TheImage[row][col+2];
         }
         col+=3;       cc += CRAS;     ss += SRAS;
      }
   }
   pthread_exit(NULL);
}
```

4.7.5   Rotate7(): Consolidating More Computations

Finally, our Rotate7() function looks at every possible computation to see if it can be done with precomputed values. The Rotate7() function in Code 4.11 achieves a run time of 161 ms, an improvement of 1.15× compared to Rotate6() that achieved 185 ms.

4.7.6   Overall Performance of imrotateMC

The execution results of all 7 Rotate functions are shown in Table 4.3. Taking CPU5 as an example, all of the improvements resulted in a 6.4× speedup on a single thread. For 8 threads, which we know is a comfortable operating point for CPU5, the speedup is 7.8×. This additional boost in speedup shows that, for core-intensive functions like Rotate(), improving the core efficiency of the program helped when more threads are launched.

TABLE 4.3 imrotateMC.c execution times for the CPUs in Table 3.1.

#Th	Func	CPU1	CPU2	CPU3	CPU4	CPU5	CPU6
1		951	1365	782	1090	1027	845
2		530	696	389	546	548	423
3		514	462	261	368	365	282
4	Rotate()	499	399	253	322	272	227
6			387	283	338	214	213
8			374	237	297	188	163
10			341	228	285	163	201
1		468	580	364	441	498	659
2		280	301	182	222	267	330
3		249	197	123	148	194	220
4	Rotate2()	280	174	126	165	137	165
6			207	127	138	101	176
8			195	138	134	84	138
10				125	141	67	
1		327	363	264	301	376	446
2		218	189	131	151	202	223
3		187	123	88	101	142	149
4	Rotate3()	202	93	106	108	101	112
6			123	97	106	75	116
8			117	101	110	59	89
10				106	92	47	
1		202	227	161	182	235	240
2		140	124	80	91	135	120
3		109	80	54	61	92	80
4	Rotate4()	109	65	73	54	69	60
8			88	62	69	37	47
10				58	55	29	
1		171	209	145	158	210	207
2		140	108	73	78	117	104
3		93	73	49	53	80	69
4	Rotate5()	93	61	69	72	61	52
6			72	51	62	44	53
8			81	56	60	36	40
10				59	48	29	
1		156	180	125	128	185	176
2		124	92	63	64	109	88
3		93	78	43	45	78	59
4	Rotate6()	93	57	63	65	55	44
6			60	43	43	37	44
8			65	51	49	30	33
1		140	155	107	110	161	156
2		109	75	53	55	97	78
3		93	52	36	37	64	52
4	Rotate7()	62	70	53	56	46	39
6			61	36	38	36	39
8			56	40	42	24	29
10				43	45	21	

4.8 CHAPTER SUMMARY

In this chapter, we looked at what happens inside the core and during the memory transfers from the CPU to the main memory. We used this information to make one example program faster. The rules we outlined are simple:

> ➤ *Stay away from sophisticated core instructions, such as* sin(), sqrt().
> *If you must use them, make sure to have a limited number of them.*
> ➤ *The ALU executes integer instructions, FPU executes floating point.*
> *Try to have a mix of these instructions in an inner loop to use both units.*
> *If you can use only one type, use integer.*
> ➤ *Avoid choppy memory accesses. Use big bulk transfers if possible.*
> *Try to take heavy computations outside the inner loops.*

Aside from these simple rules, Table 4.3 shows that if the threads we design are *thick*, we will not be able to take advantage of the multiple threads that a core can execute. In our code so far, even the improved Rotate7() function has thick threads. So, the performance falls off sharply when either the launched threads gets close to the number of the physical cores, or we saturate the memory. This brings up the concept of "whichever constraint comes first." In other words, when we change our program to improve it, we might eliminate one bottleneck (say, FPU inside the core), but create a totally different bottleneck (say, main memory bandwidth saturation).

The "whichever constraint comes first" concept will be the key ingredient in designing GPU code, since inside a GPU, there are multiple constraints that can saturate and the programmer has to be fully aware of every single one of them. For example, the programmer could saturate the *total number of launched threads* before the memory is saturated. The *constraint ecosystem* inside a GPU will be very much like the one we will learn in the CPU. More on that coming up very shortly ...

Before we jump right into the GPU world, we have one more thing to learn: *thread synchronization* in Chapter 5. Then, we are ready to take on the GPU challenge starting in Chapter 6.

## CHAPTER 5

Thread Management and Synchronization

W Hen we say *hardware architecture*, what are we talking about?
    The answer is *everything physical*: CPU, memory, I/O controller, PCI express bus, DMA controller, hard disk controller, hard disk(s), SSDs, CPU chipsets, USB ports, network cards, CD-ROM controller, DVDRW, I can go on for a while ...

How about, when we say *software architecture*, what are we talking about?
The answer is *all of the code* that runs on the hardware. Code is everywhere: your hard disk, SSDs, your USB controller, your CD-ROM, even your cheap keyboard, your operating system, and, finally, the programs you write.

The big question is: which one of these do we *care about* as a high-performance programmer? The answer is: CPU cores and threads, as well as memory in the hardware architecture and the operating system (OS) and your application code in the software architecture. Especially if you are writing high performance programs (like the ones in this book), a vast percentage of your program performance will be determined by these things. The disk performance generally does not play an important role in the overall performance because most modern computers have plenty of RAM to cache large portions of the disk. The OS tries to efficiently allocate the CPU cores and memory to maximize performance while your application code requests these two resources from the OS and uses them – hopefully efficiently. Because the purpose of this book is high-performance programming, let's dedicate this chapter to understanding the interplay between your our own code and the OS when it comes to allocating/using the cores and memory.

5.1   EDGE DETECTION PROGRAM: IMEDGE.C

In this section, I will introduce an image edge detection program, imedge.c, that detects the edges in an image as shown in Figure 5.1. In addition to providing a good template example that we will continuously improve, edge detection is actually one of the most common image processing tasks and is an essential operation that the human retina performs. Just like the programs I introduced in the preceding chapters, this program is also resource-intensive; therefore, I will introduce its variant imedgeMC.c that is memory-friendly ("M") and core-friendly ("C"). One interesting variant I will add is imedgeMCT.c, which adds the thread-friendliness ("T") concept by utilizing a *MUTEX* structure to facilitate communication among multiple threads.

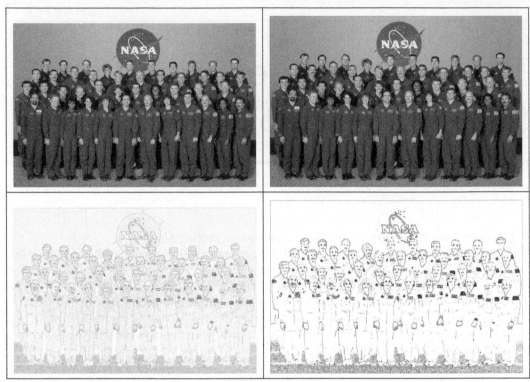

FIGURE 5.1 The imedge.c program is used to detect edges in the original image astronaut.bmp (top left). Intermediate processing steps are: GaussianFilter() (top right), Sobel() (bottom left), and finally Threshold() (bottom right).

5.1.1 Description of the imedge.c

The purpose of imedge.c is to create a program that is core-heavy, memory-heavy, and is composed of multiple independent operations (functions):

- GaussianFilter(): The initial smoothing filter to reduce noise in the original image.

- Sobel(): The edge detection kernel we apply to amplify the edges.

- Threshold(): The operation that turns a grayscale image to a binary (black & white) image (denoted B&W going forward), thereby "detecting" the edges.

imedge.c takes an image such as the one shown in Figure 5.1 (top left) and applies these three distinct operations one after the other to finally yield an image that is composed of only the edges (bottom right). To run the program, the following command line is used:

imedge InputfileName OutputfileName [1-128] [ThreshLo] [ThreshHi]

where [1–128] is the number of threads to launch as usual and the [ThreshLo], [ThreshHi] pair determines the thresholds used during the Threshold() function.

5.1.2 imedge.c: Parametric Restrictions and Simplifications

Some simplifications had to be made in the program to avoid unnecessary complications and diversion of our attention to unrelated issues. Numerous improvements are possible to this program and the readers should definitely attempt to discover them. However, the choices made in this program are geared toward improving the instructional value of this chapter, rather than the ultimate performance of the imedge.c program. Specifically,

- A much more robust edge detection program can be designed by adding other post-processing steps. They have not been incorporated into our program.

- The granularity of the processing is limited to a single row when running the "MCT" version (imedgeMCT.c) at the end of this chapter. Although a finer granularity processing is possible, it does not add any instructional value.

- To calculate some of the computed pixel values, the choice of the `double` variable type is a little overkill; however, `double` type has been employed in the program because it improves the instructional value of the code.

- Multiple arrays are used to store different processed versions of the image before its final edge-detected version is computed: `TheImage` array stores the original image. `BWImage` array stores the B&W version of this original. `GaussImage` array stores the Gaussian filtered version. `Gradient` and `Theta` arrays store the Sobel-computed pixels. The final edge-detected image is stored in `CopyImage`. Although these arrays could be combined, separate arrays have been used to improve the clarity of the code.

5.1.3   imedge.c: Theory of Operation

imedge.c edge detection program includes three distinct operations as described in Section 5.1. It is definitely possible to combine these operations to end up with a more efficient program, but from an instructional standpoint, it helps to see how we will speed up each individual operation; because these three operations have different resource characteristics, separating them will allow us to analyze each one individually. Later, when we design improved versions of it – as we did in the previous chapters – we will improve each individual operation in different ways. Now, let me describe each individual operation separately.

GaussianFilter(): You can definitely skip this operation at the expense of confusing some noise with the edges. As shown in Figure 5.1, Gaussian filter blurs the original image, which can be perceived as a useless operation that degrades image quality. However, the intended positive consequence of blurring is the fact that we get rid of noise that might be present in the image, which could be erroneously detected as edges. Also, remember that our final image will be turned into a *binary* image consisting of only two values, which will correspond to edge/no edge. A binary image only has a 1-bit depth for each pixel, which is substantially lower quality than the original image that has a 24-bit depth, i.e., 8-bits for R, G, and B each. Considering that this edge detection operation reduces the image bit-depth from 24-bits per pixel down to 1-bit per pixel anyway, the depth reduction due to the blurring operation only has a positive effect for us. By the same token, though, too much blurring might worsen the ability of our entire process by eventually melting the *actual edges* into the picture, thereby making them unrecognizable. Quantitatively speaking, GaussianFilter() turns the original 24-bit image into a blurred 8-bit-per-pixel B&W image according to the equation below:

$$\text{BWImage} = \frac{\text{TheImage}_R + \text{TheImage}_G + \text{TheImage}_B}{3} \tag{5.1}$$

where the `TheImage` is the original 24-bit/pixel image, composed of R, G, and B parts; averaging of the R, G, and B values turns this color image into an 8-bit/pixel B&W image, saved in the variable named `BWImage`.

The Gaussian filtering step is composed of a convolution operation as follows:

$$\text{Gauss} = \frac{1}{159} \begin{bmatrix} 2 & 4 & 5 & 4 & 2 \\ 4 & 9 & 12 & 9 & 4 \\ 5 & 12 & 15 & 12 & 5 \\ 4 & 9 & 12 & 9 & 4 \\ 2 & 4 & 5 & 4 & 2 \end{bmatrix} \implies \text{GaussImage} = \text{BWImage} * \text{Gauss} \qquad (5.2)$$

where `Gauss` is the filter kernel and $*$ is the *convolution* operation, which is one of the most common operations in digital signal processing (DSP). Together, the operation in Equation 5.2 *convolves* the B&W image we created in Equation 5.2 – that is contained in variable `BWImage` – with the Gauss filter mask, contained in variable `Gauss` to result in the blurred image, contained in variable `GaussImage`. Note that other filter kernels are available that result in different levels of blurring, however, for our demonstration purposes, this filter kernel is totally fine. At the end of this step, our image looks like Figure 5.1 (top right).

Sobel(): The goal of this step is to use a Sobel gradient operator on each pixel to determine the direction – and existence – of edges. From Equation 5.3, Sobel kernels are

$$G_x = \begin{bmatrix} -1 & 0 & 1 \\ -2 & 0 & 2 \\ -1 & 0 & 1 \end{bmatrix}, \qquad G_y = \begin{bmatrix} -1 & -2 & -1 \\ 0 & 0 & 0 \\ -1 & -2 & -1 \end{bmatrix} \qquad (5.3)$$

where G_x and G_x are the Sobel *kernels* used to determine the edge gradient for each pixel in the x and y directions. These kernels are convolved with the previously computed `GaussImage` to compute a gradient image as follows:

$$GX = Im * G_x, \quad GY = Im * G_y, \quad G = \sqrt{GX^2 + GY^2}, \quad \theta = \tan^{-1}\left(\frac{GX}{GY}\right) \qquad (5.4)$$

where Im is the previously computed blurred image in `GaussImage`, $*$ is the *convolution* operation, GX and GY are the edge gradients in the x and y directions. These temporary gradients are not of interest to us, whereas the magnitude of the gradient is; the G variable in Equation 5.4 is the magnitude of the gradient for each pixel, stored in variable `Gradient`. Additionally, we would like to have some idea about what the direction of the edge was (θ) at a specific pixel location; the variable `Theta` stores this information – also calculated from Equation 5.4 – and will be used in the thresholding process. At the end of the Sobel step, our image looks like Figure 5.1 (bottom left). This image is the value of the `Gradient` array, which stores the magnitudes of the edges; it is a grayscale image.

Threshold(): The thresholding step takes the `Gradient` array and turns it into a 1-bit edge/no edge image based on two threshold values: `ThreshLo` value is the threshold value below which a pixel is definitely considered a "NO EDGE." Alternatively, `ThreshHi` value is the threshold value above which a pixel is definitely considered an "EDGE." In these two cases, using the `Gradient` variable suffices, as formulated in the equation below:

$$\text{pixel at}[x, y] \implies \begin{cases} \text{Gradient}[x, y] < \text{ThreshLo}, & \text{CopyImage}[x, y] = \text{NOEDGE} \\ \text{Gradient}[x, y] > \text{ThreshHi}, & \text{CopyImage}[x, y] = \text{EDGE} \end{cases} \qquad (5.5)$$

where the `CopyImage` array stores the final binary image. In case the gradient value of a pixel is in between these two thresholds, we use the second array, `Theta`, to determine the direction of the edge. We classify the direction into one of four possible values: horizontal (EW), vertical (NS), left diagonal (SW-NE), and right diagonal (SE-NW). The idea is that

TABLE 5.1   Array variables and their types, used during edge detection.

Function to perform	Starting array variable	Destination array variable	Destination type
Convert image to B&W	TheImage	BWImage	unsigned char
Gaussian filter	BWImage	GaussImage	double
Sobel filter	GaussImage	Gradient	double
		Theta	double
Threshold	Gradient (Theta if needed)	CopyImage	unsigned char

if the Theta of the edge is pointing to vertical (that is, up/down), we determine whether this pixel is an edge or not by looking at the pixel above or below it. Similarly, for a horizontal pixel, we look at its horizontal neighbors. This well-studied method is formulated below:

$$L < \Delta[x,y] < H \implies \begin{cases} \Theta < -\frac{3}{8}\pi \quad \text{or} \quad \Theta > \frac{3}{8}\pi \quad \implies \quad \text{EW neighbor} \\ \Theta \geq -\frac{1}{8}\pi \quad \text{and} \quad \Theta \leq \frac{1}{8}\pi \quad \implies \quad \text{NS neighbor} \\ \Theta > \frac{1}{8}\pi \quad \text{and} \quad \Theta \leq \frac{3}{8}\pi \quad \implies \quad \text{SW-NE neighbor} \\ \Theta \geq -\frac{3}{8}\pi \quad \text{and} \quad \Theta < -\frac{1}{8}\pi \quad \implies \quad \text{SE-NW neighbor} \end{cases} \quad (5.6)$$

where $\Theta = \Theta[x,y]$ is the angle of edge $[x,y]$ and L and H are the low, high thresholds. The gradient is shown as Δ. The final result of the imedge.c program is shown in Figure 5.1 (bottom right). The #define values define how EDGE/NOEDGE will be colored; in this program I assigned 0 (black) to EDGE and 255 (white) to NO EDGE. This makes the edges *print-friendly*.

5.2   IMEDGE.C : IMPLEMENTATION

Table 5.1 provides the names and types of different arrays that are used during the execution of imedge.c. The image is initially read into the TheImage array, which contains the RGB values of each pixel in unsigned char type (Figure 5.1 top left). This image is converted to its B&W version according to Equation 5.1 and saved in the BWImage array (Figure 5.1 top right). Each pixel value in the BWImage array is of type double, although a much lower resolution type would suffice. Gaussian filtering takes the BWImage array and places its filtered (i.e., *blurred*) version in the Gradient and Theta arrays, which contain the gradient magnitudes and edge angles, respectively. Gradient array is shown in Figure 5.1 (bottom left), which shows how *edgy* this pixel is. The final thresholding step takes these two arrays to produce the final result of the program, CopyImage array, which contains the binary edge/no edge image (Figure 5.1 bottom right).

Although a single-bit depth would suffice for the CopyImage, each pixel is saved in CopyImage using RGB pixel values (EDGE, EDGE, EDGE) to denote an existing edge, or (NOEDGE, NOEDGE, NOEDGE) to denote a non-existent edge. This makes it feasible to use the same function to save a BMP image to save the final binary image. The default #define value of EDGE is 0, so each edge looks black, i.e., the RGB value (0,0,0). Similarly, NOEDGE is 255, which looks white, i.e., RGB value (255,255,255). The edge image in Figure 5.1 (bottom right) contains only these two RGB values. A 1-bit depth BMP file format could also be used to conserve space when saving the final edge image, which is left up to the reader as an exercise.

CODE 5.1:   imedge.c   ... main() {...

Time stamping code in main to identify partial execution times.

```c
unsigned char**  TheImage;          // This is the main image
unsigned char**  CopyImage;         // This is the copy image (to store edges)
double           **BWImage;         // B&W of TheImage (each pixel=double)
double           **GaussImage;      // Gauss filtered version of the B&W image
double           **Gradient, **Theta; // gradient and theta for each pixel
struct ImgProp   ip;
...
double GetDoubleTime()              // returns the time stamps in ms
{
   struct timeval    tnow;
   gettimeofday(&tnow, NULL);
   return ((double)tnow.tv_sec*1000000.0 + ((double)tnow.tv_usec))/1000.00;
}

double ReportTimeDelta(double PreviousTime, char *Message)
{
   double  Tnow,TimeDelta;
   Tnow=GetDoubleTime();
   TimeDelta=Tnow-PreviousTime;
   printf("\n.....%-30s ... %7.0f ms\n",Message,TimeDelta);
   return Tnow;
}

int main(int argc, char** argv)
{
   int      a,i,ThErr;          double  t1,t2,t3,t4,t5,t6,t7,t8;
   ...
   printf("\nExecuting the Pthreads version with %li threads ...\n",NumThreads);
   t1 = GetDoubleTime();
   TheImage=ReadBMP(argv[1]);   printf("\n");
   t2 = ReportTimeDelta(t1,"ReadBMP complete"); // Start time without IO
   CopyImage = CreateBlankBMP(NOEDGE);          // Store edges in RGB
   BWImage   = CreateBWCopy(TheImage);
   GaussImage = CreateBlankDouble();
   Gradient = CreateBlankDouble();
   Theta    = CreateBlankDouble();
   t3=ReportTimeDelta(t2, "Auxiliary images created");
   ...
```

5.2.1 Initialization and Time-Stamping

imedge.c time stamps each one of the three separate steps, as well as the initialization times to create the aforementioned arrays to assess program performance better. Initialization and time-stamping code is shown in Code 5.1. Each time stamp is obtained using the GetDoubleTime() helper function; the time difference between two time stamps is computed and reported using the ReportTimeDelta() function, which also uses a string to explain the achieved function at that time stamp.

CODE 5.2:   imageStuff.c   Image Initialization Functions

Initialization of the different image values.

```c
double** CreateBlankDouble()
{
   int i;
   double** img = (double **)malloc(ip.Vpixels * sizeof(double*));
   for(i=0; i<ip.Vpixels; i++){
      img[i] = (double *)malloc(ip.Hpixels*sizeof(double));
      memset((void *)img[i],0,(size_t)ip.Hpixels*sizeof(double));
   }
   return img;
}


double** CreateBWCopy(unsigned char** img)
{
   int i,j,k;
   double** imgBW = (double **)malloc(ip.Vpixels * sizeof(double*));
   for(i=0; i<ip.Vpixels; i++){
      imgBW[i] = (double *)malloc(ip.Hpixels*sizeof(double));
      for(j=0; j<ip.Hpixels; j++){  // convert each pixel to B&W = (R+G+B)/3
         k=3*j;
            imgBW[i][j]=((double)img[i][k]+(double)img[i][k+1]+(double)img[i][k+2])/3.0;
      }
   }
   return imgBW;
}


unsigned char** CreateBlankBMP(unsigned char FILL)
{
   int i,j;
   unsigned char** img=(unsigned char **)malloc(ip.Vpixels*sizeof(unsigned char*));
   for(i=0; i<ip.Vpixels; i++){
      img[i] = (unsigned char *)malloc(ip.Hbytes * sizeof(unsigned char));
      memset((void *)img[i],FILL,(size_t)ip.Hbytes); // zero out every pixel
   }
   return img;
}
```

5.2.2   Initialization Functions for Different Image Representations

ReadBMP() function (Code 2.5) reads the source image into **TheImage** using an unsigned char for each of the RGB values. CreateBlankBMP() function (Code 5.2) creates an initialized BMP image, with R=G=B=0 unsigned char initial pixel values; it is used to initialize the **CopyImage** array. CreateBWCopy() is used to initialize the **BWImage** array; it turns a 24-bit image into its B&W version (using Equation 5.1), where each pixel has a double value. CreateBlankDouble() function (Code 5.2) creates an image array, populated with 0.0 double values; it is used to initialize the **GaussImage**, **Gradient**, and **Theta** arrays.

CODE 5.3: imedge.c main() ...}

Launching multiple threads for three separate functions.

```c
int main(int argc, char** argv)
{
    ...
    pthread_attr_init(&ThAttr);
    pthread_attr_setdetachstate(&ThAttr, PTHREAD_CREATE_JOINABLE);
    for(i=0; i<NumThreads; i++){
        ThParam[i] = i;
        ThErr = pthread_create(&ThHandle[i], &ThAttr, GaussianFilter, (void *)...
        if(ThErr != 0){
            printf("\nThread Creation Error %d. Exiting abruptly... \n",ThErr);
            exit(EXIT_FAILURE);
        }
    }
    for(i=0; i<NumThreads; i++){ pthread_join(ThHandle[i], NULL); }
    t4=ReportTimeDelta(t3, "Gauss Image created");
    for(i=0; i<NumThreads; i++){
        ThParam[i] = i;
        ThErr = pthread_create(&ThHandle[i], &ThAttr, Sobel, (void *)&ThParam[i]);
        if(ThErr != 0){ ... }
    }
    for(i=0; i<NumThreads; i++){ pthread_join(ThHandle[i], NULL); }
    t5=ReportTimeDelta(t4, "Gradient, Theta calculated");
    for(i=0; i<NumThreads; i++){
        ThParam[i] = i;
        ThErr = pthread_create(&ThHandle[i], &ThAttr, Threshold, (void *)&ThParam[i]);
        if(ThErr != 0){ ... }
    }
    pthread_attr_destroy(&ThAttr);
    for(i=0; i<NumThreads; i++){ pthread_join(ThHandle[i], NULL); }
    t6=ReportTimeDelta(t5, "Thresholding completed");
    WriteBMP(CopyImage, argv[2]); printf("\n");   //merge with header. write to file
    t7=ReportTimeDelta(t6, "WriteBMP completed");
    for(i = 0; i < ip.Vpixels; i++) {               // free() image memory and pointers
        free(TheImage[i]);   free(CopyImage[i]);  free(BWImage[i]);
        free(GaussImage[i]); free(Gradient[i]);   free(Theta[i]);
    }
    free(TheImage);     ...        free(Theta);
    t8=ReportTimeDelta(t2, "Program Runtime without IO"); return (EXIT_SUCCESS);
}
```

5.2.3 Launching and Terminating Threads

Code 5.3 shows the part of the main() that launches and terminates multiple threads for three separate functions: GaussianFilter(), Sobel(), and Threshold(). Using the timestamping variables t4, t5, and t6, the execution times for these three separate functions are individually determined. This will come in handy for different analyses we will conduct, because these functions have varying core/memory resource demands.

CODE 5.4:   imedge.c   ... GaussianFilter() {...

This function converts the BWImage into its Gaussian-filtered version, GaussImage.

```
double Gauss[5][5] = { { 2,   4,   5,   4,   2 },
                       { 4,   9,  12,   9,   4 },
                       { 5,  12,  15,  12,   5 },
                       { 4,   9,  12,   9,   4 },
                       { 2,   4,   5,   4,   2 }   };
// Calculate Gaussian filtered GaussFilter[][] array from BW image
void *GaussianFilter(void* tid)
{
   long tn;           // My thread number (ID) is stored here
   int row,col,i,j;
   double G;          // temp to calculate the Gaussian filtered version
   tn = *((int *) tid);         // Calculate my Thread ID
   tn *= ip.Vpixels/NumThreads;

   for(row=tn; row<tn+ip.Vpixels/NumThreads; row++){
      if((row<2) || (row>(ip.Vpixels-3))) continue;
      col=2;
      while(col<=(ip.Hpixels-3)){
         G=0.0;
         for(i=-2; i<=2; i++){
            for(j=-2; j<=2; j++){
               G+=BWImage[row+i][col+j]*Gauss[i+2][j+2];
            }
         }
         GaussImage[row][col]=G/159.00D;
         col++;
      }
   }
   pthread_exit(NULL);
}
```

5.2.4   Gaussian Filter

Code 5.4 shows the implementation of Gaussian filtering, which applies Equation 5.2 to the BWImage array to generate the GaussImage array. The Gaussian filter kernel – shown in Equation 5.2 – is defined as a 2D global double array Gauss[][] inside imedge.c. For efficiency, each pixel's filtered value is saved in the G variable inside the two inner for loops and the final value of G is divided by 159 *only once* before being written into the GaussImage array.

Let us analyze the resource requirements of the GaussianFilter() function:

- Gauss array is only 25 double elements (200 Bytes), which can easily fit in the cores' L1$ or L2$. Therefore, this array allows the cores/threads to use the cache very efficiently to store the Gauss array.

- BWImage array is accessed 25 times for each pixel's computation, and should take advantage of the cache architecture efficiently due to this high *reuse* ratio.

- GaussImage is written once for each pixel and takes no advantage of the cache memory.

CODE 5.5:   imedge.c   ... Sobel() {...

This function converts the `GaussImage` into its `Gradient` and `Theta`.

```
double Gx[3][3] = {    { -1,    0,    1 },
                       { -2,    0,    2 },
                       { -1,    0,    1 }    };

double Gy[3][3] = {    { -1,   -2,   -1 },
                       {  0,    0,    0 },
                       {  1,    2,    1 }    };
...
// Function that calculates the Gradient and Theta for each pixel
// Takes the Gauss[][] array and creates the Gradient[][] and Theta[][] arrays
void *Sobel(void* tid)
{
   int row,col,i,j;            double GX,GY;
   long tn = *((int *) tid);   tn *= ip.Vpixels/NumThreads;

   for(row=tn; row<tn+ip.Vpixels/NumThreads; row++){
      if((row<1) || (row>(ip.Vpixels-2))) continue;
      col=1;
      while(col<=(ip.Hpixels-2)){
         // calculate Gx and Gy
         GX=0.0; GY=0.0;
         for(i=-1; i<=1; i++){
            for(j=-1; j<=1; j++){
               GX+=GaussImage[row+i][col+j]*Gx[i+1][j+1];
               GY+=GaussImage[row+i][col+j]*Gy[i+1][j+1];
            }
         }
         Gradient[row][col]=sqrt(GX*GX+GY*GY);
         Theta[row][col]=atan(GX/GY)*180.0/PI;
         col++;
      }
   }
   pthread_exit(NULL);
}
```

5.2.5 Sobel

Code 5.5 implements the gradient computation. It achieves this by applying Equation 5.3 to the `GaussImage` array to generate the two resulting arrays: `Gradient` array contains the magnitude of the edge gradients, whereas the `Theta` array contains the angle of the edges.

Resource usage characteristics of the Sobel() function are

- `Gx` and `Gy` arrays are small and should be cached nicely.

- `GaussImage` array is accessed 18 times for each pixel and should also be cache-friendly.

- `Gradient` and `Theta` arrays are written once for each pixel and take no advantage of the cache memory inside the cores.

CODE 5.6:   imedge.c   ... Threshold() {...

This function finds the edges and saves the resulting binary image in `CopyImage`.

```c
void *Threshold(void* tid)
{
   int row,col; unsigned char PIXVAL;    double L,H,G,T;
   long tn = *((int *) tid);          tn *= ip.Vpixels/NumThreads;
   for(row=tn; row<tn+ip.Vpixels/NumThreads; row++){
      if((row<1) || (row>(ip.Vpixels-2))) continue;
      col=1;
      while(col<=(ip.Hpixels-2)){
         L=(double)ThreshLo;         H=(double)ThreshHi;
         G=Gradient[row][col];       PIXVAL=NOEDGE;
         if(G<=L)   PIXVAL=NOEDGE; else if(G>=H){PIXVAL=EDGE;} else {
            T=Theta[row][col];
            if((T<-67.5) || (T>67.5)){         // Look at left and right
               PIXVAL=((Gradient[row][col-1]>H) ||
                       (Gradient[row][col+1]>H)) ? EDGE:NOEDGE;
            }else if((T>=-22.5) && (T<=22.5)){   // Look at top and bottom
               PIXVAL=((Gradient[row-1][col]>H) ||
                       (Gradient[row+1][col]>H)) ? EDGE:NOEDGE;
            }else if((T>22.5) && (T<=67.5)){      // Look at upper right, lower left
               PIXVAL=((Gradient[row-1][col+1]>H) ||
                       (Gradient[row+1][col-1]>H)) ? EDGE:NOEDGE;
            }else if((T>=-67.5) && (T<-22.5)){     // Look at upper left, lower right
               PIXVAL=((Gradient[row-1][col-1]>H) ||
                       (Gradient[row+1][col+1]>H)) ? EDGE:NOEDGE;
            }
         }
         CopyImage[row][col*3]=PIXVAL;        CopyImage[row][col*3+1]=PIXVAL;
         CopyImage[row][col*3+2]=PIXVAL;       col++;
      }
   }
   pthread_exit(NULL);
}
```

5.2.6 Threshold

Code 5.6 shows the implementation of the thresholding function that determines whether a given pixel at location $[x, y]$ should be classified as an EDGE or NOEDGE. If the gradient value is lower then `ThreshLo` or higher than `ThreshHi`, the EDGE/NOEDGE determination requires only Equation 5.5 and the `Gradient` array. Any gradient value between these two values requires a more elaborate computation, based on Equation 5.6.

The `if` condition makes the resource determination of Threshold() more complicated:

- `Gradient` has a good reuse ratio; it should therefore be cached nicely.

- `Theta` array is accessed based on pixel – and edge – values. So, it is hard to determine its cache-friendliness.

- `CopyImage` array is accessed only once per pixel, making it cache-unfriendly.

TABLE 5.2 imedge.c execution times for the W3690 CPU (6C/12T).

#Th/Func	1	2	4	8	10	12
ReadBMP()	73	70	71	72	73	72
Create arrays	749	722	741	724	740	734
GaussianFilter()	5329	2643	1399	1002	954	880
Sobel()	18197	9127	4671	2874	2459	2184
Threshold()	499	260	147	132	95	92
WriteBMP()	70	70	66	60	61	62
Total without IO	24850	12829	7030	4798	4313	3957

5.3 PERFORMANCE OF IMEDGE

Table 5.2 shows the run time results for the imedge.c program. Here are the summarized observations for each line:

- ReadBMP(): Clearly, the performance of reading the BMP image from disk does not depend on the number of threads used because the bottleneck is the disk access speed.

- **Create arrays**: Because the arrays are initialized using the efficient memset() function in Code 2.5, no notable improvement is observed when multiple threads are running. The calls to the memset() function completely saturate the memory subsystem even if a single thread is running.

- GaussianFilter(): This function seems to be taking advantage of however many threads you throw at it! This is a perfect function to take advantage of multithreading, because it has a balanced usage pattern between memory and core resources. In other words, it is both core- and memory-intensive. Therefore, added threads have sufficient work to utilize the memory subsystem and core resources at an increasing rate. However, the diminishing returns phenomenon is evident when the number of threads is increased.

- Sobel(): The characteristics of this step are almost identical to the Gaussian filter because the operations are very similar.

- Threshold(): This operation is clearly more core-intensive, so it is less balanced than the previous two operations. Because of this, the diminishing returns start at much lower thread counts; from 4 to 8 threads, there is nearly no improvement.

- WriteBMP(): Much like reading the file, writing gets no benefit from multiple threads, since the operation is I/O-intensive.

To summarize Table 5.2, the I/O intensive image-reading and image-writing functions, as well as memory-saturating image-array-initialization functions cannot benefit from multiple threads, although they consume only less than 5% of the execution time. On the other hand, the memory- and core-intensive filtering functions consume more than 95% of the execution time and can greatly benefit from multiple threads. This makes it clear that to improve the performance of imedge.c, we should be focusing on improving the performance of the filtering functions, GaussianFilter(), Sobel(), and Threshold().

5.4 IMEDGEMC: MAKING IMEDGE MORE EFFICIENT

To improve the overall computation speed of imedge.c, let us look at Equation 5.1 closely; to compute the B&W image, Equation 5.1 simply requires one to add the R, G, and B components of each pixel and divide the resulting value by 3. Later, when the Gaussian filter is being applied, as shown in Equation 5.2, this B&W value for each pixel gets multiplied by 2, 4, 5, 9, 12, and 15 and the Gaussian kernel is formed. This kernel – that is essentially a 5×5 matrix – possesses favorable symmetries, containing only six different values (2, 4, 5, 9, 12, and 15).

A close look at Equation 5.2 shows that the constant $\frac{1}{3}$ multiplier for the `BWImage` array variable, as well as the constant $\frac{1}{159}$ multiplier for the `Gauss` array matrix can be taken outside the computation and be dealt with at the very end of the computation. That way the computational burden is only experienced once for the entire formula, rather than for each pixel. Therefore, to compute Equation 5.1 and Equation 5.2, one can get away with multiplying each pixel value with simple integer numbers.

It gets better ... look at the Gauss kernel; the corner value is 2, which means that some pixel at some point gets multiplied by 2. Since the other corner value is also 2, some pixel four columns ahead (horizontally) also gets multiplied by 2. Same for four rows below and four rows and four columns below. An extensive set of such symmetries brings about the following idea to speed up the entire convolution operation:

For each given pixel with a B&W value of, say, X, why not *precompute* different multiples of X and save them somewhere *only once*, right after we know what the pixel's B&W value is. These multiples are clearly $2X$, $4X$, $5X$, $9X$, $12X$, and $15X$. A careful observer will come up with yet another optimization: instead of multiplying X by 2, why not simply add X to itself and save the result into another place. Once we have this $2X$ value, why not add that to itself to get $4X$ and continue to add $4X$ to X to get $5X$, thereby completely avoiding multiplications.

Since we saved each pixels's B&W value as a `double`, each multiplication and addition is a `double` type operation; so, saving the multiplications will clearly help reduce the core computation intensity. Additionally, each pixel is accessed only once, rather than 25 times during the convolution operation that Equation 5.2 prescribes.

5.4.1   Using Precomputation to Reduce Bandwidth

Based on the ideas we came up with, we need to store multiple values for each pixel. Code 5.7 lists the updated imageStuff.h file that stores the R, G, and B values of each pixel as an `unsigned char`, as well as the precomputed B&W pixel value as a `float` in `BW`. Instead of `double`, a `float` value has more than sufficient precision, however, curious readers are welcome to try it with `double`.

The precomputed values are stored in variables `BW2`, `BW4`, ..., `BW15`. The `Gauss`, `Gauss2`, `Theta`, and `Gradient` values are also precomputed and will be explained shortly. We are not necessarily talking about reducing memory bandwidth ("M"), but also core computational bandwidth ("C") with precomputation. First, the computation of the B&W image can be melted into the actual precomputation operation. This reduces memory accesses and core utilization substantially by avoiding the computation of the same multiplications over and over again.

CODE 5.7:   imageStuff.h   ...

Header file that includes the struct to store the precomputed pixels.

```
#define EDGE        0
#define NOEDGE      255
#define MAXTHREADS 128

struct ImgProp{
   int Hpixels;
   int Vpixels;
   unsigned char HeaderInfo[54];
   unsigned long int Hbytes;
};
struct Pixel{
   unsigned char  R;
   unsigned char  G;
   unsigned char  B;
};
struct PrPixel{
   unsigned char  R;
   unsigned char  G;
   unsigned char  B;
   unsigned char  x;     // unused. to make it an even 4B
   float          BW;
   float          BW2,BW4,BW5,BW9,BW12,BW15;
   float          Gauss, Gauss2;
   float          Theta,Gradient;
};

double** CreateBWCopy(unsigned char** img);
double** CreateBlankDouble();
unsigned char** CreateBlankBMP(unsigned char FILL);
struct PrPixel** PrAMTReadBMP(char*);
struct PrPixel** PrReadBMP(char*);
unsigned char** ReadBMP(char*);
void WriteBMP(unsigned char** , char*);

extern struct ImgProp  ip;
extern long   NumThreads, PrThreads;
extern int    ThreadCtr[];
```

5.4.2   Storing the Precomputed Pixel Values

Code 5.7 shows the new struct that contains the precomputed pixel values. RGB values are stored in the usual unsigned char, occupying 3 bytes total, followed by a "space filler" variable named x to round the storage up to 4 bytes, for efficient access by a 32-bit load operation. The precomputed B&W value of the pixel is stored in variable BW, while multiples of this value are stored in BW2, BW4, ... BW15. The Gaussian filter value and Sobel kernel precomputations are also stored in the same struct as explained shortly.

CODE 5.8: imedgeMC.c ...main() ...}
Precomputing pixel values.

```
struct PrPixel **PrImage;          // the pre-calculated image
...
int main(int argc, char** argv)
{
   ...
   printf("\nExecuting the Pthreads version with %li threads ...\n",NumThreads);
   t1 = GetDoubleTime();
   PrImage=PrReadBMP(argv[1]);  printf("\n");
   t2 = ReportTimeDelta(t1,"PrReadBMP complete"); // Start time without IO
   CopyImage = CreateBlankBMP(NOEDGE);  // This will store the edges in RGB
   t3=ReportTimeDelta(t2, "Auxiliary images created");
   pthread_attr_init(&ThAttr);
   pthread_attr_setdetachstate(&ThAttr, PTHREAD_CREATE_JOINABLE);
   for(i=0; i<NumThreads; i++){
      ThParam[i] = i;
      ThErr = pthread_create(&ThHandle[i], &ThAttr, PrGaussianFilter, ...);
      if(ThErr != 0){ ... }
   }
   for(i=0; i<NumThreads; i++){ pthread_join(ThHandle[i], NULL); }
   t4=ReportTimeDelta(t3, "Gauss Image created");
   for(i=0; i<NumThreads; i++){
      ThParam[i] = i;  ThErr = pth...(..., PrSobel, ...);  if(ThErr != 0){ ... }
   }
   for(i=0; i<NumThreads; i++){ pthread_join(ThHandle[i], NULL); }
   t5=ReportTimeDelta(t4, "Gradient, Theta calculated");
   for(i=0; i<NumThreads; i++){
      ThParam[i] = i;  ThErr = pth...(..., PrThreshold, ...);  if(ThErr != 0){ ... }
   }
   pthread_attr_destroy(&ThAttr);
   for(i=0; i<NumThreads; i++){ pthread_join(ThHandle[i], NULL); }
   t6=ReportTimeDelta(t5, "Thresholding completed");
   WriteBMP(CopyImage, argv[2]); printf("\n"); //merge with header and write to file
   t7=ReportTimeDelta(t6, "WriteBMP completed");
   // free() the allocated area for image and pointers
   for(i = 0; i < ip.Vpixels; i++) { free(CopyImage[i]); free(PrImage[i]); }
   free(CopyImage);   free(PrImage);
   t8=ReportTimeDelta(t2, "Program Runtime without IO");
   return (EXIT_SUCCESS);
}
```

5.4.3 Precomputing Pixel Values

We read the image and its precomputed pixel values in Code 5.8 using the PrReadBMP()
function. CreateBlankBMP() function was already explained in Code 5.2. Aside from that, the
only difference is the replacement of the GaussianFilter(), Sobel(), and Threshold() functions
with PrGaussianFilter(), PrSobel(), and PrThreshold() functions. The additional struct array
PrPixel has the form that was described in Code 5.7.

CODE 5.9:   imageStuff.c   PrReadBMP() {...}

Reading the image from the disk and precomputing pixel values.

```c
struct PrPixel** PrReadBMP(char* filename)
{
   int i,j,k;           unsigned char r, g, b;      unsigned char Buffer[24576];
   float R, G, B, BW, BW2, BW3, BW4, BW5, BW9, BW12, Z=0.0;
   FILE* f = fopen(filename, "rb"); if(f == NULL){ printf(...); exit(1); }
   unsigned char HeaderInfo[54];
   fread(HeaderInfo, sizeof(unsigned char), 54, f); // read the 54-byte header
   // extract image height and width from header
   int width = *(int*)&HeaderInfo[18];   ip.Hpixels = width;
   int height = *(int*)&HeaderInfo[22];  ip.Vpixels = height;
   int RowBytes = (width*3 + 3) & (~3);  ip.Hbytes = RowBytes;
   //copy header for re-use
   for(i=0; i<54; i++) { ip.HeaderInfo[i] = HeaderInfo[i]; }
   printf("\n Input BMP File name: %20s (%u x %u)",filename,ip.Hpixels,ip.Vpixels);
   // allocate memory to store the main image
   struct PrPixel **PrIm=(struct PrPixel **)malloc(height*sizeof(struct PrPixel *));
   for(i=0; i<height; i++) PrIm[i]=(struct...)malloc(width*sizeof(struct PrPixel));
   for(i = 0; i < height; i++) { // read image, pre-calculate the PrIm array
      fread(Buffer, sizeof(unsigned char), RowBytes, f);
      for(j=0,k=0; j<width; j++, k+=3){
         b=PrIm[i][j].B=Buffer[k];            B=(float)b;
         g=PrIm[i][j].G=Buffer[k+1];          G=(float)g;
         r=PrIm[i][j].R=Buffer[k+2];          R=(float)r;
         BW3=R+G+B;                           PrIm[i][j].BW  = BW  = BW3*0.33333;
         PrIm[i][j].BW2  = BW2 = BW+BW;       PrIm[i][j].BW4  = BW4 = BW2+BW2;
         PrIm[i][j].BW5  = BW5 = BW4+BW;      PrIm[i][j].BW9  = BW9 = BW5+BW4;
         PrIm[i][j].BW12 = BW12 = BW9+BW3;    PrIm[i][j].BW15 = BW12+BW3;
         PrIm[i][j].Gauss = PrIm[i][j].Gauss2 = Z;
         PrIm[i][j].Theta = PrIm[i][j].Gradient = Z;
      }
   }
   fclose(f);           return PrIm; // return the pointer to the main image
}
```

5.4.4 Reading the Image and Precomputing Pixel Values

The PrReadBMP() function reads the image from the disk, initializes each pixel with its RGB value as well as its precomputed values, as shown in Code 5.9. One interesting aspect of this function is that it overlaps the disk read with precomputation. Therefore, PrReadBMP() no longer contains an exhaustive set of I/O operations. It performs core-intensive and memory intensive operations while it is reading the data from the disk. The fact that PrReadBMP() function is core, memory, and I/O intensive hints at the possibility of speeding this function up by multithreading it. This function does not actually calculate Gauss, Gauss2, Gradient, or Theta. It strictly initializes them.

CODE 5.10: imedgeMC.c ...PrGaussianFilter() {...}

Performing the Gaussian filter using the precomputed pixel values.

```
#define ONEOVER159 0.00628931
...
// Function that takes the pre-calculated .BW. .BW2, .BW4, ...
// pixel values and compute the .Gauss value from it
void *PrGaussianFilter(void* tid)
{
   long tn;                    int row,col,i,j;
   tn = *((int *) tid);        tn *= ip.Vpixels/NumThreads;
   float G;           // temp to calculate the Gaussian filtered version

   for(row=tn; row<tn+ip.Vpixels/NumThreads; row++){
      if((row<2) || (row>(ip.Vpixels-3))) continue;
      col=2;
      while(col<=(ip.Hpixels-3)){
         G=PrImage[row][col].BW15;
         G+=(PrImage[row-1][col].BW12    + PrImage[row+1][col].BW12);
         G+=(PrImage[row][col-1].BW12    + PrImage[row][col+1].BW12);
         G+=(PrImage[row-1][col-1].BW9   + PrImage[row-1][col+1].BW9);
         G+=(PrImage[row+1][col-1].BW9   + PrImage[row+1][col+1].BW9);
         G+=(PrImage[row][col-2].BW5     + PrImage[row][col+2].BW5);
         G+=(PrImage[row-2][col].BW5     + PrImage[row+2][col].BW5);
         G+=(PrImage[row-1][col-2].BW4   + PrImage[row+1][col-2].BW4);
         G+=(PrImage[row-1][col+2].BW4   + PrImage[row+1][col+2].BW4);
         G+=(PrImage[row-2][col-2].BW2   + PrImage[row+2][col-2].BW2);
         G+=(PrImage[row-2][col+2].BW2   + PrImage[row+2][col+2].BW2);
         G*=ONEOVER159;
         PrImage[row][col].Gauss=G;
         PrImage[row][col].Gauss2=G+G;
         col++;
      }
   }
   pthread_exit(NULL);
}
```

5.4.5 PrGaussianFilter

The PrGaussianFilter() function, shown in Code 5.10, has the same exact functionality as Code 5.4, which is the version that does not use the precomputed values. The difference between GaussianFilter() and PrGaussianFilter() is that the former achieves the same computation result by adding the appropriate precomputed values for the corresponding pixels, rather than performing the actual computation.

The inner-loop simply computes the Gaussian-filtered pixel value from Equation 5.2 by using the precomputed values that were stored in the struct in Code 5.7.

CODE 5.11:   imedgeMC.c   PrSobel() {...}

Performing the Sobel using the precomputed pixel values.

```
// Function that calculates the .Gradient and .Theta for each pixel.
// Uses the pre-computed .Gauss and .Gauss2x values
void *PrSobel(void* tid)
{
    int row,col,i,j;            float GX,GY;         float RPI=180.0/PI;
    long tn = *((int *) tid);    tn *= ip.Vpixels/NumThreads;

    for(row=tn; row<tn+ip.Vpixels/NumThreads; row++){
        if((row<1) || (row>(ip.Vpixels-2))) continue;
        col=1;
        while(col<=(ip.Hpixels-2)){
            // calculate Gx and Gy
            GX  = PrImage[row-1][col+1].Gauss + PrImage[row+1][col+1].Gauss;
            GX += PrImage[row][col+1].Gauss2;
            GX -= (PrImage[row-1][col-1].Gauss + PrImage[row+1][col-1].Gauss);
            GX -= PrImage[row][col-1].Gauss2;
            GY  = PrImage[row+1][col-1].Gauss + PrImage[row+1][col+1].Gauss;
            GY += PrImage[row+1][col].Gauss2;
            GY -= (PrImage[row-1][col-1].Gauss + PrImage[row-1][col+1].Gauss);
            GY -= PrImage[row-1][col].Gauss2;
            PrImage[row][col].Gradient=sqrtf(GX*GX+GY*GY);
            PrImage[row][col].Theta=atanf(GX/GY)*RPI;
            col++;
        }
    }
    pthread_exit(NULL);
}
```

5.4.6 PrSobel

The PrSobel() function, shown in Code 5.11, has the same exact functionality as Code 5.5, which is the version that does not use the precomputed values. The difference between Sobel() and PrSobel() is that the former achieves the same computation result by adding the appropriate precomputed values for the corresponding pixels, rather than performing the actual computation.

The inner-loop simply computes the Sobel-filtered pixel value from Equation 5.3 by using the precomputed values that were stored in the struct in Code 5.7.

CODE 5.12:   imedgeMC.c   ...PrThreshold() {...}

Performing the Threshold function using the precomputed pixel values.

```c
// Function that takes the .Gradient and .Thetapre-computed values for
// each pixel and calculates the final value (EDGE/NOEDGE)
void *PrThreshold(void* tid)
{
   int row,col,col3;            unsigned char PIXVAL;        float L,H,G,T;
   long tn = *((int *) tid);    tn *= ip.Vpixels/NumThreads;
   for(row=tn; row<tn+ip.Vpixels/NumThreads; row++){
      if((row<1) || (row>(ip.Vpixels-2))) continue;
      col=1;   col3=3;
      while(col<=(ip.Hpixels-2)){
         L=(float)ThreshLo;               H=(float)ThreshHi;
         G=PrImage[row][col].Gradient;       PIXVAL=NOEDGE;
         if(G<=L){ PIXVAL=NOEDGE; }else if(G>=H){ PIXVAL=EDGE; }else{ // noedge,edge
            T=PrImage[row][col].Theta;
            if((T<-67.5) || (T>67.5)){       // Look at left and right
               PIXVAL=((PrImage[row][col-1].Gradient>H) ||
                       (PrImage[row][col+1].Gradient>H)) ? EDGE:NOEDGE;
            }else if((T>=-22.5) && (T<=22.5)){  // Look at top and bottom
               PIXVAL=((PrImage[row-1][col].Gradient>H) ||
                       (PrImage[row+1][col].Gradient>H)) ? EDGE:NOEDGE;
            }else if((T>22.5) && (T<=67.5)){   // Look at upper right, lower left
               PIXVAL=((PrImage[row-1][col+1].Gradient>H) ||
                       (PrImage[row+1][col-1].Gradient>H)) ? EDGE:NOEDGE;
            }else if((T>=-67.5) && (T<-22.5)){  // Look at upper left, lower right
               PIXVAL=((PrImage[row-1][col-1].Gradient>H) ||
                       (PrImage[row+1][col+1].Gradient>H)) ? EDGE:NOEDGE;
            }
         }
         if(PIXVAL==EDGE){ // Each pixel was initialized to NOEDGE
            CopyImage[row][col3]=PIXVAL;        CopyImage[row][col3+1]=PIXVAL;
            CopyImage[row][col3+2]=PIXVAL;
         }
         col++;      col3+=3;
      }
   }
   pthread_exit(NULL);
}
```

5.4.7 PrThreshold

The PrThreshold() function, shown in Code 5.12, has the same exact functionality as Code 5.6, which is the version that does not use the precomputed values. The difference between Threshold() and PrThreshold() is that the former achieves the same computation result by adding the appropriate precomputed values for the corresponding pixels, rather than performing the actual computation.

The inner-loop simply computes the resulting binary (thresholded) pixel value from Equation 5.5 by using the precomputed values that were stored in the struct in Code 5.7.

TABLE 5.3 imedgeMC.c execution times for the W3690 CPU (6C/12T) in ms for a varying number of threads (above). For comparison, execution times of imedge.c are repeated from Table 5.2 (below).

Function #threads ⟹	1	2	4	8	10	12
PrReadBMP()	2836	2846	2833	2881	2823	2898
Create arrays	31	32	31	36	31	31
PrGaussianFilter()	2179	1143	570	526	539	606
PrSobel()	7475	3833	1879	1141	945	864
PrThreshold()	358	193	121	107	113	107
WriteBMP()	61	60	61	61	60	61
imedgeMC.c runtime no I/O	12940	8107	5495	4752	4511	4567
ReadBMP()	73	70	71	72	73	72
Create arrays	749	722	741	724	740	734
GaussianFilter()	5329	2643	1399	1002	954	880
Sobel()	18197	9127	4671	2874	2459	2184
Threshold()	499	260	147	132	95	92
WriteBMP()	70	70	66	60	61	62
imedge.c runtime no I/O	24850	12829	7030	4798	4313	3957
Speedup	1.92×	1.58×	1.28×	1.01×	0.96×	0.87×

5.5 PERFORMANCE OF IMEDGEMC

Table 5.3 shows the run time results for the imedgeMC.c program. To be able to compare this performance to the previous version, imedge.c, Table 5.3 actually provides the performance results of both of the programs including the detailed break-down.

So, what did we improve? Let us analyze the performance results:

- The ReadBMP() function needed only ≈ 70 ms to read the BMP image file, whereas the PrReadBMP() took ≈ 2850 ms on average, due to the precomputation while reading the image. In other words, we added a large overhead to perform the precomputation.

- Creating the arrays required ≈ 700 ms less in imedgeMC.c, since the B&W image computation ended up being shifted to the precomputation phase in imedgeMC.c. In other words, our effective precomputation overhead, i.e., the non-BW-computation part, was only ≈ 2100 ms.

- Since each compute- and memory-intensive function uses the precomputed values in imedgeMC.c, they achieved healthy speed-ups: PrGaussianFilter() seems to be ≈ 2× faster than GaussianFilter() on average, while PrSobel() is ≈ 2.5× faster consistently. PrThreshold(), on the other hand, got a much lower speed-up; this is not a big deal since thresholding is a fairly small portion of the overall execution time anyway.

- The WriteBMP() function was not affected since it is strictly I/O-intensive.

Putting it all together, we achieved our most important goal of speeding up the compute-intensive functions by using precomputed values. One interesting observation, though, seems to be that the increased number of threads help a lot less with the imedgeMC.c. This is due to the fact that the threads are a lot less balanced in imedgeMC.c; due to the precomputation, there are a lot fewer memory accesses and the weight is shifted toward the core, providing a lot less work for the added threads and reducing their utility.

CODE 5.13:   imedgeMCT.c   ...main() {...}

Structure of main() that uses MUTEX and barrier synchronization.

```
int main(int argc, char** argv)
{
    ...
    t1 = GetDoubleTime();
    PrImage=PrAMTReadBMP(argv[1]); printf("\n");
    t2 = ReportTimeDelta(t1,"PrAMTReadBMP complete"); // Start time without IO
    CopyImage = CreateBlankBMP(NOEDGE);  // This will store the edges in RGB
    t3=ReportTimeDelta(t2, "Auxiliary images created");
    pthread_attr_init(&ThAttr); pthread_attr_setdetachstate(&ThAttr, PTHREAD_CR...);
    for(i=0; i<NumThreads; i++){ ... pthread_create(...PrGaussianFilter...); ... }
    for(i=0; i<NumThreads; i++){ pthread_join(ThHandle[i], NULL); }
    t4=ReportTimeDelta(t3, "Gauss Image created");
    for(i=0; i<NumThreads; i++){ ... pthread_create(...PrSobel...); }
    for(i=0; i<NumThreads; i++){ pthread_join(ThHandle[i], NULL); }
    t5=ReportTimeDelta(t4, "Gradient, Theta calculated");
    for(i=0; i<NumThreads; i++){ ... pthread_create(...PrThreshold...); }
    pthread_attr_destroy(&ThAttr);
    for(i=0; i<NumThreads; i++){ pthread_join(ThHandle[i], NULL); }
    t6=ReportTimeDelta(t5, "Thresholding completed");
    //merge with header and write to file
    WriteBMP(CopyImage, argv[2]); printf("\n");
    t7=ReportTimeDelta(t6, "WriteBMP completed");
    // free() the allocated area for image and pointers
    for(i = 0; i < ip.Vpixels; i++) { free(CopyImage[i]); free(PrImage[i]); }
    free(CopyImage);   free(PrImage);
    t8=ReportTimeDelta(t2, "Prog ... IO"); printf("\n\n--- ... -----\n");
    for(i=0; i<PrThreads; i++) {
        printf("\ntid=%2li processed %4d rows\n",i,ThreadCtr[i]);
    } printf("\n\n--- ... -----\n");     return (EXIT_SUCCESS);
}
```

5.6 IMEDGEMCT: SYNCHRONIZING THREADS EFFICIENTLY

Code 5.13 shows the final version of main() in imedgeMCT.c. This is the thread-friendly version ("T") of the imedge.c. The only change we notice is the implementation of the precomputation using the PrAMTReadBMP(), rather than the previous imedgeMC.c implementation using the PrReadBMP() function in Code 5.8.

The PrAMTReadBMP() function will allow us to introduce two new concepts: (1) the concept of barrier synchronization, and (2) usage of MUTEXes. Detailed descriptions of these two concepts will follow shortly in Section 5.6.1 and Section 5.6.2, respectively. Note that these techniques that we introduce in PrAMTReadBMP() (AMT denoting "asymmetric multithreading) can be readily applied to the other functions that use multithreading, such as PrGaussianFilter(), PrSobel(), and PrThreshold(); this is, however, left to the reader because it does not have any additional instructional value. Our focus will be on understanding the effect of these two concepts on the performance of the PrAMTReadBMP() function, which is sufficient to apply them to any function you want later.

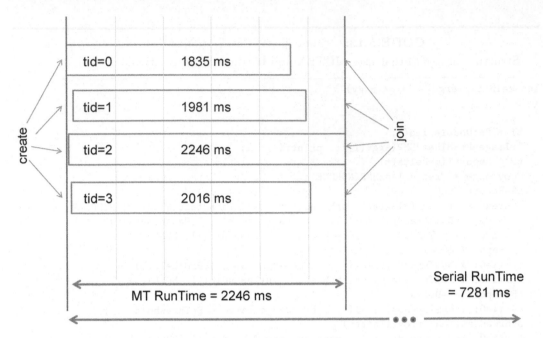

FIGURE 5.2 Example barrier synchronization for 4 threads. Serial runtime is 7281 ms and the 4-threaded runtime is 2246 ms. The speedup of 3.24× is close to the best-expected 4×, but not equal due to the imbalance of each thread's runtime.

5.6.1 Barrier Synchronization

When N threads are executing the same function, which is $\frac{1}{N}^{th}$ of the entire task, at what point do we consider the entire task *finished*? Figure 5.2 shows an example where the entire task takes 7281 ms when executed in a serial fashion. The same task, when executed using 4 threads, takes 2246 ms, which means a 3.24× speedup (81% threading efficiency); from everything we have seen previously, this 81% threading efficiency is not that bad. But, the important question is: could we have done better?

To answer this, let's dig deep into the details of this 81%. First of all, are we getting so much below 100% because of the core-intensive or the memory-intensive nature of the task? It turns out, there is another factor: the *synchronization* of threads; when you split the entire task into 4 pieces, even if the amount of work that has to be done is equal among all four tasks, there is no guarantee that they will be executed at the same time by different hardware threads. There are just too many factors that come into play at runtime, one of them being the involvement of the OS, as we saw in Section 3.4.4.

Ideally, each thread will complete its own sub-task exactly in 25% of the time that it takes to complete the entire task, i.e., $\frac{7281}{4} = 1820$ ms and the task will be completed in 1820 ms. However, in a realistic scenario that we see in Figure 5.2, although one of the threads executes its portion in an amount of time that is fairly close to the *ideal* 1820 ms interval (1835 ms), three of them are far from it (1981, 2016, and 2246 ms). In the end, we cannot deem the task *finished* until the very last thread completes in 2246 ms, thereby making the multithreaded execution time 2246 ms. Sadly, three of the threads sit idle and wait for the others to be done, reducing the efficiency.

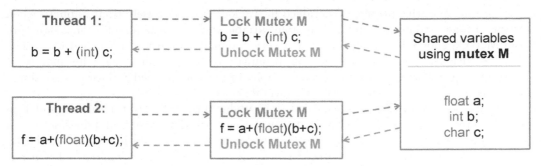

FIGURE 5.3   Using a MUTEXdata structure to access shared variables.

5.6.2   MUTEX Structure for Data Sharing

We will dedicate the imageMCT.c program to answering whether we could have done better by using better synchronization; could we have made it thread-friendly to deserve the "T" in our imageMCT.c? The answer is clearly yes, but we need to explain a few more things before we can explain the implementation of imageMCT.c.

First of all, when multiple threads are *updating* a variable (read or write), there is the danger of one thread overwriting that variable's value before another thread gets a chance to read the correct value. This can be prevented with a MUTEX structure that allows the threads to update a variable's value in a "thread-safe" manner, completely eliminating the possibility for one thread to read an incorrect value. A MUTEX structure is showing in Figure 5.3. To understand it, let us look at an analogy.

ANALOGY 5.1: *Thread Synchronization.*

Coco Town needed to harvest 1800 coconuts, for which they employed 4 farmers. The town supervisor would ask each farmer to take 450 coconuts and come back when they are harvested. Because each farmer harvested the coconuts at different speeds, the task was completed by different farmers in 450, 475, 500, and 515 minutes. The supervisor declared the entire harvesting task complete in 515 minutes.

Since the faster farmers had to wait for the slower ones to complete, they asked the supervisor to try something different next year: what if the farmers took one coconut at a time and when harvested, they came back and asked the supervisor for another one? To track how many coconuts have been harvested so far, the supervisor put up a chalk board and asked each farmer to update the coconut count on it. That year, the overall task took 482 minutes during which farmers harvested 418, 440, 464, 478 coconuts, respectively; a healthy improvement from 515 minutes.

Farmers noticed something strange when they were updating the coconut count; although rarely, sometimes they would read the coconut count *exactly* at the same time (say, count=784) and politely wait for the other farmer to finish and write the updated count (count=785) on the board. Clearly, this count was one less than what it was supposed to be, since the count ended up being updated only once instead of twice. They found the solution by introducing a red flag, which each farmer used to let the other know that he was updating the counter. The other farmer waited for the flag to be down and brought it up and read the count *after* he brought the flag up.

The count updating problem mentioned in Analogy 5.1 is precisely what happens when multiple threads attempt to update the same variable without using an additional structure like a MUTEX, shown in Figure 5.3. It is a very common bug in multithreaded code. The red flag solution is precisely what is used in multithreaded programming. The underlying idea to prevent incorrect updating is very simple: instead of a thread recklessly updating a variable (i.e., *unsafely*), a shared MUTEX variable is used.

If a variable is a MUTEX variable, it has to be updated according to the rules of updating such a variable; each thread knows that it is updating a MUTEX *variable* and does not touch the variable itself before it lets the other threads know that it is updating it. It does this by *locking the* MUTEX, which is equivalent to *bringing up the red flag* in Analogy 5.1. Once it locks the MUTEX, it is free to make any update it desires to the variable that is controlled by this MUTEX. In other words, it either excludes itself or the other threads from updating a MUTEX variable, hence the name *mutually exclusive*, or in short, MUTEX.

Before the terms get confused, let me make something clear: there is a difference between a MUTEX itself and MUTEX variables. For example, in Figure 5.3, the name of the MUTEX is M, while the variables that MUTEX M controls are f, a, b, and c. In such a setup, the multithreaded functions are supposed to lock and unlock the MUTEX M itself. After a lock has been obtained, they are free to update the four MUTEX-controlled variables, f, a, b, and c. When done, they are supposed to unlock MUTEX M.

Let me make one point very clear: although a MUTEX eliminates the incorrect-updating problem, implementing a MUTEX requires hardware-level *atomic* operations. For example, the CAS (compare and swap) instructions in the x86 Intel ISA achieve this. Luckily, a programmer enjoys the readily available MUTEX implementation functions that are a part of POSIX and this is what we will use in our implementation of imageMCT.c.

There is absolutely no mechanism that checks to see whether a thread has updated a variable *safely* by locking/unlocking a controlling MUTEX, therefore, it is also a common bug for a programmer to forget that he or she was supposed to lock/unlock a MUTEX for a variable that is being shared. In this case, exactly the same problems that were mentioned in Analogy 5.1 will creep up, presenting an impossible-to-debug problem. It is also common for a programmer to realize that a variable should have really been a MUTEX variable and declare a MUTEX for it halfway into the program development. However, declaring a MUTEX does not magically solve the incorrect-updating problem; correct locking/unlocking does. All it takes is forgetting one place where the variable is being accessed and forget to lock/unlock its corresponding MUTEX; worse yet, most incorrect update problems are infrequent and they manifest themselves as *weird intermittent problems*, keeping most programmers up at night! So, a good upfront planning is the best way to prevent these problems.

5.7 IMEDGEMCT: IMPLEMENTATION

Like I mentioned before, only the PrAMTReadBMP() function, inside imageStuff.c, will be implemented using MUTEX structures, as shown in Code 5.14. We will borrow important ideas from Analogy 5.1: instead of assigning a $\frac{1}{N}$ chunk of the entire task to N different threads, we will assign a much smaller task to each thread, namely reading and precomputing a single row of the image, and keep a counter to determine what the next row is.

PrAMTReadBMP() reads the image one row at a time and updates a MUTEX variable named `LastRowRead`, controlled by the MUTEX named `CtrMutex`. It creates N different threads that perform the precomputation using a function named AMTPreCalcRow(), shown in Code 5.15.

CODE 5.14: imageStuff.c ...PrAMTReadBMP() {...}
Structure of the asymmetric multithreading and barrier synchronization.

```c
pthread_mutex_t    CtrMutex;                        // MUTEX
int                NextRowToProcess, LastRowRead;   // MUTEX variables
// This function reads one row at a time, assigns them to threads to precompute
struct PrPixel** PrAMTReadBMP(char* filename)
{
   int i,j,k,ThErr;                    unsigned char Buffer[24576];
   pthread_t ThHan[MAXTHREADS];        pthread_attr_t attr;
   FILE* f = fopen(filename, "rb");    if(f == NULL){ ... }
   unsigned char HeaderInfo[54];
   fread(HeaderInfo, sizeof(unsigned char), 54, f); // read the 54-byte header
   // extract image height and width from header, and copy header for re-use
   int width = *(int*)&HeaderInfo[18];    ip.Hpixels = width;
   int height = *(int*)&HeaderInfo[22];   ip.Vpixels = height;
   int RowBytes = (width*3 + 3) & (~3);   ip.Hbytes = RowBytes;
   for(i=0; i<54; i++) { ip.HeaderInfo[i] = HeaderInfo[i]; }
   printf("\n Input BMP File name: %20s (%u x %u)",filename,ip.Hpixels,ip.Vpixels);
   // allocate memory to store the main image
   PrIm = (struct PrPixel **)malloc(height * sizeof(struct PrPixel *));
   for(i=0; i<height; i++) {
      PrIm[i] = (struct PrPixel *)malloc(width * sizeof(struct PrPixel));
   }
   pthread_attr_init(&attr); pthread_attr_setdetachstate(&attr, PTHRE...JOINABLE);
   pthread_mutex_init(&CtrMutex, NULL); // create a MUTEX named CtrMutex
   pthread_mutex_lock(&CtrMutex);   // MUTEX variable updates require lock,unlock
      NextRowToProcess=0;           // Set the asynchronous row counter to 0
      LastRowRead=-1;               // no rows read yet
      for(i=0; i<PrThreads; i++) ThreadCtr[i]=0; // zero every thread counter
   pthread_mutex_unlock(&CtrMutex);
   // read the image from disk and pre-calculate the PRImage pixels
   for(i = 0; i<height; i++) {
      if(i==20){ // when sufficient # of rows are read, launch threads
         // PrThreads is the number of pre-processing threads
         for(j=0; j<PrThreads; j++){
            ThErr=p..create(&ThHan[j], &attr, AMTPreCalcRow, (void *)&ThreadCtr[j]);
            if(ThErr != 0){ ... }
         }
      }
      fread(Buffer, sizeof(unsigned char), RowBytes, f); // Read one row
      for(j=0,k=0; j<width; j++, k+=3){
         PrIm[i][j].B=Buffer[k]; PrIm[i][j].G=Buffer[k+1]; PrIm[i][j].R=Buffer[k+2];
      }
      // Advance LastRowRead. While doing this, lock the CtrMutex, then unlock.
      pthread_mutex_lock(&CtrMutex); LastRowRead=i; pthread_mutex_unlock(&CtrMutex);
   }
   for(i=0; i<PrThreads; i++){ pthread_join(ThHan[i], NULL); } // join threads
   pthread_attr_destroy(&attr); pthread_mutex_destroy(&CtrMutex); fclose(f);
   return PrIm; // return the pointer to the main image
}
```

5.7.1 Using a MUTEX: Read Image, Precompute

Looking at Code 5.14, PrAMTReadBMP() looks very similar to the multi-threaded functions we have seen before, with the exception of the MUTEX variables:

- MUTEX variables are sacred! Any time you need to touch them, you need to lock the responsible MUTEX and unlock it after completing the update. Here is an example:

```
pthread_mutex_lock(&CtrMutex);    // MUTEX vars require lock,unlock
NextRowToProcess=0;               // Set the asynchronous row counter to 0
LastRowRead=-1;                   // no rows read yet
for(i=0; i<PrThreads; i++) ThreadCtr[i]=0; // zero every thread counter
pthread_mutex_unlock(&CtrMutex);
```

- The indentation of the variables inside the lock/unlock make it easy to see the variables that are being updated while the MUTEX is locked.

- We have to create and destroy each MUTEX using the following functions:

```
pthread_mutex_init(&CtrMutex, NULL); // create a MUTEX named CtrMutex
...
pthread_mutex_destroy(&CtrMutex);  // destroy the MUTEX named CtrMutex
```

- For N threads that are being launched, there are $N+2$ MUTEX variables, all controlled by the same MUTEX named CtrMutex. These variables are: NextRowToProcess, LastRowRead, and the array ThreadCtr[0]..ThreadCtr[N-1].

- PrAMTReadBMP() updates LastRowRead to indicate the last image row it read.

- NextRowToProcess tells AMTPreCalcRow() which row to preprocess next; while preprocessing each row, each thread knows its own tid; for example ThreadCtr[5] is only incremented by the thread that has $tid = 5$. Since we are expecting each thread to preprocess a different number of rows, the counts in the ThreadCtr[] array will be different, although we expect them to be relatively close to each other. In our Analogy 5.1, four farmers harvested 418, 440, 464, 478 coconuts each; that analogy is equivalent to four different threads preprocessing 418, 440, 464, and 478 rows of the image, thereby leaving the array values $ThreadCtr[0] = 418$, $ThreadCtr[1] = 440$, $ThreadCtr[2] = 464$, and $ThreadCtr[30] = 478$ when PrAMTReadBMP() joins them.

- PrAMTReadBMP() does not launch the multiple threads before it reads a few rows (20 in Code 5.14). This avoids threads being idle and continuously spinning (checking to see if there are rows available to preprocess). This number is not critical, however, I am showing it since it has some value in pointing out the concept of idle threads.

- While the reading is going on from the disk, precomputation of the last read row also continues by the launched threads. The function AMTPreCalcRow() is responsible for precomputing the last row, as shown in Code 5.15.

- Notice that PrAMTReadBMP() is itself an active I/O intensive thread, in addition to the N threads it launches. So, the execution of PrAMTReadBMP() launches $N+1$ threads on the computer. Clearly, there is no reason why PrAMTReadBMP() and the rest of the functions of the imedgeMCT program cannot be launched by a different number of threads; this is something that is left to the reader to implement.

CODE 5.15:   imageStuff.c   ...AMTPreCalcRow() {...}

Function that precomputes each row. The same `CtrMutex` MUTEX is used to share multiple variables among the threads and PrAMTReadBMP().

```
pthread_mutex_t    CtrMutex;
struct PrPixel     **PrIm;
int                NextRowToProcess, LastRowRead;
int                ThreadCtr[MAXTHREADS]; // Counts # rows processed by each thread
void *AMTPreCalcRow(void* ThCtr)
{
  unsigned char r, g, b;           int i,j,Last;
  float R, G, B, BW, BW2, BW3, BW4, BW5, BW9, BW12, Z=0.0;
  do{  // get the next row number safely
    pthread_mutex_lock(&CtrMutex);
      Last=LastRowRead;             i=NextRowToProcess;
      if(Last>=i){
        NextRowToProcess++;        j = *((int *)ThCtr);
        *((int *)ThCtr) = j+1;     // One more row processed by this thread
      }
    pthread_mutex_unlock(&CtrMutex);
    if(Last<i) continue;
    if(i>=ip.Vpixels) break;
    for(j=0; j<ip.Hpixels; j++){
      r=PrIm[i][j].R;     g=PrIm[i][j].G;     b=PrIm[i][j].B;
      R=(float)r;         G=(float)g;         B=(float)b;         BW3=R+G+B;
      PrIm[i][j].BW  = BW  = BW3*0.33333;  PrIm[i][j].BW2 = BW2 = BW+BW;
      PrIm[i][j].BW4 = BW4 = BW2+BW2;      PrIm[i][j].BW5 = BW5 = BW4+BW;
      PrIm[i][j].BW9 = BW9 = BW5+BW4;      PrIm[i][j].BW12 = BW12 = BW9+BW3;
      PrIm[i][j].BW15 = BW12+BW3;          PrIm[i][j].Gauss=PrIm[i][j].Gauss2=Z;
      PrIm[i][j].Theta= PrIm[i][j].Gradient = Z;
    }
  }while(i<ip.Vpixels);
  pthread_exit(NULL);
}
```

5.7.2   Precomputing One Row at a Time

PrAMTReadBMP() launches N threads (requested by the user) and assigns the same precomputation function AMTPreCalcRow() to all of them, shown in Code 5.15. While AMTPreCalcRow() is similar to PrReadBMP() in Code 5.9, there are major differences:

- The *granularity* of AMTPreCalcRow() is much finer; it is expected to process only a single row of the image before it updates some MUTEX variables. This contrasts with an entire $\frac{1}{N}$ of the image that each thread has to process in PrReadBMP(). This is one idea we are borrowing from the farmers in Analogy 5.1. Although the processing times of each row may be different, their influence on overall execution time is negligible.

- AMTPreCalcRow() updates a MUTEX variable named `NextRowToProcess`, by properly locking/unlocking the MUTEX that controls this variable, named `CtrMutex`. Clearly `NextRowToProcess` is being updated by all N threads, necessitating the usage of a MUTEX to avoid updating problems.

TABLE 5.4 imedgeMCT.c execution times (in ms) for the W3690 CPU (6C/12T), using the Astronaut.bmp image file (top) and Xeon Phi 5110P (60C/240T) using the dogL.bmp file (bottom).

Function #threads ⟹	1	2	4	8	10	12		
PrAMTReadBMP()	2267	1264	920	1014	1020	1078		
Create arrays	33	31	31	33	32	33		
PrGaussianFilter()	2223	1157	567	556	582	611		
PrSobel()	7415	3727	1910	1124	948	842		
PrThreshold()	341	195	119	107	99	104		
WriteBMP()	61	62	60	63	61	63		
imedgeMCT.c w/o IO	12640	6436	3607	2897	2742	2731		
PrReadBMP()	2836	2846	2833	2881	2823	2898		
Create arrays	31	32	31	36	31	31		
PrGaussianFilter()	2179	1143	570	526	539	606		
PrSobel()	7475	3833	1879	1141	945	864		
PrThreshold()	358	193	121	107	113	107		
WriteBMP()	61	60	61	61	60	61		
imedgeMC.c w/o IO	12940	8107	5495	4752	4511	4567		
Speedup (W3690)	1.02×	1.26×	1.52×	1.64×	1.64×	1.67×		

Xeon #threads ⟹	1	2	4	8	16	32	64	128
Xeon Phi 5110P no IO	3994	2178	1274	822	604	507	486	532

- Each instance of the AMTPreCalcRow() function makes no assumption on how many rows it is supposed to process, since it could differ among different instances of this function; therefore, the only terminating condition for each instance of AMTPreCalcRow() is when the `NextRowToProcess` reaches the end of the image and there is nothing more to process.

- AMTPreCalcRow() idles in the case that the `LastRowRead` variable indicates a slow hard disk read by PrAMTReadBMP(). Although, remember that the initially buffered 20 rows should avoid this in Code 5.14.

5.8   PERFORMANCE OF IMEDGEMCT

Table 5.4 shows the run time results for the imedgeMCT.c program. The top part of the code compares the runtimes of imedgeMCT.c and imedgeMC.c. Because we only redesigned the PrReadBMP() function, its redesigned asymmetric multithreaded version PrAMTReadBMP() benefits from multithreading and gets progressively faster as the number of threads increases. The impact of this on the overall performance is clearly visible.

I also ran imedgeMCT.c on a Xeon 5110P, which has 60 cores and 240 threads; because each thread is a thick thread in almost every function we are using, the additional threads did not benefit the Xeon, saturating the performance when we got closer to 60 threads. Remember from Section 3.9 that benefitting from the vast amount of threads that exist in a Xeon required meticulous engineering; unless thick threads are intermixed with thin threads as described in Section 3.2.2, no additional benefit from Xeon will be gained when the number of threads increases beyond the core count.

PART II

GPU Programming Using CUDA

## CHAPTER 6

Introduction to GPU Parallelism and CUDA

W E have spent a considerable amount of time in understanding the CPU parallelism and how to write CPU parallel programs. During the process, we have learned a great deal about how simply bringing a bunch of cores together will not result in a magical performance improvement of a program that was designed as a serial program to begin with. This is the first chapter where we will start understanding the inner-workings of a GPU; the good news is that we have such a deep understanding of the CPU that we can make comparisons between a CPU and GPU along the way. While so many of the concepts will be dead similar, some of the concepts will only have a place in the GPU world. It all starts with the monsters ...

6.1 ONCE UPON A TIME ... NVIDIA ...

Yes, it all starts with the monsters. As many game players know, many games have monsters or planes or tanks moving from here to there and interacting heavily during this movement, whether it is crashing into each other or being shot by the game player to kill the monsters. What do these actions have in common? (1) A plane moving in the sky, (2) a tank shooting, or (3) a monster trying to grab you by moving his arms and his body. The answer — from a mathematical standpoint — is that all of these *objects* are high resolution graphic elements, composed of many pixels, and moving them (such as rotating them) requires heavy floating point computations, as exemplified in Equation 4.1 during the implementation of the imrotate.c program.

6.1.1 The Birth of the GPU

Computer games have existed as long as computers have I was playing computer games in the late 1990s and I had an Intel CPU in my computer; something like a 486. Intel offered two flavors of the 486 CPU: 486SX and 486DX. 486SX CPUs did not have a built-in floating point unit (FPU), whereas 486DX CPUs did. So, 486SX was really designed for more general purpose computations, whereas 486DX made games work much faster. So, if you were a gamer like me in the late 1990s and trying to play a game that had a lot of these tanks, planes, etc. in it, hopefully you had a 486DX, because otherwise your games would play so slow that you would not be able to enjoy them. Why? Because games require heavy floating point operations and your 486SX CPU does not incorporate an FPU that is capable of performing floating point operations fast enough. You would resort to using your ALU to emulate an FPU, which is a very slow process.

This story gets worse. Even if you had a 486DX CPU, the FPU inside your 486DX was still not fast enough for most of the games. Any exciting game demanded a 20× (or even 50×) higher-than-achievable floating point computational power from its host CPU. Surely, in every generation the CPU manufacturers kept improving their FPU performance, just to witness a demand for FPU power that grew much faster than the improvements they could provide. Eventually, starting with the Pentium generation, the FPU was an integral part of a CPU, rather than an option, but this didn't change the fact that significantly higher FPU performance was needed for games. In an attempt to provide much higher scale FPU performance, Intel went on a frenzy to introduce vector processing units inside their CPUs: the first ones were called MMX, then SSE, then SSE2, and the ones in 2016 are SSE4.2. These vector processing units were capable of processing many FPU operations in parallel and their improvement has never stopped.

Although these vector processing units helped certain applications a lot — and they still do – the demand for an ever-increasing amount of FPU power was insane! When Intel could deliver a 2× performance improvement, game players demanded 10× more. When they could eventually manage to deliver 10× more, they demanded 100× more. Game players were just monsters that ate lots of FLOPS! And, they were always hungry! Now what? This was the time when a paradigm shift had to happen. Late 1990s is when the manufacturers of many plug-in boards for PCs — such as sound cards or ethernet controller — came up with the idea of a card that could be used to accelerate the floating point operations. Furthermore, routine image coordinate conversions during the course of a game, such as 3D-to-2D conversions and handling of triangles, could be performed significantly faster by dedicated hardware rather than wasting precious CPU time. Note that the actual unit element of a monster in a game is a *triangle*, not a pixel. Using triangles allows the games to associate a *texture* for the surface of any object, like the skin of a monster or the surface of a tank, something that you cannot do with simple pixels.

These efforts of the PC card manufacturers to introduce products for the game market gave birth to a type of card that would soon be called a *Graphics Processing Unit*. Of course, we love acronyms: it is a GPU ... A GPU was designed to be a "plug-in card" that required a connector such as PCI, AGP, PCI Express, etc. Early GPUs in the late 1990s strictly focused on delivering as high of a floating point performance as possible. This freed the CPU resources and allowed a PC to perform 5× or 20× better in games (or even more if you were willing to spend a lot of money on a fancy GPU). Someone could purchase a $100 GPU for a PC that was worth $500; for this 20% extra investment, the computer performed 5× faster in games. Not a bad deal. Alternatively, by purchasing a $200 card (i.e., a 40% extra investment), your computer could perform 20× faster in games. Late 1990s was the point of no return, after which the GPU was an indispensable part of every computer, not just for games, but for a multitude of other applications explained below. Apple computers used a different strategy to build a GPU-like processing power into their computers, but sooner or later (e.g., in the year 2017, the release year of this book) the PC and Mac lines have converged and they started using GPUs from the same manufacturers.

6.1.2 Early GPU Architectures

If you were playing Pacman, an astonishingly popular game in the 1980s, you really didn't need a GPU. First of all, all of the objects in Pacman were 2D — including the monster that is chasing you and trying to eat you — and the movement of all of the objects was restricted to 2D — x and y — dimensions. Additionally, there were no sophisticated computations in the game that required the use of transcendental functions, such as sin()

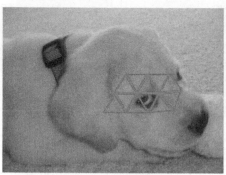

Triangle Coordinates

Tesselation

3D Object → Triangles

Texture Mapping——→

FIGURE 6.1 Turning the dog picture into a 3D wire frame. Triangles are used to represent the object, rather than pixels. This representation allows us to map a texture to each triangle. When the object moves, so does each triangle, along with their associated textures. To increase the resolution of this kind of an object representation, we can divide triangles into smaller triangles in a process called *tesselation*.

or cos(), or even floating point computations of any sort. The entire game could run by performing integer operations, thereby requiring only an ALU. Even a low-powered CPU was perfectly sufficient to compute all of the required movements in real time. However, having watched the *Terminator 2* movie a few years ago, the Pacman game was far from exciting for gamers of the 1990s. First of all, objects had to be 3D in any good computer game and the movements were substantially more sophisticated than Pacman — and in 3D, requiring every transcendental operation you can think of. Furthermore, because the result of any transcendental function due to a sophisticated object move — such as the rotation operation in Equation 4.1 or the scaling operation in Equation 4.3 — required the use of floating point variables to maintain image coordinates, GPUs, by definition, had to be computational units that incorporated significant FPU power. Another observation that the GPU manufacturers made was that the GPUs could have a significant edge in performance if they also included dedicated processing units that performed routine conversions from pixel-based image coordinates to triangle-based object coordinates, followed by texture mapping.

To appreciate what a GPU has to do, consider Figure 6.1, in which our dog is represented by a bunch of triangles. Such a representation is called a *wire-frame*. In this representation, a 3D *object* is represented using triangles, rather than an *image* using 2D pixels. The unit of element for this representation is a triangle with an associated texture. Constructing a 3D wire-frame of the dog will allow us to design a game in which the dog jumps up and down; as he makes these moves, we have to apply some transformation — such as rotation, using the 3D equivalent of Equation 4.1 — to each triangle to determine the new location of that triangle and map the associated texture to each triangle's new location. Much like a 2D image, this 3D representation has the same "resolution" concept; to increase the resolution of a triangulated object, we can use *tesselation*, in which a triangle is further subdivided into smaller triangles as shown in Figure 6.1. Note: Only 11 triangles are shown in Figure 6.1 to avoid cluttering the image and make our point on a simple figure; in a real game, there could be millions of triangles to achieve sufficient resolution to please the game players.

Now that we appreciate what it takes to create scenes in games where 3D objects are moving freely in the 3-dimensional space, let's turn our attention to the underlying

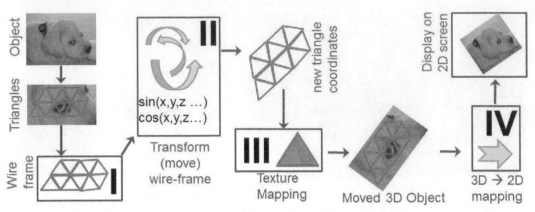

FIGURE 6.2 Steps to move triangulated 3D objects. Triangles contain two attributes: their *location* and their *texture*. Objects are moved by performing mathematical operations only on their coordinates. A final texture mapping places the texture back on the moved object coordinates, while a 3D-to-2D transformation allows the resulting image to be displayed on a regular 2D computer monitor.

computations to create such a game. Figure 6.2 depicts a simplified diagram of the steps involved in moving a 3D object. The designer of a game is responsible for creating a wire-frame of each object that will take part in the game. This wire-frame includes not only the locations of the triangles — composed of 3 points for each triangle, having an x, y, and z coordinate each — but also a texture for each triangle. This operation *decouples* the two components of each triangle: (1) the *location* of the triangle, and (2) the *texture* that is associated with that triangle. After this coupling, triangles can be moved freely, requiring only mathematical operations on the coordinates of the triangles. The texture information — stored in a separate memory area called *texture memory* — doesn't need to be taken into account until all of the moves have been computed and it is time to display the resulting object in its new location. Texture memory does not need to be changed at all, unless, of course, the object is changing its texture, as in the *Hulk* movie, where the main character turns green when stressed out! In this case, the texture memory also needs to be updated in addition to the coordinates, however, this is a fairly infrequent update when compared to the updates on the triangle coordinates. Before displaying the moved object, a *texture mapping* step fills the triangles with their associated texture, turning the wire-frame back into an object. Next, the recomputed object has to be displayed on a computer screen; because every computer screen is composed of 2D pixels, a 3D-to-2D transformation has to be performed to display the *object* as an *image* on the computer screen.

6.1.3 The Birth of the GPGPU

Early GPU manufacturers noticed that the GPUs could excel in being specialized game cards if they incorporated dedicated hardware into the GPUs that performed the following operations. Each operation is denoted with a Roman numeral in Figure 6.2:

- Ability to deal with the natural data type *triangle*, Box I in Figure 6.2

- Ability to perform heavy floating point operations on triangles (Box II)

- Ability to associate texture with each triangle and store this texture in a separate memory area called *texture memory* (Box III)

- Ability to convert from triangle coordinates back to image coordinates for display in a computer screen (Box IV)

Based on this observation, right from the first day, every GPU was manufactured with the ability to implement some sort of functionality that matched all of these boxes. GPUs kept evolving by incorporating faster Box IIs, although the concept of Box I, III, and IV never changed too much. Now, imagine that you are a graduate student in the late 1990s —in a physics department— and trying to write a particle simulation program that requires an extensive amount of floating point computations. Before the introduction of the GPUs, all you could use was a CPU that had an FPU in it and, potentially, a vector unit. However, when you bought one of these GPUs at an affordable price and realized that they could perform a much higher volume of FPU operations, you would naturally start thinking: "Hmmm... I wonder if I could use one of these GPU things in my particle simulations?" This investigation would be worth every minute you put into it because you know that these GPUs are capable of $5\times$ or $10\times$ faster FPU computations. The only problem at that time was that the functionality of Box III and Box IV couldn't be "shut off." In other words, GPUs were not designed for non-gamers who are trying to do particle simulations!

Nothing can stop a determined mind! It didn't take too long for our graduate student to realize that if he or she mapped the location of the particles as the triangle locations of the monsters and somehow performed particle movement operations by emulating them as monster movements, it could be possible to "trick" the GPU into thinking that you are actually playing a game, in which particles (monsters) are moving here and there and smashing into each other (particle collisions). You can only imagine the massive challenges our student had to endure: First, the native language of the games was OpenGL, in which objects were graphics objects and computer graphics APIs had to be used to "fake" particle movements. Second, there were major inefficiencies in the conversions from monster-to-particle and particle-back-to-monster. Third, accuracy was not that great because the initial cards could only support single precision FPU operations, not double precision. It is not like our student could make a suggestion to the GPU manufacturers to incorporate double precision to improve the particle simulation accuracy; GPUs were game cards and they were game card manufacturers, period! None of these challenges stopped our student! Whoever that student was, the unsung hero, created a multibillion dollar industry of GPUs that are in almost every top supercomputer today.

Extremely proud of the success in tricking the GPU, the student published the results ... The cat was out of the bag ... This started an avalanche of interest; if this trick can be applied to particle simulations, why not circuit simulations? So, another student applied it to circuit simulations. Another one to astro-physics, another one to computational biology, another ... These students invented a way to do general purpose computations using GPUs, hence the birth of the term GPGPU.

6.1.4   Nvidia, ATI Technologies, and Intel

While the avalanche was coming down the mountain and universities were purchasing GPUs in a frenzy to perform scientific computations — despite the challenges associated with trying to use a game card for "serious" purposes by using sophisticated mappings — GPU manufacturers were watching on the sidelines and trying to gauge the market potential of their GPUs in the GPGPU market. To significantly expand this market, they had to make modifications to their GPUs to eliminate the tedious transformations the academics had to go through to use GPUs as scientific computation cards. What they realized fairly quickly was that the market for GPGPUs was a lot wider than just the academic arena; for example, oil explorers could analyze the underwater SONAR data to find oil under water, an application that requires a substantial volume of floating point operations. Alternatively, the academic and research market, including many universities and research institutions such as NASA or Sandia National Labs, could use the GPGPUs for extensive scientific simulations. For these simulations, they would actually purchase hundreds of the most expensive versions of GPGPUs and GPU manufacturers could make a significant amount of money in this market and create an alternative product to the already-healthy game products.

In the late 1990s, GPU manufacturers were small companies that saw GPUs as ordinary add-on cards that were no different than hard disk controllers, sound cards, ethernet cards, or modems. They had no vision of the month of September 2017, when Nvidia would become a company that is worth \$112 B (112 billion US dollars) in the Nasdaq stock market (Nasdaq stock ticker NVDA), a pretty impressive 20-year accomplishment considering that Intel, the biggest semiconductor manufacturer on the planet with its five decade history, was worth \$174 B the same month (Nasdaq stock ticket INTC). The vision of the card manufacturers changed fairly quickly when the market realized that GPUs were not in the same category as other add-on cards; it didn't take a genius to figure out that the GPU market was ready for an explosion. So the gold rush started. GPU cards needed two main ingredients: (1) the GPU chips, responsible for all of the computation, (2) GPU memory, something that could be manufactured by the CPU DRAM manufacturers that were already making memory chips for the CPU market, (3) interface chips to interface to the PCI bus, (4) power supply chips that provide the required voltages to all of these chips, and (5) other semiconductors to make all of these work together, sometimes called "glue logic."

The market already had manufacturers for (2), (3), and (4). Many small companies were formed to manufacture (1), the GPU "chips," so the functionality shown in Figure 6.2 could be achieved. The idea was that GPU chip designers — such as Nvidia — would design their chips and have them manufactured by third parties — such as TSMC — and sell the GPU chips to contractor manufacturers such as FoxConn. FoxConn would purchase the other components (2,3,4, and 5) and manufacture GPU add-on cards. Many GPU chip designers entered the market just to see a massive consolidation toward the end of 1990s. Some of them bankrupted and some of them sold out to bigger manufacturers. As of 2016, only three key players remain in the market (Intel, AMD, and Nvidia), two of them being actual CPU manufacturers. Nvidia became the biggest GPU manufacturer in the world as of 2016 and made multiples pushes to enter into the GPU/CPU market by incorporating ARM cores into their Tegra line GPUs. Intel and AMD kept incorporating GPUs into their CPUs to provide an alternative to consumers that didn't want to buy a discrete GPU. Intel has gone through many generations of designs eventually incorporating Intel HD Graphics and Intel Iris GPUs into their CPUs. Intel's GPU performance improved to the point when in 2016, Apple deemed the built-in Intel GPU performance sufficient to be included in their Mac Books as the only GPU, instead of discrete GPUs. Additionally, Intel introduced the Xeon Phi cards to compete with Nvidia in the high-end supercomputing market. While this major competition was taking place in the desktop market, the mobile market saw a completely different set of players emerge. QualComm and Broadcom built GPU cores into their mobile processors by licensing them from other GPU designers. Apple purchased processor designers to design their "A" family processors that had built-in CPUs and GPUs with extreme low power consumption. By about 2011 or 2012, CPUs couldn't be thought of as the only processing unit of any computer or mobile device. CPU+GPU was the new norm.

6.2 COMPUTE-UNIFIED DEVICE ARCHITECTURE (CUDA)

Nvidia, convinced that the market for GPGPUs was large, decided to turn all of their GPUs into GPGPUs by designing the "Box II" in Figure 6.2 as a general purpose computation unit and exposing it to the GPGPU programmers. For efficient GPGPU programming, bypassing the graphics functionality (Box I, Box III, and Box IV, in which *graphics-specific* operations take place) was necessary, while Box II is the only necessary block for scientific computations. To phrase alternatively, games needed access to Boxes I, III, and IV (the "G" part) while scientific computation needs only Box II (the "PU" part). They also had to allow the GPGPU programmers to input data directly into Box II without having to go through Box I. Furthermore, *triangle* wasn't a friendly data type for the scientific computations, suggesting that the natural data types in Box II had to be the usual integers, float, and double.

6.2.1 CUDA, OpenCL, and Other GPU Languages

In addition to these hardware implications, a software architecture was necessary to allow GPGPU programmers to develop GPU code without having to learn anything about computer graphics, which uses OpenGL. Considering all of these facts, Nvidia introduced their language Computer-Unified Device Architecture (CUDA) in 2007, which was — and still is — designed strictly for Nvidia platforms. Two years later, the OpenCL language emerged that allowed GPU code to be developed for Intel, AMD, and other GPUs. While AMD initially introduced its own language CTM (Close to Metal), it eventually abandoned these efforts and went strictly with OpenCL. As of the year 2016, there are two predominant desktop GPU languages in the world: OpenCL and CUDA. I must note here that the landscape is different in the mobile market and this is not the focus of this book.

Right from the introduction, GPUs were never viewed as *processors*; they were always conceptualized as being "*co-processors*" that worked under the supervision of a host CPU. All of the data went through the CPU first before reaching the GPU. Therefore, a connection of some sort, for example, a PCI Express bus, was always necessary to interface the GPU to its host CPU. This fact completely dictated the hardware design of the GPU, as well as the programming language required for GPU coding. Both Nvidia's CUDA and its competition OpenCL developed their programming languages to include a *host side code* and a *device side code*. Instead of calling it *GPU code*, it is more general and appropriate to call it "device side code" because, for example, a device doesn't have to be a GPU in OpenCL; it can be an FPGA, DSP, or any other device that has a similar parallel architecture to the GPU. The current implementation of OpenCL 2.3 allows the same code to be used for a multitude of aforementioned devices with very minor modifications. While this generalization is great in some applications, our focus is strictly the GPU code in this book. So, my apologies if I slip and call it GPU code in some parts of the book.

6.2.2 Device Side versus Host Side Code

The initial CUDA language developers had the following dilemma: CUDA had to be a programming language that allowed the programmers to write code for both the CPU and the GPU. Knowing that two completely different processing elements (CPU and GPU) had to be programmed, how would CUDA work? Because the CPU and GPU both had their separate memory, how would the data transfers work? Programmers simply didn't want to learn a brand new language, so CUDA had to have a similar syntax on both the CPU and

GPU code side ... Furthermore, a single compiler would be great to compiler both sides' code, without requiring two separate compilations.

> ➤ *There is no such thing as GPU Programming ...*
> ➤ *GPU always interfaces to the CPU through certain APIs ...*
> ➤ *So, there is always CPU+GPU programming ...*

Given these facts, CUDA had to be based on the C programming language (for the CPU side) to provide high performance. The GPU side also had to be almost exactly like the CPU side with some specific keywords to distinguish between host versus device code. The burden to determine how the execution would take place at runtime — regarding CPU versus GPU execution sequences — had to be determined by the CUDA compiler. GPU parallelism had to be exposed on the GPU side with a mechanism similar to the Pthreads we saw in the Part I of this book. By taking into account all of these facts, Nvidia designed its nvcc compiler that is capable of compiling CPU and GPU code simultaneously. CUDA, since its inception, has gone through many version updates, incorporating an increasing set of sophisticated features. The version I use in this book is CUDA 8.0, released in September 2016. Parallel to the progress of CUDA, Nvidia GPU architectures have gone through massive updates as I will document shortly.

6.3 UNDERSTANDING GPU PARALLELISM

The reason you are reading a GPU programming book is the fact that you want to program a device (GPU) that can deliver a superior computational performance as compared to a CPU. The big questions is: why can a GPU deliver such a high performance? To understand this, let us turn our attention to an analogy.

ANALOGY 6.1: *CPU versus GPU.*

Cocotown had an annual competition for harvesting 2048 coconuts. The strongest farmer in town, Arnold, had a big reputation for owning the fastest tractor and being the strongest guy that could pick and harvest coconuts twice as fast as any other farmer in town. This year, a group of ambitious farmer brothers, Fred and Jim, challenged Arnold; they claimed that although their tractor was half the size of Arnold's, they could still beat Arnold in the competition. Arnold gladly accepted the challenge. This was going to be the most fun competition to watch for the residents who stayed on the sidelines and cheered for their team.

Just before the competition started, another farmer — Tolga, who had never competed before — claimed that he could win this. His setup was completely different: he would drive a bus, which could seat 32 people and the driver. Inside the bus, he would actually seat 32 boy and girl scouts that would help him harvest the coconuts. He wouldn't do any work, but actually give instructions to the scouts, so they could know what to do next. Additionally, he would report the results and return the harvested coconuts. The scouts had no experience, so their individual performance was a quarter of the other farmers. Additionally, Tolga faced major challenges in coordinating the instructions among the scouts. He actually had to give them a piece of rope to hold onto, so they could coordinate their movements in lock step.

Who do you think won the competition?

FIGURE 6.3 Three farmer teams compete in Analogy 6.1: (1) Arnold competes alone with his 2× bigger tractor and "the strongest farmer" reputation, (2) Fred and Jim compete together in a much smaller tractor than Arnold. (3) Tolga, along with 32 boy and girl scouts, compete together using a bus. Who wins?

Analogy 6.1 is depicted in Figure 6.3 with three alternatives: Arnold represents a single-threaded CPU that can work at 4 GHz, while Fred and Jim together in their smaller tractor represent a dual-core CPU in which each core works at something like 2.5 GHz. We have done major evaluations on the performance differences between these two alternatives in Part I of the book. The interesting third alternative in Figure 6.3 is Tolga with the 32 boy/girl scouts. This represents a single CPU core — probably working at 2.5 GHz — and a GPU co-processor composed of 32 small cores that each work at something like 1 GHz. How could we compare this alternative to the first two?

6.3.1   How Does the GPU Achieve High Performance?

First of all, looking at Figure 6.3, if Tolga was alone, his performance would be half that of the second team and probably half that of Arnold's. So, Tolga wouldn't be able to win the competition alone. But, as long as Tolga can coordinate efficiently with the 32 scouts, we can expect a significant performance from the third team. But, how much? Let's do the math. If the *parallelization overhead* was negligible, i.e., the threading efficient was 100%, translating to $\eta = 1.0$ in Equation 4.4, the theoretical maximum we expect would be $2.5 + 32*1 = 34.5$, which is close to 8× of Arnold's performance. This is a very rough estimate to provide a simple back-of-the-envelope number — by simply adding the GHz values of the different processing elements together, i.e., Tolga is at 2.5 GHz, and the scouts are at 1 GHz each — and ignores architectural differences as well as many other phenomena that contribute to reduced performance in parallel architectures. However, it definitely provides a *theoretical maximum* because any of these negative factors will *reduce* the performance further beyond this number.

In this example, even if Tolga was burdened with shuttling data from/to the scouts and didn't have time to do any real work, we are still at 32, instead of 34.5. The power of this third alternative comes from the sheer quantity of the small GPU cores; multiplying any number by 32 creates a large number. So, even if we work the GPU cores at 1 GHz, as long as we can pack 32 cores into the GPU, we can still surpass a single core CPU working at 4 GHz. However, in reality, things are different: CPUs have a lot more than two cores and so do GPUs. In 2016, an 8 core, 16-thread (8C/16T) desktop processor was common and a high-end GPU incorporated 1000–3000 cores. As the CPU manufacturers kept building an increasing number of cores into their CPUs, so did GPU manufacturers. What makes the GPUs win is the fact that the GPU cores are much simpler and they work at lower speed. This allows the GPU chip designers to build a significantly higher number of cores into their GPU chips and the lower speed keeps the power consumptions below the magic 200–250 W, which is about the peak power you can consume from any semiconductor device (i.e., "chip").

Note that the power that is consumed by the GPU is **not** proportional to the frequency of each core; instead, the dependence is something like quadratic. In other words, a 4 GHz CPU core is expected to consume 16× more power than the same core working at 1 GHz. This very fact allows GPU manufacturers to pack hundreds or even thousands of cores into their GPUs without reaching the practical power consumption limits. This is actually exactly the same design philosophy behind multicore CPUs too. A single core CPU working at 4 GHz versus a dual-core CPU in which both cores work at 3 GHz could consume similar amounts of power. So, as long as the parallelization overhead is low (i.e., η is close to 1), a dual-core 3 GHz CPU is a better alternative than a single core 4GHz CPU. GPUs are nothing more than this philosophy taken to the ultimate extreme with one big exception: while the CPU multicore strategy calls for using multiple **sophisticated** (out-of-order) cores that work at lower frequencies, this design strategy only works if you are trying to put 2, 4, 8, or 16 cores inside the CPU. It simply won't work for 1000! So, the GPUs had to go through an additional step of making each core **simpler**. Simpler means that each core is *in-order* (see Section 3.2.1), work at lower frequencies, and their L1$ memories are not coherent. Many of these details are going to become clear as we go through the following few chapters. For now, the take-away from this section should be that GPUs incorporate a lot of architectural changes — as compared to CPUs — to provide a manageable execution environment for such a high core count.

6.3.2 CPU versus GPU Architectural Differences

Although we will learn the GPU architecture in great depth in the following chapters, for now, let's analyze the most important distinctions between CPUs and GPUs conceptually by just looking at the symbolic Figure 6.3. Here are our observations:

1. In our conceptualization of the "bus" that Tolga drives with the scouts in the back, Tolga does all of the driving. Scouts never get involved in it. In a GPU, the GPU cores do all of the execution, but the work orders always come from the CPU.

2. Tolga will be the one that is getting the coconuts and distributing them to the scouts. Scouts never directly get the coconuts by themselves. They simply wait in the bus until Tolga brings them the coconuts. In the GPU case, the GPU cores never actually get the data themselves. It always comes from the CPU side and the results always go back to the CPU side. So, the GPU simply acts as a computation accelerator in the background, working for the CPU for outsourcing certain tasks.

3. This type of an architecture only works well when there is a significant amount of parallel processing units, not just two or four. Indeed, in a GPU, nothing less than 32 things get executed at any point in time. This number — 32 — is almost the replacement of the "thread" concept we saw in the CPU world, something like a "super thread." There is even a name for it in the GPU world: 32 threads glued together are called a "warp." Although GPUs call the tasks executed by a single GPU core a *thread*, there is a necessity to define this other term *warp* to indicate the fact that no less than a warp's worth of threads get executed at any point in time.

4. The existence of the *warp* concept has dramatic implications on the GPU architecture. In Figure 6.3, we never talked about how the coconuts arrive in the bus. If you brought only 5 coconuts into the bus, 27 of the scouts would sit there doing nothing; so, the data elements must be brought in to the GPU in the same bulk amounts, although the unit of these data chunks is *half warp* or 16 elements.

5. The fact that the data arrives into the GPU cores in half warp *chunks* means that the memory sub-system that is bringing the data into the GPU cores should be bringing in the data 16-at-a-time. This implies a parallel memory subsystem that is capable of shuttling around data elements 16 at a time, either 16 floats or 16 integers, etc. This is why the GPU DRAM memory is made from GDDR5, which is *parallel memory*.

6. Because the CPU cores and GPU cores are completely different processing units, it is expected that they have different ISAs (instruction set architectures). In other words, they speak a different language. So, two different sets of instructions must be written: one for Tolga, one for the scouts. In the GPU world, a single compiler — nvcc — compiles both the CPU instructions and GPU instructions, although there are two separate *programs* that the developer must write. Thank goodness, the CUDA language combines them together and makes them so similar that the programmer can write both of these programs without having to learn two totally different languages.

6.4 CUDA VERSION OF THE IMAGE FLIPPER: IMFLIPG.CU

It is time to write and analyze our first CUDA program. Our first program will have the main() and everything else we saw in the CPU programs in Part I. If I didn't show you the additional CUDA code that is stuck in the few tens of lines inside the program, you would have thought that it is CPU code. This is good news in that although the ISAs of the CPU and GPU cores are vastly different, we write code for both of them in the same C language and add a few keywords to indicate whether a specific part of the code belongs to the CPU (*host side code*) or the GPU (*device side code*).

If this is the case, let's start writing the CPU code as if we had nothing to do with the GPU, with one exception. Because we know that we will eventually incorporate the GPU code into this program, we will call it imflipG.cu — the .cu extension denoting CUDA, which is the GPU version of the imflipP.c code we developed in Section 2.1. Remember that imflipP.c flipped an image either vertically or horizontally, based on the command line option that the user specified. It had two functions, MTFlipH() and MTFlipV(), that did the actual work of flipping the pixels in the image memory. The rest of the code in main() was responsible for reading and parsing the command lines and shuttling the data from/to these functions. What the CPU does in the imflipG.cu will be surprisingly similar to imflipP.c, because, remember from Analogy 6.1 that Tolga (the CPU core) will do a lot of the *onesy-twosy work* while the GPU cores will do the *massively parallel work*. In other words, why waste the scouts' time in Figure 6.3 to harvest just one coconut; you should have Tolga do it! If you assigned that task to the scouts, 31 of them will be idle. Worse yet, one scout works at 1 GHz, while Tolga can work at 2.5 GHz. So, the general principle is that the "serial" part of the code is more suitable for the CPU, even the "parallel" part is to an extent. However, the "massively parallel" part that requires at least the execution of 32-things or 64-things at a time is a perfect match for the GPU cores. The *programmer* decides how to split the tasks among the CPU and GPU cores.

Before I start explaining the details of imflipG.cu, let's go through the conceptual steps one by one. How is this program going to work?

1. First, the CPU will read the command line arguments and will parse them and place the parsed values in the appropriate CPU-side variables. Exactly the same story as the plain-simple CPU version of the code, the imflipP.c.

2. One of the command line variables will be the file name of the image file we have to flip, like the file that contains the dog picture, dogL.bmp. The CPU will read that file by using a CPU function that is called ReadBMP(). The resulting image will be placed inside a CPU-side array named TheImg[]. Notice that the GPU does absolutely **nothing** so far.

3. Once we have the image in memory and are ready to flip it, now it is time for the GPU's sun to shine! Horizontal or vertical flipping are both massively parallel tasks, so the GPU should do it. At this point in time, because the image is in a CPU-side array (more generally speaking, in CPU memory), it has to be transferred to the device side. What is obvious from this discussion is that the GPU has *its own memory*, in addition to the CPU's own memory — DRAM — that we have been studying since the first time we saw it in Section 3.5.

4. The fact that the CPU memory versus GPU memory are completely different memory areas (or "chips") should be pretty clear because the GPU is a different plug-in device that shares none of the electronic components with the CPU. The CPU memory is soldered on the motherboard and the GPU memory is soldered on the GPU plug-in card; the only way a data transfer can happen between these two memory areas is an explicit data transfer — using the available APIs we will see shortly in the following pages — through the PCI Express bus that is connecting them. I would like the reader to refresh his or her memory with Figure 4.3, where I showed how the CPU connected to the GPU through the X99 chipset and the PCI Express bus. The X99 chip facilitates the transfers, while the I/O portion of the CPU "chip" employs hardware to interface to the X99 chip and shuttle the data back and forth between the GPU memory and the DRAM of the CPU (by passing through the L3$ of the CPU along the way).

5. So, this *transfer* must take place from the CPU's memory into the GPU's memory before the GPU cores can do anything with the image data. This transfer occurs by using an API function that looks like an ordinary CPU function.

6. After this transfer is complete, now somebody has to tell the GPU cores what to do with that data. It is the *GPU side code* that will accomplish this. Well, the reality is that you should have transferred the code before the data, so by the time the image data arrives at the GPU cores they are aware of what to do with it. This implies that we are really transferring two things to the GPU side: (1) data to process, (2) code to process the data with (i.e., compiled GPU instructions).

7. After the GPU cores are done processing the data, another GPU→CPU transfer must transfer the results back to the CPU.

Using our Figure 6.3 analogy, it is as if Tolga is first giving a piece of paper with the instructions to the scouts so they know what to do with the coconuts (GPU side code), grabbing 32 coconuts at a time (read from CPU memory), dumping 32 coconuts at a time in front of the scouts (CPU→GPU data transfer), telling the scouts to execute their given instructions, which calls for harvesting the coconuts that just got dumped in front of them (GPU-side execution), and grabbing what is in front of them when they are done (GPU→CPU data transfer in the reverse direction) and putting the harvested coconuts back in the area where he got them (write the results back to the CPU memory).

CODE 6.1:   imflipG.cu   ... main() {...

First part of main() in imflipG.cu. `TheImg` and `CopyImg` are CPU-side image pointers, while `GPUImg`, `GPUCopyImg`, and `GPUResult` are GPU-side pointers.

```
#include <cuda_runtime.h>
#include <device_launch_parameters.h>
#include <stdio.h>
#include <stdlib.h>
#include <stdint.h>
#include <string.h>
#include <iostream>
#include <ctype.h>
#include <cuda.h>

typedef unsigned char uch;
typedef unsigned long ul;
typedef unsigned int ui;

uch *TheImg, *CopyImg;                   // Where images are stored in CPU
uch *GPUImg, *GPUCopyImg, *GPUResult;  // Where images are stored in GPU
...
int main(int argc, char **argv)
{
   char InputFileName[255], OutputFileName[255], ProgName[255];
   ...
   strcpy(ProgName, "imflipG");
   switch (argc){
     case 5:  ThrPerBlk=atoi(argv[4]);
     case 4:  Flip = toupper(argv[3][0]);
     case 3:  strcpy(InputFileName, argv[1]);
              strcpy(OutputFileName, argv[2]);
              break;
     default: printf("\n\nUsage: %s InputFilename Outp ...");
         ...
   }
   ...
   TheImg = ReadBMPlin(InputFileName); // Read the input image
   CopyImg = (uch *)malloc(IMAGESIZE); // allocate space for the work copy
   ...
```

This process sounds a little inefficient due to the continuous back-and-forth data transfer, but don't worry. There are multiple mechanisms that Nvidia built into their GPUs to make the process efficient and the sheer processing power of the GPU eventually partially hides the underlying inefficiencies, resulting in a huge performance improvement.

6.4.1   imflipG.cu: Read the Image into a CPU-Side Array

OK. We are done with analogies. Time for real CUDA code. The main() function within imflipG.cu is a little long, so I will chop it up into small pieces. These pieces will look very much like the steps I just listed in the last page. First, let's see the variables involved in holding the original and processed CPU side images. Code 6.1 lists the first part of the main() function and the following five pointers that facilitate image storage in CPU and GPU memory:

- **TheImg** variable is the pointer to the memory that will be malloc()'d by the ReadBMPLin() function to hold the image that is specified in the command line (e.g., dogL.bmp) in the CPU's memory. Notice that this variable, **TheImg**, is a pointer to the CPU DRAM memory.

- **CopyImg** variable is another pointer to the CPU memory and is obtained from a separate malloc() to allocate space for a copy of the original image (the one that will be flipped while the original is not touched). Note that we have done nothing with the GPU memory so far.

- As we will see very shortly, there are APIs that we will use to allocate memory in the *GPU memory*. When we do this, using an API called cudaMalloc(), we are asking the GPU memory manager to allocate memory for us **inside the GPU DRAM**. So, what the cudaMalloc() returns back to us is *a pointer to the GPU DRAM memory*. Yet, we will take that pointer and will store it in a *CPU-side variable*, **GPUImg**. This might look confusing at first because we are saving a pointer to the GPU side inside a CPU-side variable. It actually isn't confusing. Pointers are nothing more than "values" or more specifically 64-bit integers. So, they can be stored, copied, added, and subtracted in exactly the same way 64-bit integers can be. When do we store GPU-side pointers on the CPU side? The rule is simple: *Any* pointer that you will ever use in an API that is called by the CPU must be saved on the CPU side. Now, let's ask ourselves the question: will the variable **GPUImg** ever be used by the CPU side? The answer is definitely yes, because we will need to transfer data from the CPU to the GPU using cudaMalloc(). We know that cudaMalloc() is a CPU-side function, although its responsibility has a lot to do with the GPU. So, we need to store the pointers to *both* sides in CPU-side variables. We will most definitely use the same GPU-side pointer on the GPU side itself as well! However, we are now making a copy of it at the host (CPU), so the CPU has the chance of accessing it when it needs it. If we didn't do this, the CPU would never have access to it in the future and wouldn't be able to initiate memory transfers to the GPU that involved that specific pointer.

- The other GPU-side pointers, **GPUCopyImg** and **GPUResult**, have the same story. They are pointers to the GPU memory, where the resulting "flipped" image will be stored (**GPUResult**) and another temporary variable that the GPU code needs for its operation (**GPUCopyImg**). These two variables are CPU-side variables that store pointers that we will obtain from cudaMalloc(); storing GPU pointers in CPU variables shouldn't be confusing.

There are multiple #include directives you will see in every CUDA program, which are <cuda_runtime.h>, <cuda.h>, and <device_launch_parameters.h> to allow us to use Nvidia APIs. These APIs, such as cudaMalloc(), are the bridge between the CPU and the GPU side. Nvidia engineers wrote them and they allow you to transfer data between the CPU and the GPU side magically without worrying about the details.

Note the types that are defined here, **ul**, **uch**, and **ui**, to denote the unsigned long, unsigned char, and unsigned int, respectively. They are used so often that it makes the code cleaner define them as user-defined types. It serves, in this case, no purpose other than to reduce the clutter in the code. The variables to hold the file names are **InputFileName** and **OutputFileName**, which both come from the command line. The **ProgName** variable is hard-coded into the program for use in reporting as we will see later in this chapter.

CODE 6.2:   imflipG.cu   main() {...
Initializing and querying the GPU(s) using Nvidia APIs.

```
int main(int argc, char** argv)
{
   cudaError_t cudaStatus, cudaStatus2;
   cudaDeviceProp GPUprop;
   ul SupportedKBlocks, SupportedMBlocks, MaxThrPerBlk;  char SupportedBlocks[100];
   ...
   int NumGPUs = 0;              cudaGetDeviceCount(&NumGPUs);
   if (NumGPUs == 0){
      printf("\nNo CUDA Device is available\n"); exit(EXIT_FAILURE);
   }
   cudaStatus = cudaSetDevice(0);
   if (cudaStatus != cudaSuccess) {
      fprintf(stderr, "cudaSetDevice failed! No CUDA-capable GPU installed?");
      exit(EXIT_FAILURE);
   }
   cudaGetDeviceProperties(&GPUprop, 0);
   SupportedKBlocks = (ui)GPUprop.maxGridSize[0] * (ui)GPUprop.maxGridSize[1] *
                 (ui)GPUprop.maxGridSize[2] / 1024;
   SupportedMBlocks = SupportedKBlocks / 1024;
   sprintf(SupportedBlocks, "%u %c", (SupportedMBlocks >= 5) ? SupportedMBlocks :
         SupportedKBlocks, (SupportedMBlocks >= 5) ? 'M' : 'K');
   MaxThrPerBlk = (ui)GPUprop.maxThreadsPerBlock;
   ...
```

6.4.2   Initialize and Query the GPUs

Code 6.2 shows the part of the main() that is responsible for *querying* the GPU(s), that is, gathering information about all of the existing GPUs in the system. There could be one (the most typical case), or more than one GPUs with different characteristics. The cudaGetDeviceCount() function, within the vast list of APIs that Nvidia provides us, writes the number of available GPUs in an int type variable that we defined, NumGPUs. If this variable is zero, clearly, we have no GPUs available and we can quit with a nasty message! Alternatively, if we have at least one GPU, we can choose the one that we want to execute our CUDA code on by using the API cudaSetDevice(). In this case we choose 0, which means *the first GPU*. Much like C variable indexes, GPU indexes range in 0, 1, 2... Once we choose a GPU, we can query that GPU's parameters using the cudaGetDeviceProperties() API function and place the results in a variable GPUProp (which is of type **cudaDeviceProp**). Let's look at a few features that we received from this API:

GPUProp.maxGridSize[0], GPUProp.maxGridSize[1], and GPUProp.maxGridSize[2] variables denote the maximum number of blocks that can be launched in x, y, and z dimensions, thereby allowing us to calculate the total number of blocks when you take the product of the three. We write the result into a variable named SupportedKBlocks after dividing by 1024 (to extract out the result in "kilo" terms). Similarly, divide it by 1024 again to get the result in "Mega." This is one simple example of how we can get the query answer to: *"what is the maximum number of blocks I can launch with this GPU?"* The upper limit of the number of threads you can launch in a block is given in the MaxThrPerBlk variable.

CODE 6.3: imflipG.cu   main() {...

The part of main() in imflipG.cu that launches the GPU kernels.

```
uch *TheImg, *CopyImg;                // Where images are stored in CPU
uch *GPUImg, *GPUCopyImg, *GPUResult;  // Where images are stored in GPU
#define IPHB        ip.Hbytes
#define IPH         ip.Hpixels
#define IPV         ip.Vpixels
#define IMAGESIZE   (IPHB*IPV)
...
int main(int argc, char** argv)
{
  cudaError_t cudaStatus, cudaStatus2;
  cudaEvent_t time1, time2, time3, time4;
  ui BlkPerRow, ThrPerBlk=256, NumBlocks, GPUDataTransfer;
  ...
  cudaEventCreate(&time1);        cudaEventCreate(&time2);
  cudaEventCreate(&time3);        cudaEventCreate(&time4);

  cudaEventRecord(time1, 0);  // Time stamp at the start of the GPU transfer
  // Allocate GPU buffer for the input and output images
  cudaStatus = cudaMalloc((void**)&GPUImg, IMAGESIZE);
  cudaStatus2 = cudaMalloc((void**)&GPUCopyImg, IMAGESIZE);
  if ((cudaStatus != cudaSuccess) || (cudaStatus2 != cudaSuccess)){
    fprintf(stderr, "cudaMalloc failed! Can't allocate GPU memory");
    exit(EXIT_FAILURE);
  }
  // Copy input vectors from host memory to GPU buffers.
  cudaStatus = cudaMemcpy(GPUImg, TheImg, IMAGESIZE, cudaMemcpyHostToDevice);
  if (cudaStatus != cudaSuccess) {
    fprintf(stderr, "cudaMemcpy CPU to GPU failed!");  exit(EXIT_FAILURE);
  }
  cudaEventRecord(time2, 0);  // Time stamp after the CPU --> GPU tfr is done
  BlkPerRow = (IPH + ThrPerBlk -1 ) / ThrPerBlk;
  NumBlocks = IPV*BlkPerRow;
  switch (Flip){
    case 'H': Hflip <<< NumBlocks, ThrPerBlk >>> (GPUCopyImg, GPUImg, IPH);
        GPUResult = GPUCopyImg;    GPUDataTransfer = 2*IMAGESIZE; break;
    case 'V': Vflip <<< NumBlocks, ThrPerBlk >>> (GPUCopyImg, GPUImg, IPH, IPV);
        GPUResult = GPUCopyImg;    GPUDataTransfer = 2*IMAGESIZE; break;
    case 'T': Hflip <<< NumBlocks, ThrPerBlk >>> (GPUCopyImg, GPUImg, IPH);
        Vflip <<< NumBlocks, ThrPerBlk >>> (GPUImg, GPUCopyImg, IPH, IPV);
        GPUResult = GPUImg;        GPUDataTransfer = 4*IMAGESIZE; break;
    case 'C': NumBlocks = (IMAGESIZE+ThrPerBlk-1) / ThrPerBlk;
        PixCopy <<< NumBlocks, ThrPerBlk >>> (GPUCopyImg, GPUImg, IMAGESIZE);
        GPUResult = GPUCopyImg;    GPUDataTransfer = 2*IMAGESIZE; break;
  }
  ...
```

6.4.3 GPU-Side Time-Stamping

You will know what a *block* means in detail very shortly when we get into the CUDA code execution details. For now, let's focus our attention on the CPU-GPU interaction; you can query the GPU right at the beginning of your CUDA code and based on the query results you can potentially execute your CUDA code using different parameters for optimum efficiency. Here is an analogy about how the CPU-GPU interaction works:

ANALOGY 6.2: *CPU-side versus GPU-side.*

CocoTown had their best year with more than 40 million coconuts ready to be harvested. Because they weren't used to harvesting so many coconuts, they decided to get help from their buddies at the moon; they knew that the moon city of cudaTown had a state-of-the-art technology to harvest the coconuts 10–20 times faster. This required them to send a spaceship to the moon with 40+ million coconuts in it. The folks at cudaTown were very organized; they were used to harvesting the coconuts in chunks they called "blocks" and required cocoTown to package each block in a single box, in sizes of 32, 64, 128, 256, or 1024. They requested cocoTown to inform them of the block size and how many blocks were coming in the spaceship.

CocoTown-cudaTown city officials had to come up with some ideas to communicate the block sizes and number of blocks. Additionally, cudaTown workers needed to *allocate* some space in their warehouse *before* the coconuts arrived. Luckily an engineer in cudaTown designed a satellite phone for the cocoTown engineers to call ahead and reserve space in cudaTown's warehouse, as well as to let them know about the block size and the number of blocks coming. One more thing: cudaTown was so big that the person responsible for allocating warehouse space for the coconuts had to give the "warehouse number" to the cocoTown people, so when the spaceship arrived cudaTown people would know where to store the coconuts.

This relationship was very exciting, however, cocoTown people didn't know what the right block size was and how to measure the time it took for space travel and harvesting in cudaTown. Because all of the work was going to be performed by the cudaTown folks, it was a good idea to make somebody at cudaTown responsible for timing all of these events; they hired an event manager for this. After long deliberations, cocoTown decided to ship the coconuts in a block size of 256, which required 166,656 blocks to be shipped. They put a notebook in the spaceship that includes warehouse addresses, and the other parameters they received from the satellite phone, so cudaTown people had a clear direction in harvesting the coconuts.

Analogy 6.2 actually has quite a bit of detail. Let's understand it.

- The city of cocoTown is the CPU and cudaTown is the GPU. Launching the spaceship between the two cities is equivalent to executing GPU code. The notebook they left in the spaceship contains the function parameters for the GPU-side function (for example, the Vflip()); without these parameters cudaTown couldn't execute any function.

- It is clear that the data transfer from the earth (cocoTown) to the moon (cudaTown) is a big deal; it takes a lot of time and might even marginalize the amazing execution speed at cudaTown. The spaceship is representing the data transfer engine, while the *space* itself is the PCI Express bus that is connecting the CPU and GPU.

- The speed of the spaceship is the PCI Express bus speed.

- The satellite phone represents the CUDA runtime API library for cocoTown and cudaTown to communicate. One important detail is that just because the satellite phone operator is in cudaTown, it doesn't guarantee that a copy is also saved in cudaTown; so, these parameters (e.g., warehouse number) must still be put inside the spaceship (written inside the notebook).

- Allocating the space in cudaTown's warehouse is equivalent to cudaMalloc(), which returns a pointer (the warehouse number in cudaTown). This pointer is given to cocoTown people, although it must be shipped back in the spaceship as a function parameter, as per my previous comment.

- Because the cocoTown people have no idea how things unfold as the spaceship leaves earth, they shouldn't be timing the events. cudaTown people should. I will provide a lot more detail on this below.

A GPU-side function — such as Vflip() — is termed a *kernel*. Code 6.3 shows how the CPU launches GPU kernels, which use the GPU as a co-processor; this is the most important part of the code to understand, so I will explain every piece of it in detail. Once you understand this part of code, it will be a smooth ride in the following chapters. First, much like the timing of the CPU code, we want to time how long it takes to do the CPU→GPU transfers as well as the GPU→CPU transfers. Additionally, we want to time the GPU code execution in between these two events. When we look at Code 6.3, we see that the following lines facilitate the timing of the GPU code:

```
cudaEvent_t time1, time2, time3, time4;
...
cudaEventCreate(&time1);    cudaEventCreate(&time2);
cudaEventCreate(&time3);    cudaEventCreate(&time4);
... // This is where we copy data from CPU to GPU
cudaEventRecord(time1, 0);  // Time stamp at the start of the GPU transfer
... // This is where we transfer data from the CPU to the GPU
cudaEventRecord(time2, 0);  // Time stamp after the CPU --> GPU tfr is done
...   // This is where we execute GPU code
cudaEventRecord(time3, 0);  // Time stamp after the GPU code execution
```

The variables time1, time2, time3, and time4 are all CPU-side variables that store time-stamps during the transfers between the CPU and GPU, as well as the execution of the GPU code on the device side. A curious observation from the code above is that we only use Nvidia APIs to time-stamp the GPU-related events. Anything that touches the GPU must be time-stamped with the Nvidia APIs, specifically cudaEventRecord() in this case. But, why? Why can't we simply use the good-and-old gettimeofday() function we saw in the CPU code listings?

The answer is in Analogy 6.2: We totally rely on Nvidia APIs (the people from the moon) to time anything that relates to the GPU side. If we are doing that, we might as well let them time all of the space travel time, both forward and back. We are recording the beginning and end of these data transfers and GPU kernel execution as *events*, which allows us to use *Nvidia event timing APIs* to time them, such as cudaEventRecord(). To be used in this API an event must be first *created* using the cudaEventCreate() API. Because the event recording mechanism is built into Nvidia APIs, we can readily use them to time our GPU kernels and the CPU⟷GPU transfers, much like we did with our CPU code.

In Code 6.3, we use `time1` to time-stamp the very beginning of the code and `time2` to time-stamp the point when the CPU→GPU transfer is complete. Similarly, `time3` is when the GPU code execution is done and `time4` is when the arrival of the results to the CPU side is complete. The difference between any of these two time-stamps will tell us how long each one of these *events* took to complete. Not surprisingly, the *difference* must also be calculated by using the cudaEventElapsedTime() API — shown in Code 6.4 — in the CUDA API library, because the stored time-stamps are in a format that is also a part of the Nvidia APIs rather than ordinary variables.

6.4.4 GPU-Side Memory Allocation

Now, let's turn our attention to *GPU-side memory allocation* in Code 6.3. The following lines facilitate the creation of GPU-side memory by using the API cudaMalloc():

```
// Allocate GPU buffer for the input and output images
cudaStatus = cudaMalloc((void**)&GPUImg, IMAGESIZE);
cudaStatus2 = cudaMalloc((void**)&GPUCopyImg, IMAGESIZE);
if ((cudaStatus != cudaSuccess) || (cudaStatus2 != cudaSuccess)){
   fprintf(stderr, "cudaMalloc failed! Can't allocate GPU memory");
   exit(EXIT_FAILURE);
}
```

Nvidia Runtime Engine contains a mechanism — through the cudaMalloc() API — for the CPU to "ask" Nvidia to see if it can allocate a given amount of GPU memory. The answer is returned in a variable of type cudaError_t. If the answer is cudaSuccess, we know that the Nvidia runtime Engine was able to create the GPU memory we asked for and placed the starting point of this memory area in a pointer that is named `GPUImg`. Remember from Code 6.1 that the `GPUImg` is a CPU-side variable, pointing to a GPU-side memory address.

6.4.5 GPU Drivers and Nvidia Runtime Engine

As you see from the way the Nvidia APIs work, the *Nvidia Runtime Engine* is pretty much the *Nvidia Operating System* that manages your GPU resources, much like the regular OS managing the CPU resources. In Analogy 6.2, this is the cudaTown government offices. This Nvidia OS is placed inside your *GPU drivers*, the drivers that you install when you plug in your GPU card and get your CPU OS to recognize it. After you install the drivers, an icon on your SysTray will show up in Windows 10 Pro, as you see in Figure 6.4. The GPU drivers make it seamless for you to access GPU resources through a set of easy-to-use APIs to copy data back and forth between the CPU and GPU and execute GPU code. Figure 6.4 is a screen shot from a Windows 10 Pro PC that has an Nvidia GTX Titan Z GPU installed, which registers itself as two separate GPUs, containing 2880 cores each. The version of the Nvidia driver is 369.30 in this case. Some of the driver versions may be buggy and may not work well with certain APIs and these issues will be the continuous discussion topics in Nvidia forums. In other words, some driver versions will "drive" you nuts until you downgrade them back to a stable version or Nvidia fixes the bugs. If you create an Nvidia Developer account by going to developer.nvidia.com, you can be a part of these discussions and can provide answers when you build sufficient experience. These issues are nothing different than the bugs that Windows (or Mac) OS's have. It is just part of the OS development process.

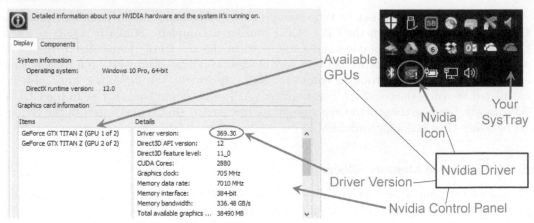

FIGURE 6.4 Nvidia Runtime Engine is built into your GPU drivers, shown in your Windows 10 Pro SysTray. When you click the Nvidia symbol, you can open the Nvidia control panel to see the driver version as well as the parameters of your GPU(s).

6.4.6 CPU→GPU Data Transfer

Once the GPU-side memory is allocated by the Nvidia runtime, we use another Nvidia API — named cudaMemcpy() — to perform the CPU→GPU data transfer. This API transfers data from a CPU-side memory area (pointed to by the **TheImg** pointer) to a GPU-side memory area (pointed to by the **GPUImg** pointer). Both of these pointers are declared in Code 6.1. The following lines perform the CPU→GPU data transfer:

```
// Copy input vectors from host memory to GPU buffers.
cudaStatus = cudaMemcpy(GPUImg, TheImg, IMAGESIZE, cudaMemcpyHostToDevice);
if (cudaStatus != cudaSuccess) {
   fprintf(stderr, "cudaMemcpy CPU to GPU failed!");
   exit(EXIT_FAILURE);
}
cudaEventRecord(time2, 0);  // Time stamp after the CPU --> GPU tfr is done
```

Much like the memory allocation API cudaMalloc(), the memory transfer API cudaMemcpy() also uses the same status type **cudaError_t**, which returns **cudaSuccess** if the transfer completes without an error. If it doesn't, then we know that something went wrong during the transfer.

Going back to our Analogy 6.2, the cudaMemcpy() API is a specialized function that the spaceship has; a way to transfer 166,656 coconuts super fast in the spaceship, instead of worrying about each coconut one by one. Fairly soon, we will see that this memory transfer functionality will become a lot more sophisticated and the transfer time will end up being a big problem that will face us. We will see a set of more advanced memory transfer functions from Nvidia to ease the pain! In the end, just because the transfers take a lot of time, cudaTown people do not want to lose business. So, they will invent ways to make the coconut transfer a lot more efficient to avoid discouraging cocoTown people from sending business their way.

6.4.7 Error Reporting Using Wrapper Functions

We can further query the *reason* for the error by using the cudaGetErrorString() API function (which returns a pointer to a string that contains a simple explanation for the error such as *"invalid device pointer"*), but this is something that is not shown in Code 6.3. A simple modification of Code 6.3 would print the reason for the failure as follows:

```
if (cudaStatus != cudaSuccess) {
    fprintf(stderr, "cudaMemcpy failed! %s",cudaGetErrorString(cudaStatus));
    exit(EXIT_FAILURE);
}
```

It is fairly common for programmers to write a *wrapper* function that wraps every single CUDA API call around some sort of error checking as shown below:

```
chkCUDAErr(cudaMemcpy(GPUImg, TheImg, IMAGESIZE, cudaMemcpyHostToDevice));
```

where the wrapper function — chkCUDAErr() — is one that we write within our C code, which directly uses the error code coming out of a CUDA API. An example wrapper function is shown below, which exits the program when a GPU runtime code is returned by any CUDA API:

```
// helper function that wraps CUDA API calls, reports any error and exits
void chkCUDAErr(cudaError_t ErrorID)
{
    if (ErrorID != CUDA_SUCCESS){
        printf("CUDA ERROR :::%\n", cudaGetErrorString(ErrorID));
        exit(EXIT_FAILURE);
    }
}
```

6.4.8 GPU Kernel Execution

After the completion of the CPU→GPU data transfer, we are now ready to *execute* the GPU kernel on the GPU side. The lines that relate to the GPU side code execution in Code 6.3 are shown below:

```
int IPH=ip.Hpixels;   int IPV=ip.Vpixels;
...
BlkPerRow = (IPH + ThrPerBlk -1 ) / ThrPerBlk;
NumBlocks = IPV*BlkPerRow;
switch (Flip){
    case 'H': Hflip <<< NumBlocks, ThrPerBlk >>> (GPUCopyImg, GPUImg, IPH);
            GPUResult = GPUCopyImg;        GPUDataTransfer = 2*IMAGESIZE;
            break;
    case 'V': Vflip <<< NumBlocks, ThrPerBlk >>> (GPUCopyImg, GPUImg, IPH, IPV);
            GPUResult = GPUCopyImg;        GPUDataTransfer = 2*IMAGESIZE;
            break;
    ...
```

The Flip parameter is set based on the command line argument the user enters. When the option 'H' is chosen by the user, the Hflip() GPU-side function is called and the three specified arguments (`GPUCopyImg`, `GPUImg`, and `IPH`) are passed onto Hflip() from the CPU side. The 'V' option launches the Vflip() kernel with four arguments, as opposed to the three arguments in the Hflip() kernel; `GPUCopyImg`, `GPUImg`, `IPH`, and `IPV`. Once we look at the details of both kernels, it will be clear why we need the additional argument inside Vflip().

The following lines show what happens when the user chooses the 'T' (transpose) or 'C' (copy) options in the command line. I could have implemented *transpose* in a more efficient way by writing a specific kernel for it; however, my goal was to show how two kernels can be launched, one after the other. So, to implement 'T', I launched Hflip followed by Vflip, which effectively transposes the image. For the implementation of the 'C' option, though, I designed a totally different kernel PixCopy().

```
switch (Flip){
    ...
    case 'T': Hflip <<< NumBlocks, ThrPerBlk >>> (GPUCopyImg, GPUImg, IPH);
              Vflip <<< NumBlocks, ThrPerBlk >>> (GPUImg, GPUCopyImg, IPH, IPV);
              GPUResult = GPUImg;        GPUDataTransfer = 4*IMAGESIZE;
              break;
    case 'C': NumBlocks = (IMAGESIZE+ThrPerBlk-1) / ThrPerBlk;
              PixCopy <<< NumBlocks, ThrPerBlk >>> (GPUCopyImg, GPUImg, IMAGESIZE);
              GPUResult = GPUCopyImg;    GPUDataTransfer = 2*IMAGESIZE;
              break;
}
```

When the option 'H' is chosen by the user, the execution of the following line is handled by the Nvidia Runtime Engine, which involves launching the Hflip() kernel and passing the three aforementioned arguments to it from the CPU side.

```
Hflip <<< NumBlocks, ThrPerBlk >>> (GPUCopyImg, GPUImg, IPH);
```

Going forward, I will use the terminology *launching GPU kernels*. This contrasts with the terminology of *calling CPU functions*; while the CPU *calls* a function within its own *planet*, say *earth* according to Analogy 6.2, this is possibly not a good terminology for GPU kernels. Because the GPU really acts as a co-processor, plugged into the CPU using a far slower connection than the CPU's own internal buses, calling a function in a far location such as *moon* deserves a more dramatic term like *launching*. In the GPU kernel launch line above, Hflip() is the *GPU kernel name*, and the two parameters that are inside the ≪ and ≫ symbols (`NumBlocks` and `ThrPerBlk`) tell the Nvidia Runtime Engine what *dimensions* to run this kernel with; the first argument (`NumBlocks`) indicates *how many blocks* to launch, and the second argument (`ThrPerBlk`) indicates *how many threads* are launched in each block. Remember from Analogy 6.2 that these two numbers are what the cudaTown people wanted to know; the number of boxes (`NumBlocks`) and the number of coconuts in each box (`ThrPerBlk`). The generalized kernel launch line is as follows:

```
GPU Kernel Name <<< dimension, dimension >>> (arg1, arg2, ...);
```

where `arg1`, `arg2`, ... are the parameters passed from the CPU side onto the GPU kernel. In Code 6.3, the arguments are the two pointers (`GPUCopyImg` and `GPUImg`) that were given to us by cudaMalloc() when we created memory areas to store images in the GPU memory and `IPH` is a variable that holds the number of pixels in the horizontal dimension of the image (`ip.Hpixels`). GPU kernel Hflip() will need these three parameters during its execution and would have no way of getting them had we not passed them during the kernel launch. Remember that the two launch dimensions in Analogy 6.2 were 166,656 and 256, effectively corresponding to the following launch line:

```
Hflip <<< 166,656, 256 >>> (GPUCopyImg, GPUImg, IPH);
```

This tells the Nvidia Runtime Engine to launch 166,656 blocks of the Hflip() kernel and pass the three parameters onto every single one of these blocks. So, the following blocks will be launched: Block 0, Block 1, Block 2, ... Block 166,655. Every single one of these blocks will execute 256 threads ($tid = 0$, $tid = 1$, ... , $tid = 255$), identical to the *pthreads* examples we saw in Part I of the book. What we are really saying is that we are launching a total of $166,656 \times 256 \approx 41$ M threads with this single launch line.

It is worth noting the difference between *Million* and *Mega*: Million threads means 1,000,000 threads, while Mega threads means $1024 \times 1024 = 1,048,576$ threads. Similarly *Thousand* is 1000 and *Kilo* is 1024. I will notate 41 Mega threads as 41 M threads. Same for 41,664 Kilo threads, being notated as 41,664 K threads. To summarize:

$$166,656 \times 256 = 42,663,936 = 41,664 \text{ K} = 40.6875 \text{ M threads} \approx 41 \text{ M threads}. \quad (6.1)$$

One important note to take here is that the GPU kernel is a bunch of GPU machine code instructions, generated by the nvcc compiler, on the CPU side. These are the instructions for the cudaTown people to execute in Analogy 6.2. Let's say you wanted them to flip the order in which the coconuts are stored in the boxes and send them right back to earth. You then need to send them instructions about how to flip them (Hflip()). Because cudaTown people do not know what to do with the coconuts once they receive them. They need the coconuts (data), as well as the sequence of commands to execute (instructions). So, the compiled instructions also travel to cudaTown in the spaceship, written on a big piece of paper. At runtime, these instructions are executed on each block independently. Clearly, the performance of your GPU program depends on the efficiency of the kernel instructions, i.e., the programmer.

Let's refresh our memory with Code 2.8, which was the MTFlipH() CPU function that accepted a single parameter named `tid`. By looking at the `tid` parameter that is passed onto it, this CPU function knew "who it was." Based on who it was it processed a different part of the image, indexed by `tid` in some fashion. The GPU kernel Hflip() has stark similarities to it: This kernel acts almost exactly like its CPU sister MTFlipH() and the entire functionality of the Hflip() kernel will be dictated by a thread ID. Let's now compare them:

- MTFlipH() function is launched with 4–8 threads, while the Hflip() kernel is launched with almost 40 million threads. I talked about the *overhead* in launching CPU threads in Part I, which was really high. This overhead is almost negligent in the GPU world, allowing us to launch a million times more of them.

- MTFlipH() expects the Pthread API call to pass the `tid` to it, while the Hflip() kernel will receive its thread ID (0...255) directly from Nvidia Runtime Engine, at runtime. As the GPU programmer, all we have to worry about is to tell the kernel *how many* threads to launch and they will be numbered automatically.

- Due to the million times higher number of the threads we launch, some sort of hierarchy is necessary. This is why the thread numbering is broken down into two values: the *blocks* are little chunks that execute, with 256 threads in each. Each block executes completely independent from each other.

CODE 6.4:   imflipG.cu   main() {...

Finishing the GPU kernel execution and transferring the results back to the CPU.

```
int main(int argc, char** argv)
{
  ...
  // cudaDeviceSynchronize waits for the kernel to finish, and returns
  // any errors encountered during the launch.
  cudaStatus = cudaDeviceSynchronize();
  if (cudaStatus != cudaSuccess) {
    fprintf(stderr, "\n\ncudaDeviceSynchronize error code %d ...\n", cudaStatus);
    exit(EXIT_FAILURE);
  }
  cudaEventRecord(time3, 0);
  // Copy output (results) from GPU buffer to host (CPU) memory.
  cudaStatus = cudaMemcpy(CopyImg, GPUResult, IMAGESIZE, cudaMemcpyDeviceToHost);
  if (cudaStatus != cudaSuccess) {
    fprintf(stderr, "cudaMemcpy GPU to CPU failed!");
    exit(EXIT_FAILURE);
  }
  cudaEventRecord(time4, 0);

  cudaEventSynchronize(time1);    cudaEventSynchronize(time2);
  cudaEventSynchronize(time3);    cudaEventSynchronize(time4);
  cudaEventElapsedTime(&totalTime, time1, time4);
  cudaEventElapsedTime(&tfrCPUtoGPU, time1, time2);
  cudaEventElapsedTime(&kernelExecutionTime, time2, time3);
  cudaEventElapsedTime(&tfrGPUtoCPU, time3, time4);

  cudaStatus = cudaDeviceSynchronize();
  if (cudaStatus != cudaSuccess) {
    fprintf(stderr, "\n Program failed after cudaDeviceSynchronize()!");
    free(TheImg);      free(CopyImg);      exit(EXIT_FAILURE);
  }
  WriteBMPlin(CopyImg, OutputFileName);   // Write the flipped image back to disk
  ...
```

6.4.9 Finish Executing the GPU Kernel

Although I explained the GPU kernel execution strictly in terms of the Hflip() kernel, almost everything is the same for the vertical flip (option 'V'), transpose (option 'T'), and copy (option 'C'). The only thing that changes is the kernel that is launched in these other cases. Vertical flip option causes the launch of the Vflip() kernel, which requires four function arguments, whereas the image transpose option simply launches both the horizontal and vertical flips, one after the other. Copy option launches another kernel PixCopy() that requires three arguments, the last one being different than the other kernels.

Code 6.4 shows the part of imflipG.cu() where we wait for the GPU to finish executing its kernel(s). When we launch one or more kernels, they continue executing until they are done. We must wait for the execution to be done using the lines below:

```
// cudaDeviceSynchronize waits for the kernel to finish, and returns
// any errors encountered during the launch.
cudaStatus = cudaDeviceSynchronize();
if (cudaStatus != cudaSuccess) {
   fprintf(stderr, "\n\ncudaDeviceSynchronize error code %d...", cudaStatus);
   exit(EXIT_FAILURE);
}
```

The cudaDeviceSynchronize() function waits for every single launched kernel to complete its execution. The result could be an error, in which case cudaDeviceSynchronize() will return an error code. Otherwise, everything is good and we move onto reporting the results.

6.4.10 Transfer GPU Results Back to the CPU

Once the execution of the kernels that we have launched is complete, the result (the flipped image) is sitting in the GPU memory area, pointed to by the pointer GPUResult. We want to transfer it to the CPU memory area, pointed to by the pointer CopyImg. We can use the cudaMemcpy() function to do this with one exception: the very last argument of cudaMemcpy() specifies the direction of the transfer. Remember from Code 6.3 that we used the same API with the cudaMemcpyHostToDevice, which meant a CPU→GPU transfer. Now we use the cudaMemcpyDeviceToHost option, which means a GPU→CPU transfer. Aside from that everything else is the same as shown below:

```
cudaEventRecord(time3, 0);
// Copy output (results) from GPU buffer to host (CPU) memory.
cudaStatus = cudaMemcpy(CopyImg, GPUResult, IMAGESIZE, cudaMemcpyDeviceToHost);
if (cudaStatus != cudaSuccess) {
   fprintf(stderr, "cudaMemcpy GPU to CPU failed!");
   exit(EXIT_FAILURE);
}
cudaEventRecord(time4, 0);
```

6.4.11 Complete Time-Stamping

We time-stamp the end of the GPU→CPU transfer with time4 and we are ready to calculate the amount of time each event took as follows:

```
cudaEventSynchronize(time1);      cudaEventSynchronize(time2);
cudaEventSynchronize(time3);      cudaEventSynchronize(time4);
cudaEventElapsedTime(&totalTime, time1, time4);
cudaEventElapsedTime(&tfrCPUtoGPU, time1, time2);
cudaEventElapsedTime(&kernelExecutionTime, time2, time3);
cudaEventElapsedTime(&tfrGPUtoCPU, time3, time4);
```

The API cudaEventSynchronize() tells Nvidia runtime to synchronize a given event to ensure that the variables time1 ... time4 have correct timing values. The cudaEventElapsedTime() API is used to calculate the difference between a pair of them, because they are not simple types. For example, the difference between time and time4 indicates how long it took for the data to depart from the CPU and arrive at the GPU, get processed by the GPU, and return back to CPU memory. The rest are easy to follow.

CODE 6.5: imflipG.cu main() ...}

Last part of main() reports the results and cleans up GPU and CPU memory areas.

```
int main(int argc, char** argv)
{
   ...
   printf("--...--\n"); printf("%s ComputeCapab=%d.%d [supports max %s blocks]\n",
       GPUprop.name,GPUprop.major,GPUprop.minor,SupportedBlocks); printf("...\n");
   printf("%s %s %s %c %u [%u BLOCKS, %u BLOCKS/ROW]\n", ProgName, InputFileName,
       OutputFileName,Flip, ThrPerBlk, NumBlocks, BlkPerRow);
   printf("-------------------- ... ------------------------------\n");
    printf("CPU->GPU Transfer = %5.2f ms ... %4d MB ... %6.2f GB/s\n",
   tfrCPUtoGPU, IMAGESIZE / 1024 / 1024, (float)IMAGESIZE / (tfrCPUtoGPU *
       1024.0*1024.0));
   printf("Kernel Execution = %5.2f ms ... %4d MB ... %6.2f GB/s\n",
   kernelExecutionTime, GPUDataTransfer / 1024 / 1024, (float)GPUDataTransfer /
       (kernelExecutionTime * 1024.0*1024.0));
   printf("GPU->CPU Transfer = %5.2f ms ... %4d MB ... %6.2f GB/s\n",
   tfrGPUtoCPU, IMAGESIZE / 1024 / 1024, (float)IMAGESIZE / (tfrGPUtoCPU *
       1024.0*1024.0));
   printf("Total time elapsed = %5.2f ms\n", totalTime);
   printf("-------------------- ... ------------------------------\n");
   // Deallocate CPU, GPU memory and destroy events.
   cudaFree(GPUImg);          cudaFree(GPUCopyImg);
   cudaEventDestroy(time1);    cudaEventDestroy(time2);
   cudaEventDestroy(time3);    cudaEventDestroy(time4);
   // cudaDeviceReset must be called before exiting, so profiling and tracing tools
   // like Parallel Nsight and Visual Profiler shows complete traces.
   cudaStatus = cudaDeviceReset();
   if (cudaStatus != cudaSuccess) {
      fprintf(stderr, "cudaDeviceReset failed!");
      free(TheImg);   free(CopyImg);   exit(EXIT_FAILURE);
   }
   free(TheImg);   free(CopyImg);
   return(EXIT_SUCCESS);
}
```

6.4.12 Report the Results and Cleanup

Code 6.5 shows how we report the results of the program. Some of these reported results are coming from our query of the GPU. For example, GPUprop.name is the name of the GPU such as "GeForce GTX Titan Z." I will go over the detailed output in Section 6.5.4.

Much like every malloc() must be cleaned up with a corresponding free(), every cudaMalloc() must be cleaned up with a corresponding cudaFree() by telling the Nvidia runtime that you no longer need that GPU memory area. Similarly, every event you created using cudaEventCreate() must now be destroyed by cudaEventDestroy(). After all of this is done, we call cudaDeviceReset(), which tells Nvidia runtime that we are done using the GPU. Then, we go ahead and free() our CPU memory areas and imflipG.cu is done!

CODE 6.6:   imflipG.cu   ReadBMPlin() {...}, WriteBMPlin() {...}

Functions that read/write a BMP file in linear indexing, rather than the x, y indexing.

```
// Read a 24-bit/pixel BMP file into a 1D linear array.
// Allocate memory to store the 1D image and return its pointer.
uch *ReadBMPlin(char* fn)
{
   static uch *Img;
   FILE* f = fopen(fn, "rb");
   if (f == NULL){ printf("\n\n%s NOT FOUND\n\n", fn); exit(EXIT_FAILURE); }
   uch HeaderInfo[54];
   fread(HeaderInfo, sizeof(uch), 54, f); // read the 54-byte header
   // extract image height and width from header
   int width = *(int*)&HeaderInfo[18];    ip.Hpixels = width;
   int height = *(int*)&HeaderInfo[22];   ip.Vpixels = height;
   int RowBytes = (width * 3 + 3) & (~3); ip.Hbytes = RowBytes;
   memcpy(ip.HeaderInfo, HeaderInfo,54);  //save header for re-use
   printf("\n Input File name: %17s (%u x %u) File Size=%u", fn,
   ip.Hpixels, ip.Vpixels, IMAGESIZE);
   // allocate memory to store the main image (1 Dimensional array)
   Img = (uch *)malloc(IMAGESIZE);
   if (Img == NULL) return Img;   // Cannot allocate memory
   // read the image from disk
   fread(Img, sizeof(uch), IMAGESIZE, f);   fclose(f);        return Img;
}

// Write the 1D linear-memory stored image into file.
void WriteBMPlin(uch *Img, char* fn)
{
   FILE* f = fopen(fn, "wb");
   if (f == NULL){ printf("\n\nFILE CREATION ERROR: %s\n\n", fn); exit(1); }
   fwrite(ip.HeaderInfo, sizeof(uch), 54, f); //write header
   fwrite(Img, sizeof(uch), IMAGESIZE, f);  //write data
   printf("\nOutput File name: %17s (%u x %u) File Size=%u", fn, ip.Hpixels,
       ip.Vpixels, IMAGESIZE);
   fclose(f);
}
```

6.4.13   Reading and Writing the BMP File

Code 6.6 shows the two functions that read and write a BMP image to/from CPU memory. The difference between these functions and the ones we saw in Code 2.4 and Code 2.5 is that they read the image in a *linear* memory area, hence the suffix "lin" at the end of the function names. Linear memory area means a single index, rather than an x, y index to store the pixels. This makes the image reading from the disk extremely simple because the image is, indeed, stored in a linear fashion on the disk. However, when we are processing it, we will need to convert it back to the x, y format using a very simple transformation:

$$\text{Pixel Coordinate Index} = (x, y) \quad \longrightarrow \quad \text{Linear Index} = x + (y \times ip.Hpixels) \quad (6.2)$$

CODE 6.7:   imflipG.cu   Vflip() {...}

The GPU kernel Vflip() that flips an image vertically.

```
// Kernel that flips the given image vertically
// each thread only flips a single pixel (R,G,B)
__global__
void Vflip(uch *ImgDst, uch *ImgSrc, ui Hpixels, ui Vpixels)
{
    ui ThrPerBlk = blockDim.x;
    ui MYbid = blockIdx.x;
    ui MYtid = threadIdx.x;
    ui MYgtid = ThrPerBlk * MYbid + MYtid;

    ui BlkPerRow = (Hpixels + ThrPerBlk - 1) / ThrPerBlk;  // ceil
    ui RowBytes = (Hpixels * 3 + 3) & (~3);
    ui MYrow = MYbid / BlkPerRow;
    ui MYcol = MYgtid - MYrow*BlkPerRow*ThrPerBlk;
    if (MYcol >= Hpixels) return;    // col out of range
    ui MYmirrorrow = Vpixels - 1 - MYrow;
    ui MYsrcOffset = MYrow      * RowBytes;
    ui MYdstOffset = MYmirrorrow * RowBytes;
    ui MYsrcIndex = MYsrcOffset + 3 * MYcol;
    ui MYdstIndex = MYdstOffset + 3 * MYcol;

    // swap pixels RGB @MYcol , @MYmirrorcol
    ImgDst[MYdstIndex] = ImgSrc[MYsrcIndex];
    ImgDst[MYdstIndex + 1] = ImgSrc[MYsrcIndex + 1];
    ImgDst[MYdstIndex + 2] = ImgSrc[MYsrcIndex + 2];
}
```

6.4.14 Vflip(): The GPU Kernel for Vertical Flipping

Code 6.7 shows the GPU kernel named Vflip(). This kernel shows what each *thread* does. As we computed a few pages ago, there are 40 million plus threads that will run this kernel, assuming that we launch it with the dimensions Vflip ≪ 166656, 256 ≫ (...). If we launch it with higher dimensions, there will be even more threads launched. Every single line of code we saw up to this point in this chapter (for that matter, in this book) was CPU code. Code 6.7 is the first GPU code we are seeing. Although Code 6.7 is written in C and everything else so far was also written in C, what we care about is the final set of CPU instructions that Code 6.7 will be compiled into. While everything we saw so far was going to be compiled into x64 — Intel 64-bit Instruction Set Architecture (ISA) — instructions, Code 6.7 will be compiled into the Nvidia GPU ISA named Parallel Thread Execution (PTX). Much like Intel and AMD extend their ISAs in every new generation of CPUs, so does Nvidia in every generation of GPUs. For example, as of late 2015, the dominant ISA was PTX 4.3 and PTX 5.0 was in development.

One big difference between a CPU ISA and GPU ISA is that while an x64 compiled CPU ISA output contains x86 instructions that will be executed one-by-one at runtime (i.e., fully compiled), PTX is actually an intermediate representation (IR). This means that it is "half-compiled," much like a Java Byte code. The Nvidia Runtime Engine takes the

PTX instructions and further *half-compiles* them at runtime and feeds the *full-compiled* instructions into the GPU cores. In Windows, all of the "Nvidia magic code" that facilitates this "further-half-compiling" is built into a Dynamic Link Library (DLL) named cudart (CUDA Run Time). There are two flavors: in modern x64 OSs, it is cudart64 and in old 32-bit OSs, it is cudart32, although the latter should never be used because all modern Nvidia GPUs require a 64-bit OS for efficient use. In my Windows 10 Pro PC, for example, I was using cudart64_80.dll (Runtime Dynamic Link Library for CUDA 8.0). This file is not something you explicitly have to worry about; the nvcc compiler will put it in the executable directory for you. I am just mentioning it so you are aware of it.

Let's compare Code 6.7 to its CPU sister Code 2.7. Let's assume that both of them are trying to flip the astronaut.bmp image in Figure 5.1 vertically. astronaut.bmp is a 7918×5376 image that takes $\approx 121\,\text{MB}$ on disk. How would their functionality be different?

- For starters, assume that Code 2.7 uses 8 threads; it will assign the flipping task of 672 lines to each thread (i.e., $672 \times 8 = 5376$). Each thread will, then, be responsible for *processing* $\approx 15\,\text{MB}$ of *information* out of the entire image, which contains $\approx 121\,\text{MB}$ of information in its entirety. Because the launch of more than 10–12 threads will not help on an 8C/16T CPU, as we witnessed over and over again in Part I, we cannot really do better than this when it comes to the CPU.

- The GPU is different though. In the GPU world, we can launch a gazillion threads without incurring any overhead. What if we went all the way to the bitter extreme and had *each thread swap a single pixel*? Let's say that each GPU thread takes a single pixel's RGB value (3 bytes) from the source image GPU memory area (pointed to by *ImgSrc) and writes it into the intended *vertically flipped* destination GPU memory area (pointed to by *ImgDst).

- Remember, in the GPU world, our unit of launch is *blocks*, which are clumps of threads, each clump being 32, 64, 128, 256, 512, or 1024 threads. Also remember that it cannot be less than 32, because "32" is the smallest amount of parallelism we can have and 32 threads are called a *warp*, as I explained earlier in this chapter. Let's say that each one of our blocks will have 256 threads to flip the astronaut image. Also, assume that we are processing one row of the image at a time using multiple blocks. This means that we need $\lceil 7918/256 \rceil = 31$ blocks to process each row.

- Because we have 5376 rows in the image, we will need to launch $5376 \times 31 = 166{,}656$ blocks to vertically flip the astronaut.bmp image.

- We observe that 31 blocks-per-row will yield some minor loss, because $31 \times 256 = 7936$ and we will have 18 threads ($7936 - 7918 = 18$) doing nothing to process each row of the image. Oh well, nobody said that massive parallelism doesn't have its own disadvantages.

- This problem of "useless threads" is actually exacerbated by the fact that not only are these threads useless, but also they have to check to see if they are the useless threads as shown in the line below:

```
if (MYcol >= Hpixels) return;   // col out of range
```

This line simply says "if my *tid* is between 7918 and 7935 I shouldn't do anything, because I am a useless thread." Here is the math: We know that the image has

7918 pixels in each row. So, the threads $tid = 0...7917$ are useful, and because we launched 7936 threads ($tid = 0...7935$), this designates threads ($tid = 7918...7935$) as useless.

- Don't worry about the fact that we do not see tid in the comparison; rather, we see MYcol. When you calculate everything, the underlying math ends up being exactly what I just described. The reason for using a variable named MYcol is because the code has to be parametric, so it works for any size image, not just the astronaut.bmp.

- Why is it so bad if only 18 threads check this? After all, 18 is only a very small percentage of the 7936 total threads. Well, this is not what happens. Like I said before, what you are seeing in Code 6.7 is what *every* thread executes. In other words, *all 7936 threads* must execute the same code and must check to see if they are useless, just to find that they aren't useless (most of the time) or they are (only a fraction of the time). So, with this line of code, we have introduced overhead to *every thread*. How do we deal with this? We will get to it, I promise. But, not in this chapter ... For now, just know that even with these inefficiencies — which are an artifact of massively parallel programming — our performance is still acceptable.

- And, finally, the __global__ is the third CUDA symbol that I am introducing here, after ≪ and ≫. If you precede any ordinary C function with __global__ the nvcc compiler will know that it is a *GPU-side* function and it compiles it into PTX, rather than the x64 machine code output. There will be a few more of these CUDA designators, but, aside from that, CUDA looks exactly like C.

6.4.15   What Is My Thread ID, Block ID, and Block Dimension?

We will spend a lot of time explaining the deep details of Vflip() in the next chapter, but for now all we want to know is how this function calculates who it is and which portion of the processing it is responsible for. Let's refresh our memory with the CPU Code 2.7:

```
void *MTFlipV(void* tid)
{
   long ts = *((int *) tid);            // My thread ID is stored here
   ts *= ip.Hbytes/NumThreads;          // start index
   long te = ts+ip.Hbytes/NumThreads-1; // end index
   ...
   for(col=ts; col<=te; col+=3){
      row=0;
      while(row<ip.Vpixels/2){
         pix.B=TheImage[row][col];  pix.G=TheImage[row][col+1];  pix.R=...
         ...
```

Here, the ts and te variables computed the starting and ending row numbers in the image, respectively. Vertical flipping was achieved by two nested loops, one scanning the columns and the other scanning the rows. Now, let's compare this to the Vflip() function in Code 6.7:

```
__global__
void Vflip(uch *ImgDst, uch *ImgSrc, ui Hpixels, ui Vpixels)
{
   ui ThrPerBlk = blockDim.x;
   ui MYbid = blockIdx.x;
   ui MYtid = threadIdx.x;
   ui MYgtid = ThrPerBlk * MYbid + MYtid;
   ui BlkPerRow = (Hpixels + ThrPerBlk - 1) / ThrPerBlk; // ceil
   ui RowBytes = (Hpixels * 3 + 3) & (~3);
```

We see that there are major similarities and differences between CPU and GPU functions. The task distribution in the GPU function is completely different because of the blocks and the number of threads in each block. So, although the GPU still calculates a bunch of indexes, they are completely different than the CPU function. GPU first wants to know how many threads were launched with each block. The answer is in a special GPU value named `blockDim.x`. We know that this answer will be 256 in our specific case because we specified 256 threads to be launched in each block (Vflip≪ ..., 256 ≫). So, each block contains 256 threads, with thread IDs 0...255. The specific thread ID of this thread is in `threadIDx.x`. It also wants to know, out of the 166,656 blocks, what is its own block ID. This answer is in another GPU value named `blockIdx.x`. Surprisingly, it doesn't care about the total number of blocks (166,656) in this case. There will be other programs that do.

It saves its block ID and thread ID in two variables named `bid` and `tid`. It then computes a global thread ID (`gtid`) using a combination of these two. This `gtid` gives a unique ID to each one of the launched GPU threads (out of the total $166,656 \times 256 \approx 41$ M threads), thereby *linearizing them*. This concept is very similar to how we linearized the pixel memory locations on the disk according to Equation 6.2. However, an immediate correlation between linear GPU thread addresses and linear pixel memory addresses is not readily available in this case due to the existence of the useless threads in each row. Next, it computes the *blocks per row* (`BlkPerRow`), which was 31 in our specific case. Finally, because the value of the number of horizontal pixels (7918) was passed onto this function as the third parameter, it can compute the total number of bytes in a row of the image ($3 \times 7918 = 23,754$ Bytes) to determine the *byte index* of each pixel.

After these computations, the kernel then moves onto computing the row and column index of the single pixel that it is responsible for copying as follows:

```
   ui MYrow = MYbid / BlkPerRow;
   ui MYcol = MYgtid - MYrow*BlkPerRow*ThrPerBlk;
   if (MYcol >= Hpixels) return;   // col out of range
   ui MYmirrorrow = Vpixels - 1 - MYrow;
   ui MYsrcOffset = MYrow      * RowBytes;
   ui MYdstOffset = MYmirrorrow * RowBytes;
   ui MYsrcIndex = MYsrcOffset + 3 * MYcol;
   ui MYdstIndex = MYdstOffset + 3 * MYcol;
```

After these lines, the source pixel memory address is in `MYsrcIndex` and the destination memory address is in `MYdstIndex`. Because each pixel contains three bytes (RGB) starting at that address, the kernel copies three consecutive bytes starting at that address as follows:

```
// swap pixels RGB @MYcol , @MYmirrorcol
ImgDst[MYdstIndex] = ImgSrc[MYsrcIndex];
ImgDst[MYdstIndex + 1] = ImgSrc[MYsrcIndex + 1];
ImgDst[MYdstIndex + 2] = ImgSrc[MYsrcIndex + 2];
```

Let's now compare this to CPU Code 2.7. Because we could only launch 4–8 threads, instead of the massive 41 M threads we just witnessed, one striking observation from the GPU kernel is that the `for` loops are gone! In other words, instead of explicitly scanning over the columns and rows, like the CPU function has to, we don't have to loop over anything. After all, the entire purpose of the loops in the CPU function was to scan the pixels with some sort of two-dimensional indexing, facilitated by the `row` and `column` variables. However, in the GPU kernel, we can achieve this functionality by using the `tid` and `bid`, because we know the precise relationship of the coordinates and the `tid` and `bid` variables.

CODE 6.8: imflipG.cu Hflip() {...}

The GPU kernel Hflip() that flips an image horizontally.

```
// Kernel that flips the given image horizontally
// each thread only flips a single pixel (R,G,B)
__global__
void Hflip(uch *ImgDst, uch *ImgSrc, ui Hpixels)
{
   ui ThrPerBlk = blockDim.x;
   ui MYbid = blockIdx.x;
   ui MYtid = threadIdx.x;
   ui MYgtid = ThrPerBlk * MYbid + MYtid;

   ui BlkPerRow = (Hpixels + ThrPerBlk -1 ) / ThrPerBlk; // ceil
   ui RowBytes = (Hpixels * 3 + 3) & (~3);
   ui MYrow = MYbid / BlkPerRow;
   ui MYcol = MYgtid - MYrow*BlkPerRow*ThrPerBlk;
   if (MYcol >= Hpixels) return;    // col out of range
   ui MYmirrorcol = Hpixels - 1 - MYcol;
   ui MYoffset = MYrow * RowBytes;
   ui MYsrcIndex = MYoffset + 3 * MYcol;
   ui MYdstIndex = MYoffset + 3 * MYmirrorcol;

   // swap pixels RGB @MYcol , @MYmirrorcol
   ImgDst[MYdstIndex] = ImgSrc[MYsrcIndex];
   ImgDst[MYdstIndex + 1] = ImgSrc[MYsrcIndex + 1];
   ImgDst[MYdstIndex + 2] = ImgSrc[MYsrcIndex + 2];
}
```

6.4.16 Hflip(): The GPU Kernel for Horizontal Flipping

Code 6.8 shows the GPU kernel function that is responsible for flipping a pixel's 3 bytes in the horizontal direction. Despite some differences in the way the indexes are calculated, Code 6.8 is almost identical to Code 6.7, in which a pixel was vertically flipped.

6.4.17 Hardware Parameters: threadIDx.x, blockIdx.x, blockDim.x

Both Code 6.7 and Code 6.8 have one thing in common: There are no explicit looping in either one because the Nvidia hardware is responsible for providing the `threadIDx.x`, `blockIdx.x`, and `blockDim.x` variables to every thread it launches. Every launched thread knows exactly what its thread and block ID is, as well as how many threads are launched in each block. So, with clever indexing, it is possible for every thread to get the `for` loops for free, as we see in both Code 6.7 and Code 6.8. Considering that each thread does such a small amount of work in GPU kernels, the savings from the looping overhead can be drastic.

We will spend a lot of time in the next chapter analyzing every single line of each function and will improve them as well as understand how they map to the GPU architecture.

CODE 6.9:   imflipG.cu   PixCopy() {...}

The GPU kernel PixCopy() that copies an image.

```
// Kernel that copies an image from one part of the
// GPU memory (ImgSrc) to another (ImgDst)
__global__
void PixCopy(uch *ImgDst, uch *ImgSrc, ui FS)
{
   ui ThrPerBlk = blockDim.x;
   ui MYbid = blockIdx.x;
   ui MYtid = threadIdx.x;
   ui MYgtid = ThrPerBlk * MYbid + MYtid;

   if (MYgtid > FS) return;        // outside the allocated memory
      ImgDst[MYgtid] = ImgSrc[MYgtid];
}
```

6.4.18 PixCopy(): The GPU Kernel for Copying an Image

Code 6.9 is how we copy an image entirely into another one. Comparing the PixCopy() function in Code 6.9 to the other two functions in Code 6.7 and Code 6.8, we see that the major difference is the usage of one-dimensional pixel indexing in PixCopy(). In other words, PixCopy() does not try to calculate which row and column it is responsible for. Each thread in PixCopy() copies only a single byte. This also eliminates the wasted threads at the edge of the rows, however, there are still wasted threads at the very end of the image. If we were to launch our program with the ('C' — copy) option, we would still have to launch it with, say, 256 threads per block. However, because the copying is linear, rather than based on a 2D indexing, we would have a different number of blocks to process it. Let's do the math: The image is a total of 127,712,256 Bytes. Each each block has 256 threads, we need a total of $\left\lceil \frac{127,712,256}{256} \right\rceil = 498,876$ blocks launched, each block containing 256 threads. In this specific

TABLE 6.1 CUDA keyword and symbols that we learned in this chapter.

CUDA Keyword	Description	Examples

__global__	precedes a device-side function (i.e., a kernel)	`__global__`
		`void PixCopy(uch *ImgDst, uch *ImgSrc, ui FS)`
		`{`
		`   ...`
		`}`

≪, ≫	Launch a device-side kernel from the host-side	`Hflip<<<NumBlocks, ThrPerBlk>>>(..., ..., ...);`
		
		`Vflip<<<NumBlocks, ThrPerBlk>>>(..., ..., ...);`
		
		`PixCopy<<<NumBlocks, ThrPerBlk>>>(..., ..., ...);`

case we got lucky. However, if we had 20 more bytes in the file, we would have 236 threads wasted out of the 256 in the very last block. This is why we still have to put the following if statement in the kernel to check for this condition as shown below:

```
if (MYgtid > FS) return;        // outside the allocated memory
```

The if statement in Code 6.9, much like the ones in Code 6.7 and Code 6.8, checks if "it is a useless thread" and does nothing if it is. The performance impact of this line is similar to the previous two kernels: although this condition will be only true for a negligible number of threads, *every* thread still has to execute it for every single byte they are copying. We can improve this, but we will save all of these improvement ideas to the upcoming chapters. For now, it is worth noting that the performance impact of this if statement in the PixCopy() kernel is far worse than the one we saw in the other two kernels. The PixCopy() kernel has a much finer granularity as it copies only a single byte. Because of this, there are only 6 lines of C code in PixCopy(), one of which being the if statement. In contrast, Code 6.7 and Code 6.8 contain 16–17 lines of code, thereby making the impact of one added line much less. Although "lines of code" clearly does not translate to "the number of cycles that it takes for the GPU core to execute the corresponding instructions" one-on-one, we can still get an idea about the magnitude of the problem.

6.4.19 CUDA Keywords

At this point, it is a good idea to summarize what makes the compiler distinguish between a C program and a CUDA program. There are a few CUDA-specific identifiers like the triple bracket pair (≪, ≫), which let the compiler know that these are CUDA function calls from the host-side. Furthermore, there are a few identifiers like __global__ to let the compiler know that the upcoming function must be compiled into PTX (GPU instruction set), not x64 (CPU instruction set). Table 6.1 lists these two most common CUDA keywords, which we learned in this chapter. There will be more we will learn in the following chapters; as we learn more, we will expand our table.

6.5 CUDA PROGRAM DEVELOPMENT IN WINDOWS

In this section, my goal was to show a working CUDA example: The imflipG.cu program, based on its command line parameters, performs one of four operations: flips an image in the (1) horizontal or (2) vertical direction, (3) copies it to another image, or (4) transposes it. The command line to run imflipG.cu is as follows:

imflipG astronaut.bmp a.bmp V 256

This vertically flips an image named astronaut.bmp and writes the flipped image into another file named a.bmp. The 'V' option is the flip direction (vertical) and 256 is the *number of threads in each block*, which is what we will plug into the second argument of our kernel dimensions with the launch parameters Vflip ≪ ..., 256 ≫ (...). We could choose 'H', 'C', or 'T' for horizontal flip, copy, or transpose operations.

6.5.1   Installing MS Visual Studio 2015 and CUDA Toolkit 8.0

To be able to develop CUDA code in a Windows PC, the best editor/compiler (Integrated Development Environment or IDE) to use is Microsoft Visual Studio 2015. I will call it VS 2015 going forward. With the release of CUDA 8.0, VS 2015 started working. Earlier versions of CUDA, such as even as late as CUDA 7.5, did not work with VS 2015. They only worked with VS 2013. In mid-2016, CUDA 8.0 was released, which allowed Pascal architecture-based GPUs to be used. In this book, VS 2015 and CUDA 8.0 will be used and notes will be made when certain features only apply to CUDA 8.0 and not the previous versions. To be able to compile CUDA 8.0 using VS 2015, you first have to install VS 2015 in your machine, followed by the Nvidia CUDA 8.0 toolkit. This toolkit will install the plug-in for VS 2015, allowing it to edit and compile CUDA code. Both of these are easy steps.

To install VS 2015, follow the instructions below:

- If you have a license for VS 2015 Professional, you can start the installation from the DVD. Otherwise, in your Internet Explorer (or Edge) browser, go to:

  https://www.visualstudio.com/downloads/

  Click "Free Download" under Visual Studio Community

- Once the download is complete, launch the installer and choose "everything" (instead of the limited set of default modules) to avoid some unforeseen issues, especially if you are installing VS 2015 for the first time.

- CUDA 8.0 will need quite a bit of the VS 2015 features. So, choosing *every* option takes up close to 40–50 GB of hard disk space. If you do not choose every option, you will get an error when you are trying to compile even the simplest CUDA code.

- I have not tried to pick and choose specific options to see which ones are absolutely needed to prevent this error. You can surely try to do it and spend a lot of time in the forums, but I can only guarantee that what is described here will work when you install every option of VS 2015.

After VS 2015 is installed, you need to install the CUDA Toolkit 8.0 as follows:

- In your Internet Explorer (or Edge) browser, go to:

  https://developer.nvidia.com/cuda-toolkit

- Click "Download" and download the toolkit if you prefer Local install (it will be about a Gigabyte). Local install option allows you to keep the installer in a temporary directory and double-click on the installer.

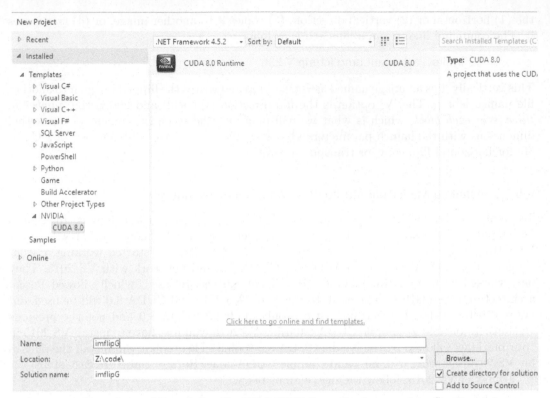

FIGURE 6.5 Creating a Visual Studio 2015 CUDA project named imflipG.cu. Assume that the code will be in a directory named Z:\code\imflipG in this example.

- You can select the Network Installer, which will install straight from the Internet. After spending 50 GB of your hard disk space on VS 2015, you will not be terribly worried about another GB. Either option is fine. I always choose the network installer, so I don't have to worry about deleting the local installer code after the installation is done.

- Click OK for the default extraction paths. The screen may go blank for a few second while the GPU drivers are being configured. After the installation is complete, you will see a new option in your Visual Studio, named "NSIGHT."

6.5.2 Creating Project imflipG.cu in Visual Studio 2015

To create a GPU program, we first clock "Create New Project" in VS 2015 and the dialog box, shown in Figure 6.5 opens up. Visual Studio wants a solution name and a name for the project. By default, it populates the same name into their respective boxes. We will not change this. To create your project, choose the name of the directory in "Location" and choose the project name as "imflipG." In the example shown in Figure 6.5, I am using Z:\code as the directory name and imflipG as the project name. Once you make these selections and click OK, VS 2015 will create a solution directory under Z:\code\imflipG.

A screen shot of the solution directory Z:\code\imflipG is shown in Figure 6.6. If you go into this directory, you will see another directory named imflipG, which is where your

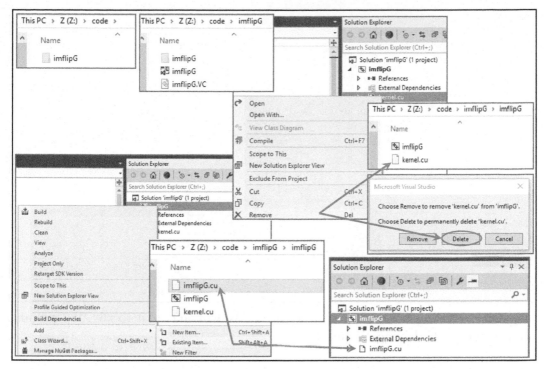

FIGURE 6.6 Visual Studio 2015 source files are in the Z:\code\imflipG\imflipG directory. In this specific example, we will remove the default file, kernel.cu, that VS 2015 creates. After this, we will add an existing file, imflipG.cu, to the project.

project source files are going to be placed by VS 2015; so, the source files will be under the directory Z:\code\imflipG\imflipG. Go into Z:\code\imflipG\imflipG; you will see a file named kernel.cu and another file we don't care about. The kernel.cu file is created in the source file directory automatically by VS 2015 by default.

At this point, there are three ways you can develop your CUDA project:

1. You can enter your code inside kernel.cu by using it as a template and delete the parts you don't want from it and compile it and run it as your only kernel code.

2. You can rename kernel.cu as something else (say, imflipG.cu) by right clicking on it inside VS 2015. You can clean what is inside the renamed imflipG.cu and put your own CUDA code in there. Compile it and run it.

3. You can remove the kernel.cu file from the project and add another file, imflipG.cu, to the project. This assumes that you already had this file; either by acquiring from someone or editing it in a different editor.

I will choose the last option. One important thing to remember is that you should never rename/copy/delete the files from Windows. You should perform any one of these operations *inside Visual Studio 2015*. Otherwise, you will confuse VS 2015 and it will try to use a file that doesn't exist. Because I intend to use the last option, the best thing to do is to actually plop the file imflipG.cu inside the Z:\code\imflipG\imflipG directory first. The screen shot after doing this is shown at the bottom of Figure 6.6. This is, for example, what you would

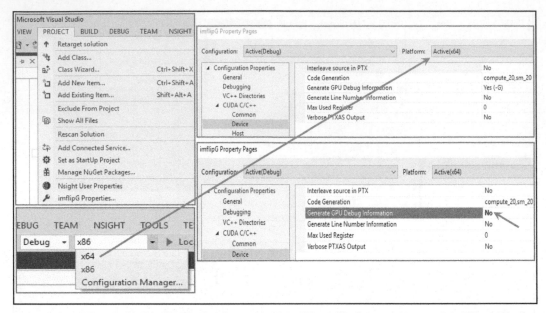

FIGURE 6.7 The default CPU platform is x86. We will change it to x64. We will also remove the GPU debugging option.

do if you are testing the programs I am supplying as part of this book. Although you will get only a single file, imflipG.cu, as part of this book, it must be properly added to a VS 2015 project, so you can compile it and execute it. Once the compilation is done, there will be a lot of miscellaneous files in the project directory, however, the source file is only a single file: imflipG.cu.

Figure 6.6 also shows the steps in deleting the kernel.cu() file. You right click and choose "Remove" first (top left). A dialog box will appear asking you whether you want to just remove it from the project, but keep the actual file (the "Remove" option) or remove it from the project and delete the actual file too (the "Delete" option). If you choose the "Delete" option, the file will be gone and it will no longer be a part of the project. This is the *graceful* way to get this file permanently out of your life, while also letting VS 2015 know about it along the way. After kernel.cu() is gone, you right click the project and this time Add a file to it. You can either add the file that we just dropped into the source directory (which is what we want to do by choosing the "Add Existing Item" option), or add a new file that doesn't exist and you will start editing (the "Add New Item" option). After we choose "Add Existing," we see the new imflipG.cu file added to the project in Figure 6.6. We are now ready to compile it and run it.

6.5.3 Compiling Project imflipG.cu in Visual Studio 2015

Before you can compile your code, you have to make sure that you choose the correct CPU and GPU platforms. As shown in Figure 6.7, the two CPU platform options are x86 (for 32-bit Windows OSs) and x64 (for 64-bit Windows OSs). I am using Windows 10 Pro, which is an x64 OS. So, I drop-down the CPU platform option and choose x64. For the GPU platform, you have to go to the project's properties by choosing PROJECT in the menu bar and selecting *imflipG Properties*. Another "imflipG Property Pages" dialog box

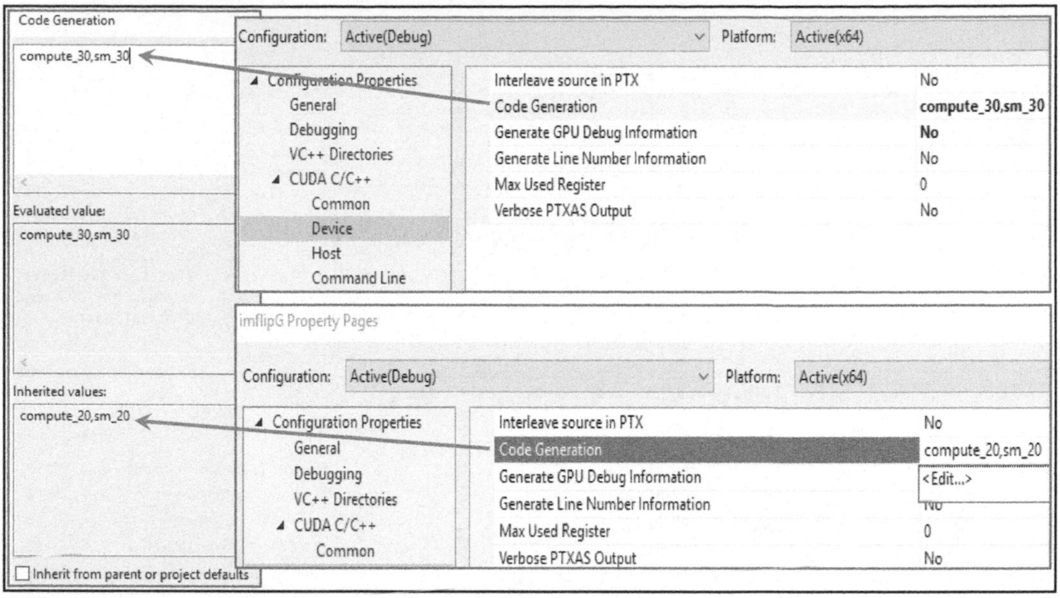

FIGURE 6.8 The default Compute Capability is 2.0. This is too old. We will change it to Compute Capability 3.0, which is done by editing *Code Generation* under *Device* and changing it to **compute_30, sm_30**.

will open, as shown in Figure 6.7. For the GPU, the first option you choose is *Generate GPU Debug Information*. If you choose "Yes" here, you will be able to run the GPU debugger, however your code will run at half the speed because the compiler has to add all sorts of break points inside your code. Typically, the best thing to do is to keep this at "Yes" while you are developing your code. After your code is fully debugged, you switch it to "No" as shown in Figure 6.7.

After you choose the GPU Debug option, you have to edit the *Code Generation* under *CUDA C/C++* → *Device* and select the *Code Generation* next, as shown in Figure 6.8. The default Compute Capability is 2.0, which will not allow you to run a lot of the new features of the modern Nvidia GPUs. You have to change this to Compute Capability 3.0. Once the "Code Generation" dialog box opens, you have to first uncheck *Inherit from parent of project defaults*. The default Compute Capability is 2.0, which the "compute_20, sm_20" string represents; you have to change it to "compute_30, sm_30" by typing this new string into the textbox at the top of the Code Generation dialog box, as shown in Figure 6.8. Click "OK" and the compiler knows now to generate code that will work for Compute Capability 3.0 and above. When you do this, your compiled code will no longer work with any GPU that *only supports 2.0 and below*. There have been major changes starting with Compute Capability 3.0, so it is better to compile for at least 3.0. Compute Capability of the Nvidia GPUs is exactly like the x86 versus x64 Intel ISA, except there are quite a few more options from Compute Capability 1.0 all the way up to 6.x (for the Pascal Family) and 7.x for the upcoming Volta family.

The best option to choose when you are compiling your code is to set your Compute Capability to the lowest that will allow you to run your code at an acceptable speed. If you set it too high, like 6.0, then your code will only run on Pascal GPUs, however you will have the advantage of using some of the high-performance instructions that are only available

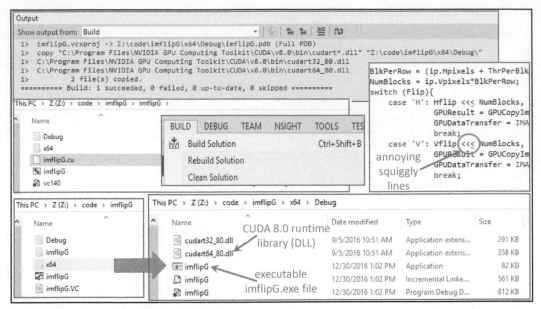

FIGURE 6.9 Compiling imflipG.cu to get the executable file imflipG.exe in the Z:\code\imflipG\x64\Debug directory.

in Pascal GPUs. Alternatively, if you use a low number, like 2.0, then your code might be exposed to the severe limitations of the early year-2000 days, when — just as a quick example — the block size limitations were so restrictive that you had to launch the kernels in a loop because each kernel launch could only have a maximum of ≈ 60,000 blocks, rather than the multibillion, starting with 3.0. This would be a huge problem even in our very first CUDA program imflipG.cu; as we analyzed in Section 6.4.15, imflipG.cu required us to launch 166,656 blocks. Using Compute Capability 2.0 would require that we somehow chop up our code into three separate kernel launches, which would make the code messy. However, using Compute Capability 3.0 and above, we no longer have to worry about this because we can launch billions of blocks with each kernel. We will study this in great detail in the next chapter. This is why the 3.0 is a good default for your projects and I will choose 3.0 as my default assumption for all of the code I am presenting in this book, unless otherwise stated explicitly. If 3.0 is continuously what you will use, it might be better to change the *Project defaults*, rather than having to change this every time you create a new CUDA program template.

Once you choose the Compute Capability, you can compile and run your code; go to *BUILD* → *Build Solution* as shown in Figure 6.9. If there are no problems, your screen will look like what I am showing in Figure 6.9 (1 succeeded and 0 failed) and your executable file will be in the Z:\code\imflipG\x64\Debug directory. If there are errors, you can click it to go to the source line of the error.

Although Visual Studio 2015 is a very nice IDE, it has a super annoying feature when it comes to developing CUDA code. As you see in Figure 6.9 (see ebook for color version), your kernel launch lines in main() — consisting of CUDA's signature ≪ and ≫ brackets — will have squiggly red lines as if they are a syntax error. It gets worse; because VS 2015 sees them as dangerous aliens trying to invade this planet, any chance it gets, it will try to separate them into double and single brackets: "≪" will become "≪<". It will drive you

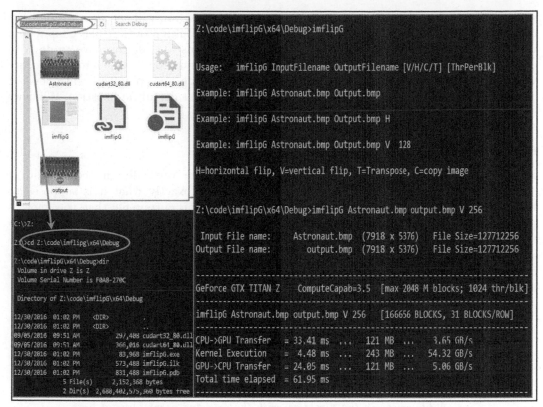

FIGURE 6.10 Running imflipG.exe from a CMD command line window.

nuts when it separates them and you connect them back together and, in a minute, they are separated again. Don't worry. You will figure out how to handle them in time and you will get over it. I no longer have any issues with it. Ironically, even after being separated, nvcc will actually compile it correctly. So, the squiggly lines are nothing more than a nuisance.

6.5.4 Running Our First CUDA Application: imflipG.exe

After a successful compilation, your executable file imflipG.exe will be in the Z:\code\ imflipG\x64\Debug directory, as shown in Figure 6.10.

The best way to run this file is to open a CMD (command line interpreter) in Windows and type the following commands to run the application:

    C:\> Z:

    Z:\> CD Z:\code\imflipG\x64\Debug

    Z:\code\imflipG\x64\Debug> imflipG Astronaut.bmp Output.bmp V 256

As seen in Figure 6.10, if you have File Explorer open, you can browse this executable code directory and when you click the location dropdown box, the directory name will be highlighted (Z:\code\imflipG\x64\Debug), allowing you to copy it using Ctrl-C. You can then type "CD" inside your CMD window and paste that directory name after "CD", which eliminates the need to remember that long directory name. The program will require the source file Astronaut.bmp that we are specifying in the command line. If you try to run it without the Astronaut.bmp in the code executable directory, you will get an error message, otherwise the program will run and place the expected resulting output file, Output.bmp in this case, in the same directory. To visually inspect this file, all you have to do is to open a browse — such as Internet Explorer or Mozilla — and drop the file into the browser window. Even simpler, you can double click on the image and Windows will open the associated application to view it. If you want to change that default application, Windows will typically give you an option to do so.

6.5.5 Ensuring Your Program's Correctness

In Figure 6.10, we see that the Output.bmp file is the vertically flipped version of Astronaut.bmp, however, there is no guarantee that it is exactly what it is supposed to be, pixel by pixel. This is why it is a good idea to have an output file that you know is perfect such as one that you wrote — and 100% confirmed — using the CPU version of the program. You can then use some file comparison tools to see if they are exactly the same. Such a file is named *golden truth* or *ground truth*. However, from my experience, your chances of having a correctly functioning program are pretty high if you go through a list of *common sense* checks:

➢ *Use these as the "minimum common-sense check" rules:*

➢ *Your program is highly likely to be functioning correctly, if*
  - *i) You didn't get an error, causing the program to terminate,*
  - *ii) The program didn't take an unusual amount of time to complete,*
  - *iii) Your computer didn't start acting weird or sluggish after the program finished running and everything appeared to be running fine,*
  - *iv) A file exists in the directory with the expected name* Output.bmp,
  - *v)   The file size of* Output.bmp *is identical to what you expect,*
  - *vi) A visual inspection of* Output.bmp *shows no signs of problems.*

If everything checks out OK in this list after the execution of the program is complete, then your program may be fine. After these checks the only remaining issues are subtle ones. These issues do not manifest themselves as errors or crashes; they may have subtle effects that are hard to tell through the checklist above, such as the image being one pixel shifted to the right and the one row on the left being blank (e.g., white). You wouldn't be able to tell this problem with the simple visual check, not even when you drag and drop the file into a browser. The one horizontal column of blank pixels would be white, much like the background color of the browser, thereby making the two difficult for you to distinguish between the browser background versus image column. However, a trained eye knows to be suspicious of everything and can spot the most subtle differences like this. In any event, a simple file checker will clear up any doubt that you have in mind for these kinds of problems.

Just as computer programmer trivia, I can't stop myself from mentioning a third kind of a problem: everything checks out fine, and the golden and output files compare fine. However, the program gradually makes computer performance degrade. So, in a sense, although your program is producing the expected output, it is *not running properly*. This is the kind of problem that will really challenge an intermediate programmer, even an experienced one. But, more than likely, an experienced programmer will not have these types of bugs in his or her code; yeah right! Examples of these bugs include ones that allocate memory and do not free it or ones that write a file with the wrong attributes, preventing another program from modifying it, assuming that the intent of the program is to produce an output that can be further modified by another program, etc. If you are a beginner, you will develop your own experience database as time goes on and will be proficient in spotting these bugs. I can make a suggestion for you though: be suspicious of everything! You should be able to detect any anomalies in performance, output speed, the difference between two different runs of the same code, and more. When it comes to computer software bugs — and, for that matter even hardware design bugs — it might be a good time to repeat Intel's former CEO and legend, late Andy Grove's words: *only the paranoid survive.*

6.6 CUDA PROGRAM DEVELOPMENT ON A MAC PLATFORM

Because a Mac OS has almost an identical structure to Unix, the instructions for Mac and general Unix will be extremely similar. Mac owners are encouraged to read everything in the Unix section too after reading this section.

6.6.1 Installing XCode on Your Mac

CUDA Toolkit, regardless of its version, requires a command line tool — such as gcc — to work. To be able to get gcc into your Mac, you have to install Xcode.

Installation instructions for Xcode are as follows:

- You must have an Apple developer account. Create one if you don't. It is free.

- In your Safari browser, go to:

    https://developer.apple.com/xcode/

- Click "Download" and download the Xcode IDE. It includes an environment for building any Mac, iPhone, iPad app, even Apple Watch or Apple TV apps. Xcode 8 is the current version at the time of the writing of this book.

- When Xcode is initially installed, a command line tool will not be a part of it.

- Go to Xcode → Preferences → Downloads → Components

- Select and install the *Command Line Tools* package. This will bring gcc into your Apple; you can now launch gcc from your terminals.

- Instead of this GUI method, you have another option to install gcc directly from your terminal with the following commands:

    xcode-select –install

    /usr/bin/cc/help

- The last line is to confirm that the command line tool chain is installed.

- The biggest difference between a Windows and a Mac — or in general, all of Unix — environments is that Windows has a strictly IDE-dictated structure for storing executable and source files, whereas Unix platforms do not create a total mess on your hard drive. One example of this mess is the database file that MS Visual Studio 2015 creates, which takes up 20 MB, etc. So, for a 20 KB imflipG.cu source file, your Mac project directory could be 100 KB including all of the executables, whereas the Windows project directory could be 20 MB!

6.6.2 Installing the CUDA Driver and CUDA Toolkit

Once you have gcc installed, you will need to install the CUDA driver and the CUDA toolkit [18], which will bring the nvcc compiler into your Mac; with this compiler, you will be able to compile the imflipG.cu file and the executable will be in the same directory. To install the CUDA driver and the CUDA toolkit, follow these instructions:

- You must have an Intel-CPU based Mac, and a supported Mac OS version (Mac OS X 10.8 or higher), as well as a CUDA-supported Nvidia GPU.

- You must have an Nvidia developer account. Create one if you don't. It is free.

- The CUDA toolkit will install a driver for you unless you have installed the standalone driver before the CUDA toolkit.

- After the CUDA toolkit installation is complete, you will have all of your CUDA stuff in /usr/local/cuda and all of your Apple Developer account in the /Developer/NVIDIA/CUDA-8.0 directory. These names might change slightly depending on the versions you are installing. There might actually be multiple CUDA directories with different versions. If you choose to, the CUDA toolkit will install all sorts of useful samples for you in the /Developer/NVIDIA/CUDA-8.0/samples directory. Once you develop a sufficient understanding of the operation/programming of the GPUs, you can look at the samples to get advanced CUDA programming ideas.

- In order to be able to run the nvcc compiler along with many other tools, set up your environment variables:

  export PATH=/Developer/NVIDIA/CUDA-8.0/bin:$PATH

  export DYLD_LIBRARY_PATH=/Dev...8.0/lib:$DYLD_LIBRARY_PATH

- If you have a Mac Book Pro that has an Nvidia, as well as an Intel-CPU-integrated GPU, the laptop will try to use the Intel GPU as much as it can to conserve energy. The Nvidia GPU is termed a *discrete GPU* — and, it indeed is a separate *card* that plugs into some slot inside your laptop, allowing you to change it in the future — and the Intel GPU is termed the *integrated GPU*, which is built-into your CPU's VLSI Integrated Circuit and cannot be upgraded unless you upgrade the CPU itself. Nvidia's Optimus technology [16] allows your laptop to switch between the integrated and discrete GPUs, however, you have to tell OS to do so by following the steps below (instructions are taken from [18], which is an online document for Getting Stared with CUDA on your Mac OS X):

  Uncheck System Preferences → Energy Saver → Automatic Graphic Switch

  Choose *Never* in the Computer Sleep bar.

6.6.3 Compiling and Running CUDA Applications on a Mac

Once you have Xcode installed, you can either let Xcode compile it for you — much like what we saw in the case of Visual Studio — or go to a terminal and type the following command line to compile it. Once this compilation succeeds, you can type the name of the executable to run it:

  nvcc -o imflipG imflipG.cu

  imflipG

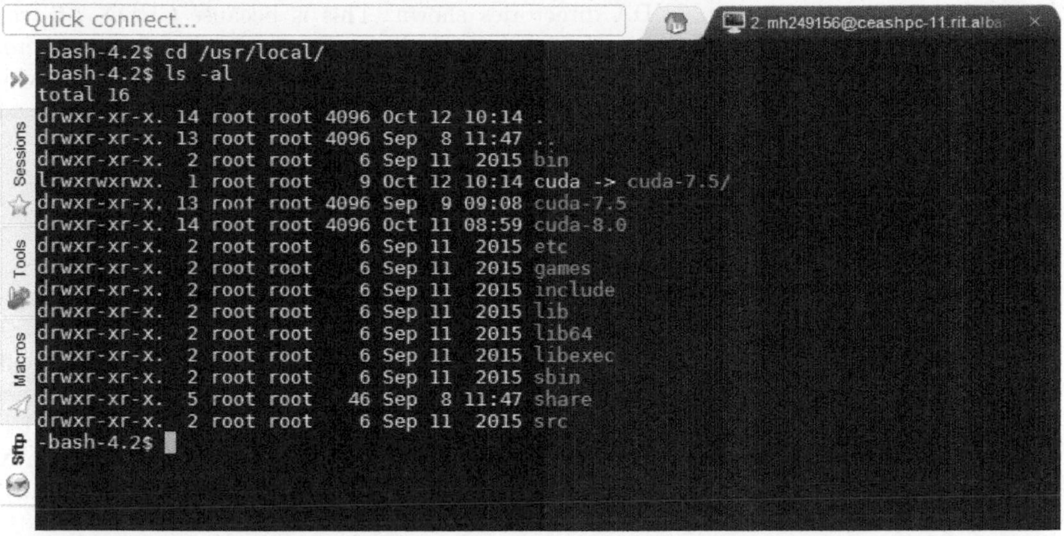

```
Quick connect...                                              2. mh249156@ceashpc-11.rit.alba    ×
-bash-4.2$ cd /usr/local/
-bash-4.2$ ls -al
total 16
drwxr-xr-x. 14 root root 4096 Oct 12 10:14 .
drwxr-xr-x. 13 root root 4096 Sep  8 11:47 ..
drwxr-xr-x.  2 root root    6 Sep 11 2015 bin
lrwxrwxrwx.  1 root root    9 Oct 12 10:14 cuda -> cuda-7.5/
drwxr-xr-x. 13 root root 4096 Sep  9 09:08 cuda-7.5
drwxr-xr-x. 14 root root 4096 Oct 11 08:59 cuda-8.0
drwxr-xr-x.  2 root root    6 Sep 11 2015 etc
drwxr-xr-x.  2 root root    6 Sep 11 2015 games
drwxr-xr-x.  2 root root    6 Sep 11 2015 include
drwxr-xr-x.  2 root root    6 Sep 11 2015 lib
drwxr-xr-x.  2 root root    6 Sep 11 2015 lib64
drwxr-xr-x.  2 root root    6 Sep 11 2015 libexec
drwxr-xr-x.  2 root root    6 Sep 11 2015 sbin
drwxr-xr-x.  5 root root   46 Sep  8 11:47 share
drwxr-xr-x.  2 root root    6 Sep 11 2015 src
-bash-4.2$ 
```

FIGURE 6.11 The /usr/local directory in Unix contains your CUDA directories.

Actually, in a Windows platform, this is precisely what Visual Studio does when you click the "Build" option. You can view and edit the command line options that VS 2015 will use when compiling your CUDA code by going to *PROJECT* → *imflipG Properties* on the menu bar. Xcode IDE is no different. Indeed, the Eclipse IDE that I will describe when showing the Unix CUDA development environment is identical. Every IDE will have an area to specify the command line arguments to the underlying nvcc compiler. In Xcode, Eclipse, or VS 2015, you can completely skip the IDE and compile your code using the command line terminal. The CMD tool of Windows also works for that.

6.7 CUDA PROGRAM DEVELOPMENT IN A UNIX PLATFORM

In this section, I will give you the guidelines for editing, compiling, and running your CUDA programs in a Unix environment. In Windows, we used the VS 2015 IDE. On a Mac, we used the Xcode IDE. In Unix, the best IDE to use is Eclipse. So, this is what we will do.

6.7.1 Installing Eclipse and CUDA Toolkit

The first step — much like Windows and Mac — is to install the Eclipse IDE, followed by the CUDA 8.0 toolkit. Before you can compile and test your CUDA code, you have to add the environment variables into your path, just like we saw in Mac. Edit your .bashrc by using an editor like gedit, vim (that's real old school), or emacs. Add this line into your .bashrc:

$ export PATH=$PATH:/usr/local/cuda-8.0/bin/

If you do not want to close and open your terminal again, just source your .bashrc by typing

$ source .bashrc

This will make sure that the new path made it into your PATH environment variable. If you browse your /usr/local directory, you should see a screen like the one in Figure 6.11.

Here, there are two different CUDA directories shown. This is because CUDA 7.5 was installed first, followed by CUDA 8.0. So, both of the directories for 7.5 and 8.0 are there. The /usr/local/cuda symbolic link points to the one that we are currently using. This is why it might be a better idea to actually put this symbolic link in your PATH variable, instead of a specific one like cuda-8.0, which I showed above.

6.7.2 ssh into a Cluster

You can either develop your GPU code on your own Unix-based computer or laptop, or, as a second option, you can login to a GPU cluster using an X terminal program. You cannot login by using a simple text-based terminal. You have to either have an xterm, or another program that allows you to do "X11 Forwarding." This forwarding means that you can run graphics applications remotely and display the results on your local machine. A good program is MobaXterm, although if you installed Cygwin to run the code in Part I, you might have also installed Cygwin-X, which will have an xterm for you to use as an X Windows-based terminal and run CUDA programs and display the graphics output locally.

To run a program remotely and display the result locally, follow these instructions:

- ssh into the remote machine with the X11 forwarding flag, "-X".

      $ ssh -X username@clustername

- Potentially, the best thing to do is to open a second terminal strictly for file transfers.

- On the second terminal, transfer the source BMP files into the cluster

      $ sftp username@clustername

- Use put and get commands, found in sftp, to transfer files back and forth. Alternatively, you can use scp for secure copy.

- Compile your code on the first terminal and make sure that the resulting output file is there.

- Transfer the output file back to your local machine.

- Alternatively, you can display the file remotely and let X11 forwarding show it on your local machine, without having to worry about transferring the file back.

6.7.3 Compiling and Executing Your CUDA Code

Instead of running nvcc on a command line, you can use the Eclipse IDE to develop and compile your code, much like the case in Visual Studio 2015. Assuming that you configured your environment correctly by following the instructions in Section 6.7.1, type this to run the Eclipse IDE:

      $ nsight &

A dialogue box opens asking you for the workspace location. Use the default or set it to your preferred location and press OK. You can create a new CUDA project by choosing *File → New → CUDA C/C++ Project*, as shown in Figure 6.12.

Build your code by clicking the hammer icon and run it. To execute a compiled program on your local machine, run it as you would any other program. However, because we are

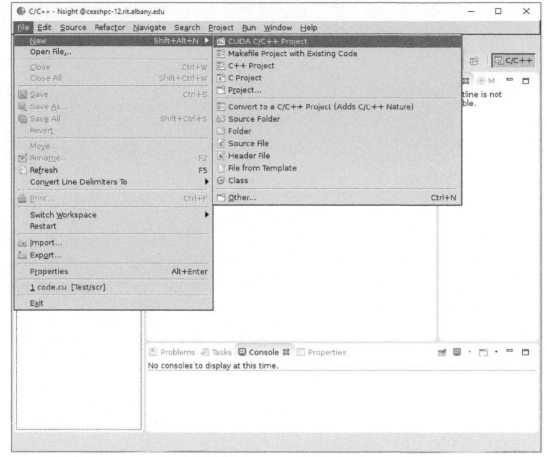

FIGURE 6.12 Creating a new CUDA project using the Eclipse IDE in Unix.

normally going to be passing files and command line arguments to the program, you will probably want to put the cd into the directory with the binary and run it from there. You could specify the command line arguments from within your IDE, but this is somewhat tedious if you are changing them frequently. The binaries generated by IDEs generally appear in some subfolder of your project (Eclipse puts them in the Debug and Release folders). As an example, to run the "Release" version of an application in Linux that we developed in Eclipse/Nsight, we may type the following commands:

      cd ~/cuda-workspace/imflipG/Release

     ./imflipG

This will run your CUDA code and will display the results exactly like in Windows.

## CHAPTER 7

CUDA Host/Device Programming Model

G PU is a co-processor, with almost no say in how the CPU does its work. So, although we will study the GPU architecture in great detail in the following chapters, we will focus on the CPU-GPU interaction in this chapter. The programming model of a GPU is a host/device model, in which the host (CPU) issues co-processor like commands to the device (GPU) and has no idea what the GPU is doing during their execution, although it can use a rich set of query commands to determine its status during the execution. To the CPU, all that matters is that the GPU finishes its execution and the results show up somewhere that it can access. GPU, on the other hand, receives commands from the CPU through a set of API functions — provided by Nvidia — and executes them. So, although CPU programming can be learned independently, GPU programming should be learned in conjunction with CPU programming, hence the reason for the organization of this book.

> ➤ *There is no such thing as GPU programming; there is only CPU+GPU*
> *programming ... You can't learn "just GPU programming."*
> ➤ *When you were learning how to ride a bicycle, what did you do?*
> *Did you learn "just how to pedal" and ignore steering?*
> *No, you either learned both or you didn't know how to ride a bike.*
> ➤ *The CPU code dictates what the GPU does.*
> *So, you need to learn their programming together; you can't just learn one.*

In this chapter, we will focus on the parts of the GPU programming that involves both the CPU and GPU, which are the launch dimensions of a GPU kernel, PCI Express bandwidth and its impact on the overall performance, and the memory bandwidth of the CPU and the GPU.

7.1 DESIGNING YOUR PROGRAM'S PARALLELISM

In this section, before we worry about what the GPU parameters — be it launch dimensions or GPU core structure — we will worry about the parallelization of the task. Remember from Section 6.4.8 that we could launch millions and millions of threads to run the GPU code. This amazing luxury means that we will do a lot less *looping* during our kernel execution, however, it introduces an interesting dilemma: how can we take advantage of this massive parallelism in the best way possible? So, the first step in GPU programming is to determine how the task should be parallelized to map to the GPU hardware perfectly. In the case of GPU parallelism, our options were fairly limited; with the potential of being able to run only a few threads (say, 8), we simply chopped up the image into 8 pieces and had each thread process a portion of $\frac{1}{8}$ of the image. However, when there are millions of threads we can launch, a slew of very different considerations come up. Let's first conceptualize our parallelism without writing a single line of GPU code.

7.1.1 Conceptually Parallelizing a Task

I am writing this section with the assumption that we are vertically flipping the 121 MB Astronaut.bmp image — shown in Figure 5.1 — using the imflipG.cu program. The first step in designing this program is to determine where the source and destination images will be stored in both the CPU and GPU side memory. We saw in Section 6.4.14 that the Vflip() kernel (Code 6.7) is the GPU kernel that actually performs this function. The main() function allocated the CPU and GPU memory areas using malloc and cudaMalloc(), respectively. The pointers that came from this allocation were as follows (shown in Code 6.1):

```
uch *TheImg, *CopyImg;          // Where images are stored in CPU
uch *GPUImg, *GPUCopyImg;       // Where images are stored in GPU
```

The pointer `TheImg` points to the original CPU image array, which is then copied onto the GPU image array, pointed to by the `GPUImg` pointer. When the GPU kernel executes, it takes the image that is stored in the array that is pointed to by `GPUImg`, flips it, and stores the flipped image in the GPU memory, pointed to by `GPUCopyImg`. This image, then, is transferred back to the CPU's second image array, pointed to by the `CopyImg` pointer.

If this code was being executed using 8 CPU threads, each thread would be responsible for copying an eighth of the image, but GPU parallelism differs despite major similarities. Here are the basic rules for GPU parallelism:

- The GPU code is written using *threads*, exactly like the CPU code, so a GPU kernel is the code for each *thread*. In that sense, a thread can be thought of as being a *building block of the task*. So, the programmer designs the GPU program based on threads.

- Because no less than 32 threads (a warp) executes at any given point in time, a *warp* can be thought of as being the *building block of code execution*. So, the GPU executes the program based on warps.

- In many cases, a warp is too small of an execution unit for the GPU. So, we launch our kernels in terms of *blocks*, which are a bunch of warps, clumped up together. In that sense, a *block* can be thought of as being a *building block of code launch*. So, the programmer launches his or her kernels in terms of blocks. The notion of the warp is more like a good trivia; a programmer conceptualizes everything in terms of threads/block. In all of our GPU kernels, we will always ask ourselves the following question: "How many threads should our blocks have?" Rarely we will worry about warps (if ever). The size of a warp for Nvidia GPUs never changed in the past two decades of Nvidia GPU designs, but we have to keep the possibility in mind that Nvidia might decide to change the warp size to something other than 32 in future GPU generations. But, the notion of *how many threads are in a block?* will never change. The programmer doesn't really think of the block size in terms of warps.

- Common block sizes are 32, 64, 128, 256, 512, or 1024 threads/block.

7.1.2 What Is a Good Block Size for Vflip()?

Based on the information in Section 7.1.1, let's try to determine what the best block size is for vertically flipping astronaut.bmp, which is a 7918×5376 image. There are many options to parallelize this, but the most important aspect of the Vflip() GPU kernel is that it is memory intensive. Although — at this point — we don't know exactly what a *memory intensive GPU program* is supposed to look like, we know that Vflip() consists of nothing more than a tornado of memory transfers within the GPU itself. That being the case, we can rely on one thing: the GPU memory likes very similar access patterns to the CPU memory, as we saw in Section 3.5.3. So, it is better to access the memory in a sequential horizontal pattern, which will map to big chunks of consecutive data being pulled out of GPU memory, thereby obeying the DRAM rules that we saw in Section 3.5.3. As described in Chapter 6, we will have multiple blocks to copy one row. If you chose 256 threads/block as our *block size* — which is the number of threads per each launched block — let's see how many block each row will require:

$$\text{Hpixels} = 7918, \quad \text{NumThreads} = 256 \quad \Longrightarrow \quad \text{BlkPerRow} = \left\lceil \frac{7918}{256} \right\rceil = 31 \qquad (7.1)$$

For coherency, variables names in Equation 7.1 are the same as Code 6.3. So, each GPU thread will copy one pixel (3 bytes) and a total of $31 \times 256 = 7936$ threads will be launched to copy one row, thereby leading to a waste of $7936 - 7918 = 18$ threads per row. Based on this computation, we will need to launch $31 \times 5376 = 166{,}656$ blocks to flip the entire image, leading to a wasted thread count of $18 \times 5376 \approx 95\,\text{K}$ threads out of the total $7918 \times 5376 \approx 41\,\text{M}$ threads launched. This is an extremely small percentage loss and can be completely ignored.

Now, let's see what happens if we use the same idea — multiple blocks per row — on a 1920×1200 image. Each row will have 8 blocks launched (each block with 256 threads) and the waste will be 128 threads/row, totaling $150\,\text{K}$ total wasted threads out of the total $2.2\,\text{M}$ launched. Although still a small percentage, you can see that it is dependent on the dimensions of the image. If you, for example, change the block size to 128 threads/block instead of 256, the waste will be less, however, you will be launching twice as many blocks to execute the same code. Which one is better? Instead of guessing, it is better to run the same code with a few different block sizes to see how the block size affects the performance.

7.1.3 imflipG.cu: Interpreting the Program Output

An example output of the imflipG.cu, when run with the 'V' option — for vertical flip — and a block size of 256, using the astronaut.bmp image, is shown in Figure 6.10. Here is the information displayed:

- The GPU was GeForce GTX Titan Z, which supports $2048\,\text{M}$ blocks (≈ 2 billion).

- We choose `ThePerBlk`=256, which means that we need 166,656 blocks to flip the image (`NumBlocks`=166,656). This is well under the $2048\,\text{M}$ blocks supported by this GPU.

- We display `BlkPerRow`=31 to confirm what we computed in Equation 7.1.

- The Compute Capability of the TITAN Z is 3.5. So, because we compiled our code with the 3.0 option (as detailed Section 6.5.3), this code will run fine, because 3.5 means that a GPU supports 3.5 and anything below.

- The program reports the completion time of three different events, as described in Section 6.4.3: (a) The CPU→GPU transfer time is $30.89\,\text{ms}$, (b) the kernel execution time of Vflip() is $4.21\,\text{ms}$, and (c) the GPU→CPU transfer time is $28.80\,\text{ms}$.

TABLE 7.1 Vflip() kernel execution times (ms) for different size images on a GTX TITAN Z GPU.

ThrPerBlk	astronaut.bmp (121 MB) 7918 × 5376			mars.bmp (241 MB) 12140 × 6940
	NumBlocks	BlkPerRow	time (ms)	time (ms)
32	1333248 ≈ 1.27 M	248	12.04	22.93
64	666624 = 651 K	124	6.58	12.63
128	333312 ≈ 326 K	62	4.24	7.90
256	166656 ≈ 163 K	31	4.48	8.15
512	86016 = 84 K	16	4.64	8.53
1024	43008 = 42 K	8	5.33	9.22

Different block sizes (**ThrPerBlk**) correspond to different total number of blocks launched (**NumBlocks**) and the total number of blocks needed to process each row (**BlkPerRow**), as tabulated. The CPU⟷GPU data transfer times are not included in this table.

7.1.4 imflipG.cu: Performance Impact of Block and Image Size

After this example run, we naturally ask ourselves the question: How would different values of `ThrPerBlk` affect the performance of the program? Furthermore, would it make a difference whether the image was large or small? To partially answer both of these questions, let's run the same program on two different images — the 121 MB astronaut.bmp (7918 × 5376) and the 241 MB mars.bmp (12,140 × 6940) — with a varying number of `ThrPerBlk` values. Table 7.1 tabulates the execution time of the Vflip() kernel for different `ThrPerBlk` values, entered from the command line argument. The CPU→GPU and GPU→CPU data transfer times are not shown in Table 7.1 because we know that the `ThePerBlk` parameter wouldn't affect them. Only the Vflip() kernel execution times are shown (case 'V' in Code 6.3).

From Table 7.1, we observe the following:

- There is an *optimum* `ThrPerBlk` value that gives us the shortest execution time, which is 128 in this specific example.

- For both images, the performance degrades significantly for `ThrPerBlk`<128.

- For both images, the performance degrades slightly for `ThrPerBlk`>128.

- This behavior is identical, regardless of the size of the image.

Although a very detailed analysis of this concave performance curve that we see in Table 7.1 will be provided later in this chapter, for now, let's think *simple* and see what the reasons could be. Here is our logic:

- This program is memory intensive; each thread copies 3 bytes and the `ThrPerBlk` parameter we choose has a direct impact on *how many bytes each block copies*.

- Knowing this, let's calculate how different `ThrPerBlk` values translate to *number of bytes copied per block*. (i) For `ThrPerBlk`=32, each one of the ≈1.27 M blocks copies 96 bytes, while (ii) for `ThrPerBlk`=64, each one of the 651 K blocks copies 192 bytes. We know that these two are the two *bad* options.

- Alternatively, (iii) for `ThrPerBlk`=128, each one of the ≈326 K blocks copies 384 bytes, making it our *optimum* option, followed by the remaining *not so terrible* `ThrPerBlk` options of 256, 512, and 1024, yielding results that are slightly worse then the optimum.

The most intuitive explanation for performance degradation for `ThrPerBlk`<128 is that above a certain threshold (more than 384 bytes copied per block in this specific case) the GPU memory's massive bandwidth can be taken advantage of, thereby yielding the best results. This is very similar to what we witnessed in Section 3.5.3, where the CPU DRAM didn't like "choppy" access; it preferred big consecutive chunks. The question is: How do we explain the slight performance degradation *above* the — `ThrPerBlk`>128 — threshold? Also, how can we be sure that bigger blocks translate to "consecutive memory access" inside the GPU memory? These questions will take a few more sections to appreciate. Let's keep studying the GPU to find answers to these questions.

7.2 KERNEL LAUNCH COMPONENTS

While Section 7.1.4 gave us a lot of food for thought, the details that contribute to the concave performance curve in Table 7.1 are way too many; we need to understand the entire mechanism in how a GPU receives and executes its kernels for a more accurate conclusion. In this section, I will only focus on the *kernel execution time*. Do not let the numbers in Section 7.1.3 trick you. It looks like the CPU→GPU and GPU→CPU data transfer times are dominating the kernel execution time, but this is just a made-up example and the mix of these different times will change significantly in the upcoming examples. So, for now, let's just focus on the kernel execution time and ignore the transfer times.

7.2.1 Grids

We observe from Table 7.1 that the total number of GPU threads we launched is always 41 M — for the astronaut.bmp image — regardless of what the size of each block is. This is because we conceptualized each thread as being responsible for one pixel (3 bytes) and the size of the image is 121 MB. So, not surprisingly, we need to launch a number of threads that is $\frac{1}{3}$ of the image size. Actually, it is a little more than that because of the useless threads that I mentioned in Section 7.1.2. So, although the 41 M pixels became a fact the second we decided to adopt the "one pixel per thread" strategy, we still have a choice in how we chop these 41 M threads up into blocks: If our blocks have 128 threads each, then we will have ≈326 K blocks. Alternatively, a much smaller block size of 32 threads will result in ≈1.27 M blocks, per Table 7.1.

1D Grids: Nvidia has a name for these army of blocks: a *grid*. A grid is a bunch of blocks, arranged in a 1D or 2D fashion. Let's choose the `ThrPerBlk`=128 as our block size option in Table 7.1, which we know yields the *optimum* kernel execution time. In this case, we are choosing to launch a grid that has 333,312 blocks. In this grid, the blocks are numbered from Block 0 to Block 333,311. Alternatively, if you launch the same grid with `ThrPerBlk`=256, you will have 166,656 blocks, numbered from Block 0 to Block 166,655.

- The grid dimension is 166,656, so block IDs will range in 0...166,655.

2D Grids: The grids do not have to be in a 1D array. They can most definitely be in a 2D array. For example, instead of launching 166,656 blocks in a 1D fashion — which will have block numbers 0 through 166,655 — you can launch them in a 2D fashion; in this case you still need to choose the size of each dimension: you have options such as 256 × 651 or 768 × 217. As an example, if you chose the 768 × 217 option, your blocks will now have a 2D block numbering (i.e., x and y block IDs) as follows:

- The x grid dimension is 768, so block IDs in the x dimension will range in 0...767,

- The y grid dimension is 217, so block IDs in the y dimension will range in 0...216.

TABLE 7.2 Variables available to a kernel upon launch.

Variable Name	Description
gridDim.x	# of blocks in each grid (x dimension)
gridDim.y	# of blocks in each grid (y dimension)
gridDim.z	# of blocks in each grid (z dimension)
blockIdx.x	Which block ID am I in the grid (in x dimension)?
blockIdx.y	Which block ID am I in the grid (in y dimension)?
blockIdx.z	Which block ID am I in the grid (in z dimension)?
Total number of launched blocks = gridDim.x × gridDim.y × gridDim.z	
blockDim.x	# of threads in each block (x dimension)
blockDim.y	# of threads in each block (y dimension)
blockDim.z	# of threads in each block (z dimension)
threadIdx.x	Which thread ID am I in the block (in x dimension)?
threadIdx.y	Which thread ID am I in the block (in y dimension)?
threadIdx.z	Which thread ID am I in the block (in z dimension)?
Total number of launched threads per block = blockDim.x × blockDim.y × blockDim.z	

3D Grids: Starting with Compute Capability 2.x, Nvidia GPUs support 3D grids of blocks. However, there is a fine print; just because a GPU supports a 3D grid of blocks does not guarantee the *total number of blocks that can be launched* with a single kernel launch. As we will see in Section 7.7.3, the GT630 GPU can support three dimensions, but its x dimension is limited to 65,535 blocks and the total number of blocks that can be launched with a single kernel launch is only ≈190 K. In other words, the product of the three dimensions cannot exceed this number, which is an architectural limitation.

One important note here is that grids are always multidimensional; so, if you launch a 1D grid, the only available dimension will be considered as the "x" dimension. One of the most beautiful features of the GPUs is that grids are automatically numbered by the hardware and passed on to the kernels. Table 7.2 shows the kernel variables that the GPU hardware passes on to the kernel. In the example I just gave, if we launch a 1D grid of 166,656 blocks, the GPU hardware will have the `gridDim.x=166656` available in every single kernel that is launched. In the case of a 2D grid launch (say, with the 768×217) dimensions, the GPU hardware will pass `gridDim.x=768` and `gridDim.y=217` to every single kernel. The kernels can take these values and use them in their calculation. Because they are hardware-generated values, the cost of obtaining them is zero. This means that whether you launch a 1D or 2D grid you do not pay any performance penalty.

7.2.2 Blocks

As I described before, blocks are the *unit element of launch*. The way a GPU programmer conceptualizes a program is that a giant tasks gets chopped up into blocks *that can execute independently*. In other words, no block should be depending on each other to execute, because this is something that will "serialize" the execution. Each block should be so independent from each other in terms of resource requirement, execution, and result reporting that you should be able to run block 10,000 and block 2 at the same time without causing any problems. As an another 2D case, block (56,125) and (743,211) should be able to run without requiring any of the others to have completed execution. It is only then you can take advantage of massive parallelism. If there is any dependency between, say, block (something, something) and block (something+1, something) or in any other way, you are hurting the very first requirement of massive parallelism:

> ➤ *For the GPU to do its job (massive parallelism), a great responsibility falls on the GPU programmer's shoulders: The GPU programmer should divide the execution into a "massively independent" set of blocks.*
> ➤ *Each block should have no resource dependence with other blocks.*
> ➤ *Massively parallel execution is only possible with massively independent blocks.*
> ➤ *Any hint of dependence among blocks will "serialize" the execution.*

Assuming that you launch your kernels using a 2D grid with dimensions 768×217, it is as if you are launching 166,656 blocks in two nested `for` loops; these 166,656 blocks would each have a unique block ID that is passed on to every single one of your kernels using the `blockIdx.x` and `blockIdx.y` variables shown in Table 7.2. So, the `for` loops would be symbolically look like this:

```
for(blockIdx.x=0; blockIdx.x<=767, blockIdx.x++){
    for(blockIdx.y=0; blockIdx.y<=216; blockIdx.y++){
        // Execute the block(blockIdx.x,blockIdx.y) here.
        // Each block will have the gridDim.x=768 and gridDim.y=217
        // parameters available during its execution.
    }
}
}
```

Alternatively, if you decided to launch the blocks in a 1D grid, like we did in Section 7.1.3, this would correspond to launching the blocks in a single `for` loop as follows:

```
for(blockIdx.x=0; blockIdx.x<=166655, blockIdx.x++){
    // Execute the block(blockIdx.x) here.
    // Each block will have the gridDim.x=166656 parameter available
    // during its execution.
}
```

7.2.3 Threads

You must have the intuition at this point that the block dimensions do not have to be strictly 1D either. They can be 2D or 3D. For example, if you are launching 256 threads in each block, you can launch them in a 2D thread array of size 16×16 or a 3D thread array of size $8 \times 8 \times 4$. In these cases, your thread IDs will range as follows:

- In a 1D thread array of size 256, `blockDim.x=256`, `blockDim.x=1`, `blockDim.x=1`,

  Thread IDs range in `threadIdx.x=0...255`, `threadIdx.y=0`, `threadIdx.z=0`.

- In a 2D thread array of size 16×16, `blockDim.x=16`, `blockDim.y=16`,

  Thread IDs range in `threadIdx.x=0...15`, `threadIdx.y=0...15`, `threadIdx.z=0`.

- In a 3D thread array of size $8 \times 8 \times 4$, `blockDim.x=8`, `blockDim.y=8`, `blockDim.z=4`,

  Thread IDs range in `threadIdx.x=0...7`, `threadIdx.y=0...7`, `threadIdx.z=0...3`.

In all three of these cases you launch 256 threads in each block. What changes — among these three cases — is that instead of the single for loop I just showed, a 2D or 3D thread array corresponds to executing the threads within two or three nested for loops, respectively, and passing the for loop variables (threadIdx.x, threadIdx.y, and threadIdx.z) into the kernel. This implies that the programmer doesn't have to worry about the for loops pertaining to the threads. This is the functionality of the GPU hardware. To put it in different terms, *you get free for loops*. However, this doesn't mean that the looping becomes free instantly. Each kernel still has to check this big list of variables to see who it is.

Continuing our 2D grid example with x and y grid dimensions 768×217, a complete block executes within the inner loop, having block IDs (blockIdx.x and blockIdx.y). Remember that the *execution of one block* means the *execution of every thread within that block*. The number of threads within each block is also in another parameter, blockDim.x, which is 256 in this specific case. Each one of these threads executes the same kernel function Vflip() (shown in Code 6.7), so it is as if you are running the Vflip() function in, yet another, for loop — with 256 iterations — as follows:

```
for(blockIdx.x=0; blockIdx.x<=767, blockIdx.x++){
  for(blockIdx.y=0; blockIdx.y<=216, blockIdx.y++){
    // Execute the block(blockIdx.x,blockIdx.y) here.
    // This block will have the gridDim.x=768 and gridDim.y=217 parameters
    // available and will pass it on to all of the 256 threads within it.
    // Executing this block means executing a total of 256 threads (8 warps).
    // Block dimensions are available in blockDim.x, blockDim.y, and blockDim.z
    for(threadIdx.x=0; threadIdx.x<=7, threadIdx.x++){
      for(threadIdx.y=0; threadIdx.y<=7, threadIdx.y++){
        for(threadIdx.z=0; threadIdx.z<=3, threadIdx.z++){
          // Execute one thread within this block with 8*8*4=256 threads
          // this thread will also inherit gridDim.x, gridDim.y
          // this thread will also inherit blockIdx.x, blockIdx.y
          // this thread will also inherit blockDim.x, blockDim.y, blockDim.z
          // all GPU kernel parameters are passed onto the function below
          Vflip(...);
        }
      }
    }
  }
}
```

This pseudo-code demonstrates the scenario where the kernel is launched as a 2D array of blocks, in which every block will have a 2D index. Furthermore, each block is composed of a 3D array of 256 threads, arranged as an $8 \times 8 \times 4$ array of 3D threads, each with a 3D thread ID. Observe that in this scenario, we get five for() loops for free because the Nvidia GPU hardware generates all five loop variables as it launches the blocks and threads.

7.2.4 Warps and Lanes

When we study CUDA assembly language (PTX) we will find that a thread is called a *lane* in the PTX language. However, unless we are going to use inline assembly (which we will briefly see within a few chapters), we do not need to worry too much about this term.

Alternatively, the term *warp* is important to a GPU programmer. This term will pop up in quite a few places. So, for now, it is important to understand that when we say *our block size is 256 threads*, we mean *our block size is 8 warps*. Warps are the *unit element of execution*, as opposed to blocks, which are *unit element of launch*. Warps are always 32 threads. This argument makes it clear that just because we launched the blocks in 256 thread clumps, it doesn't mean that they will be executed instantaneously, with all 256 threads executing and finishing in a flash. Instead, the block execution hardware inside the GPU will execute them in 8 warps, warp0, warp1, warp2, ... warp7.

Although each warp will contain this warp ID that I just showed, we only worry about this ID if we are writing low-level PTX assembly language, which makes this warp ID available to the kernel. Otherwise, we simply worry about the blocks at the high level CUDA language. The significance of a warp will become clear when we go through some simple PTX examples, however, in general, a programmer can conceptualize everything in terms of blocks and the code will run fine.

7.3 IMFLIPG.CU: UNDERSTANDING THE KERNEL DETAILS

Now that we understand the details in launching a kernel with different block dimensions, let's go back to Code 6.3, where one of the three kernels (Vflip(), Hflip(), or PixCopy()) are launched, based on the block dimensions specified by the user on the command line.

7.3.1 Launching Kernels in main() and Passing Arguments to Them

Figure 6.10 shows a specific example execution of imflipG.exe (the compiled CUDA code from imflipG.cu), in which the *threads per block* is chosen to be 256 and the source image file is astronaut.bmp, with dimensions 7918 × 5376. Code 6.3 looks something like this:

```
#define IPHB      ip.Hbytes
#define IPH       ip.Hpixels
#define IPV       ip.Vpixels
#define IMAGESIZE  (IPHB*IPV)
int main(int argc, char** argv)
{
   ...
   cudaEventRecord(time2, 0);  // Time stamp after the CPU --> GPU tfr is done
   BlkPerRow = (IPH + ThrPerBlk -1 ) / ThrPerBlk;
   NumBlocks = IPV*BlkPerRow;
   switch (Flip){
      case 'H': Hflip <<< NumBlocks, ThrPerBlk >>> (GPUCopyImg, GPUImg, IPH);
         GPUResult = GPUCopyImg;      GPUDataTransfer = 2*IMAGESIZE;
         break;
      case 'V': Vflip <<< NumBlocks, ThrPerBlk >>> (GPUCopyImg, GPUImg, IPH, IPV);
         GPUResult = GPUCopyImg;      GPUDataTransfer = 2*IMAGESIZE;
         break;
      case 'T': Hflip <<< NumBlocks, ThrPerBlk >>> (GPUCopyImg, GPUImg, IPH);
         Vflip <<< NumBlocks, ThrPerBlk >>> (GPUImg, GPUCopyImg, IPH, IPV);
         GPUResult = GPUImg;          GPUDataTransfer = 4*IMAGESIZE;
         break;
      case 'C': NumBlocks = (IMAGESIZE+ThrPerBlk-1) / ThrPerBlk;
         PixCopy <<< NumBlocks, ThrPerBlk >>> (GPUCopyImg, GPUImg, IMAGESIZE);
         GPUResult = GPUCopyImg;      GPUDataTransfer = 2*IMAGESIZE;
      break;
   }
   ...
```

The following variable values will be available or calculated in main():

- IPH=ip.Hpixels=7918   IPV=ip.Vpixels=5376   Flip='V'

- ThrPerBlk=256   BlkPerRow=31

- NumBlocks=166656 for the Vflip() and Hflip() kernels

- NumBlocks=498834 for the PixCopy() kernel

- GPUDataTransfer=either 243 MB or twice that, depending on the kernel. This value is used to compute the amount of data transferred within the GPU global memory. The result is used in reporting the global memory bandwidth with the assumption that *this is a 100% memory intensive GPU program*. For example, in Figure 6.10, the reported 54.32 GB/s is far below the maximum achievable global memory bandwidth.

- TheImg, CopyImg=pointer to the original (and copy) image in CPU memory.

- GPUImg, GPUCopyImg=pointer to the original (and copy) image in GPU memory.

- GPUResult=extra GPU memory pointer to implement the transpose operation easily.

During the execution of the Vflip() kernel, gridDim.x=166,656 and the block IDs range in blockIDx.x=0...166,655. Because the Hflip() and Vflip() kernels both require the NumBlocks to be computed in exactly the same way, NumBlocks is computed before the switch statement. Both Hflip() and Vflip() kernels copy one pixel (3 bytes) per thread, therefore the 'H' and 'V' cases of the switch will compute NumBlocks=166,656, thereby launching 166,656×256 ≈ 41 M threads total for the entire imflipG.cu code.

7.3.2   Thread Execution Steps

Observing that ThePerBlk is a 1D value for all three kernels, the thread array in each block is a 1D array, in which threadIdx.x ranges in 0...255. For all three kernels, blockDim.x=256. Additionally, because NumBlocks is also a 1D value, it is clear that this CUDA program is launching a 1D grid of blocks for all three kernels; however, the dimension of the grid changes based on the kernel, as detailed below. Observe that 6 arguments are being passed onto the GPU kernel Vflip() (5 for the other kernels) from the CPU side as follows:

- Four arguments, GPUCopyImg, GPUImg, IPH, and IPV are passed *explicitly* as function arguments to Vflip(). Explicit means that the arguments are passed onto the function in the stack, as function call arguments, as in the case of any regular C function call.

- Two arguments, NumBlocks and ThrPerBlk, are passed *implicitly* via the Nvidia Runtime Engine because they will eventually make their way into the kernel through the hardware-generated variable names blockDim.x and gridDim.x (shown in Table 7.2), respectively. Nothing prevents a programmer from passing these values explicitly also, but there are limited uses for this. After all, why not use something that is *free*?

- From the standpoint of our code, all that matters is that these parameters are *available* for the kernel to use, regardless of how they got there. However, the reason for my emphasis of these two different types of arguments is to make the readers aware of the mechanisms in which the parameters get there. One interesting immediate impact of this is that *if something is already going to be passed onto your kernel for free, you should know it and not pass it as an explicit argument*. After all, passing arguments is an additional computational overhead that will slow down your program.

Here are general guidelines for the structure of a GPU (or CPU) thread. We will look at the details of the GPU kernels based on the following guidelines:

> *Every thread of the multithreaded CPU and GPU code goes through three stages of operation.*

> *These three stages are:*

   1. **Who am I?** *This is where the kernel finds out about its own ID.*
   2. **What is my task?** *This is where the kernel determines which part of the data it is supposed to process, based on its ID.*
   3. **Do it...** *and, it does what it is supposed to do.*

7.3.3 Vflip() Kernel Details

Let's apply these guidelines to determining the distinct steps in Vflip():

   1. **Who am I?** Here are the first lines of the Vflip() kernel in Code 6.7:

```
__global__
void Vflip(uch *ImgDst, uch *ImgSrc, ui Hpixels, ui Vpixels)
{
   ui ThrPerBlk = blockDim.x;
   ui MYbid = blockIdx.x;
   ui MYtid = threadIdx.x;
   ui MYgtid = ThrPerBlk * MYbid + MYtid;
```

This is where the Vflip() kernel extracts its block ID, thread ID, and the `ThrPerBlk` value. Just for clarity, the kernel is using the same variable name `ThrPerBlk` as the main(), but this is a local variable to this kernel, so the name could have been anything else. Because our example assumes that 41 M kernels are being launched, the Vflip() function above represents only a single thread out of these 41 M kernels. Therefore, as its first task, this kernel computes which one of the 41 M kernels it is; this *global thread ID* is placed in a variable named `MYgtid`. This computation "linearizes" the thread index (or, alternatively the thread ID), departing from the block ID (in the `MYbid` variable) and the thread ID (in the `MYtid` variable).

   2. **What is my task?** I am using the term *linearized thread ID* to refer to `MYgtid`. After determining its `MYgtid`, Vflip() continues as follows:

```
   ui BlkPerRow = (Hpixels + ThrPerBlk - 1) / ThrPerBlk; // ceil
   ui RowBytes = (Hpixels * 3 + 3) & (~3);
   ui MYrow = MYbid / BlkPerRow;
   ui MYcol = MYgtid - MYrow*BlkPerRow*ThrPerBlk;
   if (MYcol >= Hpixels) return;    // col out of range
   ui MYmirrorrow = Vpixels - 1 - MYrow;
   ui MYsrcOffset = MYrow      * RowBytes;
   ui MYdstOffset = MYmirrorrow * RowBytes;
   ui MYsrcIndex = MYsrcOffset + 3 * MYcol;
   ui MYdstIndex = MYdstOffset + 3 * MYcol;
```

The concept of *linearization* refers to turning a 2D index into a 1D index here, as we saw in Equation 6.2, where we computed the linear memory address of a pixel based on its x and y pixel coordinates. The similarity here — in the case of `MYgtid` — is that the launch pattern of the 41 M threads here are indeed 2D, the blocks occupying the x dimension and the threads in each block occupying the y dimension. Therefore, *linearization* in this case allows the thread to determine a globally unique ID — MYgtid — among all 41 M threads, something that cannot be determined from the MYbid or MYtid alone.

In determining what its task is, Vflip() first determines which row (MYrow) and which column (MYcol) pixel coordinates it is supposed to copy, as well as the mirroring row that this will be copied onto (MYmirrorrow) with the same column index. After computing its column index, it *quits* if it realizes that it is one of the useless threads, as I described in Section 6.4.14. Next, it turns row and column information into source and destination GPU memory addresses (MYsrcIndex and MYdstIndex). Note that it uses the ImgSrc and ImgDst pointers that were passed onto this kernel after being allocated using cudaMalloc() inside main().

3. Do it ... After computing the source and destination memory addresses, all that is left is to copy a pixel's three consecutive bytes from the source to the destination GPU memory.

```
ImgDst[MYdstIndex] = ImgSrc[MYsrcIndex];
ImgDst[MYdstIndex + 1] = ImgSrc[MYsrcIndex + 1];
ImgDst[MYdstIndex + 2] = ImgSrc[MYsrcIndex + 2];
```

7.3.4 Comparing Vflip() and MTFlipV()

What we see so far with the Vflip() kernel is similar to the corresponding steps of the CPU version of the function, MTFlipV() in Code 2.7; the MTFlipV() function also went through these three steps, however, each step contained a totally different set of lines as follows:

1. Who am I? Determining its thread ID was not a big deal for the CPU thread, with only a single line of code — shown below — because the thread ID (tid) was not hardware-generated; rather, it came from the for loop we use in the main() as shown below:

```
for(i=0; i<NumThreads; i++){
    ThParam[i] = i;
    ThErr = pthread_create(&ThHandle[i], &ThAttr, MTFlipFunc,
        (void *)&ThParam[i]);
    if(ThErr != 0){...exit(EXIT_FAILURE);}
}
}
```

The &ThHandle[i] was substituted as tid in the thread function as follows:

```
void *MTFlipV(void* tid)
{
    long ts = *((int *) tid);        // My thread ID is stored here
```

2. What is my task? Determining the start and end indexes is also extremely easy for the CPU; it was just a better simple formula to compute the part of the image that this thread was responsible for as follows:

```
ts *= ip.Hbytes/NumThreads;        // start index
long te = ts+ip.Hbytes/NumThreads-1;  // end index
```

3. Do it ... However, the execution was much longer as follows:

```
for(col=ts; col<=te; col+=3){
    row=0;
    while(row<ip.Vpixels/2){
        pix.B = TheImage[row][col];
        ...
        TheImage[ip.Vpixels-(row+1)][col+2] = pix.R;
        row++;
    }
}
pthread_exit(NULL);
}
```

Let's compare Vflip() and MTFlipV():

- There is a significant shift of work when comparing the two. This shouldn't come as a surprise. In the CPU case, we are talking about 8–16 threads, while in the GPU case, we are talking about tens of millions of threads. When you have so many threads, the computation of the indexes gets complicated, with the added dimensions, etc.

- In the GPU, the amount of time spent in determining thread identification (step 1) and the required indexes to do the work (step 2) dominates the amount of time required to do the actual work (step 3).

- In the CPU, the first two steps are trivial and the actual work dominates the execution time. This is because when you have tens of millions of threads, the "thread index management" ends up being something that takes a lot of time.

- We see no `for` loops in the GPU code, whereas two nested `for` loops are required for the CPU code.

- So, as a summary, although managing the thread indexes ends up being a task of its own in the GPU code, we get hardware managed indexes for free, easing most of the pain.

7.3.5   Hflip() Kernel Details

Execution of Hflip() requires the same number of blocks to be launched as Vflip(); so, `gridDim.x`=166,656. Block IDs have exactly the same range, `blockIDx.x`=0...166,655. These two kernels are written in such a way that the horizontal versus vertical flip is achieved by using the same code and only reversing the row versus column order. Therefore, the Hflip() kernel, originally shown in Code 6.8, is not repeated here. By looking at Code 6.8, it is fairly easy to identify the corresponding three thread execution steps. Furthermore, the CPU version of the horizontal flipper, MTFlipH() is identical to MTFlipV() with only the row and column order reversed.

7.3.6   PixCopy() Kernel Details

Because the PixCopy() kernel copies one "byte" per thread rather than 3 bytes per thread, imflipG.cu has to launch three times as many threads when the user chooses the 'C' option (Copy) at the command line. In this specific example, for a 121 MB image, 121 M threads will be launched, corresponding to gridDim.x=498,834. Each block will have IDs in the range blockIDx.x=0...498,833. Let's look at the thread execution steps.

1. Who am I? Here is the PixCopy() kernel, originally shown in Code 6.9. The global index is still computed, just like the other two kernels.

```
__global__
void PixCopy(uch *ImgDst, uch *ImgSrc, ui FS)
{
    ui ThrPerBlk = blockDim.x;
    ui MYbid = blockIdx.x;
    ui MYtid = threadIdx.x;
    ui MYgtid = ThrPerBlk * MYbid + MYtid;
```

2. What is my task? A *range check* is done in this step, although it is a little different; this thread is useless only if the global thread ID is bigger than the size of the image. This would only happen if the size of the image is not perfectly divisible by the number of threads in a block. As an example, astronaut.bmp is $7918 \times 5376 \times 3 = 127,701,504$ bytes and if we are using 1024 threads per block, we would need 124,708.5 blocks; so, we would launch 124,709 blocks, wasting half a block (i.e., 512 threads). Although this is a minuscule amount as compared to 121 MB, it still forces us to use the line to check to see if this global thread ID is out of range.

```
    if (MYgtid > FS) return;            // outside the allocated memory
```

3. Do it ... Because each thread only copies a single pixel, the global thread ID (MYgtid) is directly correlated to the source and destination GPU memory addresses. Actually, more than that: it *is* the same addresses, making it unnecessary to compute any other index. Therefore, the line that does the actual work is only a single line.

```
    ImgDst[MYgtid] = ImgSrc[MYgtid];
}
```

All this line does is to copy a single byte from one part of GPU global memory (pointed to by *ImgSrc) to another part of the GPU global memory (pointed to by *ImgDst).

The fact that the PixCopy() kernel was so much simpler than the other two should make you scratch your heads at this point. The questions you should be asking yourself are

- Does the fact that PixCopy() was so much simpler mean that these simplifications can be applied to *all* GPU kernels?

- In other words, was the complicated nature of the Vflip() and Hflip() kernels a fact of life for GPUs or something that we have caused by choosing the design of parallelization in Section 7.1?

- Does this, then, mean that we could have designed the Vflip() and Hflip() kernels much better?

- Will the PixCopy() kernel work *faster* just because it looks simpler?

- If *simple* doesn't mean *faster*, should we — ever — care about making code look simpler or should we strictly care about the execution time?

- What would happen if we tried to copy 3 bytes in the PixCopy() kernel?

- What would happen if we tried to copy only a single byte in the Vflip() and Hflip() kernel?

Phew! These are all excellent questions to ask. And, the answers will take a few chapters to appreciate. For now, let me remind you of my golden rules in designing good code:

➢ *A good programmer writes fast code.*
➢ *A really good programmer writes even faster code.*
➢ *An excellent programmer writes super fast code.*

➢ *An exceptional programmer doesn't worry about writing fast code;*
  he or she only worries about understanding the fundamental cause(s)
  for low performance and redesigning the code to avoid them.
  It is highly likely that he or she will end up writing the fastest code in the end.

The moral of the story is that your take-away from this book should be understanding the causes of low performance by bombarding your brain with questions like the ones above. When I showed some of the examples in this book to my students, I received reactions such as *"Wow, I could have done better, this code is really slow."* My answer is: *Exactly! That's my point.* In this book, I went to extremes to find a sequence of improvements that can demonstrate a performance improvement at each step, which clearly implied starting with fairly pathetic code! So, I did that. The most important thing I want to demonstrate in this book is the *incremental performance improvement* achieved by adding/removing certain lines. Just to let you know, the Vflip(), Hflip(), and PixCopy() kernels are so **exceptionally bad** that I will write many improved versions of them in the upcoming chapters, each demonstrating a healthy performance gain due to a particular architectural reason.

Your goal in reading this book should be to thoroughly — painfully and obsessively — understand the *reasons* behind the improvements. Once you understand the reasons, you have control over them and you can be an *exceptional programmer* per my comments above. Otherwise, do not ever try to improve code without understanding the reasons of the improvement; more than likely, each one will have an architectural correlate. GPU programming differs majorly from CPU programming in that the nice architectural improvements such as *out of order execution* and *cache coherency among all of L1$'s* are missing. This pushes a lot of responsibility to the shoulders of the programmer. So, if you do not understand the underlying reasons for performance degradations, you will not be able to work the 3000–5000 cores that you have in this beast to their fullest potential!

7.4   DEPENDENCE OF PCI EXPRESS SPEED ON THE CPU

My home computer is about 4–5 years old; it is a home-grown PC with an Intel DX79SR motherboard [7]. A few years back, I got excited about the new Kepler family GPUs that supported PCI Express 3.0. Before that, all Fermi family GPUs supported only PCI Express 2.0. At that time, I had an i7-3820 CPU [8] in my DX79SR motherboard. Like any other DIY enthusiast, I got excited about the extra GPU power I would get, as well as twice the PCI Express throughput. The specifications of the DX79SR motherboard said that it supports PCI Express 3.0 on Intel's website [7]. With the old Fermi GPU, my Nvidia Control Panel was telling me that I had a PCIe 2.0 bus. My thinking was: "Of course, although the motherboard supports PCIe 3.0, the GPU doesn't support more than PCIe 2.0. So, I will get the worse of the two. This is why my Nvidia Control Panel was showing PCIe 2.0."

I plugged in the fancy Kepler GPU (GTX Titan Z, the one reported in Table 7.3 inside Box V). Guess what? My Nvidia Control Panel still reported PCI Express 2.0. Of course, like any other rational computer scientist or electrical engineer would, I panicked and rebooted my PC multiple times. Nope! Still PCIe 2.0. What could be the problem? After my "denial" period passed, I started digging through the Intel website just to find that the i7-3820 CPU [8] did *not* support PCIe 3.0. It simply does not have a PCIe 3.0 controller built into it. It only supports PCIe 2.0; in other words, I got the worst of the three! I kept reading and eventually found that only the Xeon E5-2680 or Xeon E5-2690 [13] support PCIe 3.0 on that specific motherboard. Although it is very expensive normally, I was able to find a used Xeon E5-2690 at a quarter of the price and plugged it in. Xeon E5-2690 is an 8C/16T CPU and I have been using it ever since.

The moral of the story here is that all *three* components have to support a specific PCIe speed: (1) motherboard, (2) CPU, and (3) the GPU. If one of them supports anything less, you get the worst of the three. Another example is my Dell laptop (shown as Box II in Table 7.3). Although this laptop should support PCIe 3.0, my Nvidia control panel reports PCIe 2.0. What could it be? (1) The motherboard should be PCIe 3.0 because Dell's website says so, (2) the CPU is i7-3740QM, which should support PCIe 3.0, and (3) the GPU, Nvidia Quadro K3000M, is a Kepler with PCIe 3.0 support. What else could it be? There is, yet, another possibility. The PCIe 3.0 support might be disabled in the BIOS.

Well, there will be another round of Googling on my end ...

7.5 PERFORMANCE IMPACT OF PCI EXPRESS BUS

In this section, I will analyze the I/O performance of the imflipG.cu code. Before I do this, let me introduce some important terms. Mixing up these terms is not technically correct, so I would like to provide a clear technical background on them.

7.5.1 Data Transfer Time, Speed, Latency, Throughput, and Bandwidth

When explaining data movement from one point to another (e.g., CPU→GPU), I will be using four different — but highly related — metrics: *transfer time*, *transfer speed*, *transfer latency*, and *transfer throughput*. I will describe each one individually below:

Transfer time is the amount of time that it takes data to be transferred from point A to point B. It doesn't make any reference to the "amount" of data. As an example, let's assume that our transfer time is 0.4 μs (0.4 microseconds = 0.4×10^{-6} seconds).

Transfer speed is the amount of time it took *per unit amount of data*. Assume that we transferred 1 KB of data (1024 Bytes) in 0.4 μs. Then, our computed transfer speed is $\left(\frac{1024 \text{ Bytes}}{0.4 \times 10^{-6} \text{ seconds}} \right) = 2.56 \times 10^9$ Bytes per second. Or, we can divide this by 1024^3 to get GBps; we can say that we transferred a 1 KB data chunk at the speed of 2.38 GBps.

Transfer latency refers to the arrival time of the *first* packet, if the transfer involves a continuous batch of packets, which is the case for all network communications, as well as PCI Express operation. This concept of sending packets one after the other is called *pipelining*.

Transfer throughput is the *average* transfer speed of many data packets over a long term. Assume that we are transferring the 121 MB Astronaut.bmp image over PCI Express in a pipelined fashion, i.e., as a continuous stream of packets. Each data packet will definitely be smaller in size than 121 MB; assume that each packet is 1 KB, as we calculated a little while ago. The first packet will arrive in 0.4 μs, but the other might be a little slower (say, 0.41 μs), and the next one might be a little faster (say, 0.33 μs), depending on the random things in the computer that you cannot control. In the end, we do not care too much about the little differences between consecutive packets. All we care about is the "average" transfer speed over a long term, say, hundreds of packets. Measuring a single packet's transfer speed as we computed above (2.38 GBps) might cause a huge error over the long term. However, if we know the total amount of time for 121 MB (say, 39.43 ms), we can calculate the transfer throughput as $\left(\frac{121 \times 1024^2}{39.43 \times 10^{-3}} \frac{\text{Bytes}}{\text{seconds}} \right) \approx 3$ GBps. We observe that we would be making a large error if we used only a single packet to measure the speed (2.56 GBps). However, this final 3 GBps is a much more accurate throughput number because all of the plus and minus errors average out to lose their influence over a longer interval.

Bandwidth of a data transfer medium (e.g., memory bus or PCI Express bus) is the maximum throughput it supports. For example, the bandwidth of a PCI Express 2.0 bus is 8 GBps. This means that we cannot expect a throughput that is higher than 8 GBps when transferring data over a PCIe 2.0 bus. However, generally the throughput we achieve will be less because there are a lot of OS-dependent reasons that prevent reaching this peak.

Upstream bandwidth is the expected bandwidth in the CPU→GPU direction, whereas **Downstream** bandwidth refers to the GPU→CPU direction. One nice feature of PCIe is that it supports simultaneous data transfers in both directions.

7.5.2 PCIe Throughput Achieved with imflipG.cu

There are so many questions about what contributes to the kernel performance that it is best, at this point, to go ahead and run the imflipG.cu code using two different functional options, 'V' and 'C', as well as multiple threads/block options. This will allow us to gauge the performance of the two kernels, Vflip() and PixCopy(). The other two kernels do not carry too much additional information because Hflip() is way too similar to Vflip() and the 'T' options simply run one after the other. Table 7.3 tabulates the results on six computer configurations (noted as Box I ... Box VI), each with a different CPU and a GPU.

First, let's strictly look at the CPU→GPU and GPU→CPU data transfer times in Table 7.3. Because there are a lot of details contributing to the PCI Express performance (written *PCIe* in short), we will try to make high-level observations instead of detailed ones. Here is a summary of the observations from Table 7.3:

- In every computer that has a PCIe Gen2 bus (short for Generation 2, including PCIe 2.0 and PCIe 2.1), the CPU→GPU data transfer throughput is ≈2.5–3 GBps (Giga Bytes per second) in almost every case (Box I, Box II, and Box III).

- In the first two computers that have a PCIe Gen3 bus (Box IV and Box V), the CPU→GPU throughput is ≈5 GBps, while Box VI (the Dell Cluster server with dual Xeon E5-2680v4 [12] CPUs) has a much higher ≈7 GBps throughput. This is because of the advanced Xeon inside the Dell server (the only server board in Table 7.3), allowing a better throughput due to its server-class (i.e., improved throughput) architecture.

- In the reverse direction (GPU→CPU), Boxes I–V have an identical throughput to their CPU→GPU throughput. Alternatively, Box VI shows a very different characteristic, with almost half the throughput achieved. So, we should not assume that the upstream and downstream throughputs will be identical.

- Note that when a data transfer is happening over the PCIe bus, whether in the CPU→GPU or GPU→CPU direction, the CPU is continuously involved in this transfer, through virtual memory paging. We will see in the following sections how we can prevent this to speed up the PCIe bus transfers.

TABLE 7.3 Specifications of different computers used in testing the imflipG.cu program, along with the execution results, compiled using *Compute Capability 3.0*.

Feature	Box I	Box II	Box III	Box IV	Box V	Box VI
CPU	i7-920	i7-3740QM	W3690	i7-4770K	i7-5930K	2xE5-2680v4
C/T	4C/8T	4C/8T	6C/12T	4C/8T	6C/12T	14C/28T
Memory	16GB	32GB	24GB	32GB	64GB	256GB
BW GBps	25.6	25.6	32	25.6	68	76.8
GPU	GT640	K3000M	GTX 760	GTX 1070	Titan Z	Tesla K80
Engine	GK107	GK104	GK104	GP104-200	2xGK110	2xGK210
Cores	384	576	1152	1920	2x2880	2x2496
Compute Cap	3.0	3.0	3.0	6.1	3.5	3.7
Global Mem	2GB	2GB	2GB	8GB	2x12GB	2x12GB
Peak GFLOPS	691	753	2258	5783	8122	8736
DGFLOPS	29	31	94	181	2707	2912
Data transfer speeds & throughput over the PCI Express bus						
CPU→GPU ms	39.43	52.37	34.06	23.07	33.41	17.36
GBps	3.09	2.33	3.58	5.28	3.65	7.01
GPU→CPU ms	40.46	52.68	35.45	25.03	24.05	42.72
GBps	3.01	2.31	3.44	4.87	5.06	2.85
PCIe Bus	Gen2	Gen2	Gen2	Gen3	Gen3	Gen3
BW GBps	8.00	8.00	8.00	15.75	15.75	15.75
Achieved (%)	(39%)	(29%)	(45%)	(31–34%)	(23–32%)	(18–45%)
Vflip() kernel run time (ms) 'V' command line option						
V 32	71.88	63.65	20.62	4.42	12.0	16.59
V 64	38.27	33.42	10.97	2.19	6.58	8.85
V 128	23.00	20.02	6.68	2.19	4.24	5.49
V 256	23.71	20.58	7.06	2.23	4.48	5.63
V 512	26.75	22.73	7.48	2.36	4.64	6.00
V 768	39.34	35.21	11.19	2.98	6.12	7.74
V 1024	28.34	23.88	8.09	2.48	5.33	6.51
GM BW GBps	28.5	89	192	256	336	240
Achieved GBps	10.59	11.84	36.47	111.35	57.39	44.39
(%)	(37%)	(13%)	(19%)	(43%)	(17%)	(18%)
PixCopy() kernel run time (ms) 'C' command line option						
C 32	102.69	86.54	27.65	7.42	14.45	19.90
C 64	52.22	43.66	13.89	3.67	7.71	10.24
C 128	27.37	22.78	7.35	2.38	4.65	5.64
C 256	27.81	22.81	7.36	2.33	4.33	5.58
C 512	28.59	23.71	7.87	2.37	4.30	5.77
C 1024	30.48	25.50	8.18	2.42	4.56	6.25
GM BW GBps	28.5	89	192	256	336	240
Achieved GBps	8.90	10.69	33.16	104.33	56.59	43.66
(%)	(31%)	(12%)	(17%)	(41%)	(17%)	(18%)

The asronaut.bmp image was used with 'V' and 'C' options and different block sizes (32..1024). CPU→GPU and GPU→CPU are reported only for a block size that achieved the best transfer speed.

TABLE 7.4 Introduction date and peak bandwidth of different bus types.

Bus Type	Peak Bandwidth	Introduction Date	Common Uses
Industry Standard Architecture (ISA)	< 20 MBps	1981	8–16 b Peripherals
VESA Local Bus (VLB)	< 150 MBps	1992	32 b High-End Peripherals
Peripheral Component Interconnect (PCI)	266 MBps	1992	Peripherals, Slow GPUs
Accelerated Graphics Port (AGP)	2133 MBps	1996	GPUs
PCIe Gen1 x1	250 MBps	2003	Peripherals, Slow GPUs
PCIe Gen1 x16	4 GBps		GPUs
PCIe Gen2 x1	500 MBps	2007	Peripherals, Slow GPUs
PCIe Gen2 x16	8 GBps		GPUs
PCIe Gen3 x1	985 MBps	2010	Peripherals
PCIe Gen3 x16	15.75 GBps		GPUs
Nvidia NVlink Bus	80 GBps	April 2016	Nvidia GPU Supercomputers
PCIe Gen4 x1	1.969 GBps	final specs expected in 2017	Peripherals
PCIe Gen4 x16	31.51 GBps		GPUs

AGP is totally obsolete today, while legacy PCI is still provided on some motherboards. Nvidia introduced the NVlink bus in mid-2016 for use in GPU-based supercomputers, with almost 5× higher bandwidth than the then-available PCIe Gen3.

In general, PCIe bus is a huge bottleneck for GPUs, placing a significant limitation on the CPU$\longleftrightarrow$GPU data throughput. To alleviate this problem, PCIe standards have continuously been improved in the past two decades, starting with the PCIe 1.0 standard in 2003 that intended to replace the then-standard Accelerated Graphics Port (AGP). The AGP standard used a 32-bit bus, which transferred the data in 32-bit chunks at a time and achieved a maximum throughput of $\approx$2 GBps. The PCIe standard reduced the bus size to a single bit, which could work at 250 MBps (1-bit instead of 32-bits, found in AGP). Although this intuitively sounds like a degradation of the standard, it is not true; synchronizing 32 bits is highly susceptible to *phase delays among 32 bits*, degrading the performance of this *parallel* transfer of the 32-bits. Alternatively, turning those 32 parallel bits into individual 32 single-bit data entities allows us to send all 32 of them in separate *PCIe lanes* without worrying about the phase delays during the transfer and synchronizing them at the receiver end. As a result, PCI Express can achieve a much better data throughput.

A list of different bus types is provided in Table 7.4, showing their introduction dates chronologically. As shown in Table 7.4, PCIe concept has another huge advantage: the PCIe x16 is downward compatible with x8, x4, and x1. Therefore, you can use different peripherals — such as network cards or sound cards — on the same PCIe bus, without requiring different standards for different cards. This allowed PCIe to take over all of the previous standards such as AGP, PCI, ISA, and possibly more that I don't even remember. This serial-transfer structure of PCIe 1.0 allowed it to deliver a peak 4 GBps throughput on 16 lanes, beating the AGP. Sixteen PCIe lanes are denoted as "PCIe x16," which is a typical number of lanes used for GPUs. Slower cards, such as Gigabit NICs, use PCIe 1x or 4x. Further revisions of PCIe kept increasing the transfer throughput; PCIe 2.0 was specified at 8 GBps and PCIe 3.0 was specified at 15.75 GBps. What is coming in the future is the PCIe 4.0 standard that will allow a 2× higher throughput than PCIe 3.0. Furthermore, Nvidia has just introduced its own NVlink bus that is designed for high-end server boards to eliminate the bottlenecks due to the PCIe standard. Starting with the Pascal family, Nvidia is offering both the PCI 3.0 as an option, as well as NVlink on high-end servers, housing Pascal GPUs.

The family names I am mentioning (Fermi, Kepler, Pascal) are different generations of GPUs designed by Nvidia and I will discuss them in more detail in Section 7.7.1 and Section 8.3. Within each family, there are different GPU engine designs; for example, **GK** family denotes the Kepler family engines, **GP** is for Pascal engines, and **GF** is for Fermi engines. In Table 7.3, Box I contains a Fermi architecture GPU, working on the PCIe 2.0 x16 bus, while Boxes II, III, V, and VI are Kepler engine GPUs, the former two working on a PCIe 2.0 bus and the latter two working on a PCIe 3.0 bus. Box IV is the only Pascal engine GPU, which works on a PCIe 3.0 bus. If we look at two different Kepler GPUs, within Box V and Box VI, they have the GK110 and GK210 engines. Therefore, although they belong to the same family, they could have significant performance differences, as we will thoroughly study throughout the book.

For now, a clean observation can be made from Table 7.3 that the CPU→GPU and GPU→CPU data transfer times are directly correlated with PCIe bus speeds. However, the fact that there is an asymmetry between the two different directions for Box I and Box VI is curious. Keep reading the book. The answers will eventually pop out.

7.6 PERFORMANCE IMPACT OF GLOBAL MEMORY BUS

Now that we understand the impact of the PCIe bus throughput (or more generally the *I/O bus throughput*) on the data transfer times in both the CPU→GPU and GPU→CPU directions, let's look at the impact of the memory bus throughputs. Figure 7.1 depicts a symbolic representation of the computer in Figure 4.1, which has an i7-5930K CPU [10] (with an internal CPU architecture is shown in Figure 4.4). The same computer is tabulated as Box V in Table 7.3, which has a GTX Titan Z GPU. As shown in Figure 7.1, the only way to transfer data from the CPU main memory (64 GB DDR4) to the GPU global memory (12 GB GDDR5) — or vice versa — is through the PCI Express bus, facilitated by the Nvidia API library. The cudaMemcpy() API function call we saw in Code 6.3 is one such API function. Figure 7.1 shows three different bus speeds:

- The PCIe Gen3 bus has a theoretical peak throughput of ≈16 GBps. Let's look at Table 7.3 again. For the upstream and downstream PCIe transfers, we achieved a throughput of ≈5 GBps for Box V, which is far below the theoretical peak of the PCIe Gen3 bus (noted in Table 7.3 as 23–32% of the theoretical peak).

- The memory bus that is connecting the CPU to its own DDR4-based DRAM memory has a theoretical peak throughput of 68 GBps. We can only test this with a program that measures the data transfers within the CPU cores and CPU DRAM memory, such as memcpy(). This was done in Part I and is not the focus of this chapter.

- The GPU also has its own internal memory, which is called *global memory* and this memory is connected to the cores with a bus that has a peak bandwidth of 336 GBps. Vertical flipping took 4.24 ms — in its best case — to transfer 121 MB of data from the global memory to the cores and from the cores back to a different area of global memory; so, the total data transfer amount was 2×121 MB. This corresponds to a transfer throughput of $2 \times 121/1024/0.00424s \approx 57.39$ GBps, which is significantly lower than the 336 GBps peak (noted as 17% of peak in Table 7.3), suggesting that there is major room for improvement.

One note to make here is that Box V in Table 7.3 shows a GTX Titan Z GPU, which is really 2 GPUs in one GPU; it has a total of 2880 cores in each GPU, with a total of 5760 for the GTX Titan Z GPU card. The K80 GPU inside the Dell Server (Box VI) is designed exactly the same way; 4992 total GPU cores, separated into two GPUs as 2×2496 cores. The GTX Titan Z *GPU card* in Box V is connected to a single PCIe Gen3 slot; there are two GPUs receiving and sending data through this single connection to the PCIe bus. To generate the results in Table 7.3, I choose GPU ID = 0, thereby telling Nvidia that I would like to use the first one of the two GPUs on that card. So, Table 7.3 can be interpreted as if the results were obtained on a single GPU. One other important note from Figure 7.1 is the drastic difference in the sizes of the cache memory of the CPU versus GPU. Indeed, GPU doesn't even have an L3$. It only has an L2$, which is a tenth of the size of the CPU, yet it feeds 2880 cores, rather than the six cores of the CPU. These quantities shouldn't be surprising. The L2$ of the GPU is significantly faster than the L3$ of the CPU and is designed to feed the GPU cores at a speed that is much higher than the CPU's bus speed. Therefore, the VLSI technology can only allow Nvidia to design an architecture with a 1.5 MB of L2$.

The L3$ of the CPU and the L2$ of the GPU share the same functionality of *Last Level Cache (LLC)*. Typically, the LLC is the only cache that directly interfaces the actual memory of the device and is responsible for being the first line of defense against data starvation. The LLC is designed for size — not for speed — because the more LLC you have, the less likely you are to starve for data. Lower level cache memories are generally built right into the cores. For example, in the case of the CPU, each core has a design with

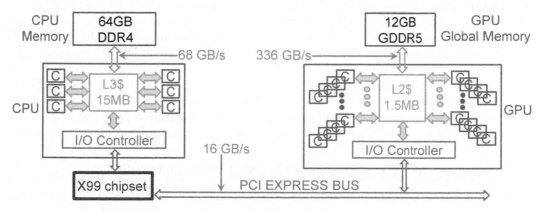

FIGURE 7.1 The PCIe bus connects for the host (CPU) and the device(s) (GPUs). The host and each device have their own I/O controllers to allow transfers through the PCIe bus, while both the host and the device have their own memory, with a dedicated bus to it; in the GPU this memory is called *global memory*.

TABLE 7.5 Introduction date and peak throughput of different CPU and GPU memory types.

Memory Type	Peak Throughput	Introduction Date	Common Uses
Synchronous DRAM (SDRAM)	<2000 MBps	1993	CPU Main Memory, Peripheral Card Memory, Peripheral Device Memory
Double Data Rate (DDR) SDRAM	3200 MBps	2000	
DDR2 SDRAM	8533 MBps	2003	
DDR3 SDRAM	17066 MBps	2007	
DDR4 SDRAM	19200* MBps	2014	
GDDR3	10–30 GBps	2004	GPU Main Memory
GDDR5	40–350 GBps	2008	
GDDR5X	300–500 GBps	2016	
High Bandwidth Memory (HBM, HBM2)	500–2000* GBps	2016	

DDRx family is used commonly in peripherals, as well as CPU main memory. Both the DDRx memory and the GPU GDDRx family designs have advanced continuously over the past two decades, delivering increasing peak throughputs.

32+32 KB L1\$ and a 256 KB L2\$. In the case of the GPU, we will see that a 64 KB or 96 KB L1\$ is shared by quite a few cores, while the LLC is the L2\$ you see in Figure 7.1 and an L3\$ does not exist in any GPU in Table 7.3. As a summary, in both of the LLC architectures, all that the architects care about is that when a core needs data it can find that data without waiting for an extended period of time, which will hurt performance. We will get deep into the details of the GPU internal architecture in Chapter 8.

A list of different CPU and GPU memory types is provided in Table 7.5, with their introduction dates chronologically. We see that further generations of CPU memory designs have offered increasing bandwidths, albeit at the expense of increased access latency, as I initially pointed out in Section 4.3.4. On an alternate progression path, GDDR family GPU memory designs have taken advantage of the advances in regular DRAM standards; for example, GDDR5 design borrowed heavily from the DDR3 standard. Today, the advanced GDDR5X standard is used in high-end Pascal GPUs, such as GTX1080, while the HBM2 standard is used in high-end GPU accelerators, such as the P100. Note that because the DDR4 and HBM2 standards are still evolving, I put down tentative peak rates (indicated with a *) in Table 7.5, which are not confirmed — but are reasonably accurate — numbers.

7.7 PERFORMANCE IMPACT OF COMPUTE CAPABILITY

When I compared the runtime results of imflipG.cu on different GPUs in Table 7.3, I used the executable version that was compiled with the "compute_30, sm_30" option in MS Visual Studio 2015, which generates the GPU executable using Compute Capability 3.0, as I described in Section 6.5.3. In Mac or any other Unix computer, this Compute Capability (compute_30, sm_30) is a command line parameter for nvcc. I will use "CC" to denote "Compute Capability" going forward to shorten the text.

When you compile your code with CC 3.0, for example, you are guaranteeing that the executable application (imflipG.exe in Windows and typically imflipG in Mac and Unix) can only run with GPUs that support CC 3.0 or higher. A built-in feature of imflipG.cu queries the GPU and outputs the supported highest CC. Looking at Table 7.3, we observe that every GPU in this table supports CC 3.0 or higher (specifically, 3.0, 3.5, 3.7, and 6.1).

7.7.1   Fermi, Kepler, Maxwell, Pascal, and Volta Families

Using CC 3.0 allows the code to take advantage of the CC 3.0, however it will not work on GPUs that support CC 2.0 or 2.1. Fermi family Nvidia GPUs (engine names starting with **GF**) support CC 2.x, while the Kepler family GPUs (engine names starting with **GK**) support CC 3.x and higher. Maxwell family (engine names starting with **GM**) supports CC 5.x. Pascal family (engine names starting with **GP**) supports CC 6.x. The upcoming Volta family (I assume that engine names starting with **GV**) will support CC 7.x. I have no clue what happened to CC 4.x. I assume that the engine design got abandoned by Nvidia, however, this is just my humble opinion.

In Table 7.3, five out of the six GPUs belong to the Kepler family (with GK engine names). The only Pascal family GPU (with a GP engine name) supports CC 6.1; unfortunately, our compiled executable will not be able to take advantage of the additional instructions in the Pascal GPU because we chose a command line that restricts the code execution to CC 3.0. Any higher CC is backwards compatible with CC 3.0, so the Pascal GPU will happily execute this application, however, had I chosen "compute_61, sm_61" to compile this code, the executable wouldn't even run on the other five GPUs. However, chances are that it would execute faster in the Pascal GPU. So, when you choose a specific CC, you are instructing the compiler to use only the instructions within that CC to compile the GPU code. In a sense, if you choose CC 3.0, you are choosing something like *Kepler family or later generation families*, which would mean that the code would run on Kepler, Maxwell, Pascal, and Volta, which support CC 3.x, 5.x, 6.x, and 7.x, respectively.

In many cases, if you know that your code will only execute on Pascal GPUs (or higher), it would actually be a good idea to choose CC 6.0 to compile it (even 6.1 if you want to get that aggressive), to take advantage of the additional Pascal-and-beyond instructions. One important point to understand is that when Nvidia designs a new engine family, they design it to perform better in the new CC and *anything below that CC*. In other words, even if you choose CC 3.0 to compile a code and run it on a Pascal GPU, it will potentially perform better than its older friends. Because the two improvements Nvidia makes in every generation are: (1) the introduction of a *new set of instructions* that perform operations that couldn't have been performed in the previous CC, and (2) performance improvements for all of the instructions that were introduced in the previous CC generations.

Just to provide a few examples, CC 3.x started supporting what is known as *Unified Memory*, something that didn't exist in CC 2.x. This was a major enhancement going from the Fermi family to the Kepler family. Furthermore, CC 5.3 and above started supporting half precision floating point numbers, something I will elaborate on in Section 9.3.10.

7.7.2   Relative Bandwidth Achieved in Different Families

Looking at Table 7.3, we see a good proof of what I just described. Box IV is the only Pascal GPU in the table and shows the following characteristics:

- The PCI Express performance of Box IV does not show any noticeable difference from the other boxes. This is because PCIe performance depends mostly on the CPU

I/O sub-system and the i7-4770K [9] CPU — in box IV — does not offer anything more than the other workstation-grade (i.e., non-Xeon) CPUs. Alternatively, the Dell Server (Box VI) performs somehow better in certain cases, thanks to the improved I/O throughput of a Xeon-based system. This Xeon-driven improved I/O speed is also noticeable in Box III, which includes a Xeon W3690 CPU [15].

- Box IV reaches a much better *relative global memory throughput* ($\approx$ 40% of the bandwidth), as compared to the execution on the Kepler family GPUs, which reach only $\approx$13–19% of their bandwidth. This is due to the architectural improvements inside the Pascal GPU that allows legacy CC 3.0 instructions (such as byte access) to be executed with much better efficiency, whereas the older generations did not perform well when the data access size was not the natural 32-bits.

- Of course, there always has to be a case that requires additional explanation. Box I was able to reach 31–37% of its bandwidth. Why? This doesn't disprove my discussion about the GK versus GP engines. If you look carefully at the bandwidth of box I's GPU (GT 640), it is a mere 28.5 GBps, which is nearly a tenth of the Pascal GPU's 256 GBps. So, when it comes to the discussion about achieving a percentage of the global memory bandwidth, we should be fair to the GPUs that have a much higher bandwidth. So, for now, it does not make sense to focus on the percentage for Box I.

- Speaking in absolute numerical terms, running the same code using the same CC 3.0, low-end Kepler engines (in Boxes I and II) achieved a global memory throughput of $\approx$10 GBps, higher-end Kepler engines (in Boxes III, V, and VI) achieved $\approx$30–60 GBps, while the newest generation Pascal GPU (Box IV) achieved $\approx$110 GBps.

To summarize:

➤ *Choose the lowest Compute Capability (CC) when compiling your GPU code, which will run at a satisfactory performance.*
➤ *Nvidia GPU families are: Fermi, Kepler, Maxwell, Pascal, and Volta.*
➤ *They support CC 2.x, 3.x, 5.x, 6.x, and 7.x, respectively.*
➤ *For example, if you choose CC 3.0, you are restricting your executable to: "Kepler or higher" engines. Similarly, 6.0 means "Pascal or higher."*

➤ *If you choose, say, 3.0 on a Pascal GPU, you won't be able to take advantage of the additional instructions introduced between CC 4.x and 6.x. However, the code will most probably take advantage of the architectural improvements, built-into the Pascal family.*

7.7.3 imflipG2.cu: Compute Capability 2.0 Version of imflipG.cu

The fact that the Pascal GPU achieved a better performance cannot really be generalized to the entire list of applications we will work with. This happened because my artificially fabricated imflipG.cu program happened to include a lot of byte access instructions, something that Pascal family improved substantially in their architecture design. We will see other applications where the advantage of Pascal will not be so obvious.

After seeing the runtime results of imflipG.cu, one question that comes to mind is *"what happens if we compare Kepler and Fermi engines?"* To answer this, in this section I will run a scaled-down version of imflipG.cu on multiple Fermi and a few Kepler engines, after compiling the code with CC 2.0. I will call this program imflipG2.cu, which is designed to display the maximum number of blocks supported when executed. For example, the GT630 GPU with a GF108 Fermi engine supports CC 2.1 and 65,535 blocks in the x dimension, but does not support a total of more than ≈190 K blocks for the entire kernel launch. Running our imflipG.cu on a GT630 would crash and quit, because, for example, from Table 7.1, we see that we need more than 300 K blocks launched for certain options.

The simplest workaround for this is to add another loop around the kernel launch. In other words, we can restrict the number of blocks (NumBlocks) to something like 32,768 and launch multiple kernels to execute the same exact code. For example, if we need to launch 166,656 blocks (example in Section 7.3), we launch 6 different kernels, first 5 with 32,768 blocks ($5 \times 32,768 = 163,840$) and the last one with 2816 blocks ($163,840 + 2816 = 166,656$). So, in addition to the block IDs that Nvidia will assign each block at runtime, we will also have to use another ID, let's say, loop ID.

An alternative to this would be to use a "real" dimension that Nvidia supports, such as the y dimension of the grid (controlled by the gridDim.y and blockIdx.y variables) or the z dimension (gridDim.z and blockIdx.z variables), as I explained in Section 7.2.1. We see from Table 7.2 that a 3D grid of blocks is allowed in CC 2.0 and above. The only problem is that the product of these three grid dimensions, i.e., *the total number of blocks that can be launched with each kernel* is limited to ≈190 K. This upper limit might differ among different cards. So, we are, in a sense, *emulating* the *fourth* dimension of the grid with the loop ID that surrounds the kernel launch, as we will see shortly in Section 7.7.4. The GPU is queried in imflipG.cu to determine this upper limit as follows:

```
cudaGetDeviceProperties(&GPUprop, 0);
SupportedKBlocks = (ui) GPUprop.maxGridSize[0] * (ui) GPUprop.maxGridSize[1] *
    (ui )GPUprop.maxGridSize[2]/1024;
SupportedMBlocks = SupportedKBlocks / 1024;
```

GPUprop.maxGridSize[1], GPUprop.maxGridSize[2], and GPUprop.maxGridSize[3] are the x, y, and z dimension limits.

> ➤ *When I started teaching my GPU classes in 2011, I used the GTX480 cards. GTX480 cards have a Fermi GF100 engine.*
> ➤ *Fermi block dimension limit is $2^{16} - 1 = 65,535$, which is a very ugly number. 65,536 (2^{16}) would be great, but 65,535 was a disaster! 65,535 is not a power of 2, and doesn't work well with anything! So, my students continuously resorted to using 32,768 blocks. Programmers of Kepler should appreciate not having that limitation.*
> ➤ *Kepler block dimension limit is $2^{31} - 1 \approx 2048$ M. You will see this reported in Figure 6.10. With Kepler and above, you never have to emulate that extra dimension.*

The newly designed program — which uses a loop to emulate the additional dimension — is called imflipG2.cu. Because it is only for experimental purposes, I designed it to only work with the 'V' and 'C' command line arguments; it does not support the 'T' or 'H' command line options. As of 2017, the year of the publication of this book, every Nvidia GPU in the market is Kepler or above, so there is no point in talking about CC 2.0 any further than this section. For the rest of the book, I will be focusing on CC 3.0 and higher. But, let's run the code using CC 2.0 to satisfy our curiosity.

CODE 7.1: imflipG2.cu main() {...

The portion of the main() that requires changes to support smaller max. blocks in Computer Capability 2.0. VfCC20() and PxCC20() kernels are designed to vertical flip and copy images by using a maximum of 32,768 blocks per launch.

```
int main(int argc, char **argv)
{
   ...
   cudaStatus = cudaMemcpy(GPUImg, TheImg, IMAGESIZE, cudaMemcpyHostToDevice);
   if (cudaStatus != cudaSuccess) { ... }
   cudaEventRecord(time2, 0);  // Time stamp after the CPU --> GPU tfr is done
   BlkPerRow = (IPH + ThrPerBlk -1 ) / ThrPerBlk;
   ui NumLoops, L;
   switch (Flip){
      case 'C': NumBlocks = (IMAGESIZE + ThrPerBlk - 1) / ThrPerBlk;
              NumLoops = CEIL(NumBlocks, 32768);
              for (L = 0; L < NumLoops; L++) {
                  PxCC20 <<< 32768, ThrPerBlk >>> (GPUCopyImg, GPUImg, IMAGESIZE, L);
              }
              GPUResult = GPUCopyImg;    GPUDataTransfer = 2*IMAGESIZE;
              break;
      case 'V': NumBlocks = IPV*BlkPerRow;
              NumLoops = CEIL(NumBlocks,32768);
              for (L = 0; L < NumLoops; L++) {
                  VfCC20 <<< 32768, ThrPerBlk >>> (GPUCopyImg, GPUImg, IPH, IPV, L);
              }
              GPUResult = GPUCopyImg;    GPUDataTransfer = 2*IMAGESIZE;
              break;
   }
   cudaStatus = cudaDeviceSynchronize();
   if (cudaStatus != cudaSuccess) {... exit(EXIT_FAILURE);  }
   cudaEventRecord(time3, 0);
   cudaStatus = cudaMemcpy(CopyImg, GPUResult, IMAGESIZE, cudaMemcpyDeviceToHost);
   if (cudaStatus != cudaSuccess) { ... }
   ...
}
```

7.7.4 imflipG2.cu: Changes in main()

The main() function of imglipG2.cu is shown in Code 7.1. The steps to transfer data in the CPU→GPU and CPU→GPU are identical to imglipG.cu by using the cudaMemcpy() function. Furthermore, the time-stamping process is also identical. The only difference is in the way the kernel is launched with a fixed number of 32,768 blocks. I made as few changes as possible, so it is easy to compare this code to imflipG.cu. Let's document our observations about the changes between imflipG.cu and imflipG2.cu.

- There is no change in the way the BlkPerRow variable is calculated. Because we will never get close to the 65,535 in computing this value, there is no point in making changes here.

- The computation of the `NumBlocks` is left intact. For the 'C' and 'V' options, they are calculated exactly the same way as Code 6.3. `NumBlocks` would be computed as 166,656 for the 'V' option (as in Section 6.4.14) and 498,876 for the 'C' option (as in Section 6.4.18).

- However, this number — `NumBlocks` — is not used in the kernel launch; instead, it is used to compute the *number of loops* needed (`NumLoops`) around the kernel launch. The "ceiling" function (CEIL) ensures that the last loop (potentially with < 32,768 blocks) is not forgotten.

```
NumLoops = CEIL(NumBlocks,32768);
```

- The loop that surrounds the actual kernel launch is as follows:

```
for (L = 0; L < NumLoops; L++) {
    PxCC20 <<< 32768, ThrPerBlk >>> (GPUCopyImg, GPUImg, IMAGESIZE, L);
}
```

- Here, the PxCC20() kernel is the CC 2.0 version of the PixCopy(), shown in Code 7.2.

CODE 7.2:   imflipG2.cu   PxCC20() {...

The PxCC20() kernel is identical to the PixCopy() with the exception of adding another loop ID "dimension" to the existing block ID dimension.

```
// Copy kernel with small block sizes (32768). Each thread copies 1 byte
__global__
void PxCC20(uch *ImgDst, uch *ImgSrc, ui FS, ui LoopID)
{
  ui ThrPerBlk = blockDim.x;
  ui MYbid = (LoopID * 32768) + blockIdx.x;
  ui MYtid = threadIdx.x;
  ui MYgtid = ThrPerBlk * MYbid + MYtid;
  if (MYgtid > FS) return;         // outside the allocated memory
  ImgDst[MYgtid] = ImgSrc[MYgtid];
}
```

7.7.5 The PxCC20() Kernel

The only noticeable change in the PxCC20() kernel in Code 7.2 is computation of the `MYbid` variable by incorporating the `LoopID` into the computation of the block ID. In a sense, we are *linearizing* the block IDs according to the formula in Equation 6.2; we are treating the `LoopID` as the added second dimension, in this case, while the existing `blockIDx.x` is the first dimension. After the linearization, we only care about the resulting one-dimensional `MYbid`. Once we perform this linearization, we can continue with the program like nothing changed. The rest of the code is identical to PixCopy(). After adding another dimension, you have to make sure that you are not exceeding the limits of your image (through out-of-bounds memory accesses). Luckily, the `if(...)` statement already checks to see if such a violation occurs, making it unnecessary for us to make any further modifications.

CODE 7.3: imflipG2.cu VfCC20() {...

The VfCC20() kernel is identical to the Vflip() with the exception of adding another loop ID on top of the block ID.

```
// Vertical flip kernel that works with small block sizes (32768)
// each thread only flips a single pixel (R,G,B)
__global__
void VfCC20(uch *ImgDst, uch *ImgSrc, ui Hpixels, ui Vpixels, ui LoopID)
{
    ui ThrPerBlk = blockDim.x;
    ui MYbid = (LoopID * 32768) + blockIdx.x;
    ui MYtid = threadIdx.x;
    ui MYgtid = ThrPerBlk * MYbid + MYtid;
    ui BlkPerRow = (Hpixels + ThrPerBlk - 1) / ThrPerBlk; // ceil
    ui RowBytes = (Hpixels * 3 + 3) & (~3);
    ui MYrow = MYbid / BlkPerRow;
    ui MYcol = MYgtid - MYrow*BlkPerRow*ThrPerBlk;
    if (MYcol >= Hpixels) return;    // col out of range
    if (MYrow >= Vpixels) return;    // row out of range
    ui MYmirrorrow = Vpixels - 1 - MYrow;
    ui MYsrcOffset = MYrow      * RowBytes;
    ui MYdstOffset = MYmirrorrow * RowBytes;
    ui MYsrcIndex = MYsrcOffset + 3 * MYcol;
    ui MYdstIndex = MYdstOffset + 3 * MYcol;

    // swap pixels RGB @MYcol , @MYmirrorcol
    ImgDst[MYdstIndex] = ImgSrc[MYsrcIndex];
    ImgDst[MYdstIndex + 1] = ImgSrc[MYsrcIndex + 1];
    ImgDst[MYdstIndex + 2] = ImgSrc[MYsrcIndex + 2];
}
```

7.7.6 The VfCC20() Kernel

The VfCC20() kernel is shown in Code 7.3. The code is almost identical to the Vflip() kernel with two noticeable differences:

- The *linearization* of the block ID is done exactly the same way as the PxCC20() kernel using the following line of code:

```
    ui MYbid = (LoopID * 32768) + blockIdx.x;
```

which combines the loopID-based dimension and the actual blockIdx.x-based dimension to end up with a single dimension, which is the *global thread ID*, using the variable MYgtid.

- An additional *out-of-bounds index check* had to be incorporated to determine whether MYrow exceeded the number of columns in the image as follows:

```
    if (MYcol >= Hpixels) return;    // col out of range
    if (MYrow >= Vpixels) return;    // row out of range
```

We notice that the bottom check for MYrow was unnecessary in the Vflip() kernel because we were launching precisely the same number of blocks as the number of vertical pixels. However, in VfCC20(), we are dividing the number of pixel rows by 32,768, which can very well give us a residual that makes the last block of threads try to access a memory area beyond the image.

- To exemplify this, assume that we are vertically flipping the Astronaut.bmp using the VfCC20() kernel. This image requires launching 166,656 blocks, which corresponds to $\left\lceil \frac{166,656}{32,768} \right\rceil = 6$ loops to launch 32,768 blocks within each loop iteration. So, there will actually be a total number of $32,768 \times 6 = 196,608$ blocks launched. The *useful* blocks will have MYgtid values in 0...166,655, while the *useless blocks* will have MYgtid values in 166,656..196,607. The extra MYrow check line prevents the useless blocks from executing.

- This means that the number of wasted blocks is quite high (29,952 to be exact), which will reduce the program's performance.

- Of course, you could have tried to launch a much smaller number of blocks within each loop iteration, such as 4096, which would have reduced the number of wasted blocks. I am leaving it up to the reader to pursue this if needed. I am not elaborating on this any further because all of the existing GPUs in the market today support CC 3.0 or higher, making this a non-issue in the new Nvidia generations.

- Yet, another possibility is to launch a variable number of blocks within each kernel to reduce the number of wasted blocks down to almost zero. For example, for 166,656, you could have used 6 loops with 27,776 blocks in each iteration, which would have dropped the number of wasted blocks down to zero (not almost, but *exactly* zero). If you can guarantee that there are no wasted blocks, there is no need for the range check on the MYrow variable.

- The story goes on and on ... Could you have even eliminated the range check on the MYcol variable? The answer is: YES. If you did not have any wasted blocks in the column dimension, there wouldn't be a need for even the first check. Say you did that. Then, potentially, the computation of the indexes might get a little more complicated. Is it worth complicating a part of the code to make another part less complicated?

- Welcome to the world of choices, choices, choices... CUDA programming is not all about strict rules. There are many ways you can get the same functionality from a program. The question is which one works faster, and which one is more readable?

➢ *You will realize that in CUDA, there are many choices of parameters, each with pros/cons. Which option should you choose?*

Here are a few rules that might help:

➢ *If there are two options, with the same final performance.*
Choose the one that is easy to understand; simple is almost always better.
Difficult-to-understand code is prone to bugs.

➢ *If only highly sophisticated techniques will result in high performance code, document it extremely well, especially non-obvious parts.*
Don't do it for others, do it for yourself! You will look at it later.

7.8 PERFORMANCE OF IMFLIPG2.CU

To test imflipG2.cu and compare it to the imflipG.cu results in Table 7.3, I used the first three boxes from Table 7.3 (Box I, II, and III), which represent three low-end Kepler GPUs by 2017's standards. I also used three Fermi GPUs (GT630, GT640, and GTX550Ti), within Box I. In other words, I unplugged the GT 640 Kepler GPU that was in Box I and replaced it with GT 630, and then unplugged it again and replaced it with GT520, and then GTX 550Ti and reported the numbers in Table 7.6.

One interesting feature of the GT520 is that it is a PCIe x1 card, plugged into the x1 port of PCI Express. So, we expect it to be able to achieve only a 0.5 GBps bandwidth for I/O transfers. This type of card is perfect if you are out of x16 slots and still need any card that can display something, without worrying about the performance of the card. For example, some motherboards have a single PCIe x16 slot and multiple other x1 slots. If you wanted to plug in two GPUs, one for high performance computing and the other for just plain-simple display, this x1 card is perfect. You might have noticed that there is a GT series and GTX. In general, GT series are the low-end cards, while the GTX series is designed to be higher performance.

Table 7.6 tabulates the results of the same exact program (imflipG2.cu) that was compiled with the CC 2.0 option. Here are the observations we can make:

- Changes in the performance for the three Kepler cards (GT640, K3000M, and GTX760) if barely noticeable. Due to the addition of the additional dimension, there is a slight slow-down for almost every block size. This is attributable to the wasted blocks and the additional few instructions that to be inserted into the kernels.

- The PCIe transfer speed of the GT 520 is extremely low, as expected; however, this is the only card that reaches a very high percentage (76%) of the PCIe bandwidth. The reason for this is the fact that the CPU has to be actively involved in this transfer and due to the actual transfer throughput being substantially lower than the throughput at which the CPU has to pull the data from its DRAM and place it on the PCIe bus, the CPU is minimally burdened and can easily reach 76% of the PCIe bandwidth.

- The GPU family does not have any visible effect on the PCIe transfers because this is a process that is strictly dependent on the CPU I/O functionality. Because I used the same CPU for the four low-end cards, PCIe transfers are at an identical percentage.

- The higher end cards reach progressively higher *achieved throughput* values in absolute terms, GTX 550Ti beating the GT 630, etc. However, in percentage terms, GTX550Ti achieves a lower percentage because the VfCC20() kernel puts a lot of pressure on the cores, as well as memory. The 192 cores that GTX550Ti has is not enough to supply the 99 GBps this GPU has, thereby making it a lot harder for the GPU to saturate its internal global memory bandwidth. A much more detailed analysis of this will be done in the following chapters.

7.9 OLD-SCHOOL CUDA DEBUGGING

I introduced the concept of old school debugging within the context of CPU programming in Section 1.7.2, where we used our best friend, printf(), to tell us what is going on inside the program. We used some other nice old school tools, such as assert() and commenting

TABLE 7.6 Results of the imflipG2.cu program, which uses the VfCC20() and PxCC20() kernels and works in Compute Capability 2.0.

Feature	Box I				Box II	Box III
CPU	i7-920				i7-3740QM	W3690
C/T	4C/8T				4C/8T	6C/12T
Memory	16GB				32GB	24GB
BW GBps	25.6				25.6	32
GPU	GT520	GT630	GTX550Ti	GT640	K3000M	GTX 760
Engine	GF119	GF108	GF116	GK107	GK104	GK104
Cores	48	96	192	384	576	1152
Compute Cap	2.1	2.1	2.1	3.0	3.0	3.0
Global Mem	0.5GB	1GB	1GB	2GB	2GB	2GB
Peak GFLOPS	155	311	691	691	753	2258
DGFLOPS	–	–	–	29	31	94
Data transfer speeds & throughput over the PCI Express bus						
CPU→GPU ms	328.29	39.19	39.18	38.89	52.28	34.38
GBps	0.37	3.11	3.11	3.13	2.33	3.54
GPU→CPU ms	319.42	39.07	39.36	39.66	52.47	36.00
GBps	0.38	3.12	3.09	3.07	2.32	3.38
PCIe Bus	Gen2 x1	Gen2	Gen2	Gen2	Gen2	Gen2
BW GBps	0.5	8.00	8.00	8.00	8.0	8.0
Achieved (%)	(76%)	(39%)	(39%)	(39%)	(29%)	(44%)
VfCC20() kernel run time (ms) 'V' command line option						
V 32	220.55	110.99	43.34	72.67	63.98	20.98
V 64	118.88	59.29	23.37	39.25	34.22	11.30
V 128	72.68	35.43	14.55	24.04	21.02	7.04
V 256	69.19	34.61	14.70	25.88	22.54	7.56
V 512	70.66	35.03	14.77	28.52	23.80	7.94
V 768	73.31	36.04	15.20	40.77	36.39	11.86
V 1024	124.00	62.31	25.49	36.23	31.05	10.34
GM BW GBps	14.4	28	99	28.5	89	192
Achieved GBps	3.52	7.04	16.74	10.13	11.59	34.59
(%)	(24%)	(25%)	(17%)	(36%)	(13%)	(18%)
PxCC20() kernel run time (ms) 'C' command line option						
C 32	356.30	186.36	69.03	102.39	86.68	27.42
C 64	179.82	93.79	34.56	51.67	43.05	13.69
C 128	97.20	49.31	18.23	27.18	22.48	7.28
C 256	69.54	35.70	13.31	28.04	23.48	7.67
C 512	74.03	37.71	13.84	29.16	24.37	7.88
C 768	84.63	42.32	15.69	42.88	35.22	11.38
C 1024	116.93	60.34	22.59	31.44	25.86	8.34
GM BW GBps	14.4	28	99	28.5	89	192
Achieved GBps	3.50	6.82	18.30	8.96	10.84	33.45
(%)	(24%)	(24%)	(18%)	(31%)	(12%)	(17%)

All Fermi GPUs are tested only on Box I (listed in Table 7.3). The astronaut.bmp image was used with 'V' and 'C' options and different block sizes (32..1024).

lines, but printf() has a different place in every old school programmer's heart. Good news: printf() — along with a bunch of other other old school concepts — will be our best friend in the CUDA world too. What I will show you in this section actually debugs a surprisingly good number of bugs. Of course, I will show you the "new school" tools too, like the fancy schmancy CUDA debugger named nvprof (or the even-fancier GUI version nvvp), but the old school concepts are surprisingly powerful because they can get you to debug your code much faster, rather than going through the entire process of running nvvp, blah blah blah.

Before we look at how to old school debug, let's look at what type of bugs are common:

> ➤ *If they found life on a new planet and wanted to let the new planet's folks know about the most common type of a computer bug on earth, and due to a low-bandwidth satellite connection, I was allowed one — and only one — word to let them know, I would say **pointers**.*
> ➤ *Yes, you can survive all sorts of bugs, but bad memory pointers will kill you!*

Yes, pointers are a giant problem in C; yet, they are one of the most powerful features of C. For example, there are no explicit pointers in Python, which is a rebel against the pointer problems in C. However, you can't live without pointers in CUDA programming. Aside from bad-pointer-bugs, there are other common bugs, too; I am listing a bunch of them in the following section. I will show you how old school debugging can eliminate a good portion of them. Remember, this section is strictly about CUDA programming. Therefore, our focus is on CUDA bugs and how to debug them in a CUDA environment, which means either using Nvidia's built-in tools or our old school concepts that utilize simple functions, etc. and work everywhere. Even more specifically, we will *not* focus on problems within your CPU code in a typical CUDA program. We will strictly focus on the bugs inside the *CUDA kernels*. Any bug inside the actual CPU code can be fixed with the old school debugging tools in Section 1.7.2 or the nice CPU debugging tools like gdb and valgrind, which I showed in Section 1.7.1 and Section 1.7.3, respectively.

7.9.1 Common CUDA Bugs

Let's list some common bugs:

- **Memory pointer bugs** deserve the #1 spot. Because stepping out of the allocated memory area by even a single byte is a No-No for an OS, these bugs will give you a Segmentation Fault immediately. A Segmentation Fault means that the OS caught your access request to a disallowed memory area and issued a shut down for your application. Your program will terminate abruptly. In CUDA programs, your OS is the CUDA Runtime Engine, as I pointed out in Section 6.4.5, and it has a very similar mechanism for catching memory address violations. However, the message you get will be different. We will see examples very shortly. A surprising number of these bugs happen due to the pointers being just one count outside the range. For example, look at this CUDA code below:

```
// inside main()
unsigned char *ImagePtr=(unsigned char *)cudaMalloc(IMAGESIZE);
// inside the GPU kernel
for(a=0; a<IMAGESIZE; a++) { *(ImagePtr+a)=76;... }
for(a=0; a<=IMAGESIZE; a++) { *(ImagePtr+a)=76; ... }
```

The main() part of the program allocates memory in the GPU global memory through the use of the CUDA API function cudaMalloc(), and the two for loops are executed within the CUDA kernel and attempt to access the GPU global memory. The bottom for loop will completely crash your program because it is stepping one count beyond the image that is stored in the GPU global memory. This will be caught by the Nvidia runtime and will totally terminate our CUDA application. The top for loop is perfectly within the range of the image memory area, so it will work totally fine.

- **Incorrect array indexes** are nothing different than bad memory pointers. An array index is just a convenient short-hand notation for the underlying memory pointer computation. Incorrect index computations have exactly the same effect as incorrect memory accesses. Check out the example below:

```
int SomeArray[20];
for(a=0; a<20; a++)        { SomeArray[a]=0; ... }
SomeArray[20]=56;
```

The top for loop will run perfectly fine, initializing the entire array to zero (from index 0 to index 19). However, the bottom assignment will crash, because SomeArray[20] is outside the array, which spans indexes SomeArray[0]...SomeArray[19].

- **Infinite loops** are loops with messed up loop variables, having an incorrect termination condition. Here is a quick example:

```
int y=0;
while(y<20){
    SomeArray[y]=0;
}
```

Where is the part that updates y? The programmer intended to create a loop that initializes the array, but forgot to put a line to update the y variable. There should have been a line that reads   y++;   after   SomeArray[y]=0;   to avoid the termination condition (y<20) never being satisfied.

- **Uninitialized variable values** are also a common bug, resulting from you declaring a variable and not initializing them. As long as these variables will be assigned a value before they are used, you are fine. But, if you use them before initializing them, more than likely, they will crash your program. Although the source of the problem is an *uninitialized variable value*, it is possibly the *consequence* of that — such as a bad index or a pointer — will possibly cause the crash. Here is an example below:

```
int SomeArray[20];
int a=0;
int b,c;
for(x=19; x>=0; x++)        { SomeArray[x-a]=0; ... }
for(x=19; x>=0; x++)        { SomeArray[x-b]=0; ... }
for(x=19; x>=0; x++)        { SomeArray[x]=x/c; ... }
```

The top for loop will never exceed the index range of [19...0], while the bottom for loop might or might not. You can never make the assumption that any value you declared will have an initial value (0 or anything else). At runtime, the memory area for variable a is created and the value 0 is explicitly written into that memory address. Alternatively, the memory area for variable b is created and nothing is written to that area, thereby leaving whatever value was there in the computer memory before the allocation of the variable. This value can be anything; if we assume that it was 50, as an example, it is clear that it will make you exceed the index range. The third for loop is a likely candidate to give you a *division by zero* error because the value of the c variable is not initialized and can very well be zero.

- **Incorrect usage of C language syntax** happens to inexperienced programmers. Common examples include confusing = with == or & with && and many other variants of this. Check out the example below:

```
int a=5;
int b=7;
int d=20;
if(a=b) d=10;
```

What is the result of d? Although the programmer meant to type if(a==b), there is now a bug. The way it is written, this translates to the following lines:

```
int a=5;
int b=7;
int d=20;
a=b;
if(a) d=10;
```

In other words, if(a=b) means *set a to b's value and if the result is TRUE* ... In C language, any non-zero value is translated to its single-bit equivalent TRUE and zero is translated to FALSE. Therefore, the final value that a received is TRUE, forcing the execution of the follow-up statement d=10; and producing a wrong result. Let's see how we can debug these using old school CUDA debugging.

7.9.2 return Debugging

You would be surprised to hear that inserting a return() in some parts of a CUDA kernel is the most powerful tool, even as compared to our all-time hero, printf(). The idea behind this is that assume that the bug is one of the ones listed in Section 7.9.1. Most of them would crash the CUDA application. If we have a 10-line CUDA kernel, why not stick a return; right after the first 5 lines? What does it mean if it doesn't crash anymore? More than likely, the bug was between lines 6 and 10. OK, so, what if we move the return to line 8 and it still doesn't crash? Well, the bug might very well be on line 9 or line 10.

This might allow you to pinpoint the line that relates to the bug real fast, but you can speed up this process. Usually you are *suspicious* of a single line of code or some point in the kernel after which you are suspecting to have a bug. Why not put the return somewhere you are suspecting? By moving the return around, you could pinpoint the bug within just a few tries. This is great, but moving the return within the kernel will change the result and if the program behavior depends on the values produced during the execution of the code, it will be difficult for you to determine why it doesn't crash anymore. This is when you might try other tricks that I will show shortly. However, I have fixed quite a few bugs with the return trick. Let's take the PixCopy() kernel as an example in Code 6.8, which is repeated below.

```
__global__
void PixCopy(uch *ImgDst, uch *ImgSrc, ui FS)
{
    ui ThrPerBlk = blockDim.x;
    ui MYbid = blockIdx.x;
    ui MYtid = threadIdx.x;
    ui MYgtid = ThrPerBlk * MYbid + MYtid;
    if (MYgtid > FS) return;        // outside the allocated memory
    ImgDst[MYgtid] = ImgSrc[MYgtid];
}
```

Imagine now that you didn't have the if()... line in your code. It would crash because of accessing a tiny bit outside the image area. To analyze this precisely, let's remember the example numbers from Section 6.4.18. For the astronaut.bmp image, our image size was 127,712,256 Bytes and when we launched the PixCopy() kernel with 256 threads/block, we ended up launching 498,876 blocks. Each block copies a single byte, so PixCopy() kernel threads accessed addresses from 0 to 127,712,255, staying perfectly within the global memory address range. Therefore, we didn't even need the if()... line if we knew that this program would be used strictly with the astronaut.bmp image and always with 256 threads/block.

What happens when we use a different image, say, $1966 \times 1363 = 8,038,974$ Bytes. Let's also assume that we want to launch it with 1024 threads/block. We would need to launch $\left\lceil \frac{8,038,974}{1024} \right\rceil = 7851$ blocks. This would launch $7851 \times 1024 = 8,039,424$ threads for your entire CUDA application. If we didn't have the if()... statement, threads with gtid values in the range (0...8,038,973) would access perfectly allowed global memory areas. However, threads with gtid values in the range (8,038,974...8,039,423) would access an unauthorized memory range, thereby crashing your CUDA application.

How could you debug this? All of the assignments look pretty harmless, although there is a single line that has a memory access. Remember what I said at the beginning of Section 7.9: be suspicious of the pointers before anything else. Assuming that your initial code looked like this without the if()... statement,

```
__global__
void PixCopy(uch *ImgDst, uch *ImgSrc, ui FS)
{
    ...
    ui MYgtid = ThrPerBlk * MYbid + MYtid;
    ImgDst[MYgtid] = ImgSrc[MYgtid];   // suspicious !!!
}
```

You could insert the `return` statement right before the memory access, as follows:

```
__global__
void PixCopy(uch *ImgDst, uch *ImgSrc, ui FS)
{
    ...
    ui MYgtid = ThrPerBlk * MYbid + MYtid;
    return;  // this line skips the execution of the line below (DEBUG)
    ImgDst[MYgtid] = ImgSrc[MYgtid];    // suspicious !!!
}
```

and determine that the problem lies with that last line of the kernel. After a few iterations, you would realize that you are accessing an unauthorized memory range. Our idea is that if we place a `return` statement just before the suspected line, we are skipping its execution. If this fixes the problem, then we analyze its cause deeper. But, let's not kid ourselves; this problem would only happen in cases where your image size has exactly the right value (more like *exactly the wrong value*) to make your number of threads go beyond the value of the number of bytes. So, this could almost be considered in the *sneaky bug* category.

7.9.3   Comment-Based Debugging

Comment-based debugging is very similar to `return` debugging. Instead of inserting a `return` statement that shuts off the entire rest of the kernel after the `return`, you comment out one of the lines. This works better when the bug is not so severe. An application of this to Code 6.8 (the Hflip() kernel) is shown below, where two out of the last three lines of the code are commented out to determine if they are the cause of the bug:

```
__global__
void Hflip(uch *ImgDst, uch *ImgSrc, ui Hpixels)
{
    ...
    // swap pixels RGB @MYcol , @MYmirrorcol
    //ImgDst[MYdstIndex] = ImgSrc[MYsrcIndex];        COMMENTED
    //ImgDst[MYdstIndex + 1] = ImgSrc[MYsrcIndex + 1]; COMMENTED
    ImgDst[MYdstIndex + 2] = ImgSrc[MYsrcIndex + 2];
}
```

7.9.4   printf() Debugging

While commenting and the `return` trick can give you an idea about *where* the bug is, they don't tell you *how* the bug happens quantitatively. We can insert printf() functions anywhere in the code to display the values of variables, in an attempt to determine the precise cause of the bug. Even better, we can insert *conditional* printf statements to avoid executing the same printf() all the time. Conditional printf()'s will make their use more practical in CUDA programs; while CPUs run a few threads and the printf() outputs coming from just a few threads are readable, you will not be able to manage the results pouring at you from 40+ million threads! So, a common trick I use is to restrict the printf() output to only a few initial blocks or a few initial threads of a known block.

Here is a demonstration of this method using the PixCopy() kernel. Assume that you did the `return` trick and realized that the bug is on this line. Now what?

```
__global__
void PixCopy(uch *ImgDst, uch *ImgSrc, ui FS)
{
    ...
    ui MYgtid = ThrPerBlk * MYbid + MYtid;
        // this line is for debugging. Suspecting a memory pointer bug
        if(MYgtid==1000000) printf("MYgtid=%u\n",MYgtid); ////DEBUG
    ImgDst[MYgtid] = ImgSrc[MYgtid];
}
```

You are trying to narrow down the range of bugs you could have. At this point, an experienced programmer would start smelling a *memory pointer bug*. If this theory is correct, the code should work for certain values of MYgtid and crash for others. The best thing to do is to print the value of MYgtid *only if it has reached a certain value*. Something like this:

I inserted the conditional printf() line that will print the output "MYgtid=1000000" only if MYgtid has reached that value. If the program crashes before printing that line, you know that the problem is when MYgtid¡1000000, or vice versa. For a 121 MB image, you could start with 100 million and if it works increase it immediately to 120 million, followed by the size of the image minus one. At some point it will crash and will give you the *gotcha* clue.

A few final words of wisdom from an old school debugger:

> ➤ *As you see above, when I add debug lines into my code, I make them salient, with annoying notes — like DEBUG — and comment lines. I also indent them to be able to tell them apart from the actual code. This is because it is common for you to go through a long and tedious debug session and forget some useless code in there — used only for debugging — and to realize later that some of your future bugs are due to that extra junk code you forgot in there.*
>
> ➤ *The last thing you want is for your debugging to inject bugs into your code that didn't exist before you started debugging.*
>
> ➤ *Another advice I can give you is: DO NOT DELETE CODE during debugging. If you are debugging a line of code, simply make a copy of that line and keep modifying the copy. If that line is fine, uncomment the original and move on.*
>
> ➤ *Nothing is more frustrating than to delete a line during debugging and not be able to get it back, because you never made a backup of that line. If you made a copy of the line, modify the copy all you want; the original is safe.*

7.10 BIOLOGICAL REASONS FOR SOFTWARE BUGS

Much like a CPU or GPU uses *electrical energy*, the human body uses *biological energy*, extracted from food. Just like a CPU or a GPU can overheat and malfunction, the human CPU (the brain) — as well as other parts of the body — can also overheat and malfunction. Almost every computer motherboard has sensors to determine whether your CPU is running too hot, etc. The human body has similar types of mechanisms to warn you about your overworked condition. Your brain communicates with you through signals like "pain" and "irritation," which are facilitated through chemical messengers (hormones, neurotransmitters, etc.), circulated throughout your blood and detected/processed by your brain in some way. From the standpoint of this chapter, why we care about this bio stuff is that when you get tired, you write buggy code! So, it is a good idea for a programmer to read these two pages and understand the human metabolism, the functionality of the brain, and how this affects your code development performance.

7.10.1   How Is Our Brain Involved in Writing/Debugging Code?

There are two major categories of bugs, which are the *sneaky* bugs and *obnoxious* bugs.

Sneaky bugs hide themselves and it takes a lot of "conscious" effort to find them. They are usually a product of bad programming logic. They are very annoying because you might think that you've got them one day just to have them show their ugly face a few weeks later when different data is input into the program. They are the ones that might be difficult to catch with old school debugging. They might require you to run the code step by step, sometimes tens or even hundreds of steps in a loop until you hit a value that gives you the *gotcha* output. I found that the best way to eliminate them is to stop working on the program and start the next day with a fresh mind. As they say: there is no point in *pushing it*. Clearly you won't be able to solve this problem today.

Obnoxious bugs are good bugs. They immediately crash your program, so it is easy to identify them. Memory pointer bugs are in this category, when you malloc() a memory area and try to step even a single byte out of that area, the OS will shut you down. Preventing one program from writing into another program's memory area is about one of the most important functions of the OS, so your buggy program will give you a Segmentation fault or another type of memory error and will exit abruptly. The bad news about this type of a bug is that the program stops working, so it never reaches the many printf()'s you put into the program.

When you cannot debug your code, your brain might get stuck in a *local minimum* and cannot converge to a global value. *Attention* is a limited brain resource and you will run out of it if you keep working on the same problem [30]. Everything I just described actually has a neuro-scientific reason behind it. Here is how the human brain works:

> ➤ *The human brain consists of two parts: (1) conscious and (2) motor [29].*
> ➤ *The conscious part is responsible for deliberate action and requires continuous attention. Clearly, this part wasn't able to debug your code and pushing it won't help today. All you will do is drain your blood glucose, which is demonstrated to be the food of your brain and your entire nervous system. The motor part does more automated work, not requiring constant attention. Go to sleep ... Try again tomorrow ...*
> *The motor part of the human brain is known to process a difficult problem continuously, even in your sleep — actually especially in your sleep — because it is not bothered with other daily tasks.*
> ➤ *It is common for an experienced programmer to wake up and start working on a program that had one of these sneaky bugs and fix it in 5 minutes.*
> ➤ *So, I have two Tolga's in my brain; Motor-Tolga and Conscious-Tolga. When I am in front of my computer, C-Tolga is debugging; in my sleep, M-Tolga is. I don't care who debugs the code first, because they are both "me."*

7.10.2   Do We Write Buggy Code When We Are Tired?

A good programmer should understand how his or her brain works. Our brain is the CPU that we use during programming and debugging; it is different than a Xeon CPU or a GTX Titan GPU because its functionality is affected by many factors that involve the entire body. The brain gets *tired*, a CPU or a GPU doesn't. Every bug will likely come back to have a cause related to one of the tiredness states listed next. Let's go a little deeper into the factors that make a programmer tired (more specifically, *a programmer's brain* tired) and increase his or her program's chances of having more bugs. There are different types of *tiredness* states that you must be aware of. Although all of them will reduce your performance, their sources are completely different; so are their remedies.

7.10.2.1 Attention

Attention is a limited resource in your brain. Unfortunately, after millions of years of development, our brain still works as a single-threaded CPU, being able to process only a single thing heavily. We cannot focus on more than one — and only one — thing heavily at any point in time. The conscious part of our brain, described in Section 7.10.1, is solely responsible for our attention; it performs heavy processing during code development and especially debugging. If your attention is diverted to somewhere else during code development, your programming and debugging performance will fall off a cliff.

7.10.2.2 Physical Tiredness

Physical tiredness is related to your muscles, which are spread throughout your entire body. During your daily physical activity, your muscles turn ATP (Adenosine Tri Phospate) into ADP (Adenosine Di Phosphate) and AMP (Mono Phosphate), and eventually Adenosine. The ATP→ADP→AMP→Adenosine degradation is through breaking a phosphate bond, which releases the bonding energy of that phosphate bond, thereby powering up your muscles. The eventual product, *Adenosine*, is basically *burnt fuel*. So, if there was a detector in your body to detect the increasing Adenosine levels, it could be used to warn you about the decreasing ATP levels, much like the fuel gauge in your car and the yellow warning light telling you that you are getting close to running out of fuel. Guess what? The brain is that detector [30]. The brain detects the increasing Adenosine levels and *warns you* to avoid fully depleting your ATP resources. This warning could be by telling you to sleep or eat. This *"eat because you are hungry"* message is a function of the *leptin* and *ghrelin* hormones, whereas the *"sleep because you are tired"* is controlled by the *melatonin* hormone. The remedy against ATP depletion is to eat proteins, fats, and carbohydrates, which will produce ATP during their break-down [39].

7.10.2.3 Tiredness Due to Heavy Physical Activity

A nice work-out at the gym is another reason for physical tiredness. You have two types of muscles in your body: (1) fast twitch muscles that use *anaerobic* metabolism (*anaerobic* means "without requiring oxygen"), and your (2) slow twitch muscles that use *aerobic* processes (*aerobic* means "requiring oxygen"). When you need to perform heavy physical activity, your heart cannot supply oxygen fast enough to the slow twitch muscles, thereby requiring you to resort to using your fast twitch muscles, which do not need oxygen to work. Both of these muscles use ATP as their energy source, but they produce a different output; although fast twitch muscles produce burst energy, they produce lactate and your brain has a detection mechanism for lactate build-up in these muscles; it will try to slow you down — by communicating with you through the feeling of *pain* — to avoid damage to your body because too much lactate build up will eventually hurt your muscles. The pain that your brain produces as a result of heavy workout will have a negative impact on your programming performance. The only remedy for this is to rest until the pain goes away, because your *attention* will be diverted away from your code toward the pain. When the lactate is out of the muscles, pain is gone and you are back to work.

7.10.2.4 Tiredness Due to Needing Sleep

You get tired when you don't get enough sleep because there is a small part of your brain that is responsible for detecting the 24-hour circadian cycle. It is your brain's clock, which is responsible for releasing the *melatonin* hormone to prepare you for sleep and increasing *cortisol* levels when it is time to wake you up. During sleep, your ATP levels are replenished and you are ready to burn them back to Adenosine again. The only remedy for sleep deprivation is to sleep! There is no way to fight the brain! Just get a good night's sleep and you will be the best debugger again tomorrow.

7.10.2.5 Mental Tiredness

Mental tiredness is related to your neurons, which are spread throughout your entire central nervous system (CNS). Neurons burn glucose when they work. To state simplistically, the *brain food* is *sugar*. So, when your blood glucose levels go lower, your neurons do not have immediate energy. This is no different than your computer consuming so many Watts per FLOP. For example, GTX 1070 consumes 110 W of power when it works at its peak, although it consumes way less than that during normal operation. The human brain consumes an equivalent of 20 W on average. However, during a heavy-duty debugging session, when you are trying super hard to find a bug in your code, I am sure that this number goes way up, causing a higher rate of depletion in glucose.

More debugging → more neural activity → more glucose consumption

There is the other side of the story. Your neurons use *neurotransmitters* to carry neural signals. The leftover neurotransmitters are mopped up and placed back into your system when you *rest*. So, if you work very hard, your neurons will keep using sugar and the leftover neuro-garbage must be collected, which takes time. So, a nice walk along a river can have a mentally replenishing effect, although not necessarily comparable to consuming sugary foods! This is more like your neurons malloc() energy with sugar and the rest of the system has to free() the garbage out of your system, otherwise your brain will issue a **Segmentation fault**. This is when you stare at the screen and nothing happens! You need a reboot. Go for a nice walk...

## CHAPTER 8

Understanding GPU Hardware Architecture

I N the previous two chapters we looked at the structure of a CUDA program, learned how to edit, compile, and run a CUDA program, and analyzed the performance of the compiled executable on different generations of GPUs, which have different Compute Capabilities (CCs). We noted that a CPU is good for *parallel computing*, which has the building block named *thread*; it is usual to expect the CPU-based parallel programs to execute 10, 20, 100, even 1000 threads at any point in time. However, a CUDA program (more generally, a GPU program) is suitable for problems that can take advantage of *massively parallel computing*, which implies the execution of hundreds of thousands or even millions of threads at a time. To allow the execution of such an enormous number of threads, GPUs had to add two additional hierarchical organizations of threads:

1. A **Warp** is a clump of 32 threads, which is really the minimum number of threads you can break your tasks down to; in other words, nothing less than 32 threads executes in a GPU. If you need to execute 20 threads, too bad, you will be wasting 12 threads, because a GPU is not designed to execute such a "small" number of threads.

2. A **Block** is a clump of 1 to 32 warps. In other words, you launch your program as 1, 2, 3, ..., or 32 warps, corresponding to anywhere from 32 to 1024 threads. A block must be designed as an isolated set of threads that can execute independently from the other blocks. If you design your program to have such smooth *separation* (or, *independence*) from other blocks, you will achieve blazing parallelism and will take advantage of the GPU parallelism.

Clearly not every problem is amenable to massive-parallelism. Image processing, or more generally, digital signal processing (DSP) problems are natural candidates for GPU massive parallelism, because (1) you apply the same exact computation to different pixels (or pixel groups), where one pixel can be computed independently from another, and (2) there are typically hundreds of thousands or millions of pixels in an image we encounter in today's digital world. Following this understanding from the previous CUDA chapters, in this chapter, we now want to understand how the GPU achieves this parallelism in hardware.

In this chapter, we will introduce the GPU edge detection program (imedgeG.cu) and run it to observe its performance. We will relate this performance to the building blocks of the GPU, such as the GPU cores and streaming multiprocessors (SM), which are the execution units that house a bunch of these GPU cores. We will also study the relationship among SM, GPU cores, and the things we have just learned, thread, warp, and block over the course of a few chapters, during which we will learn the *CUDA Occupancy Calculator*, which is a simple tool that tells us how "occupied" our GPU is, i.e., how busy we are keeping it with our program. This will be our primary tool for crafting efficient GPU programs. The good news is that although writing efficient GPU programs is an art as much as a science, we are not alone! Throughout the following few chapters, we will learn how to use a few such useful tools that will give us a good view of *what is going on inside the GPU*, as well as during the transfers between the CPU and the GPU. It is only with this understanding of the hardware that a programmer can write efficient GPU code.

8.1 GPU HARDWARE ARCHITECTURE

So far, we wrote GPU code and tweaked the *threads per block* parameter to observe the changes in performance. In this chapter, we will look at the hardware at a high level to get an idea about how the cores and memory are organized and how the data flows inside the GPU. Remember from Analogy 6.1 (Figure 6.3) that we viewed the GPU as a school bus with 32 scouts in it; the scouts represented the slower GPU cores who got more work done than any single- or dual-core (or even quad-core) CPU architecture due to their shear quantity, as long as the application was massively parallel computing friendly.

One very important note about Figure 6.3 is that we had somebody, **Tolga**, to organize things inside this bus, which was such an intense task that Tolga couldn't do any useful work; he was dedicated to organizing things. Let's remember: *What did Tolga organize?* The answer is: *data flow and task distribution to scouts.* Because the scouts are inexperienced executers, somebody had to tell them exactly what to do and put the data right in front of them to do it. The big question is, we do not see anything too impressive being done with only 32 scouts; we need thousands of them to beat the CPU performance. In real-life GPUs, as exemplified in our Table 8.3, GTX 1070 has 1920 GPU cores and the GTX Titan Z has 5760 cores. Is it even possible to expand our scout analogy to cover such a huge number of cores? The answer is Yes, but major additional organizational structures will be needed.

8.2 GPU HARDWARE COMPONENTS

Analogy 8.1 (Figure 8.1) is exactly what is required to be able to process Gigabytes of data using thousands of cores. None of these components can be missing because the execution will be a mess otherwise! Let us now see what each person (and object) corresponds to inside the GPU. Note: I gave a little bit of a hint in the caption of Figure 8.1.

8.2.1 SM: Streaming Multiprocessor

In Analogy 8.1, each school bus, including the small barrel and the box (to receive the instructions) is equivalent to an SM (streaming multiprocessor) inside the GPU. This was the term that Nvidia used in their older GPUs; for example, in the Fermi family, the GTX550Ti GPU (with 192 cores, as shown in Table 7.6) had 32 GPU cores in each SM. Later, Nvidia changed this term to SMX in the Kepler family, SMM in the Maxwell family, and back to SM in the Pascal family, however, Pascal introduced another hierarchical structure named GTC, as we will see very shortly.

ANALOGY 8.1: *GPU Architecture.*

Cocotown's great success in their coconut competition last year encourages the town officials to significantly expand the competition to include two million coconuts. While Tolga successfully drove the school bus and organized the scouts in last year's competition, he realizes that it is impossible to use the same school bus for such a large competition. So, he asks the school to lend him four school buses this year (with a capacity for 48 or 64 scouts) and puts his brothers Tony, Tom, and Tim in charge of managing their own scouts in the other three school buses. They also put their neighbor Gina (a skilled manager) in charge of chopping each big job into *blocks* (each block containing instructions about how to process 128 coconuts) and assigning each block to a different school bus, one at a time by placing them in a box (right side of Figure 8.1). Four guys get their instructions from the box in the same order Gina assigned them and they internally distribute them to their scouts. Gina's instructions are something like "process coconuts with numbers 0...127" on them, "... 128...255 on them", ... Gina assumes that each coconut has a unique number associated with it.

They realize that somebody actually should "get" the coconuts with specific numbers (the way Gina described them in her tasks) from the jungle. This task in itself is a big deal, so they put Melissa in charge of strictly getting big chunks of coconuts from the jungle, numbering each one, and dumping them into a huge barrel in a nice and organized form. Leland, on the other hand, is made responsible for distributing these coconuts into the smaller barrels outside each school bus. Leland brings the coconuts with numbers exactly described in Gina's instructions. Tolga's four sisters, Laura, Linda, Lilly, and Libby, are made responsible for grabbing the coconuts from the smaller barrels and giving them to the scouts in exactly the order they need them.

8.2.2 GPU Cores

These are the scouts (GPU cores). We will dedicate an entire chapter (Chapter 9) to studying the inside of each core and how we can get the maximum performance out of each core. For now, it suffices to say that the ALU and FPU are inside each core.

8.2.3 Giga-Thread Scheduler

This is Gina in our Analogy 8.1. Assume that you launched a kernel with 2000 blocks. Giga thread scheduler's responsibility is to assign the execution of Block 0 through 1999 to SM0, SM1, ... SM5. It assigns each block based on which SM is available at the time of the assignment. For example, assume that Block 0 is assigned to SM0, Block 1→SM1, ... Block 5 →SM5. Now what? Possibly, Gina will keep assigning them in a round robin fashion, Block 6→SM0, Block 7→SM1, ..., Block 11→SM5, ... However, the assignment is a super fast operation, while executing a block is much slower. This means that Gina will be done assigning all of the blocks super fast, and because each SM is limited to how many blocks it can get from Gina before it blocks Gina from assigning any more, at some point, Gina will pause and wait for another SM to be ready. In other words, in Figure 8.1, the boxes next to the school buses have a limit of how many tasks (blocks) they can hold. Beyond that Gina has to wait until a box has room freed up. This will continue until Gina assigns Block 1999, which is when she is done. The kernel launch with 2000 blocks will be complete when

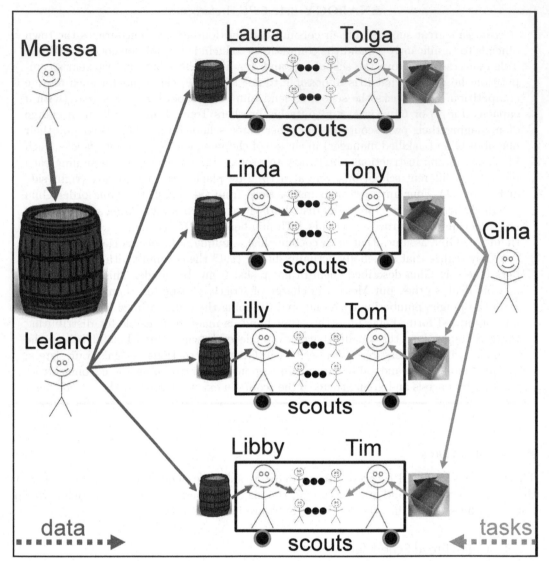

FIGURE 8.1 Analogy 8.1 for executing a massively parallel program using a significant number of GPU cores, which receive their instructions and data from different sources. Melissa (*Memory controller*) is solely responsible for bringing the coconuts from the jungle and dumping them into the big barrel (*L2$*). Larry (*L2$ controller*) is responsible for distributing these coconuts into the smaller barrels (L1$) of Laura, Linda, Lilly, and Libby; eventually, these four folks distribute the coconuts (*data*) to the scouts (*GPU cores*). On the right side, Gina (*Giga-Thread Scheduler*) has the big list of tasks (*list of blocks to be executed*); she assigns each block to a school bus (*SM* or *streaming multiprocessor*). Inside the bus, one person — Tolga, Tony, Tom, and Tim — is responsible to assign them to the scouts (*instruction schedulers*).

every single block has been completed, which will be a lot later than when Gina finishes her assignments. If another kernel is launched with a different number of blocks, Gina's job is to assign those to the buses, before even the execution of the previous ones are done. A very important note here is that Gina is also responsible for assigning a block ID to each block that she assigns (from 0 to 1999 in this specific case). It is important to note that out of all of the variables in Table 7.2, Gina's responsibility is to make a note of the `gridDim`, `blockIdx`, and `blockDim` variables on the papers she is preparing. In other words, it is the Giga Thread Scheduler that passes these variables onto the GPU cores that will execute the corresponding blocks. Alternatively, the assignment of the `threadIdx` variables has nothing to do with Gina. They will end up being the responsibility of Tolga (and his brothers) when he is assigning tasks to the scouts.

8.2.4   Memory Controllers

Melissa is the DRAM controller and she is responsible for bringing big chunks of data from global memory (GM) into the L2$. Much like the L1$ memory in each SM, L2$ is also parallel, however, there is a big difference between their operation: while L2$ is *coherent*, L1$ memory areas are not coherent with each other. In other words, the memory addresses inside the L2$ all refer to exactly the same memory areas, while the L1$ memory areas in each SM are disconnected from each other. This is due to the fact that making cache memory coherent substantially slows it down; the slowdown is acceptable in the L2$, because frequently used data eventually makes its way into the L1$, but high-performance is preferred in L1$ instead of coherence. This also contrasts with the CPU cache sub-system, in which every cache is coherent.

8.2.5   Shared Cache Memory (L2$)

This is the only shared memory in each GPU and it is the Last Level Cache (LLC). The GPU inside the LLC is surprisingly small compared to the CPU cache memory. While our i7-5930K had a 15 MB LLC, the GTX550Ti only has a 768 MB LLC. The reason is the VLSI design constraints that do not allow such large cache memories to be manufactured that can supply parallel copies of data fast enough to the GPU cores. Much like what we saw when studying the CPU architecture, L2$ works exactly the same: nothing goes from the GM directly into the cores; everything has to go through LLC (in this specific case, L2$). This allows commonly used data elements to be cached for later use.

8.2.6   Host Interface

This is the controller inside the GPU that is responsible for interfacing to the PCIe bus. This is what allows the GPU to shuttle data back and forth between the CPU and itself. It is very similar to the I/O controller we saw inside the CPU.

Figure 8.2 shows the inside of a GTX550Ti GPU, which looks exactly like our analogy except with six school buses. Each small barrel in Figure 8.1 next to a school bus represents the L1$ of that SM. Therefore, there is only one L1$ for 32 cores. This contrasts with the CPU architecture in a big way because each CPU core has its own dedicated L1$ (as well as a nice L2$). Although this looks like a major performance disadvantage for the GPU, don't be tricked; each L1$ in the SM of the GPU is a *parallel cache memory*, which is capable of supplying 16 cores data in parallel. The red boxes next to the cores in Figure 8.2 (see ebook for color version) represent the Load/Store queues designed to facilitate this parallel data read. So, as long as your program reads data in a massively parallel way (which we will study in detail in Chapter 10), this L1$ is a lot more powerful than the CPU's L1$ and

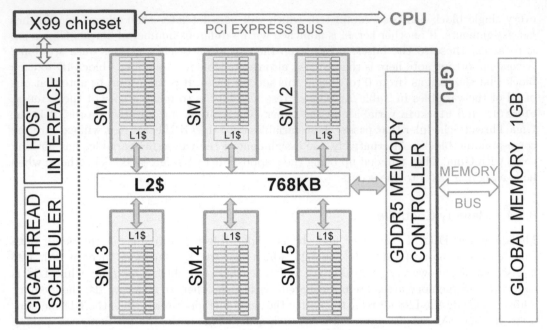

FIGURE 8.2 The internal architecture of the GTX550Ti GPU. A total of 192 GPU cores are organized into six streaming multiprocessor (SM) groups of 32 GPU cores. A single L2$ is shared among all 192 cores, while each SM has its own L1$. A dedicated memory controller is responsible for bringing data in and out of the GDDR5 global memory and dumping it into the shared L2$, while a dedicated host interface is responsible for shuttling data (and code) between the CPU and GPU over the PCIe bus.

L2$ combined. This is the reason there are only two types of cache memory in the GPU: L1$ is inside the SM and L2$ is shared among every SM.

8.3 NVIDIA GPU ARCHITECTURES

In the past two decades of GPU evolution, one fact never changed: in every generation, Nvidia, as well as other GPU designers, kept increasing the number of cores in their GPUs. For example, the GTX 550Ti we just mentioned included 192 cores in it, therefore, it was easy to design it as 6 SMs with 32 cores in each SM. In that Fermi generation of GPUs, the maximum number of cores a Fermi GPU could have was 512, thereby requiring 16 SMs. However, as the number of cores increased beyond this, Nvidia (and other GPU manufacturers) had to come up with different ways to organize the cores because increasing the number of SMs significantly increased the chip area and does not necessarily provide the same performance improvement as, for example, increasing the number of cores in each SM to 64. Going back to our Analogy 8.1, in which we had 128 scouts (32 in each school bus), what if you wanted to increase the number of scouts to 256? What is the best solution? (1) Get 8 school buses, or (2) put 64 scouts in each bus. The latter solution would eliminate the need for 4 extra barrels (L1$), and four more task schedulers, four more everything ... So, it might be a much better idea to redesign the inside of school buses to seat 64 scouts and make Tolga and Laura do a little more work. This might potentially make things a tiny bit slower in each school bus, but saves a lot of resources, which in turn allows us to get more work done as a result, and this is all we care about.

With each new generation of Nvidia GPUs, their Parallel Thread Execution (PTX) Instruction Set Architecture (ISA) evolved. You can think of the new ISAs as follows: in new generation GPUs, the scouts in each school bus got more experienced and they had *new instructions* to peel the coconuts more efficiently, maybe using more advanced tools. This allowed them to get things done faster by doing more average work with each instruction. There is a direct correlation between the PTX ISA and the Compute Capability (CC), as we saw in Section 7.7. Each new CC effectively requires a new ISA. So, every generation of Nvidia GPUs not only introduced a new CC, which is what the programmer cares about, but also a new PTX ISA. Now, let us study each generation. A summary of some key parameters for each generation is provided in Table 8.1.

8.3.1 Fermi Architecture

Although obsolete today, Fermi is a good microarchitecture to study from an instructional standpoint, due to its simplicity (by today's standards). I provided run time results for some old Fermi GPUs (like GTX 550Ti) in Table 7.6. Fermi microarchitecture was designed to incorporate a maximum of 512 GPU cores; each SM contained 32 cores, so Nvidia manufactured different models of Fermi-based GPUs by simply changing the number of SMs. For example, GT440 had 3 SMs (96 cores). GTX480 had 15 SMs (with 480 cores). After phasing out the 4xx series, Nvidia introduced the 5xx series, still based on the Fermi architecture; 5xx series was more power efficient than its predecessor 4xx and incorporated 16 SMs (512 cores).

8.3.2 GT, GTX, and Compute Accelerators

For each generation of Nvidia GPUs, you will find the "GT" and "GTX" model names for the consumer market and some other fancier name for the compute accelerator market. For example, in the year 2011, GT530 was a low-end and GTX580 was a high-end consumer market GPU; GT530 only incorporated 96 cores, while GTX580 boasted 512 cores. None of these GPUs had significant double-precision floating point computational power. They were priced for the consumer (especially the *game*) market ($50–$700), which didn't care about double-precision performance. Furthermore, GTX 580 could have a maximum of 3 GB GPU memory (GDDR5). However, if you are trying to do scientific computing, double-precision computational capability is extremely important, as well as a higher amount of memory; Nvidia manufactured a totally different line of products named *compute accelerators* (e.g., the C2075 compute module back in the Fermi days) that had 6 GB memory and much higher double-precision performance, albeit at a much higher price ($2000–$3000), geared toward the scientific computing market.

While GTX580 single-precision performance was 1581 GFLOPS and double-precision performance was ≈50 GFLOPS, the C2075 had a single versus double precision performance of 1030 versus 515 GFLOPS. In other words, you could pay something like 5× more to get 5× higher double-precision performance. For scientific applications that cannot live without double-precision, this is almost like having five GPUs in one GPU, while only using a single slot for it. These accelerators also included ECC memory support, which is important for the corporate market. On the flip side, their "graphic" capability was inferior to the GT or GTX family, because they didn't really need to *display* anything; they just needed to *compute.* Some accelerators barely had a single monitor output (e.g., C2075), while some of the earlier models (e.g., C1060) had no monitor output. Nvidia still continues to offer its products in these two categories, although its higher-end GPUs started offering higher double-precision performance.

8.3.3 Kepler Architecture

Kepler was an exciting improvement over Fermi and there are still plenty of Kepler cards in the corporate and consumer computers today. Nvidia redesigned the SMs to cram 192 cores into each and called them SMX. Going from Compute Capability 2.x to 3.x allowed Nvidia to substantially improve the internal parametric limitations; we just saw one dramatic example of it in Table 7.6, where the upper limit of the *x-dimension number of threads per block* was increased from 65,535 to 2 billion. Furthermore, the SM structure was redesigned (as shown in Table 8.1) to allow a much higher number of cores (512 in Fermi vs. 2880 in Kepler). To support such a large number of cores, the L2$ was increased to 1.5 MB from 768 KB.

Based on the Kepler engine, Nvidia introduced the K10, K20, K20X, K40, and K80 computer accelerators, which had ECC support and high double-precision capability advantages, much like the C2075 in the Fermi days. Remember from our discussion of C2075 that it had inferior graphics capabilities; K10–K80 family was no different. In these cards, the design priority was a huge double-precision power among some other extra features like ECC. Nvidia didn't even manufacture Kxx compute accelerators with any monitor output, which allowed it to use the extra chip area for even improved double-precision and more cores.

This was the time Nvidia realized two things: (1) nobody really used the monitor output in compute units, like the K80, because they were built into servers and accessed remotely (using ssh or something); these servers also had a very simple video output anyway (VGA or DVI or something), and (2) there was a growing demand for personal GPUs with excellent graphics capabilities and K80-like double-precision performance. This is when the GTX Titan was born. GTX Titan GPU was a $1000 card that had a respectable double-precision floating point performance (1500 GFLOPS) and excellent single-precision floating performance that is necessary for gaming (4500 GFLOPS). The GTX Titan Z is a two-GPUS-in-one-GPU, which has an 8122 GFLOPS single-precision and 2707 GFLOPS double-precision floating point performance. Compare that with the K80, with almost the same (8736 vs. 2912) performance. Not surprisingly, though, the price of the GTX Titan Z was ≈$3K, which was comparable to the K80. To summarize, one was designed to be used in a server and the other in a regular personal computer. I provided a lot of performance results with the Kepler GTX Titan Z and will be providing multiple results in the following chapters using the GTX Titan X (Pascal).

8.3.4 Maxwell Architecture

In the Maxwell architecture, Nvidia crammed a significantly higher number of cores into each SM (which they called SMM). They also increased the L2$ size to 2 MB. One very interesting difference between Maxwell and any other Nvidia GPU is that even the M10, M40, and M60 compute accelerators had very low double-precision performance, although they had pretty good single-precision performance. Their M60 (also a two-in-one GPU) had a 9650 versus 301 GFLOPS single versus double-precision performance, respectively. For someone who requires a high double-precision performance, this was very disappointing. However, for

TABLE 8.1   Nvidia microarchitecture families and their hardware features.

Family	Intro Year	Comp Cap.	Cores /SM	Total SMs	Total cores	L1$ (KB)	L2$ (KB)	Model (GTX)
Fermi	2009	2.x	32	16 SM	512	48	768	4xx 5xx
Kepler	2012	3.x	192	15 SMX	2880	48	1536	6xx 7xx
Maxwell	2014	5.x	128	24 SMM	3072	96	2048	7xx 8xx 9xx
Pascal	2016	6.x	64	60 SM	3840	64	4096	10xx
Volta	2018	7.x						

applications like deep-learning, where double-precision is not needed, the double-precision disadvantage is not a big deal.

8.3.5   Pascal Architecture and NVLink

Pascal engine incorporates 4 MB L2$ and is designed toward smaller but resource-rich SMs. Due to the high number of SMs, each SM is actually designed to be a part of a larger organizational unit named Graphics Processing Clusters (GPC). A GPC contains 10 SMs and each SM contains 32 cores. A much more detailed architectural diagram will be presented when we study the GPU cores and memory.

With the introduction of the Pascal engine, Nvidia introduced its own bus, NVLink, to address the PCIe bandwidth issue. Looking back to 2012, PCIe bandwidth never improved (stayed at 3.0), while Nvidia introduced three generations of GPUs (Kepler, Maxwell, and Pascal) with increasing internal global memory bandwidth. NVLink bus is designed to provide throughputs up to 80 GBps between the host and the device, which is much better than the 15.75 GBps of PCIe 3. However, the NVLink will not be available in regular PCs; it will only be available in supercomputers and potentially high-end graphics stations.

8.4   CUDA EDGE DETECTION: IMEDGEG.CU

In Section 5.1, we introduced the imdedge.c program, which was intended to be implemented using CPU multithreaded code. Distinct operations in this program (Gaussian, Sobel, etc.) took the image data from one array and wrote the computed version in a different array. We will do exactly the same thing in the GPU version of the program, imedgeG.cu.

8.4.1   Variables to Store the Image in CPU, GPU Memory

Starting with the original image in the CPU memory, we will transfer it to the GPU (device) side and will apply the same exact operations as the CPU version and the data will move from one memory area to another, as shown in Table 8.2. Note that we are avoiding using the term "array" for reasons that will become clear shortly. Now, let's look at the GPU memory areas responsible for holding the results of the separate operations.

8.4.1.1   *TheImage* and *CopyImage*

These two are unsigned char pointers and **TheImage** points to the original image that is read from the disk, while **CopyImage** points to the image that is processed by the GPU and will be written to disk as the resulting (processed) image. It is easier to refer to their sizes by choosing an example image, e.g., astronaut.bmp, which is ≈121 MB. Therefore, both of these image files are 121 MB and are allocated as follows, including the code to read in the original image:

```
TheImg = ReadBMPlin(InputFileName); // Read the input image ...
if (TheImg == NULL){ ... }
CopyImg = (uch *)malloc(IMAGESIZE);
if (CopyImg == NULL){ ... }
```

8.4.1.2 *GPUImg*

As shown in Table 8.2, `GPUImg` is the pointer to the GPU memory area where the original image will reside. Based on our example, it takes up 121 MB and the original CPU image is copied into this area as follows:

```
cudaStatus = cudaMemcpy(GPUImg, TheImg, IMAGESIZE, cudaMemcpyHostToDevice);
if (cudaStatus != cudaSuccess) { ... }
```

8.4.1.3 *GPUBWImg*

Because our algorithm needs only a B&W version of the image, the original image is immediately turned into its B&W version using the BWKernel(), which is saved in the memory area pointed to by `GPUBWImg`. The size of this area is one `double` per pixel, which stores the B&W value of the pixel. For our example astronaut.bmp image, it is 325 MB. In Table 8.2 the "Data Move" column indicates the total amount of data that each kernel is responsible for *moving*. For example, for the BWKernel(), this is indicated as 446 MB (i.e., 121+325). The reason for that is the fact that this kernel has to read all of the `GPUImg` area (121 MB), compute the resulting B&W image, and write the resulting B&W image into `GPUBWImg` area (325 MB), therefore moving around a total of 446 MB worth of data. Both of these memory areas are in global memory (GM); reading and writing from GM do not necessarily take the same amount of time; however, for the sake of simplicity, it doesn't hurt our argument to assume that they are equal and calculate some meaningful bandwidth metrics.

8.4.1.4 *GPUGaussImg*

The GaussKernel() in our GPU code will take the B&W image (`GPUBWImg`) and compute its Gaussian-filtered version and write it into the memory area that is pointed to by `GPUGaussImg`. For each B&W pixel (type `double`), one Gaussian-filtered pixel (also type `double`) is computed; therefore, the size of both of these memory areas is the same (325 MB). So, the total amount of data that the GaussKernel() has to move is 650 MB, as indicated in Table 8.2.

8.4.1.5 *GPUGradient* and *GPUTheta*

The SobelKernel() is responsible for computing a gradient (`GPUGradient`) and angle (`GPUTheta`) for each pixel, that are of type `double`. Therefore, the total size that this kernel must move is 325+325+325=975 MB. We must note here that although `GPUGradient`

TABLE 8.2 Kernels used in imedgeG.cu, along with their source array name and type.

| Kernel Name | Source | | | Destination | | | Data Move |
	Name	Type	Size	Name	Type	Size	
BWKernel	GPUImg	uc	121	GPUBWImg	double	325	446
GaussKernel	GPUBWImg	double	325	GPUGaussImg	double	325	650
SobelKernel	GPUGaussImg	double	325	GPUGradient GPUTheta	double	650	975
Threshold Kernel	GPUGradient GPUTheta	double	325	GPUResultImg	uc	121	446
						Total »»»	2517

The amount of data that each kernel manipulates for the astronaut.bmp file is also shown. All sizes are in MB. "uc" denotes unsigned char.

is always computed, GPUTheta is only computed in some cases, as we discussed previously. Therefore, the actual amount of data this kernel moves is less than 975 MB.

8.4.1.6 GPUResultImg

The ThresholdKernel() is responsible for computing the B&W version of the image and write it into the memory area pointed to by GPUResultImg, which will eventually be copied into the CPU's CopyImage. This resulting image writes 0-0-0 in pixels that will be black (edge) and 255-255-255 for pixels that will be white (no-edge). Note that the color used for edge versus no-edge can be changed using the #define, as described in the CPU version of the program. Sometimes you want black to indicate no-edge. This is useful to print the edge-detected version of the image versus display it on a compute monitor.

8.4.2 Allocating Memory for the GPU Variables

In Section 8.4.1, we went over the GPU-side memory areas we need to compute the final edge-detected image. These six areas, pointed to by the pointers shown in Table 8.2, are allocated in a single cudaMalloc() using the following lines of Code 8.1:

```
#define IMAGESIZE  (ip.Hbytes*ip.Vpixels)
#define IMAGEPIX (ip.Hpixels*ip.Vpixels)
   ...
   GPUtotalBufferSize = 4 * sizeof(double)*IMAGEPIX + 2 * sizeof(uch)*IMAGESIZE;
   cudaStatus = cudaMalloc((void**)&GPUptr, GPUtotalBufferSize); if(...);
```

The quantity 2*sizeof(uch)*IMAGESIZE computes the area needed for the initial (GPUImg) and the final (GPUResultImg) images, while the intermediate results require 4*sizeof((double))*IMAGEPIX to store GPUBWImg, GPUGaussImg, GPUGradient, and GPUTheta. In the C programming language, as long as you know the types of the variables, you can address these areas as if they were arrays, which is what is done in imedgeG.cu. The trick is to set the type of the pointers to the type of the variables that the pointer is pointing to. This is why the initial cudaMalloc() is of type void*, which allows infinite flexibility in setting each individual pointer to whatever type memory pointer our heart pleases in Code 8.1 as follows:

```
GPUImg       = (uch *)GPUptr;
GPUResultImg = GPUImg + IMAGESIZE;
GPUBWImg     = (double *)(GPUResultImg + IMAGESIZE);
GPUGaussImg  = GPUBWImg + IMAGEPIX;
GPUGradient  = GPUGaussImg + IMAGEPIX;
GPUTheta     = GPUGradient + IMAGEPIX;
```

Look at the first line:

```
GPUImg       = (uch *)GPUptr;
```

This is nothing more than a copy of two 64-bit variables, however, the big deal is casting the pointer type to uch; this lets the compiler know that the pointer on the left side (GPUImg) is now of type uch * (correctly pronounced "uch pointer" or "pointer to unsigned character" or even more precisely "pointer to an array where each array element is an unsigned char and occupies a single byte"). Remember that the right-side pointer (GPUptr) was of type void*, which is a way of saying "it is a 64-bit integer that represents a typeless pointer." Performing the casting serves a crucial purpose: now that the compiler knows that the left-side pointer is of type unsigned char, it will perform *pointer arithmetic* based on that. Look at the next line:

```
GPUResultImg = GPUImg + IMAGESIZE;
```

One thing you cannot take for granted here is the meaning of the "+" operator on two different types; GPUImg is an uch* and IMAGESIZE is some type of an integer. How would you perform the addition in this case? The answer comes from prescribed *pointer arithmetic rules* in the C language. Because of the casting, the compiler knows that GPUImg is pointing to an array which has elements of size 1 byte. Therefore, each integer indeed adds 1 and GPUResultImg points to a GPU memory area that is IMAGESIZE bytes apart from GPUImg. In our example, when we are processing astronaut.bmp, each area is 121 MB; therefore, GPUImg is at the very beginning of the allocated area (call it offset=0) and GPUResultImg is offset 121 MB. This was easy because both of these pointers are the same type; therefore, a simple addition — without casting — suffices.

CODE 8.1: imedgeG.cu ... main() {...

First part of main() in imedgeG.cu; `time2BW`, `time2Gauss`, and `time2Sobel` are used for time-stamping the execution time of four separate GPU kernels.

```
#define CEIL(a,b)   ((a+b-1)/b)
#define IPH         ip.Hpixels
#define IPV         ip.Vpixels
#define IMAGESIZE   (ip.Hbytes*ip.Vpixels)
#define IMAGEPIX (ip.Hpixels*ip.Vpixels)
uch *TheImg, *CopyImg, *GPUImg, *GPUResultImg; // CPU, GPU image pointers
int  ThreshLo=50, ThreshHi=100;    // "Edge" vs. "No Edge" thresholds
double *GPUBWImg, *GPUGaussImg, *GPUGradient, *GPUTheta;
...
int main(int argc, char **argv)
{
   cudaEvent_t time1, time2, time2BW, time2Gauss, time2Sobel, time3, time4;
   ui BlkPerRow, ThrPerBlk=256, NumBlocks, GPUDataTransfer;
   ...
   TheImg = ReadBMPlin(InputFileName);  if(TheImg==NULL){ ... }
   CopyImg = (uch *)malloc(IMAGESIZE);  if(CopyImg == NULL){ ... }
   cudaGetDeviceCount(&NumGPUs);   if (NumGPUs == 0){ ... }
   cudaGetDeviceProperties(&GPUprop, 0);
   ...
   cudaEventCreate(&time1); ...(&time2BW); ...(&time2Gauss); ...(&time2Sobel); ...
   cudaEventRecord(time1, 0);  // Time stamp at the start of the GPU transfer
   GPUtotalBufferSize = 4 * sizeof(double)*IMAGEPIX + 2 * sizeof(uch)*IMAGESIZE;
   cudaStatus = cudaMalloc((void**)&GPUptr, GPUtotalBufferSize); if(...);
   GPUImg      = (uch *)GPUptr;
   GPUResultImg = GPUImg + IMAGESIZE;
   GPUBWImg    = (double *)(GPUResultImg + IMAGESIZE);
   GPUGaussImg = GPUBWImg + IMAGEPIX;
   GPUGradient = GPUGaussImg + IMAGEPIX;
   GPUTheta    = GPUGradient + IMAGEPIX;

   cudaStatus = cudaMemcpy(GPUImg, TheImg, IMAGESIZE, cudaMemcpyHostToDevice);
   if(...);   cudaEventRecord(time2, 0); // Time stamp after CPU->GPU tfr

   BlkPerRow=CEIL(IPH, ThrPerBlk);            NumBlocks=IPV*BlkPerRow;

   BWKernel <<< NumBlocks, ThrPerBlk >>> (GPUBWImg, GPUImg, IPH);
   if(...); cudaEventRecord(time2BW, 0); // Time stamp after BW image calculation
   GaussKernel <<< NumBlocks, ThrPerBlk >>> (GPUGaussImg, GPUBWImg, IPH, IPV);
   if(...); cudaEventRecord(time2Gauss, 0); // after Gauss image calculation
   SobelKernel <<< Num...,... >>> (GPUGradient, GPUTheta, GPUGaussImg, IPH, IPV);
   if(...); cudaEventRecord(time2Sobel, 0); // after Gradient, Theta computation
   ThresholdKernel <<< NumBlocks, ThrPerBlk >>> (GPUResultImg, GPUGradient,
       GPUTheta, IPH, IPV, ThreshLo, ThreshHi);
   if(...); cudaEventRecord(time3, 0); // after threshold
   cudaStatus=cudaMemcpy(CopyImg, GPUResultImg, IMAGESIZE, cudaMemcpyDeviceToHost);
   cudaEventRecord(time4, 0); // after GPU-> CPU tfr
   ...
```

Let's continue the pointer computations, the next line is:

```
GPUBWImg    = (double *)(GPUResultImg + IMAGESIZE);
```

which is fairly straightforward: IMAGESIZE the amount of area we need for the `GPUResultImg` array, which can be added to this variable using uch * pointer arithmetic, however, the resulting pointer must be cast to `double*` because the new area that follows `GPUResultImg` is `GPUBWImg` and will point to an array of `double` type elements with size 64-bits (8 bytes). Once the pointer `GPUBWImg` is computed, the remaining three pointers are also type `double*` and computed one `double*` from another is a matter of adding "how many double elements apart" as follows:

```
GPUGaussImg = GPUBWImg + IMAGEPIX;
```

which means "IMAGEPIX elements apart, where each element is of type `double`" or alternatively "IMAGEPIX*8 bytes apart" or if you want to be even more technically correct, "IMAGEPIX*`sizeof(double)` bytes apart." The remaining pointer computations follow exactly the same logic:

```
GPUGradient = GPUGaussImg + IMAGEPIX;
GPUTheta    = GPUGradient + IMAGEPIX;
```

8.4.3 Calling the Kernels and Time-Stamping Their Execution

In Code 8.1, we start by copying the image that is read from the disk into the CPU memory (`TheImg`) into the allocated GPU memory area (`GPUImg`) using the following lines:

```
cudaStatus = cudaMemcpy(GPUImg, TheImg, IMAGESIZE, cudaMemcpyHostToDevice);
if(...);    cudaEventRecord(time2, 0); // Time stamp after CPU->GPU tfr
```

We use the variable `time2` to time-stamp the beginning of the remaining lines, which will call the GPU kernels to perform edge detection. The first kernel is BWKernel(), which computes the B&W version of the original RGB image:

```
BlkPerRow=CEIL(IPH, ThrPerBlk);              NumBlocks=IPV*BlkPerRow;

BWKernel <<< NumBlocks, ThrPerBlk >>> (GPUBWImg, GPUImg, IPH);
if(...); cudaEventRecord(time2BW, 0); // Time stamp after BW image calculation
```

Here, `BlkPerRow` and `ThrPerBlk` variables are needed for every kernel, so they are calculated before launching any of the kernels. The BWKernel() takes the original image that just got transferred into GPU memory (`GPUImg`) and writes its B&W version into the GPU memory area that holds the B&W image (`GPUBWImg`), consisted of `double` elements. The time it finishes its execution is time-stamped with the `time2BW` variable.

The other two kernels are called and time-stamped in the same way:

```
GaussKernel <<< NumBlocks, ThrPerBlk >>> (GPUGaussImg, GPUBWImg, IPH, IPV);
if(...); cudaEventRecord(time2Gauss, 0); // after Gauss image calculation
SobelKernel <<< Num...,... >>> (GPUGradient, GPUTheta, GPUGaussImg, IPH, IPV);
if(...); cudaEventRecord(time2Sobel, 0); // after Gradient, Theta computation
ThresholdKernel <<< NumBlocks, ThrPerBlk >>> (GPUResultImg, GPUGradient,
    GPUTheta, IPH, IPV, ThreshLo, ThreshHi);
if(...); cudaEventRecord(time3, 0); // after threshold
```

The time when all kernels finish executing is time-stamped with the `time3` variable. The result resides in the GPU memory area (`GPUResultImg`) and is transferred into CPU memory as follows:

```
cudaStatus=cudaMemcpy(CopyImg, GPUResultImg, IMAGESIZE, cudaMemcpyDeviceToHost);
cudaEventRecord(time4, 0); // after GPU-> CPU tfr
```

8.4.4 Computing the Kernel Performance

We have gone through an extensive description of the edge detection algorithm in Section 5.1 and we will not repeat it here. We detailed its CPU implementation in Section 5.2, which involved four separate functions; we measured their execution time and realized that similar modifications in each one of these four functions resulted in different performance improvements. This was because of the nature of each function being different; some are core-intensive, some are memory-intensive. The GPU version of the program (imedgeG.cu) will perform exactly the same functions using kernels that we will improve independently to observe the effect of each improvement. Much like their CPU version, each kernel can be memory-intensive, core-intensive, or a little bit of both. Most importantly, because of the differences in CPU and GPU memory sub-systems, the memory-intensiveness of a given GPU kernel (e.g., the GaussKernel()) can be different than its CPU counterpart (the Gauss() function in this specific case). Regardless of this fact, we can measure one parameter with great accuracy: the utilized bandwidth, according to Equation 4.6. This equation states that the memory bandwidth that a function utilizes is the ratio of the amount of data it moves during its execution by the amount of time it takes. This is the equation we will use to compute how close each one of the GPU kernels is getting to saturating the bandwidth of the global memory. We will run the same kernel on multiple GPUs to observe its behavior. We will also use this metric as a baseline for assessing our improvements; as we will see shortly, this metric will help even if the kernel is core-intensive.

8.4.5 Computing the Amount of Kernel Data Movement

Based on Equation 4.6, all we need to determine a kernel's performance is its execution time and the amount of data it moves. We know that the time stamps already take care of keeping track of individual kernel execution times. So, all that is left is determining the amount of data each kernel moves. It is worth repeating here that we are not distinguishing between memory reads and memory writes, which definitely matters. However, from the standpoint of assessing the *relative performance improvement* between two versions of the same kernel, this crude method (i.e., lumping together memory reads and writes and calling them *memory movements*) is good enough. Code 8.2 shows the part of main() that is responsible for keeping track of the amount of data movement for each kernel. For the first three kernels, this is what we have:

```
BWKernel <<< ... >>> (...);   if (...);      cudaEventRecord(time2BW, 0);
GPUDataTfrBW = sizeof(double)*IMAGEPIX + sizeof(uch)*IMAGESIZE;
GaussKernel <<< ... >>> ();   if (...);      cudaEventRecord(time2Gauss, 0);
GPUDataTfrGauss = 2*sizeof(double)*IMAGEPIX;
SobelKernel <<< ... >>>();   if (...);      cudaEventRecord(time2Sobel, 0);
GPUDataTfrSobel = 3 * sizeof(double)*IMAGEPIX;
```

The variables `GPUDataTfrBW`, `GPUDataTfrGauss`, and `GPUDataTfrSobel` correspond to the example values shown in Table 8.2 and will end up with 446 MB, 650 MB, and 975 MB, respectively for the astronaut.bmp image. These three kernels share a characteristic that makes it easy to determine the amount of data they move with 100% certainty: the data they move does not depend on any condition, i.e., they always move the same amount of data for a given image. However, the same cannot be said about ThresholdKernel(). While the `GPUGradient` is always computed, the `GPUTheta` entries are written only if necessary. These are the lines that compute the amount of data movement in ThresholdKernel():

```
ThresholdKernel <<< ... >>> (...);   if (...);
GPUDataTfrThresh=sizeof(double)*IMAGEPIX + sizeof(uch)*IMAGESIZE;
```

Clearly, this is the lower estimate for how much data is moving, while the upper estimate would require us to add another **IMAGEPIX**, i.e., if every single `GPUTheta` value is calculated, as shown in Table 8.2. In case of such a scenario, the best course of action is to avoid making critical decisions about kernel performance based on this specific kernel. The other three kernels will allow us to judge the performance impact of our improvements more than sufficiently anyway, so this hurdle will not hold us back from making accurate judgments. Finally, the following lines are needed to compute the total amount of data movement:

```
GPUDataTfrKernel=GPUDataTfrBW+GPUDataTfrGauss+GPUDataTfrSobel+GPUDataTfrThresh;
GPUDataTfrTotal =GPUDataTfrKernel + 2 * IMAGESIZE;
```

Note that the addition of 2*IMAGESIZE takes into account the original and final image that will be transported from/to the CPU.

CODE 8.2: imedgeG.cu   ... main() ...}

Time reporting part of main() in imedgeG.cu. The amount of data each kernel manipulates is in the `GPUDataTfrBW`, `GPUDataTfr...` variables.

```
#define MB(bytes)      (bytes/1024/1024)
#define BW(bytes,timems) ((float)bytes/(timems * 1.024*1024.0*1024.0))
...
int main(int argc, char **argv)
{
   ...
   BWKernel <<< ... >>> (...);   if (...);       cudaEventRecord(time2BW, 0);
   GPUDataTfrBW = sizeof(double)*IMAGEPIX + sizeof(uch)*IMAGESIZE;
   GaussKernel <<< ... >>> ();   if (...);       cudaEventRecord(time2Gauss, 0);
   GPUDataTfrGauss = 2*sizeof(double)*IMAGEPIX;
   SobelKernel <<< ... >>>();  if (...);     cudaEventRecord(time2Sobel, 0);
   GPUDataTfrSobel = 3 * sizeof(double)*IMAGEPIX;
   ThresholdKernel <<< ... >>> (...);  if (...);
   GPUDataTfrThresh=sizeof(double)*IMAGEPIX + sizeof(uch)*IMAGESIZE;
   GPUDataTfrKernel=GPUDataTfrBW+GPUDataTfrGauss+GPUDataTfrSobel+GPUDataTfrThresh;
   GPUDataTfrTotal =GPUDataTfrKernel + 2 * IMAGESIZE;
   cudaEventRecord(time3, 0);
   ...
   printf("\n\n--------------------\n");
   printf("%s   ComputeCapab=%d.%d [max %s blocks; %d thr/blk] \n",
   GPUprop.name, GPUprop.major, GPUprop.minor, SupportedBlocks, MaxThrPerBlk);
   printf("\n\n--------------------\n");
   printf("%s %s %s %u %d %d [%u BLOCKS, %u BLOCKS/ROW]\n", ProgName,
       InputFileName, OutputFileName, ThrPerBlk, ThreshLo, ThreshHi, NumBlocks,
       BlkPerRow);
   printf("\n\n---         -----------\n");
   printf("          CPU->GPU Transfer =...\n", tfrCPUtoGPU, MB(IMAGESIZE),
       BW(IMAGESIZE,tfrCPUtoGPU));
   printf("          GPU->CPU Transfer =...\n", tfrGPUtoCPU, MB(IMAGESIZE),
       BW(IMAGESIZE, tfrGPUtoCPU));
   printf("\n\n--------------------\n");
   printf("     BW Kernel Execution Time =...\n", kernelExecTimeBW,
       MB(GPUDataTfrBW), BW(GPUDataTfrBW, kernelExecTimeBW));
   printf("   Gauss Kernel Execution Time =...\n", kernelExecTimeGauss,
       MB(GPUDataTfrGauss), BW(GPUDataTfrGauss, kernelExecTimeGauss));
   printf("   Sobel Kernel Execution Time =...\n", kernelExecTimeSobel,
       MB(GPUDataTfrSobel), BW(GPUDataTfrSobel, kernelExecTimeSobel));
   printf("Threshold Kernel Execution Time =...\n", kernelExecTimeThreshold,
       MB(GPUDataTfrThresh), BW(GPUDataTfrThresh, kernelExecTimeThreshold));
   printf("\n\n--------------------\n");
   printf("       Total Kernel-only time =...\n", totalKernelTime,
       MB(GPUDataTfrKernel), BW(GPUDataTfrKernel, totalKernelTime));
   printf("  Total time with I/O included =...\n", totalTime, MB(GPUDataTfrTotal),
       BW(GPUDataTfrTotal, totalTime));
   printf("\n\n--------------------\n");
   ...
}
```

```
Z:\code>imedgeG astronaut.bmp output.bmp 256 50 100

 Input File name:      astronaut.bmp  (7918 x 5376)    File Size=127712256
Output File name:        output.bmp  (7918 x 5376)    File Size=127712256

----------------------------------------------------------------------------
GeForce GTX TITAN Z    ComputeCapab=3.5 [max 2048 M blocks; 1024 thr/blk]
----------------------------------------------------------------------------
imedgeG astronaut.bmp output.bmp 256 50 100  [166656 BLOCKS, 31 BLOCKS/ROW]
----------------------------------------------------------------------------
            CPU->GPU Transfer =   30.93 ms  ...   121 MB  ...    3.85 GB/s
            GPU->CPU Transfer =   28.51 ms  ...   121 MB  ...    4.17 GB/s
----------------------------------------------------------------------------
      BW Kernel Execution Time =    8.95 ms  ...   446 MB  ...   48.72 GB/s
   Gauss Kernel Execution Time =   48.11 ms  ...   649 MB  ...   13.19 GB/s
   Sobel Kernel Execution Time =   39.59 ms  ...   974 MB  ...   24.03 GB/s
Threshold Kernel Execution Time =    7.51 ms  ...   446 MB  ...   58.04 GB/s
----------------------------------------------------------------------------
         Total Kernel-only time =  104.16 ms  ...  2516 MB  ...   23.60 GB/s
    Total time with I/O included =  163.60 ms  ...  2760 MB  ...   16.48 GB/s
----------------------------------------------------------------------------
```

FIGURE 8.3 A sample output of the imedgeG.cu program executed on the astronaut.bmp image using a GTX Titan Z GPU. Kernel execution times and the amount of data movement for each kernel is clearly shown.

8.4.6 Reporting the Kernel Performance

The lines at the end of Code 8.2 report the performance of each kernel, including the amount of data each kernel moved as shown in Figure 8.3; in this figure, excluding the ThresholdKernel() from our analysis, the GaussKernel() achieves the lowest bandwidth utilization, while the BWKernel() achieves the highest. Excluding the I/O time, the total amount of data movement is 2516 MB, which is consistent with Table 8.2 and the overall bandwidth utilization for edge detection is 23.60 GBps, which is substantially below the 336 GBps that GTX Titan Z can offer. Therefore, intuitively, there may be a lot of room for improvement in our code.

8.5 IMEDGEG: KERNELS

In Section 8.6, we will run this program on multiple GPUs and observe their performance. For now, let us look at the details of each kernel.

8.5.1 BWKernel()

Code 8.3 provides a listing of the BWKernel(), which computes the B&W version of the image according to Equation 5.1. The computations required for this kernel are so simple that it almost looks like we shouldn't even bother with a detailed explanation; however, let's definitely dig deep into Code 8.3 and see how it is implemented and whether we could have done a better job.

CODE 8.3:   imedgeG.cu   ... BWKernel() {...}

The BWKernel() calculates the B&W image from an original color image.

```
// Kernel that calculates a B&W image from an RGB image
// resulting image has a double type for each pixel position
__global__
void BWKernel(double *ImgBW, uch *ImgGPU, ui Hpixels)
{
   ui ThrPerBlk = blockDim.x;
   ui MYbid = blockIdx.x;
   ui MYtid = threadIdx.x;
   ui MYgtid = ThrPerBlk * MYbid + MYtid;
   double R, G, B;

   ui BlkPerRow = CEIL(Hpixels, ThrPerBlk);
   ui RowBytes = (Hpixels * 3 + 3) & (~3);
   ui MYrow = MYbid / BlkPerRow;
   ui MYcol = MYgtid - MYrow*BlkPerRow*ThrPerBlk;
   if (MYcol >= Hpixels) return;   // col out of range

   ui MYsrcIndex = MYrow * RowBytes + 3 * MYcol;
   ui MYpixIndex = MYrow * Hpixels + MYcol;

   B = (double)ImgGPU[MYsrcIndex];
   G = (double)ImgGPU[MYsrcIndex + 1];
   R = (double)ImgGPU[MYsrcIndex + 2];
   ImgBW[MYpixIndex] = (R+G+B)/3.0;
}
```

Why not start with the very few lines at the beginning of the kernel? Let's skip the first three lines that are simple assignments and look at any line that involves some sort of a *computation*:

```
   ui MYgtid = ThrPerBlk * MYbid + MYtid;
   double R, G, B;

   ui BlkPerRow = CEIL(Hpixels, ThrPerBlk);
   ui RowBytes = (Hpixels * 3 + 3) & (~3);
   ui MYrow = MYbid / BlkPerRow;
   ui MYcol = MYgtid - MYrow*BlkPerRow*ThrPerBlk;
   if (MYcol >= Hpixels) return;   // col out of range
```

Do you see anything wrong? This is the problem with writing nice and organized code. CEIL is a macro that is defined at the very beginning of imedgeG.cu and contains the following few lines:

```
#define CEIL(a,b)   ((a+b-1)/b)
```

Now do you see anything even more wrong? I do! We stressed in Section 4.7.3 that integer divisions can be *weapons of mass destruction* for kernel performance, and, yet, you have TWO OF THEM for EACH PIXEL: one in computing MYrow and one in the CEIL macro.

So, for the astronaut.bmp image, you are forcing this kernel to perform 84 million integer divisions for 42 million pixels, worse yet, this is only to compute the B&W version of the image. We still have the Gauss, Sobel, etc. We are not done. Look at the way we compute the B&W pixel from its RGB components:

```
B = (double)ImgGPU[MYsrcIndex];
G = (double)ImgGPU[MYsrcIndex + 1];
R = (double)ImgGPU[MYsrcIndex + 2];
ImgBW[MYpixIndex] = (R+G+B)/3.0;
```

Do you see anything wrong? There is yet another division and it is of type `double`. So, this division is definitely slower than the previous integer type divisions. Additionally, the RGB pixel values are originally `unsigned char` and they are converted to `double`. Each such case is as bad as a regular `double` operation, which ties up double-precision computational resources. Could we have performed this computation differently? YES! Wait until we reach Chapter 9, when we will come up with many ideas to improve the performance of this kernel (and the others).

8.5.2 GaussKernel()

Code 8.4 provides a listing of the GaussKernel(), which computes the Gaussian-filtered image from its B&W version according to Equation 5.2. A close observation of this kernel shows the same two integer divisions on top and a single division — of type `double` on the bottom. However, these operations get lost in the wind when you look at the substantial amount of additions and multiplications performed in the nested `for` loops, not to mention the fact that every single one of these addition, multiplications are of type `double`. In the CPU version of imedge.c, we didn't have to pay too much attention to whether a floating point operation was `float` or `double`, i.e., single-precision or double-precision. Because all modern CPUs are 64-bits and the 64-bit length of double precision is their native data size. So, their performance to compute double-precision is almost the same as single precision, at least for the simple additional and multiplication operations. This contrasts substantially with GPUs; the native data size of all of the GPUs (at least up to the Pascal family) is 32 bits. Therefore, their performance tanks when you perform too many double precision floating point operations. As we will deeply investigate in Chapter 9, the performance difference can be a couple of orders of magnitude! Therefore, when we do not need to, we shouldn't use the double-precision floating point variables so generously.

Aside from this, there is so much more to talk about that we will go over all of it one by one. Let's look at the definition of the Gaussian filter constants:

```
__device__
double Gauss[5][5] = { { 2,  4,  5,  4,  2 },
                       { 4,  9, 12,  9,  4 },
                              ...
```

I didn't repeat all of it. The __device__ prefix we put in front of the array means that this array is a *device-side array*. In other words, we are letting the compiler decide where it goes. We don't really specify where it goes. We just know that it is on the device-side, not the host-side. Of course, this has to be the case, because every addition and multiplication in the kernel is requiring one of these 25 values. Inside the two nested `for` loops, there are about 150 integer and 75 `double` operations to compute a single pixel. You can see why we are not really terribly concerned with the 3 divisions anymore. We have bigger problems.

CODE 8.4: imedgeG.cu ... GaussKernel() {...}

The GaussKernel() calculates the Gaussian-blurred image from the B&W image, that just got calculated by the BWKernel().

```
__device__
double Gauss[5][5] = { { 2,    4,    5,    4,    2 },
                       { 4,    9,   12,    9,    4 },
                       { 5,   12,   15,   12,    5 },
                       { 4,    9,   12,    9,    4 },
                       { 2,    4,    5,    4,    2 } };
   // Kernel that calculates a Gauss image from the B&W image
   // resulting image has a double type for each pixel position
__global__
void GaussKernel(double *ImgGauss, double *ImgBW, ui Hpixels, ui Vpixels)
{
   ui ThrPerBlk = blockDim.x;
   ui MYbid = blockIdx.x;
   ui MYtid = threadIdx.x;
   ui MYgtid = ThrPerBlk * MYbid + MYtid;
   int row, col, indx, i, j;
   double G=0.00;

   //ui NumBlocks = gridDim.x;
   ui BlkPerRow = CEIL(Hpixels, ThrPerBlk);
   int MYrow = MYbid / BlkPerRow;
   int MYcol = MYgtid - MYrow*BlkPerRow*ThrPerBlk;
   if (MYcol >= Hpixels) return;     // col out of range

   ui MYpixIndex = MYrow * Hpixels + MYcol;
   if ((MYrow<2) || (MYrow>Vpixels - 3) || (MYcol<2) || (MYcol>Hpixels - 3)){
      ImgGauss[MYpixIndex] = 0.0;
      return;
   }else{
      G = 0.0;
      for (i = -2; i <= 2; i++){
         for (j = -2; j <= 2; j++){
            row = MYrow + i;
            col = MYcol + j;
            indx = row*Hpixels + col;
            G += (ImgBW[indx] * Gauss[i + 2][j + 2]);
         }
      }
      ImgGauss[MYpixIndex] = G / 159.00;
   }
}
```

8.5.3 SobelKernel()

Code 8.5 provides a listing of the SobelKernel(), which computes the Sobel-filtered image from its Gaussian-filtered version according to Equation 5.3 and Equation 5.4. This code looks very similar to its CPU sister, Code 5.5. So, from the implementation standpoint, we can feel comfortable that once we have the CPU version of something, it is pretty easy to develop its GPU version, right? NO! Not so fast!!! IF you learned one thing in this book, it is the fact that the code you are reading says nothing about the underlying hardware it will be mapped to. In other words, if the CPU and GPU hardware were identical, you would expect Code 5.5 (CPU Sobel) and Code 8.5 to give you identical performance; however, there will be so many differences that we will dedicate an entire chapter to how GPU cores work (Chapter 9) and another full chapter to how the GPU memory structure works (Chapter 10). The biggest difference between CPU and GPU cores is that GPUs are designed to fit 1000, 2000, 3000 cores inside them, while CPUs only have 4–12 cores. So, each core in the CPU is a lot faster, more capable, and can execute 64-bits natively, while the GPU cores are much simpler and they are designed to achieve their performance from their sheer quantity, rather than their sophisticated design. So, starting this chapter, the best practice for the readers will be to watch what strengths GPU has and try to use these strengths in their programs.

The SobelKernel() has two nested loops with 3 values in each loop, so it is 9 iterations total. We see plenty of integer operations:

```
row = MYrow + i;
col = MYcol + j;
indx = row*Hpixels + col;
```

and also double-precision multiplications and additions:

```
GX += (ImgGauss[indx] * Gx[i + 1][j + 1]);
GY += (ImgGauss[indx] * Gy[i + 1][j + 1]);
```

As you remember, we spent an entire Section 4.7 going over each mathematical operation and trying to understand how "bad" each one of these operations was in terms of computational requirement. Operations such as division, sin(), sqrt() were nothing but bad news for the CPU cores. Considering that GPU cores are even simpler, we do not expect them to have less damage to GPU performance. In SobelKernel, the most alarming lines are the following:

```
ImgGrad[MYpixIndex] = sqrt(GX*GX + GY*GY);
ImgTheta[MYpixIndex] = atan(GX / GY)*180.0 / PI;
```

Besides the less harmful double-precision addition and multiplication, these lines contain double-precision division, square root, and arc tangent. If the CPU equivalent of SobelKernel is any indication (the Sobel() function in Code 4.7), we expect these lines to execute very slow. The only good news is that these lines are only executed *once*, while the double-precision addition and multiplications inside the two for loops are executed many times. It is good to keep an eye on all of these details, but, in the end, we will be able to quantify all of these when we present runtime results for these kernels in the following pages.

CODE 8.5:   imedgeG.cu   ... SobelKernel() {...}

The SobelKernel() calculates the Gradient and Theta images from the Gaussian-blurred image. These will be used in deciding the magnitudes and angles of the edges.

```
__device__
double Gx[3][3] = { { -1, 0,  1 },
                    { -2, 0,  2 },
                    { -1, 0,  1 } };
__device__
double Gy[3][3] = { { -1, -2, -1 },
                    {  0,  0,  0 },
                    {  1,  2,  1 } };
// Kernel that calculates Gradient, Theta from the Gauss image
// resulting image has a double type for each pixel position
__global__
void SobelKernel(double *ImgGrad, double *ImgTheta, double *ImgGauss, ui Hpixels,
    ui Vpixels)
{
   ui ThrPerBlk = blockDim.x;
   ui MYbid = blockIdx.x;
   ui MYtid = threadIdx.x;
   ui MYgtid = ThrPerBlk * MYbid + MYtid;
   int row, col, indx, i, j;
   double GX,GY;

   //ui NumBlocks = gridDim.x;
   ui BlkPerRow = CEIL(Hpixels, ThrPerBlk);
   int MYrow = MYbid / BlkPerRow;
   int MYcol = MYgtid - MYrow*BlkPerRow*ThrPerBlk;
   if (MYcol >= Hpixels) return;    // col out of range

   ui MYpixIndex = MYrow * Hpixels + MYcol;
   if ((MYrow<1) || (MYrow>Vpixels - 2) || (MYcol<1) || (MYcol>Hpixels - 2)){
      ImgGrad[MYpixIndex] = 0.0;
      ImgTheta[MYpixIndex] = 0.0;
      return;
   }else{
      GX = 0.0;  GY = 0.0;
      for (i = -1; i <= 1; i++){
         for (j = -1; j <= 1; j++){
            row = MYrow + i;
            col = MYcol + j;
            indx = row*Hpixels + col;
            GX += (ImgGauss[indx] * Gx[i + 1][j + 1]);
            GY += (ImgGauss[indx] * Gy[i + 1][j + 1]);
         }
      }
      ImgGrad[MYpixIndex] = sqrt(GX*GX + GY*GY);
      ImgTheta[MYpixIndex] = atan(GX / GY)*180.0 / PI;
   }
}
```

CODE 8.6: imedgeG.cu ... ThresholdKernel() {...}

The ThresholdKernel() calculates from the Gradient and Theta images (magnitudes and angles, respectively) of the edges.

```
// Kernel that calculates the threshold image from Gradient, Theta
// resulting image has an RGB for each pixel, same RGB for each pixel
__global__
void ThresholdKernel(uch *ImgResult, double *ImgGrad, double *ImgTheta, ui
    Hpixels, ui Vpixels, ui ThreshLo, ui ThreshHi)
{
  ui ThrPerBlk=blockDim.x;          ui MYbid=blockIdx.x;     ui MYtid=threadIdx.x;
  ui MYgtid=ThrPerBlk*MYbid+MYtid;  double L,H,G,T;          uc PIXVAL;
  ui BlkPerRow=CEIL(Hpixels,ThrPerBlk);
  int MYrow = MYbid / BlkPerRow;     ui RowBytes=(Hpixels*3+3) & (~3);
  int MYcol = MYgtid - MYrow*BlkPerRow*ThrPerBlk;
  if (MYcol >= Hpixels) return;    // col out of range
  ui MYresultIndex=MYrow*RowBytes+3*MYcol;
  ui MYpixIndex=MYrow*Hpixels+MYcol;
  if ((MYrow<1) || (MYrow>Vpixels-2) || (MYcol<1) || (MYcol>Hpixels-2)){
    ImgResult[MYresultIndex]=NOEDGE;        ImgResult[MYresultIndex+1]=NOEDGE;
    ImgResult[MYresultIndex+2]=NOEDGE;      return;
  }else{
    L = (double)ThreshLo;        H = (double)ThreshHi;
    G = ImgGrad[MYpixIndex];     PIXVAL=NOEDGE;
    if (G <= L){                 PIXVAL=NOEDGE;         // no edge
    }else if (G >= H){           PIXVAL = EDGE;         // edge
    }else{
      T = ImgTheta[MYpixIndex];
      if ((T<-67.5) || (T>67.5)){
        // Look at left and right: [row][col-1] and [row][col+1]
        PIXVAL=((ImgGrad[MYpixIndex-1]>H) || (ImgGrad[MYpixIndex+1]>H)) ? EDGE :
            NOEDGE;
      }else if ((T >= -22.5) && (T <= 22.5)){
        // Look at top and bottom: [row-1][col] and [row+1][col]
        PIXVAL=((ImgGrad[MYpixIndex-Hpixels]>H) ||
            (ImgGrad[MYpixIndex+Hpixels]>H)) ? EDGE : NOEDGE;
      }else if ((T>22.5) && (T <= 67.5)){
        // Look at upper right, lower left: [row-1][col+1] and [row+1][col-1]
        PIXVAL=((ImgGrad[MYpixIndex-Hpixels+1]>H) ||
            (ImgGrad[MYpixIndex+Hpixels-1]>H)) ? EDGE : NOEDGE;
      }else if ((T >= -67.5) && (T<-22.5)){
        // Look at upper left, lower right: [row-1][col-1] and [row+1][col+1]
        PIXVAL=((ImgGrad[MYpixIndex-Hpixels-1]>H) ||
            (ImgGrad[MYpixIndex+Hpixels+1]>H)) ? EDGE : NOEDGE;
      }
    }
    ImgResult[MYresultIndex]=PIXVAL;
    ImgResult[MYresultIndex+1]=PIXVAL;
    ImgResult[MYresultIndex+2]=PIXVAL;
  }
}
```

8.5.4 ThresholdKernel()

Code 8.6 provides a listing of the ThresholdKernel(), which computes the thresholded (i.e., edge-detected) version of the image from its Sobel-filtered version according to Equation 5.5 and Equation 5.6. An interesting observation of Code 8.6 is that its performance is highly dependent on the image itself; this concept is called *data dependence of the performance*. There is a deep chain of if statements that can vastly vary the performance based on which one of them are TRUE versus FALSE. In a situation like this, only a range of performance numbers can be given, rather than a single value. This wasn't the case for the other three kernels, because all of the pixels required the computation of exactly the same functions. Because of the "data dependence" we will not use this function to gauge the success for some of the improvements we will suggest. Another interesting characteristic of this function is that it lacks the nested loops (or even a single loop for that matter); therefore, it is expected to execute substantially faster than the other kernels, especially because it doesn't even contain any of the transcendental functions (e.g., sin(), sqrt()).

8.6 PERFORMANCE OF IMEDGEG.CU

Table 8.3 tabulates six computer configurations that I used to test the performance of the imedgeG.cu program, both in terms of kernel performance and PCIe bus utilization (i.e., throughput during the CPU→GPU and GPU→CPU transfers). As we always do, we will first look at the performance for different families of GPUs just to develop a sense and in the chapters that follow, we will try to understand the reasons behind these numbers.

TABLE 8.3 PCIe bandwidth results of imedgeG.cu on six different computer configurations.

Feature	Box I	Box II	Box III	Box IV	Box V	Box VI
CPU	i7-920	i7-3740QM	W3690	i7-4770K	i7-5930K	2xE5-2680v4
C/T	4C/8T	4C/8T	6C/12T	4C/8T	6C/12T	14C/28T
Memory	16GB	32GB	24GB	32GB	64GB	256GB
BW GBps	25.6	25.6	32	25.6	68	76.8
GPU	GT640	K3000M	GTX 760	GTX 1070	Titan Z	Tesla K80
Engine	GK107	GK104	GK104	GP104-200	2xGK110	2xGK210
Cores	384	576	1152	1920	2x2880	2x2496
Compute Cap	3.0	3.0	3.0	6.1	3.5	3.7
Global Mem	2GB	2GB	2GB	8GB	2x12GB	2x12GB
GM BW GBps	28.5	89	192	256	336	240
Peak GFLOPS	691	753	2258	5783	8122	8736
DGFLOPS	29	31	94	181	2707	2912
Data transfer speeds & throughput over the PCI Express bus						
CPU→GPU ms	39.76	51.40	39.49	23.31	32.25	18.08
GBps	2.99	2.31	3.01	5.10	3.69	6.58
GPU→CPU ms	39.18	37.58	36.53	24.60	18.82	42.42
GBps	3.04	3.16	3.26	4.84	6.32	2.80
PCIe Bus	Gen2	Gen2	Gen2	Gen3	Gen3	Gen3
BW GBps	8.00	8.00	8.00	15.75	15.75	15.75
Achieved (%)	(38%)	(29–40%)	(41%)	(32%)	(23–40%)	(18–42%)

The asronaut.bmp image was used with ThreshLo=50 and ThresHi=100.

8.6.1   imedgeG.cu: PCIe Bus Utilization

First, let us focus on the PCIe transfers. Table 8.3 lists how fast each one of the computers completed their CPU→GPU and GPU→CPU transfers that were launched in the following lines of Code 8.1:

```
cudaStatus = cudaMemcpy(GPUImg, TheImg, IMAGESIZE, cudaMemcpyHostToDevice);
if(...);    cudaEventRecord(time2, 0); // Time stamp after CPU->GPU tfr
...
cudaStatus=cudaMemcpy(CopyImg, GPUResultImg, IMAGESIZE, cudaMemcpyDeviceToHost);
if(...);    cudaEventRecord(time4, 0); // after GPU-> CPU tfr
```

What we observe from Table 8.3 (for the imedgeG.cu program) is pretty much a repeat of what we observed in Table 7.3, when we were analyzing the performance of the imflipG.cu program. Aside from minor deviations, we see that these programs exhibit identical PCIe throughput behavior because they use exactly the same API for transfers: cudaMemcpy(). They both reach ≈20–45% of the peak throughput of PCIe and the performance is not symmetric for GPU→CPU versus CPU→GPU transfers in certain cases. Without getting into details, this is a more than sufficient conclusion for us.

8.6.2   imedgeG.cu: Runtime Results

Table 8.4 is the results for the execution of imedgeG.cu (using the same six configurations as Table 8.3). In the imflipG.cu program, we had only two kernels, Hflip() and Vflip(), which did nothing more than shuttle data from one place to another. However, imedgeG.cu is a lot more exciting. We not only have four kernels, but also each kernel has quite different characteristics when it comes to memory versus core operations. This is why we will analyze each kernel separately and try to improve each kernel's performance using different *tricks*.

Much like the imflipG.cu program, imedgeG.cu allows us to launch kernels using the *threads per block* as a command line parameter; Table 8.4 provides the run time results of each one of the four kernels using different *threads per block* values. In this section, we are not interested in how to find the best value of this parameter; instead, we are interested in what the best performance of each kernel would be, had we — magically — known the optimum value for it. Because of this reason, the best performance for each kernel for a given GPU is indicated using red ink, while the values that are close enough to it are indicated in blue ink. In general, the reader can assume that the "colored" values are optimum (or, good enough).

The "Achieved GBps" row indicates the corresponding throughput of this kernel, according to Equation 4.6. For example, when we ran the BWKernel() on a Pascal engine GPU GTX1070, we achieved 34% for the throughput. This is because the BWKernel() moved 446 MB data in 4.95 ms in the best case (64 threads/block), which corresponds to ≈88 GBps. Looking at the "Feature" section of Table 8.3 (on top), we see that the global memory peak bandwidth of the GTX 1070 GPU is 256 GBps. Therefore, we compute that our BWKernel() reached ≈34% of this peak. We also observe that different GPUs were able to utilize their global memory in vastly varying ways for different kernels. Although there are clear winners in many cases, there is no GPU that wins every category.

These performance differences cannot be explained in just a few sentences or even paragraphs. It will take us two more chapters to get a good sense. Let us start somewhere. First, in the following section, we need to understand the internal architecture that glues the GPU cores and GPU memory together. The GPU memory structure is a lot more sophisticated than the CPU memory, so one full chapter (Chapter 10) will be dedicated to it.

TABLE 8.4 imedgeG.cu kernel runtime results; red numbers are the best option for the *number of threads* and blue are fairly close to the best option (see ebook for color version).

Feature	Box I	Box II	Box III	Box IV	Box V	Box VI
GPU	GT640	K3000M	GTX 760	GTX 1070	Titan Z	Tesla K80
Compute Cap	3.0	3.0	3.0	6.1	3.5	3.7
GM BW GBps	28.5	89	192	256	336	240
Peak GFLOPS	691	753	2258	5783	8122	8736
DGFLOPS	29	31	94	181	2707	2912
# Threads	BWKernel() kernel run time (ms)					
32	82.02	72.84	23.83	5.38	12.90	16.43
64	55.69	**49.33**	**16.35**	4.95	9.69	9.23
128	52.99	48.68	16.26	**5.00**	7.62	**6.79**
256	**53.28**	**48.97**	**16.36**	4.97	**7.02**	6.76
512	56.76	51.59	17.17	5.18	7.14	7.07
768	64.02	57.30	18.96	5.74	7.46	8.60
1024	60.16	54.10	17.98	5.20	6.90	7.37
Achieved GBps	8.23	8.96	26.82	88.09	63.18	64.48
(%)	(29%)	(10%)	(14%)	(34%)	(19%)	(27%)
# Threads	GaussKernel() kernel run time (ms)					
32	311.92	246.74	81.84	30.97	48.12	63.19
64	212.85	188.18	62.74	30.71	48.14	62.34
128	**208.15**	**186.69**	**62.24**	30.64	39.87	62.34
256	206.87	183.94	61.26	30.28	**35.71**	62.34
512	**207.90**	**185.86**	**62.00**	30.47	35.64	62.34
768	223.05	193.67	64.55	29.52	35.67	62.37
1024	212.47	188.55	62.89	**29.84**	35.74	62.41
Achieved GBps	3.07	3.45	10.35	21.48	17.80	10.18
(%)	(11%)	(4%)	(5%)	(8%)	(5%)	(4%)
# Threads	SobelKernel() kernel run time (ms)					
32	289.86	255.17	84.63	**31.50**	43.64	44.32
64	253.82	**231.04**	76.77	31.29	40.37	31.36
128	**254.30**	**231.76**	**77.00**	**31.31**	33.33	31.36
256	**255.11**	231.13	76.76	**31.32**	29.28	31.36
512	260.81	232.67	77.47	**31.31**	30.00	31.36
1024	287.05	256.06	85.37	**31.38**	32.32	31.46
Achieved GBps	3.75	4.12	12.40	30.41	32.49	30.34
(%)	(13%)	(5%)	(6%)	(12%)	(10%)	(13%)
# Threads	ThresholdKernel() kernel run time (ms)					
32	99.40	85.74	27.89	4.13	12.74	20.24
64	61.13	53.45	17.54	**3.46**	8.37	12.00
128	46.73	42.36	14.08	**3.31**	6.16	7.73
256	**47.27**	**42.73**	**14.20**	**3.45**	5.77	**7.75**
512	53.29	47.46	15.49	3.93	**5.97**	9.42
768	68.52	60.80	18.65	4.90	7.12	13.01
1024	62.95	54.89	17.95	4.66	6.72	11.46
Achieved GBps	9.33	10.29	30.98	131.60	75.55	56.43
(%)	(33%)	(12%)	(16%)	(51%)	(22%)	(24%)

TABLE 8.5 Summarized imedgeG.cu kernel runtime results; runtime is reported for 256 threads/block for every case.

Feature	Box I	Box II	Box III	Box IV	Box V	Box VI
GPU	GT640	K3000M	GTX 760	GTX 1070	Titan Z	Tesla K80
Engine	Kepler	Kepler	Kepler	Pascal	Kepler	Kepler
Compute Cap	3.0	3.0	3.0	6.1	3.5	3.7
GM BW GBps	28.5	89	192	256	336	240
Peak GFLOPS	691	753	2258	5783	8122	8736
DGFLOPS	29	31	94	181	2707	2912
BWKernel()	Memory intensive. Mixed integer/double precision					
Runtime (ms)	53.28	48.97	16.36	4.97	7.02	6.76
Achieved GBps %	29%	10%	14%	34%	19%	27%
GaussKernel()	Double-precision and memory intensive					
Runtime (ms)	206.87	183.94	61.26	30.28	35.71	62.34
Achieved GBps %	11%	4%	5%	8%	5%	4%
SobelKernel()	Double-precision and memory intensive					
Runtime (ms)	255.11	231.13	76.76	31.32	29.28	31.36
Achieved GBps %	13%	5%	6%	12%	10%	13%
ThresholdKernel()	Double-precision, transcendental function intensive					
Runtime (ms)	47.27	42.73	14.20	3.45	5.77	7.75
Achieved GBps %	33%	12%	16%	51%	22%	24%

The "Achieved GBps %" column shows the *relative* bandwidth achieved (% of reported peak bandwidth). The "Peak GFLOPS" lists the peak single-precision floating and "Peak DGFLOPS" lists the peak double-precision computational capability of the GPU.

8.6.3 imedgeG.cu: Kernel Performance Comparison

Table 8.5 repeats the run time results shown in Table 8.4 under the assumption that we do not have a crystal ball and we wouldn't know which value of *threads per block* would result in the best performance. Therefore, 256 is chosen for every runtime result in Table 8.5, which yields the optimum runtime for some cases and almost every value shown in the table has some type of a "color," implying that choosing 256 for threads/block is close to optimum in many cases. This shouldn't be taken as a suggestion to use this value without thinking about its implications. We will study this number really hard in the following chapters. For now, our goal is to make some high-level observations from Table 8.5, regarding different family of GPUs. Here are the observations:

- Pascal family GPU seems to beat the Kepler in most of the memory-intensive operations. This seems to be the highest performance GPU in almost every kernel.

- Pascal does well even when we *unfairly* run the GaussKernel() on it using 256 threads/ block, which we know isn't optimum. This *forgiveness for bad programming* is a general characteristic of advanced architectures, by allowing the hardware to optimize things at runtime. More on that later ...

- The impressive performance of Pascal is surprising because as we will see in Chapter 9, GTX 1070 does not have great double-precision capability. This can also be seen in Table 8.5, under the "Peak DGFLOPS" column; the peak double-precision capability of GTX 1070 is only 181 GFLOPS, while its single-precision capability is 5783 GFLOPS. Remember from Code 8.4 that the GaussKernel() is packed with double-precision.

- For now, it suffices for us to propose a hypothesis that the reason for this surprising performance is the inefficiency of the code, which Pascal is able to partially compensate for. However, in Chapter 9, we will significantly tweak these kernels and get them close to optimal, in terms of core operation. In that case, the double-precision capable GPUs should start shining; for example, we expect K80 to scream in this case.

- To state the previous comments, we are making Pascal work as a "bad code fixer-upper" rather than a good "computational unit." However, when we optimize the code, we will see who the good computational heros are. I already ruined the surprise partially. I don't want to talk more about this. We have an entire Chapter 9 to talk about *the GPU cores*.

- Without thinking about the double-precision issue too much, almost every Kepler GPU seems to have a comparable *relative* performance, with minor exceptions. This means that the more cores Nvidia stuffs into their GPUs, the higher they make the global memory bandwidth; otherwise, adding more cores would cause *data starvation*. Although this is true when you look at these numbers at a higher level, details show interesting hidden trends. However, we cannot pick out these trends on such inefficient code. Currently, the inefficiencies in the kernels shown in Table 8.4 are masking the real performance numbers; in the following chapters, when we make the code core- and memory-friendly, much more interesting trends will emerge.

8.7 GPU CODE: COMPILE TIME

In Section 8.7, we will go over how a CUDA program is developed, followed by what happens when you launch a CUDA program (Section 8.8) and how it is executed (Section 8.9). Understanding all three steps is crucial in building efficient CUDA code.

8.7.1 Designing CUDA Code

First, we need to understand how the CUDA code developer designs a CUDA program. CUDA is nothing more than a C program, with the exception of the few symbols to indicate one of the following types of code:

1. **Plain simple CPU code with CPU-side variables**: This part is nothing but C language and can be compiled with gcc or MS Visual Studio, because it has nothing to do with the GPU. Furthermore, all of the variables used in this code are CPU-side variables, which refer to the CPU memory region. In Code 8.1, the following lines are an example of this category:

```
GPUImg       = (uch *)GPUptr;
GPUResultImg = GPUImg + IMAGESIZE;
GPUBWImg     = (double *)(GPUResultImg + IMAGESIZE);
GPUGaussImg  = GPUBWImg + IMAGEPIX;
GPUGradient  = GPUGaussImg + IMAGEPIX;
GPUTheta     = GPUGradient + IMAGEPIX;
```

I expect the readers to loudly reject that the above lines are *plain simple CPU code*. Even the variable names say GPU-something! The fact is that they are completely CPU code and the variables are CPU-side variables. Unless you use these in any conjunction with the GPU, you can call your variables whatever your heart desires. The most important thing about these previous lines of code is that they are compiled into CPU x64 instructions, and they use CPU memory and don't know if a GPU even exists.

2. **Kernel launch code**: This is pure CPU code, but its entire purpose is to launch GPU code. Everything has to initiate at the CPU, because CPU is the host. So, the only way to get access to the GPU side execution is these kernel launch lines, as shown below (from Code 8.1):

```
BWKernel <<< NumBlocks, ThrPerBlk >>> (GPUBWImg, GPUImg, IPH);
```

The lines above are nothing but a shortcut for an API function that is provided by Nvidia in facilitating a kernel launch. Instead of the ≪ and ≫ shortcut symbols, you can simply use that API and nothing will be different, with the exception that you will not get the annoying squiggly red lines that MS Visual Studio gives you. One important aspect of this category is that the variables that are passed onto the kernel are either constant values or GPU-side variables, such as GPU memory pointers.

3. **GPU→CPU and CPU→GPU Data Transfer APIs**: This is also pure CPU code. These APIs are a part of the library Nvidia provides CUDA developers and they are not different than the ≪ and ≫; they are simply an API that calls a function on the CPU side, which let's the CUDA runtime know to do something on the GPU side. By definition, they include CPU-side and GPU-side in their function call. Here is an example from Code 8.1:

```
cudaMemcpy(GPUImg, TheImg, IMAGESIZE, cudaMemcpyHostToDevice);
```

4. **Pure CUDA code**: This part is totally written in C language, however, a few symbols preceding the C code tell the compiler that what is coming is CUDA code. Here is an example from Code 8.3:

```
__global__
void BWKernel(double *ImgBW, uch *ImgGPU, ui Hpixels)
{
   ui ThrPerBlk = blockDim.x;
   ui MYbid = blockIdx.x;
   ui MYtid = threadIdx.x;
   ui MYgtid = ThrPerBlk * MYbid + MYtid;
   double R, G, B;

   ui BlkPerRow = CEIL(Hpixels, ThrPerBlk);
```

Clearly, there is no way to compile this part of the code with gcc because it is pure GPU code and the expected compiled output is the PTX, the GPU assembly language. There are two clues in this code that would completely give away the secret that it is GPU code: (1) the `__global__` identifier and (2) the specialized CUDA variables, `blockDim.x`, etc.

8.7.2 Compiling CUDA Code

The Nvidia CUDA compiler, nvcc, has two jobs:

- **Compile CPU Code**: The compiler has to not only determine where the CPU code is, but also compile it to the x64 CPU instructions.

- **Compile CUDA Code**: The compiler has to also determine which part of the CUDA code belongs to the GPU side and compile it to the GPU instruction set, as specified in the command line. For example, if you specified compute_20,sm_20, this means that you are asking it to compile to Compute Capability 2.0. This corresponds to a specific PTX instruction set that the compiler must generate the output for.

8.7.3 GPU Assembly: PTX, CUBIN

From the programmer's standpoint, it is not that important to know the details of PTX; however, it is good to know which file does what. When you run nvcc, both the x64 CPU instructions and the CUDA PTX instructions are compiled and made ready. PTX is an Intermediate Representation (IR) and it is not device-specific. However, you can force the compiler to create the ultimate CUDA Binary (referred to as cubin), which contains device-specific machine instructions. The nvcc compiler takes the options --ptx to produce PTX code and --cubin to produce the CUBIN code. I urge the reader to be curious about it and print the PTX or CUBIN code to see what GPU assembly looks like. Its cool stuff.

8.8 GPU CODE: LAUNCH

Assume that you compiled the imedgeG.cu into imedgeG.exe (in Windows) or another executable name in UNIX. I will explain this using the Windows platform. To launch your CPU code, you simply type the name of the program in Windows, or double click on the icon. Remember that this is a CUDA executable, so it will have CUDA-side instructions and plain-simple x64 instructions in it somehow.

8.8.1 OS Involvement and CUDA DLL File

As I mentioned before, everything starts on the CPU side. The very first few lines of your imedgeG.exe executable would include CPU x64 instructions to parse the arguments, etc., all of which is pure-CPU code. In Code 8.1, your code would execute all of the CPU-side instructions until it hits the following line, which is the very first line that has anything to do with the GPU:

```
cudaGetDeviceCount(&NumGPUs);
```

What does this code do? The OS runs this file as a usual EXE executable. As far as the OS is concerned, this is nothing more than a function call for which the executable is included in the cudart64_80.dll file. If you didn't have this file accessible to the OS (by setting the correct path), your program would crash right here and would tell you that it cannot locate the compiled binaries of the function cudaGetDeviceCount(). So, let's assume that you have this file accessible.

8.8.2 GPU Graphics Driver

The graphics driver you installed when you first plugged in your GPU has a lot more than just display setting adjustments, etc. It is the glue that holds the CPU and GPU together. When the `cudaGetDeviceCount()` is called from the CPU side, the compiled function inside your DLL file immediately cooperates with the graphics driver to use the driver in establishing the link between the CPU and the GPU. While the DLL file contains nothing more than compiled code, the driver is the runtime manager. Any communication between the CPU and GPU has to go through the NRE (Nvidia Runtime Engine), which is inside the graphics driver. So, the NRE would immediately determine what to do to ask the proper questions to the GPU, so it can get the answer to the *number of GPUs installed in the system*, which is the intended result of the `cudaGetDeviceCount()` API. The answer must be brought into the CPU side, because it will be saved in the CPU-side variable, `NumGPUs`.

8.8.3 CPU⟷GPU Memory Transfers

At some point, your CPU instructions will encounter an API function call as follows:

```
cudaStatus = cudaMalloc((void**)&GPUptr, GPUtotalBufferSize); if(...);
```

Who handles the `cudaMalloc()`? This is a function that requires full cooperation from the CPU and GPU side. So, the graphics driver has to have access to both. GPU is not a problem because it is the driver's own territory. However, accessing the CPU memory requires the graphics driver to be nice and apologetic, and obey the master that controls the CPU territory (the OS). Let us now go through a few situations, where things can go wrong:

- **Improper CPU-Side Memory Pointers:** This is the OS territory and anybody who uses improper (out-of-bounds) CPU-side pointers gets slapped in the face, execution is taken away from them and they are thrown on the street! An incorrect CPU-side memory pointer access would cause a *Segmentation Fault* in Unix and something equivalent of that in Windows.

- **Improper GPU-Side Memory Pointers:** It is highly likely that your program will have improper GPU-side memory accesses, but not CPU-side. If you violated the CPU territory, the OS already slapped you in the wrist and the execution of your CUDA program is over. However, the OS has no way of accessing the GPU side. Who is going to protect you from GPU-side bad memory pointers? The answer is simple: NRE (Nvidia Runtime Engine). So, NRE is the police who rules the GPU town! Any bad memory pointers would cause the NRE to detect this and decide to terminate the program. But, wait... Your program is really a CPU program. So, only the OS can terminate you. Well, two different town's cops cooperate to catch the bad guy in this case. The NRE tells the OS that this program has done something that it wasn't supposed to; and the OS listens to its cop friend in the GPU-town and terminates the program's execution. The message you get will be different than the ones you encounter with the CPU-side pointer issues, but, still, you are caught trespassing and can't run anymore!

- **Improper GPU-Side Parameters:** Bad GPU memory pointers are not the only reason for the NRE to terminate you. Let's assume that you are trying to run a kernel with 2048 threads per block and the GPU engine you are using cannot support it. The compiler would have no idea about this at compile time. It is only when you run the GPU code that your NRE would detect this and tell the OS to shut you down, because there is no way to continue the execution.

8.9 GPU CODE: EXECUTION (RUN TIME)

OK, so, what happens when everything is right? No issues with anybody trying to shut you down and the execution of your CUDA code continues peacefully. How does your CUDA code execute?

8.9.1 Getting the Data

First, let's look at the data transfers between the CPU and GPU. Look at the following code:

```
cudaMemcpy(GPUImg, TheImg, IMAGESIZE, cudaMemcpyHostToDevice);
```

In Figure 8.2, the *Host Interface* is responsible for shutting the data back and forth between the CPU and GPU. So, clearly, the NRE uses some internal APIs to facilitate this. This requires reading the data from the CPU memory, and uses the X99 chipset to transfer it to the GPU; this data is welcomed by the Host Interface and makes it into the L2$ and goes into its final destination in the global memory on the right side of Figure 8.2.

8.9.2 Getting the Code and Parameters

Data alone is not sufficient for the GPU to execute a kernel. What happens when we launch a kernel as follows:

```
BWKernel <<< NumBlocks, ThrPerBlk >>> (GPUBWImg, GPUImg, IPH);
```

which translates to the following quantities at runtime, when the values of the `NumBlocks` and `ThrPerBlk` variables are plugged in:

```
BWKernel <<< 166656, 256 >>> (GPUBWImg, GPUImg, IPH);
```

My specific question is: aside from launching the kernel itself, what additional information has to go to the GPU side, so the GPU can execute this thing internally? The answer is one of two different sets of parameters as follows:

- **Launch Parameters:** These parameters (`166656, 256`) are strictly related to the dimensions of the blocks and grids; they have nothing to do with the internal operation of the kernel. So, the only part of the GPU that cares about this is the *Giga Thread Scheduler (GTS)*. All that the GTS does is, like its name suggests, *schedule* the blocks; the only criteria for its scheduling is that an SM is *willing* to take on another block. If it is, the GTS schedules it and its job is done. Of course, it has to communicate the threads/block to the SM before the SM can decide whether it can handle this extra job. If it can, it gets the job!

- **Kernel Parameters:** The parameters passed onto the kernel (`GPUBWImg, GPUImg, IPH`) are strictly needed for the execution of the kernel by the GPU's cores and SMs and is of no concern to the GTS. Once the GTS schedules the block, that block is the SM's problem after that point.

8.9.3 Launching Grids of Blocks

Let's look at Figure 8.2 to see how the scheduling would work in our specific case, where the Giga Thread Scheduler (GTS) would have 16,656 blocks to schedule. Because there are only 6 SMs in this case, clearly, according to the Pigeon Hole principle, each SM would get more than one (way more than one) block. Each SM can only *execute one block* at any given point in time, however, this doesn't mean that it cannot accept and buffer more than one block in its internal queue. Therefore, when it comes to the job of GTs, there are three parameters at play:

- **Total Number of Blocks**: This is the entirety of the task that the GTS is responsible for. In our specific case, the GTS has 166,656 blocks to schedule to 6 SMs. Let's assume that the gird dimension is only one, i.e., only the `blockIdx.x` parameters will be passed onto the SMs to which the blocks are scheduled.

- **Number of Streaming Multiprocessors (SMs)**: When you write your CUDA code, you generally do not make any assumptions on which GPU it is running on. The GPU can have 1, 2, 3, ...16, or even 60 SMs, as evidenced in Table 7.6. Your program should work regardless of the number of SMs in the GPU it is running the program on. The number of SMs is a property you find out when you query the GPU using cudaGetDeviceProperties(). In our specific example, it is 6, so I will give you the example below using this number.

- **Maximum Number of Blocks that an SM Can Receive**: This is another parameter you get with cudaGetDeviceProperties(). This is the maximum number the GTS can assign to an SM before the SM continuously says *it can no longer accept any more blocks*. For example, this number is 8 for Fermi GPUs, 16 for Kepler, and 32 for Pascal. So, for GTX550Ti, in Table 7.6, it is 8.

Just to satisfy your curiosity, here are some relevant parameters:

- `cudaDeviceProp.multiProcessorCount` tells you the number of SMs in the device,
- `cudaDeviceProp.maxThreadsPerBlock`=1024 for Fermi, Kepler, Maxwell, Pascal,
- `cudaDeviceProp.maxGridSize[3]` gives you the maximum size of a grid (in number of blocks) in x, y, z dimensions, and
- `cudaDeviceProp.maxThreadsDim[3]` gives you the maximum size of a block (in number of threads) in x, y, z dimensions.

8.9.4 Giga Thread Scheduler (GTS)

From the facts in Section 8.9.3, we deduce the following run time activity:

- 166,656 blocks must be scheduled by the GTS,
- `gridDim.x`=166656, `gridDim.y`=1, and `gridDim.z`=1 for every single block that GTS schedules for this specific grid,
- `blockDim.x`=256, `blockDim.y`=1, and `blockDim.z`=1 for every single block, because our example assumed a single-dimensional block of threads with 256 threads/block,
- `blockIdx.x` = [0...166655] for the blocks that are scheduled,
- `blockIdx.y` = 1 and `blockIdx.z` = 1 for every block.

- Every single one of these 166,656 blocks gets an exact copy of the CUDA binary code for the CUDA kernel that it is supposed to execute. This is in *cubin* format and is already just-in-time compiled from *PTX* into *cubin*, because *cubin* is the native GPU core language, where PTX is an intermediate representation that must be translated into *cubin*.

- Clearly, there is a way to make this more efficient by avoiding repetition, but, from the standpoint of our understanding, note that every SM that receives a specific block as a task also receives its code somehow and caches it in its own instruction cache (inside the SM).

- None of the SMs can get more than 8 blocks in its queue, although it can only execute a single one at any point in time. So, while one is executing, 7 are queued up, waiting for execution.

8.9.5   Scheduling Blocks

The question is: if you were the GTS, what would you do? Well, you can only schedule one block at a time, so you would try to schedule `blockDim.x=0` first. For the sake of reducing the clutter during my explanation, let's call this *Block0*. So, in an attempt to schedule *Block0*, you would broadcast its internal launch parameters to all six of your candidates, *SM0...SM5* and see who responds. Let's say that nothing is executing inside any SM at the moment and everyone is free; it is highly likely that all six SMs will say "gimme gimme gimme." Who do you choose? Well, you've got 166,656 total blocks to go and it is so nice that everyone is so willing. So, the most meaningful and simple scheduling algorithm is round robin: if multiple SMs respond to you and ask you to assign them the block, you prefer the one with the smaller number, which is *SM0* in this case. After all of this drama, *Block0* got assigned to *SM0*. Well, remember, the other five SMs are also hoping that you will give them something to do. So, why not do all of the assignments below:

- *Block0→SM0,*   *Block1→SM1,*   *Block2→SM2,*
  Block3→SM3,   *Block4→SM4,*   *Block5→SM5.*

OK, now you got rid of 6, and still have 166,650 to go. Now what? Remember that our example is for a Fermi GPU and Fermi's can grab 8 blocks before they are full. Also remember that the instant you assigned *Block0→SM0*, it started executing *Block0*; so, it is highly likely that it will finish executing *Block0* earlier than the other SMs. After these six blocks, your request for more SMs is still broadcasting and each SM now has a single block they started executing and can take 7 more. So, they will still be telling you to assign them more. You, again, start with the lower indexed one and keep assigning six more as follows:

- *Block6→SM0,*   *Block7→SM1,*   *Block8→SM2,*
  Block9→SM3,   *Block10→SM4,*   *Block11→SM5.*

Now, each SM got 2 blocks assigned and they can still take 6 more. In short, you can assign 48 blocks lightning fast to 6 SMs, where each SM soaks in 8 blocks before letting you know that it can no longer receive more blocks to execute. So far, you got rid of 48 of the 166,656 blocks and have 166,608 more to go. At this moment, your "HELP WANTED" sign is still on the window but everyone is so busy that they are not volunteering for more work. Each SM has a control logic in it that checks two things to see if it can accept the incoming job: (1) its maximum block count. For Fermi, this is 8. So, if a Fermi SM has 8 blocks inside it, it wouldn't even consider taking another block, (2) if it can still take more, it compares the parameters of the kernel you are advertising to its own parameters and sees if it has "resources" to take this new block. Resources include a lot of things, such as cache memory, register file, among many others as we will see in Chapter 9.

Back to our example: By this time, each SM absorbed the following blocks:

- *SM0* ⟹ [*Block0*, *Block6*, *Block12*, *Block18*, *Block24*, *Block30*, *Block36*, *Block42*]

- *SM1* ⟹ [*Block1*, *Block7*, *Block13*, *Block19*, *Block25*, *Block31*, *Block37*, *Block43*]

    ...

- *SM5* ⟹ [*Block5*, *Block11*, *Block17*, *Block23*, *Block29*, *Block35*, *Block41*, *Block47*]

Now that you have scheduled 48 blocks (*Block0...Block47*), you have to wait for somebody to be free again to continue with the other 166,608 blocks (*Block48 ... Block166655*). It is not necessarily true that the SMs will finish the execution of the blocks assigned to them in exactly the same order of assignment. At this moment, you have each SM executing 8 blocks, but they can only execute one at a time, and put 7 to sleep temporarily. When a block accesses resources that will take a while to get (say, some data from global memory), it has the option to switch to another block to avoid staying idle. This is why you stuff 8 blocks to the SM and give it 8 options to choose from to keep itself busy. This concept is identical to why assigning two threads to the CPU helped it *do more work* on average, although it could not execute more than one of those threads at a time. In the case of the SM, it grabs a bunch of blocks, so it can switch to another one when one comes to a standstill.

Now, let's fast forward the time a little bit. Say, *SM1* got finished with *Block7* before anybody else. It would immediately raise its hand and volunteer to take in another block. Having gotten rid of 0...47, your next block to schedule is 48. So, you would make the following scheduling decision: *Block48→SM1*. After this assignment, *SM1* would clean up all of the resources it needed for *Block7* and replace it with *Block48*. So, *SM1*'s queue of 8 blocks is looking like this now:

- *SM1* ⟹ [*Block1*, **Block48**, *Block13*, *Block19*, *Block25*, *Block31*, *Block37*, *Block43*]

Let's say that *SM5* finished *Block23* next and raised its hand; you would assign your next block (*Block49*) to it, which will change its queue to the following:

- *SM5* ⟹ [*Block5*, *Block11*, *Block17*, **Block49**, *Block29*, *Block35*, *Block41*, *Block47*]

This would continue until you finally assigned *Block166655*. When you assign this very last block, GTS's responsibility is over. It might take a while to finish what is in the queue of each SM after this very last assignment, but, as far as the GTS is concerned, job is done!

8.9.6 Executing Blocks

Now that we understand how the blocks are scheduled to SMs for execution, let's understand how they are executed. Assume that you are *SM5* and *Block49* is scheduled to you for execution. Here is what you receive from the GTS:

- `gridDim.x`=166656, `gridDim.y`=1, `gridDim.z`=1,

- `blockDim.x`=256, `blockDim.y`=1, `blockDim.z`=1,

- `blockIdx.x`=49, `blockIdx.y`=0, `blockIdx.z`=0.

This is enough for you to understand that, out of the 166,656 blocks, you are #49. Because each block consists of 256 threads, you must execute this block using 256 threads. So, the SMs responsibility is to execute Block49 using 256 — single-dimensional — threads, numbered `threadIdx.x`=0...255. This means that it will facilitate the execution of 256 threads, where each thread gets the exact same parameters above; additionally, they also get their `threadIdx.x` computed and passed onto them. So, if you are thread #75 out of the 256 threads, this is what is passed on to you when you are executing:

- `gridDim.x`=166656, `blockDim.x`=256, `blockIdx.x`=49, `threadIdx.x`=75.

I omitted the parameters with values "1" because the programmer will not even look at them and they will not be used during the execution anyway. To summarize, the responsibility of the GTS ends when the block is scheduled and the responsibility of the SM itself starts, which is further numbering the threads before starting the execution of the block. Of course, in addition to assigning the thread ID, the shared resources in an SM, such as cache memory, register file, and more have to be allocated. More on that later. In this chapter, all we care about is the scheduling functionality at the block level.

8.9.7 Transparent Scalability

As you see, the only responsibility of the programmer is to write the CUDA code in such a way that each block is a highly independent code with no dependence on the other blocks. If the code is written this way, the hardware details are transparent to the programmer and the more SMs a GPU has, the faster your code can execute. This is called *transparent scalability*. After all, the last thing a GPU user wants is to purchase a GPU with more cores (which really corresponds to more SMs) and have his or her program barely benefit from the additional SMs. Ideally, the speed-up should be linear with the increasing number of cores.

## CHAPTER 9

Understanding GPU Cores

I N THE previous chapters, we looked at the GPU architecture at a high level; we tweaked the *threads per block* parameter and observed its impact on the performance. The readers should have a clear understanding by now that the unit of kernel launch is a *block*. In this chapter, we will go much deeper into how the GPU actually executes the blocks. As I mentioned at the very beginning of Part II, the unit of execution is not a block; rather, it is a *warp*. You can think of the block as the big task to do, which can be chopped down into much smaller sub-tasks named warps. The significance of the warp is that a smaller unit of execution than a warp makes no sense, because it is too small considering the vast parallelism the GPU is designed for.

In this chapter, we will understand how this concept of warp ties to the design of the GPU cores and their placement inside a streaming multiprocessor (SM). With this understanding, we will design many different versions of the kernels inside the imflipG.cu and imedgeG.cu programs, run them, and observe their performance. We will run these experiments in four different GPU architecture families: Fermi, Kepler, Maxwell, and Pascal. With each new family, a new instruction set and compute capability have been introduced; to accommodate these new instructions sets, the cores and other processing units had to be designed differently. Because of this fact, some of the techniques we will learn in this chapter will be broadly applicable to every family, while some of them will only work faster in the new generations, due to the utilization of the more advanced instructions, available only in the newer generations such as Pascal.

While we "guessed" the *threads per block* parameter in the previous chapters, we will learn how to use a tool named *CUDA Occupancy Calculator* in the next chapter, which will allow us to establish a formal methodology for determining this important parameter, along with many other critical parameters that ensure optimum utilization of the SM resources during kernel execution.

9.1 GPU ARCHITECTURE FAMILIES

In Table 8.1, we briefly introduced the architectural components of each family. In this chapter, we are interested in the details of each family's internal architecture and how we can utilize this knowledge to write higher performance GPU code. First, we want to know the organization of the SMs inside each family.

9.1.1 Fermi Architecture

The GF110 engine, in the Fermi family, is shown in Figure 9.1. Figure 8.2 depicted the GTX 550Ti, which has the GF116 Fermi engine with only 192 cores; therefore, although each SM is identical to the GF110 containing 32 cores, it only has 6 SMs to house 192 cores. If you wrote a kernel with 166,656 blocks and running it on a GTX 580 (with 16 SMs) versus GTX

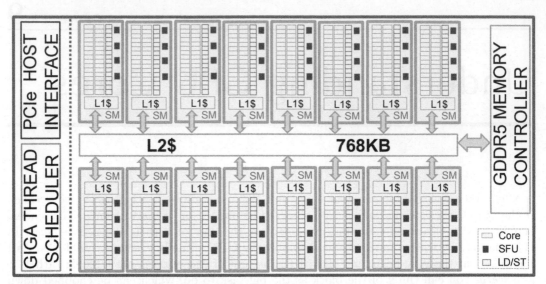

FIGURE 9.1 GF110 Fermi architecture with 16 SMs, where each SM houses 32 cores, 16 LD/ST units, and 4 Special Function Units (SFUs). The highest end Fermi GPU contains 512 cores (e.g., GTX 580).

550Ti (with 6 SMs), each SM would get assigned an average of 10,416 versus 27,776 blocks on average. So, it is not unreasonable to expect a ≈2.7× higher performance from GTX580 for *core-intensive* code. How about memory? The GTX550Ti has a global memory bandwidth of 98.5 GBps, which corresponds to about 0.51 GBps per core. Alternatively, GTX580's GM bandwidth is 192.4 GBps, which corresponds to 0.38 GBps per core. So, we might want to consider the possibility that the GTX 550Ti might beat the GTX 580 in relative terms for *memory-intensive* code by at least 20–30%. However, there are many more factors to affect the memory performance, such as the cache memory built into the SMs, which can ease the pressure on the GM with intelligent programming.

Figure 9.1 depicts a "PCI Express Host Interface," which is a PCIe 2.0 controller. Remember that the Fermi family did not support PCIe 3.0. Every family after that did. This host interface works in concert with the I/O controller in the GPU→CPU data transfers. Giga Thread Scheduler is the unit that is responsible for Block→ SM assignments, as we detailed in Section 8.9.5. The memory controller can either be a GDDR3 or GDDR5 controller and it is composed of multiple controllers, as most of the books will show. Here, I am only showing a single memory controller to depict the component that is responsible for the data transfers, Global Memory ⟷ L2$. The 768 KB L2$ is the Last Level Cache (LLC) and it is *coherent* and shared among all of the cores. This is where the GPU caches the contents of the GM. Alternatively, the L1$ inside each SM is not coherent and it is strictly used as a local cache during the processing of individual blocks. More details on the internal structure of each SM are given in the following subsection.

9.1.2 Fermi SM Structure

Fermi SM structure is shown in Figure 9.2, which contains 32 cores (one warp). There are two *warp schedulers* that turn the block — just received from the Giga Thread Scheduler

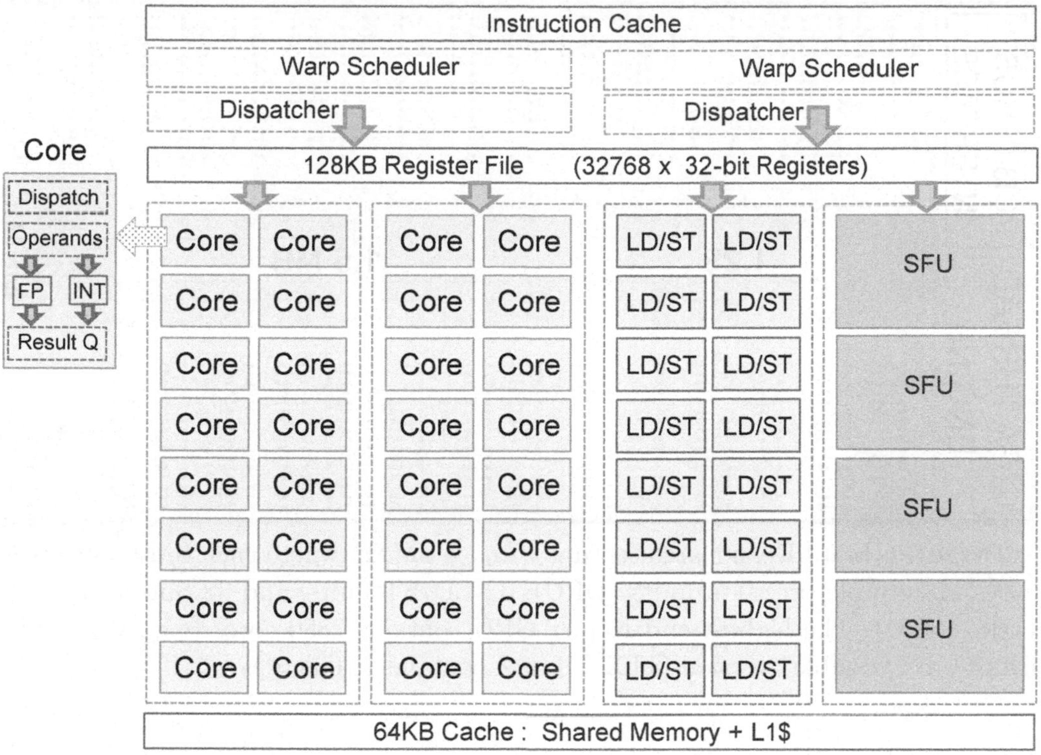

FIGURE 9.2 GF110 Fermi SM structure. Each SM has a 128 KB register file that contains 32,768 (32 K) registers, where each register is 32-bits. This register file feeds operands to the 32 cores and 4 Special Function Units (SFU). 16 Load/Store (LD/ST) units are used to queue memory load/store requests. A 64 KB total cache memory is used for L1$ and shared memory.

(GTS) — into a set of warps and schedules them to be executed by the execution units inside this SM, which are the cores, Load/Store queues, and Special Function Units (SFUs). When memory read/write instructions need to be executed, these memory requests are queued up in the Load/Store queues and when they receive the requested data, they make it available to the requesting instruction. Each core has a Floating Point (FP) and an Integer (INT) execution unit to execute float or int instructions, as shown on the left side of Figure 9.2. Instructions need to access a lot of registers; rather than giving a register file to each individual core (like in a CPU), the cores inside an SM share a large register file. In the SM shown in Figure 9.2, a 128 KB register file (RF) is shared among the 32 cores. Each register is a 32-bit (4 byte) unit, therefore making the RF a 32 K-register one. The SFU is responsible for executing transcendental functions (e.g., sin(), cos(), log()). The *Instruction Cache* holds the instructions within the block, while the *L1$ Cache* is responsible for caching commonly used data, which is also shared with another type of cache memory named *Shared Memory*. This 64 KB cache is split between the two as either (16 KB+48 KB) or (48 KB+16 KB).

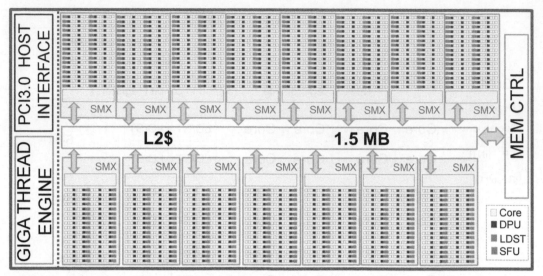

FIGURE 9.3 GK110 Kepler architecture with 15 SMXs, where each SMX houses 192 cores, 48 double precision units (DPU), 32 LD/ST units, and 32 Special Function Units (SFU). The highest end Kepler GPU contains 2880 cores (e.g., GTX Titan Black); its "double" version GTX Titan Z contains 5760 cores.

9.1.3 Kepler Architecture

Kepler architecture is shown in Figure 9.3. There are major differences as compared to Fermi:

- The L2$ memory size is doubled to 1536 KB (1.5 MB); this is a little troubling — at first look — considering that there are twice as many cores to feed and only a doubled L2$ is used,

- The host interface now supports PCIe 3.0,

- Each SM is now called SMX and it has a completely different internal structure than Fermi, as shown in Figure 9.4,

- The Giga Thread Scheduler is now called Giga Thread Engine (GTE). Intuitively, because there are almost 6× as many cores as Fermi, one would expect the Kepler to execute things a lot faster, which in turn requires a faster GTE to keep up with scheduling the blocks,

- The SMX contains a different type of execution unit, named a double precision unit (DPU), which specializes in execution `double` data types efficiently,

- Because the Kepler is designed to hold almost 6 times as many cores as Fermi (512 vs. 2880), each SMX is structured to hold a significantly higher number of cores than Fermi (192), although there is one less SMX (15 vs. 16). Having such heavily populated SMX units has interesting performance implications, as we will detail in the future sections, because the cores inside an SM (or SMX) share their L1$, register file, and another type of cache that is introduced, the *Read Only Cache*.

Instruction Cache			
Warp Scheduler	Warp Scheduler	Warp Scheduler	Warp Scheduler
Dispatcher Dispatcher	Dispatcher Dispatcher	Dispatcher Dispatcher	Dispatcher Dispatcher

256KB Register File   (65536 x 32-bit Registers)

Core	Core	Core	DPU	Core	Core	Core	DPU	LDST	SFU	Core	Core	Core	DPU	Core	Core	Core	DPU	LDST	SFU
Core	Core	Core	DPU	Core	Core	Core	DPU	LDST	SFU	Core	Core	Core	DPU	Core	Core	Core	DPU	LDST	SFU
Core	Core	Core	DPU	Core	Core	Core	DPU	LDST	SFU	Core	Core	Core	DPU	Core	Core	Core	DPU	LDST	SFU
Core	Core	Core	DPU	Core	Core	Core	DPU	LDST	SFU	Core	Core	Core	DPU	Core	Core	Core	DPU	LDST	SFU
Core	Core	Core	DPU	Core	Core	Core	DPU	LDST	SFU	Core	Core	Core	DPU	Core	Core	Core	DPU	LDST	SFU
Core	Core	Core	DPU	Core	Core	Core	DPU	LDST	SFU	Core	Core	Core	DPU	Core	Core	Core	DPU	LDST	SFU
Core	Core	Core	DPU	Core	Core	Core	DPU	LDST	SFU	Core	Core	Core	DPU	Core	Core	Core	DPU	LDST	SFU
Core	Core	Core	DPU	Core	Core	Core	DPU	LDST	SFU	Core	Core	Core	DPU	Core	Core	Core	DPU	LDST	SFU
Core	Core	Core	DPU	Core	Core	Core	DPU	LDST	SFU	Core	Core	Core	DPU	Core	Core	Core	DPU	LDST	SFU
Core	Core	Core	DPU	Core	Core	Core	DPU	LDST	SFU	Core	Core	Core	DPU	Core	Core	Core	DPU	LDST	SFU
Core	Core	Core	DPU	Core	Core	Core	DPU	LDST	SFU	Core	Core	Core	DPU	Core	Core	Core	DPU	LDST	SFU
Core	Core	Core	DPU	Core	Core	Core	DPU	LDST	SFU	Core	Core	Core	DPU	Core	Core	Core	DPU	LDST	SFU
Core	Core	Core	DPU	Core	Core	Core	DPU	LDST	SFU	Core	Core	Core	DPU	Core	Core	Core	DPU	LDST	SFU
Core	Core	Core	DPU	Core	Core	Core	DPU	LDST	SFU	Core	Core	Core	DPU	Core	Core	Core	DPU	LDST	SFU
Core	Core	Core	DPU	Core	Core	Core	DPU	LDST	SFU	Core	Core	Core	DPU	Core	Core	Core	DPU	LDST	SFU
Core	Core	Core	DPU	Core	Core	Core	DPU	LDST	SFU	Core	Core	Core	DPU	Core	Core	Core	DPU	LDST	SFU

48KB Read-Only Cache
64KB Cache : Shared Memory + L1$

FIGURE 9.4 GK110 Kepler SMX structure. A 256 KB (64 K-register) register file feeds 192 cores, 64 Double-Precision Units (DPU), 32 Load/Store units, and 32 SFUs. Four warp schedulers can schedule four warps, which are dispatched as 8 half-warps. Read-only cache is used to hold constants.

9.1.4 Kepler SMX Structure

Kepler SMX structure is shown in Figure 9.4. Probably the most noticeable difference — as compared to Fermi — is the introduction of the double precision units (DPU). The idea behind this is to have high performance not only in computing float, but also double data types. While a game player may not care about double-precision performance, a scientific program does. A detailed explanation of the DPU will be provided in Section 9.2.2.

As far as the cache structure goes, aside from adding a Read-Only cache, the entire rest of the cache memory units are the same; there is still an instruction cache, capable of caching the instructions in a block and still the same amount of L1$ and shared memory. Chapter 10 will be dedicated to what these memory units do (specifically the *shared memory*), so I am not spending a lot of time on them here.

To accommodate such an increased number of cores, there are twice as many warp schedulers, twice as many LD/ST units, and a register file that is twice as big. It might seem a little strange that increasing the number of cores by $6\times$, but only doubling the RF, LD/ST units, and the warp schedulers suffices to feed the hungry cores with data and operands, when you compare Kepler to Fermi. When Nvidia architects design these architectures, they run an extensive set of benchmarks (simulations) to see which arrangements yield the best performance results on average, based on different arrangements of the cores, DPUs, LD/ST units in the SMX. This must have yielded the best results to make it to the final design. Also, remember: the goal is to have good performance in both float and double, which would mean that Nvidia ran a set of scientific and game applications to come up with this SMX structure.

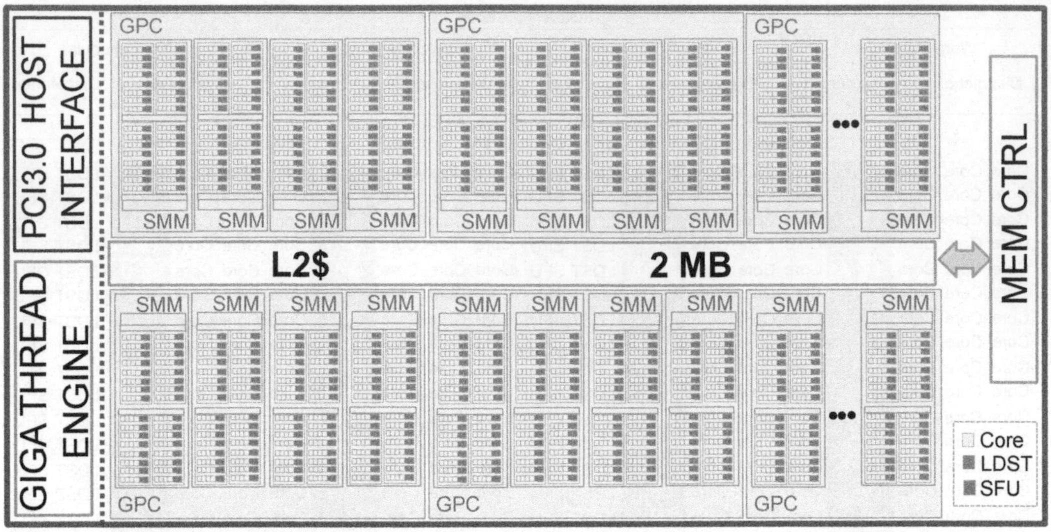

FIGURE 9.5 GM200 Maxwell architecture with 24 SMMs, housed inside 6 larger GPC units; each SMM houses 128 cores, 32 LD/ST units, and 32 Special Function Units (SFU), *does not contain* double-precision units (DPUs). The highest end Maxwell GPU contains 3072 cores (e.g., GTX Titan X).

9.1.5 Maxwell Architecture

Maxwell architecture is shown in Figure 9.5, which includes 6 Graphics Processing Clusters (GPCs), where each GPC houses 4 SMMs (this is the Maxwell version of SM); therefore, the highest-end Maxwell (GTX Titan X) houses 24 SMMs in total, each containing 128 cores, for a total of 3072 cores. This is barely an increase in the core count, as compared to Kepler's 2880 max. It is clear that the design priority in Maxwell was not to increase the core count; instead, Nvidia aimed to make every core more capable by giving them more resources inside the SMM.

9.1.6 Maxwell SMM Structure

Maxwell SMM structure is shown in Figure 9.6. As we discussed in Section 9.1.4, the nagging question with the design of Kepler (Figure 9.3 and Figure 9.4) was how the architecture would accommodate six times as many cores with twice the resources inside each SMX, not to mention only a 2× L2$. Maxwell addresses this issue by making SMMs smaller and increasing the resources available to each core. Maxwell cache infrastructure is different too: an *Instruction Cache* is shared among four identical "sub-units" that contain 32 cores, 8 LD/ST queues, and 8 SFUs, although each one of these sub-units buffers their instructions in their own *Instruction Buffer*, because they may be executing different blocks, containing different instructions. Although the total RF size is the same as SMX (256 KB), it is divided into four 64 KB RFs that are a part of each sub-unit. Also, the organization of the L1$ cache memory and shared memory allows the cores to access data easier. To summarize, similar resources are shared by 128 cores+32 SFUs in Maxwell, rather than 192 cores+64 DPUs+32 SFUs in Kepler. However, `double` instructions take 32× longer than `float` instructions in Maxwell, because they are executed in the cores, rather than the lacking DPUs.

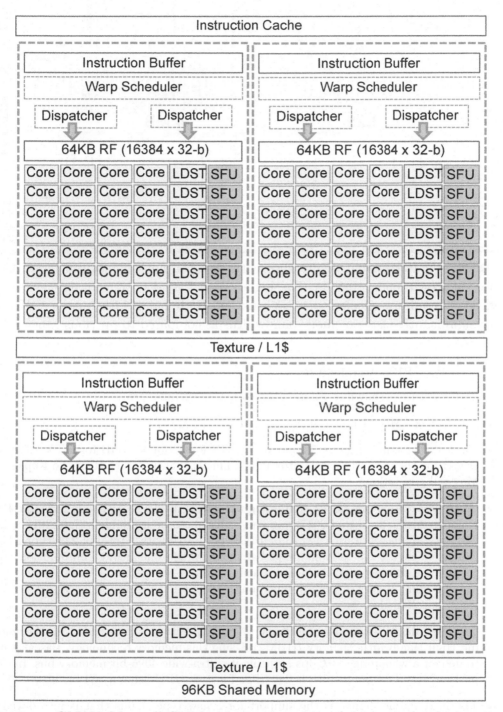

FIGURE 9.6 GM200 Maxwell SMM structure consists of 4 identical sub-structures with 32 cores, 8 LD/ST units, 8 SFUs, and 16 K registers. Two of these sub-structures share an L1\$, while four of them share a 96 KB shared memory.

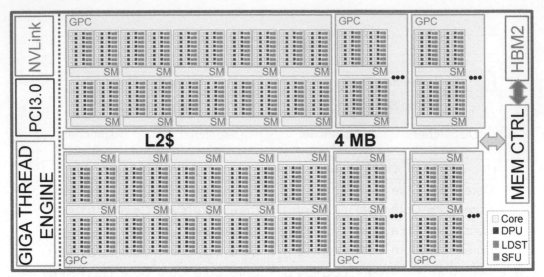

FIGURE 9.7 GP100 Pascal architecture with 60 SMs, housed inside 6 larger GPC units, each containing 10 SMs. The highest end Pascal GPU contains 3840 cores (e.g., P100 compute accelerator). NVLink and High Bandwidth Memory (HBM2) allow significantly faster memory bandwidths as compared to previous generations.

9.1.7 Pascal GP100 Architecture

Pascal architecture is shown in Figure 9.7. Noticeable changes are

- The L2$ is 4 MB, although it is only 2 MB in GTX 1070.

- The total amount of global memory is 12 GB for Titan X, 8 GB for GTX 1070, and 16 GB for the compute accelerator P100. This is slightly larger than the Kepler and Maxwell generations.

- K80 was a two-in-one GPU, containing two K40's inside it. Although a double-P100 is not available as of mid-2017, you can expect it to be available in late 2017, which will possibly incorporate a 12 GB or 16 GB per GPU, with a total of 24 GB or 32 GB. Having such a large amount of global memory helps in heavy computations that require the storage of a significant amount of data inside the GPU. For example, in the emerging *Deep Learning* application, a large GPU memory is necessary to store the "neural" nodes.

- The most drastic change in the Pascal microarchitecture is the support for the emerging High Bandwidth Memory (HBM2). Using this memory technology, Pascal is able to deliver a whopping 720 GBps by using its ultra-wide 4096-bit memory bus. In comparison, K80 GM bandwidth is 240 GBps for each GPU using a 384-bit bus.

- The next major change is the support for a new type of bus that is introduced by Nvidia: the NVLink bus, which boasts a 80 GBps transfer rate. This is much faster than the 15.75 GBps the PCIe 3.0 bus and alleviates the CPU⟷GPU data transfer bottleneck. However, NVLink is only available on high-end servers.

Instruction Cache															

Instruction Buffer								Instruction Buffer							
Warp Scheduler								Warp Scheduler							
Dispatcher				Dispatcher				Dispatcher				Dispatcher			
128KB RF (32768 x 32-b)								128KB RF (32768 x 32-b)							
Core	Core	DPU	Core	Core	DPU	LDST	SFU	Core	Core	DPU	Core	Core	DPU	LDST	SFU
Core	Core	DPU	Core	Core	DPU	LDST	SFU	Core	Core	DPU	Core	Core	DPU	LDST	SFU
Core	Core	DPU	Core	Core	DPU	LDST	SFU	Core	Core	DPU	Core	Core	DPU	LDST	SFU
Core	Core	DPU	Core	Core	DPU	LDST	SFU	Core	Core	DPU	Core	Core	DPU	LDST	SFU
Core	Core	DPU	Core	Core	DPU	LDST	SFU	Core	Core	DPU	Core	Core	DPU	LDST	SFU
Core	Core	DPU	Core	Core	DPU	LDST	SFU	Core	Core	DPU	Core	Core	DPU	LDST	SFU
Core	Core	DPU	Core	Core	DPU	LDST	SFU	Core	Core	DPU	Core	Core	DPU	LDST	SFU
Core	Core	DPU	Core	Core	DPU	LDST	SFU	Core	Core	DPU	Core	Core	DPU	LDST	SFU

Texture / L1$

64KB Shared Memory

FIGURE 9.8 GP100 Pascal SM structure consists of two identical sub-structures that contain 32 cores, 16 DPUs, 8 LD/ST units, 8 SFUs, and 32 K registers. They share an instruction cache, however, they have their own instruction buffer.

9.1.8 Pascal GP100 SM Structure

Pascal SM structure is shown in Figure 9.8. Noticeable changes from Maxwell are

- The register file is twice the size of Maxwell, giving kernels a lot more registers to work with.

- The core-to-DPU ratio is improved in comparison to Kepler; one DPU is used for every two cores in Pascal, instead of one DPU for three cores (in Kepler). So, we can expect Pascal to do better in double-precision computations. Unfortunately, this is a little bit of an illusion; Pascal does much worse in double-precision for the models that do not incorporate the DPUs. We will talk about this in detail in Section 9.1.9.

- A very important difference between Pascal cores and previous generation cores is the support for half-precision floating point, with the data type `half`. We will detail this in Section 9.3.10. While scientific applications require double-precision power, the emerging Deep Neural Networks (DNNs) do not require such high precisions. Introducing this new data type (`half`) allows the application to trade-off lower precision for higher throughput; the P100 delivers 5.3 TFLOPS in double-precision (FP64), 10.6 TFLOPS in single-precision (FP32), and 21.2 TFLOPS in half-precision (FP16), practically achieving a 2× performance in applications that do not need the accuracy of FP32 or FP64.

- To generalize the previous comment, Pascal cores handle smaller data types such as 16-bit integer (INT16), 8-bit integer (INT8), and 16-bit floating point (FP16) much more efficiently.

9.1.9 Family Comparison: Peak GFLOPS and Peak DGFLOPS

A comparison of the peak computational power (in GFLOPS) of different Nvidia generations is provided in Table 9.1. The peak Giga-Floating-Point-Operations (GFLOPS) of a GPU is determined as follows:

$$\text{Peak GFLOPS} = \begin{cases} f_{shader} \times n \times 2 = (f \times 2) \times n \times 2 & \text{for Fermi models} \\ f \times n \times 2 & \text{beyond Fermi} \end{cases} \quad (9.1)$$

where f is the base core clock for a CUDA core and n is the total number of CUDA cores. In the Fermi generation, the concept of a "CUDA core" was a little different; cores were called SPs (Streaming Processors) and a *shader clock* was defined to be 2× the base core clock. This is why the computation of the peak power is different with Fermi; for example the GTX 580's core clock is 772 MHz and its shader clock (f_{shader}) is 1544 MHz. So, its peak output is $772 \times 2 \times 512 \times 2 = 1581$ GFLOPS.

Starting with the Kepler generation, Nvidia called the cores *CUDA cores*. For example, for the GTX 780, the core clock is 863 MHz and there are 2304 CUDA cores. Therefore, GTX 780 peak compute power is $863 \times 2304 \times 2 = 3977$ GFLOPS (single precision).

Double precision peak compute power is calculated as follows:

$$\text{Peak DGFLOPS} = \begin{cases} \dfrac{\text{Peak GFLOPS}}{24} & \text{Kepler GPUS with no DPU} \\[2mm] \dfrac{\text{Peak GFLOPS}}{3} & \text{Kepler GPUs with a DPU} \\[2mm] \dfrac{\text{Peak GFLOPS}}{32} & \text{Maxwell GPUS} \\[2mm] \dfrac{\text{Peak GFLOPS}}{32} & \text{Pascal GPUS with no DPU} \\[2mm] \dfrac{\text{Peak GFLOPS}}{2} & \text{Pascal GPUs with a DPU} \end{cases} \quad (9.2)$$

As an example of Equation 9.2, let us calculate the peak DGFLOPS of GTX 1070 that we used in our previous results; GTX 1070 cores run at a clock frequency of 1506 MHz and the GTX 1070 is a "Pascal GPU with no DPUs," so, with its 1920 cores, GTX 1070's single-precision peak output is computed as $1506 \times 1920 \times 2 = 5783$ GFLOPS from Equation 9.2 and its double-precision peak output is computed as $5783 \div 32 = 181$ DGFLOPS from Equation 9.2.

It is important to note here that these computations assume the case where each core is non-stop delivering FLOPS with no inefficiencies. If we learned anything in this book, it is the fact that such a perfect scenario only occurs if-and-only-if the programmer designs the CUDA kernels with infinite efficiency. So far, with the kernels we saw, we barely hit 20, 30, or may be 40% of this peak. In this chapter, our goal is to get as close to 100% as possible. Also note that the alternative problem is being able to saturate the global memory bandwidth. In other words, a memory-intensive program might saturate the memory bandwidth far before it saturates the cores. We will use a tool named *CUDA Occupancy Calculator* toward the end of this chapter to give us an idea about which one might occur first (i.e., core saturation of memory saturation), before even we launch our kernels.

The reason behind Equation 9.1 and Equation 9.2 is that single-precision versus double-precision peaks have everything to do with the CUDA core-to-DPU ratio inside the SM, SMX, and SMM units. For example, in a Kepler SMX unit (Figure 9.4), we can clearly see that there is a DPU for every three CUDA cores; therefore, we divide the GFLOPS by 3 to

TABLE 9.1 Nvidia microarchitecture families and their peak computational power for single precision (GFLOPS) and double-precision floating point (DGFLOPS).

Family	Engine	GTX Model	# SMs	# cores	# DPUs	# SFUs	Peak Compute Power		
							float	double	Ratio
Fermi	GF110	550 Ti	4	192		16	691		
		GTX 580	16	512		64	1581		
Kepler	GK110	GTX 780	12	2304	0	384	3977	166	24×
	GK110	Titan	14	2688	896	448	4500	1500	3×
	2×GK110	Titan Z	30	5760	1920	960	8122	2707	3×
	2×GK210	K80	26	4992	1664	832	8736	2912	3×
Maxwell	GM200	980 Ti	22	2816	0	704	5632	176	32×
		Titan X	24	3072	0	768	6144	192	32×
Pascal	GP104	GTX 1070	15	1920	0	480	5783	181	32×
	GP102	Titan X	28	3584	0	896	10157	317	32×
	GP100	P100	56	3584	1792	896	9519	4760	2×
Volta									

get the DGFLOPS. Similarly, for the Pascal SM units, this ratio is 2, which is reflected in Equation 9.2. However, it is less obvious why the 32× ratio is the fact for Pascal's with no DPU (such as Titan X - Pascal Edition and GTX 1070). This has to do with the design of the CUDA cores inside Pascal; they can execute double-precision operations, although they take 32× more time to do so, as compared to single-precision operations. For Kepler's with no DPUs (such as GTX 780), this ratio was a less dramatic 24×.

Another interesting observation from Table 9.1 is that the *number of cores per SM* does not match for the GTX 1070 and P100 GPUs; while the P100 has 64 cores per SM, GTX 1070 and GTX Titan X both have 128 cores per SM. This is because P100 is designed to be a double-precision engine, while the other two are single-precision engines with a 32× lower performance in double-precision. Because of this, their SM architectures are completely different. We will be studying the GP104 engine Section 9.3.14 when we are going over the different data types that Pascal supports.

9.1.10 GPU Boost

There is another important note to make about core performance, which is a performance boost technology termed *GPU Boost*, by clocking the GPU cores at a higher frequency temporarily. Every CUDA core is designed to work at a *base clock frequency* (minimum) and a *boost clock frequency* (maximum). For example, the GTX Titan Z has a base clock frequency of 705 MHz and a boost clock frequency of 875 MHz, which is 24% higher. Its peak compute power of 8122 GFLOPS is computed according to Equation 9.2 with the base clock (705 MHz) in mind. If it could continuously work at its boost frequency of 875 MHz, it could deliver a peak of ≈10 TFLOPS. Most GPU manufacturers (e.g., ASUS, Gigabyte, PNY) provide a software to enable you to boost GPU performance by increasing the clock and voltage simultaneously (called OC – overclocking). This is made possible by Nvidia's clock frequency management that is built into their architecture.

There is a catch with *GPU Boost* though! An Integrated Circuit (IC) needs a higher voltage to be able to run at a higher frequency. So, to be able to clock the cores at 875 MHz, the GPU internal voltage circuitry would have to increase the core voltage. Then, what happens to the power consumption? The power consumption formula is as follows:

$$P \propto V^2 \cdot f \tag{9.3}$$

where f is the frequency of the core and V is the operating voltage. Although the details are not very important for this qualitative argument, it is clear that when you increase the frequency of the core by 24%, the power consumption of the GPU goes up a lot more than 24%; if you are really trying to come up with a number, call it something like 50%.

9.1.11 GPU Power Consumption

What we are saying here is the following: assume that your GTX Titan Z GPU was consuming 300 W at its normal base clock frequency (705 MHz) and you kicked it into higher gear (875 MHz) with the GPU Boost. It is highly likely that it will start consuming >400 W to be able to run at this frequency and deliver the 10 TFLOPS you are hoping to get. This has multiple implications:

- Higher power consumption means higher temperature. Every IC is designed for a specific Thermal Design Point (TDP); for example, the TDP is 80°C for GTX Titan Z, which would not be exceeded at the base clock frequency. However, with the GPU Boost, it might reach >100°C; the internal GPU hardware would detect this thermal overload and would clock the cores back down until the GPU has cooled back down to the TDP.

- Water-based cooling units are excellent for taking away the excess heat and would prevent the temperature from rising beyond its specified TDP; however, they cannot solve the power consumption issue. You will still consume the same amount of power, although you will be able to quickly remove the heat that is generated as a consequence of this power consumption.

- Additional power consumption puts strain on your PCIe connectors. Six pin PCIe power connectors support 75 W max and 8-pin connectors support 150 W max. GTX Titan Z has two 8-pin connectors, which handles a peak of 300 W. Exceeding this power limit would cause excessive heating on the PCIe connectors.

9.1.12 Computer Power Supply

Additional power consumption can also overload your power supply. If you are temporarily reaching 400 W even for very short periods of time (say, a few seconds), your computer's total power consumption is possibly something like 500–600 W because a high-powered Extreme Intel i7 CPU can consume up to 150 W and motherboard chipset, etc. can easily consume another 70 W. Additionally, power supplies are not 100% efficient. Assuming that you are using a high-quality Corsair AX 760 power supply, here is what your power consumption looks like:

- CPU Power Consumption = 150 W

- GPU Power Consumption = 400 W

- Motherboard Power Consumption = 70 W

- Total Power Consumption = 620 W

- Total Load on the Corsair AX 760 (760 W) Power Supply = 82%

- Power Supply Efficiency = 90%

Most power supplies work great at 50% load and their efficiency goes down at high loads. AX 760 has a 92% efficiency at 50% load (380 W), which goes down to 88% at 100% load (760 W). To summarize, although GPU boost is a great way to temporarily boost your performance, you should be planning to own a really good power supply and a cooling unit to be able to reach such high performance peaks. Otherwise, you will damage the GPU and your power supply. This situation is a lot worse with low-quality power supplies, most of which have only 70–80% efficiency at high loads. At an 80% efficiency, a low quality supply is burning 20% of the power it delivers, which is > 100 W; this generates a lot of heat just inside the power supply and quickly kills every component inside your computer. If you want to create a personal supercomputer, here are some suggestions:

> *Invest in an extremely high quality (and high efficiency) power supply.*
  My suggestion: Corsair AX760 (760 W). Its efficiency is ≈90%,
      which means only 10% of the delivered power is turned into heat.
      It has six 8-pin PCIe connectors, which is good for multiple GPUs.
> *Specify your power supply very generously.*
      Ideally, your peak power consumption should be 50–60% of the peak.
> *AX 760 (760 W) is good for the following configurations:*
      A GTX Titan Z (350 W) and an i7-6950X (140 W)
      Dual GTX Titan X Pascal (250+250 W) and an i7-6950X (140 W)
> *HX 1200i (1200 W) is good for the following configurations:*
      Dual GTX Titan Z (350+350 W) and an i7-6950X (140 W)
      Dual P100 Pascal (300+300 W) and dual Xeon E5-2699V4 (145 W)
> *Invest in a liquid-cooled block for the GPU to exhaust heat quickly.*
> *Invest in an extremely high-quality computer chassis with lots of cooling.*
      They should have multiple 120 mm or 140 mm fans.
> *Get an Extreme Intel CPU or a Xeon E5.*
> *Invest in a Liquid CPU Cooler. My suggestion: Corsair H90.*

9.2   STREAMING MULTIPROCESSOR (SM) BUILDING BLOCKS

Now that we understand the building blocks of an SM at a high level, we will learn their operational details in this section.

9.2.1   GPU Cores

As shown in Figure 9.2, each CUDA core has an integer and a single-precision floating point unit. These are the equivalent of an ALU and FPU inside a CPU core. Each FPU is capable of double-precision operations, but they have to spend multiple cycles to do so; this means that using the GPU cores for double-precision operations is like using a bicycle to go on a Nascar rally, where everyone else is using a 200 mph race car. Each GPU core has a dispatch port through which it receives the next instruction and an operand collector port through which it receives its operands (from the register file). It also has a result queue, to which it writes its result — to be committed to the register file eventually.

9.2.2 Double Precision Units (DPU)

One may wonder why the FPU inside each core cannot execute the `double` data type operations. It sure can ... but, the question is *how fast*? Without a dedicated DPU, each core is forced to use its FPU to perform double-precision computations in multiple (e.g., 24 or 32) cycles; so it takes a core 24 or 32 as much time to compute a single double-precision operation as a single-precision computation. There are versions of the Kepler engines that are missing the DPUs; for example, the high-end home-market GPU GTX780 Ti does not have DPUs, while the GTX Titan Z we use in this book does. When we look at their peak GFLOPS specifications, GTX 780 Ti has a 5046 GFLOPS single-precision and 210 DGFLOPS double-precision performance, which is a 24× disparity. Alternatively, the GTX Titan Z has a 8122 GFLOPS single-precision and a 2707 DGFLOPS double-precision peak processing power, which is only a 3× disparity. In other words, having the DPUs drops the single÷double ratio (i.e., DGFLOPS÷GFLOPS ratio) from 24 to 3.

Figure 9.4 depicts each DPU as being slightly physically larger than a core. This is not an optical illusion, but a VLSI IC design fact. Each DPU takes up a much larger area than a core inside the IC, because the physical area required to design a multiplier goes up quadratically with the number of bits; a single-precision floating point number has a 23-bit mantissa, while a double precision has a 52-bit mantissa, which intuitively implies a 4× larger chip area for a DPU, as compared to an FPU. However, the core has more than just the FPU in it, implying that the ratio is less than 4× in reality. We will get into floating point arithmetic details in Section 9.3.8.

9.2.3 Special Function Units (SFU)

Although the FPU and DPU are good for simple arithmetic, such as addition and multiplication, *transcendental* functions, such as sin(), cos(), exp(), sqrt(), and log(), require specialized units to execute them efficiently. Although the choice of whether to put DPUs in a given family is an option based on the market segment that the GPU is targeted for, Nvidia never omitted the SFUs in any of the GPU generations we have seen; from Fermi to Pascal (and definitely the upcoming Volta), SFUs have been an integral part of Nvidia GPUs. Without the ability to perform these *transcendental* arithmetic operations efficiently, a GPU loses its ability to cater well to both the scientific and game market. Applications in both of these markets need the ability to rotate and manipulate 3D objects efficiently by using sophisticated transcendental functions, which makes SFUs indispensable in a GPU architecture.

9.2.4 Register File (RF)

To understand how the RF works, let's take a simple kernel as an example, such as the BWKernel() in Code 8.3. This kernel declares the following variables:

- `double`   R, G, B
- `unsigned int` ThrPerBlk, MYbid, MYtid, MYgtid,
- `unsigned int` BlkPerRow, RowBytes, MYrow, MYcol,
- `unsigned int` MYsrcIndex, MYpixIndex

So, altogether, there are 3 `double` type variables and 10 `unsigned int` type variables. The compiler should map all of these variables to *registers*, because they are the most efficient for the GPU cores to execute instructions with. Considering that each register is a 32-bit value, double types eat away two registers worth of space. So, we need 10 registers for the unsigned int types and 6 to store the double types, for a total of 16. This is clearly not enough. The compiler needs to store at least one or two 32-bit temporary values and some 64-bit temporary values. So, it is realistic to assume that the BWKernel() in Code 8.3 requires 20–24 registers total to execute. Let's call it 24 to be on the safe side.

Assume that we are running the BWKernel() on a Pascal GP100 GPU, for which the SM structure is shown in Figure 9.8. Also assume that we launched these kernels with 128 threads per block. In this case, every single one of these 128 threads in each block needs 24 registers; so each block requires $128 \times 24 = 3072 = 3\,K$ registers in the RF to be even scheduled into that SM. Recall from Section 8.9.3 that a Pascal SM can accept up to 32 blocks. If the Giga Thread Engine (GTE) ends up scheduling 32 blocks into this SM, each SM will require $3 \times 32 = 96\,K$ registers to accept all of these 32 blocks. However, as we see in Figure 9.8, the Pascal SM only has $32\,K$ registers (with a total storage of $128\,KB$ in terms of bytes). So, the GTE would actually only be able to schedule 10 blocks into this SM before the SM cannot accept any more blocks due to lack of register space in the RF. What would happen if I decided to launch these kernels with 64 threads per block instead of 128? The answer is: each block would now require $1.5\,K$ registers and the GTE would now be able to schedule 20 blocks. Even if I drop the threads/block down to 32, I can still schedule only 30 blocks, which is less than the maximum number possible.

This highlights the fact that the programmer should write kernels that use as few registers as possible to avoid register starvation in the SM. Using too many registers has another interesting implication: the maximum number of registers a thread can have is limited by the CUDA hardware. This number was 63 until Compute Capability 3.0 and Nvidia increased it to 255 beyond CC 3.0. It stayed there since then. Increasing it to beyond this makes no practical sense, because kernels with a really high number of registers exhaust the RF so fast that any performance benefit that can come from these large kernels is negated by the inability to schedule them inside the SMs in large quantities.

The summary about the register file is that:

➢ *A programmer should write kernels that require as few registers as possible. Otherwise, the kernel will have "register pressure."*
➢ *The maximum number of registers in each kernel is a CUDA parameter. It was 63 until CC 3.0; later, it went up to 255.*

9.2.5 Load/Store Queues (LDST)

Load/Store queues are used to bring data from the cores into the memory and vice versa. Any core that requires a memory load or store queues up its request in the LD/ST queues and waits until the request is fulfilled. During the wait, another warp is scheduled to execute. If the load takes a long time, the scheduling logic might have cycled through a lot of other warps; these wait times for data to come from the memory do not hurt the performance as long as the SM has other work to do. This is why being able to schedule quite a few blocks into the SM helps keep it busy.

9.2.6 L1$ and Texture Cache

L1$ is the *hardware-controlled cache* inside each SM, which is governed by cache replacement algorithms such as *Least Recently Used (LRU)*. In other words, the programmer has no control over which data elements the L1$ keeps and which ones it evicts. In determining what to keep inside the L1$, the SM cache controller looks for very simple patterns in data usage. However, the underlying idea is that nobody knows the data better than the programmer himself or herself. The texture cache is where the GPU keeps the textures of the objects used in computer games.

9.2.7 Shared Memory

Shared memory is the *software-controlled cache*; in other words, through the CUDA program, the programmer tells the SM hardware precisely how to cache the data. The hardware has almost no control over how this memory is used. This is one of the most powerful tools of a GPU hardware and more than two-thirds of Chapter 10 will be dedicated to the usage of the shared memory.

9.2.8 Constant Cache

This is another very important type of cache, as it caches values that are *immutable*. This includes constants, which are used in programs to feed constant values to the threads, such as filter coefficients that do not change throughout the entire program. The operation of this cache differs significantly in that this cache is responsible for repeating and supplying the same value to multiple (e.g., 16) cores, as opposed to 16 different values to 16 cores.

9.2.9 Instruction Cache

This cache holds the instructions of a block that an SM is executing. Every time the GTE assigns a block to an SM, it also fills its cache with the instructions that make up this block.

9.2.10 Instruction Buffer

This is the local instruction cache of each SM, which copies instructions from the Instruction cache. In this sense, the Instruction cache acts as the L2I$, while the Instruction buffer acts as the L1I$.

9.2.11 Warp Schedulers

Recall from Section 8.9.5 that the Giga Thread Engine (GTE) scheduled the blocks to SMs until there was no SM left that could accept additional blocks due to either (1) lack of resources such as register file or shared memory or (2) due to exceeding architectural parameters, such as the maximum number of blocks per SM. In this section, let us assume that everything went right and the assignment $Block0{\to}SM0$ has succeeded. How will SM0 execute this block? Before we answer this, we should refresh our memory from Section 8.9.5 that the GTE sent the instructions for this block down to the Instruction Cache of SM0, along with the following parameters as part of the kernel launch:

- `gridDim.x`=166656, `gridDim.y`=1, and `gridDim.z`=1,

- `blockDim.x`=256, `blockDim.y`=1, and `blockDim.z`=1,

- `blockIdx.x` = 0, `blockIdx.y` = 0, `blockIdx.z` = 0.

These block-level parameters are necessary information, but what is missing is the thread IDs, which will be needed once the block starts executing. Who will assign them? This is where the warp schedulers come into the game; their job is to turn each block into a set of warps and schedule them to be individually executed. Using the same parameters in Section 8.9.5, we launched 256 threads/block, which effectively corresponds to 8 warps/block. Therefore, the execution of *Block0* in our example requires warp0...warp7 in this block, which corresponds to `threadIdx.x` values in the range 0...255. To accomplish this, here is what the warp schedulers schedule:

- schedule **warp0**: `gridDim.x=166656`, `blockDim.x=256`, `blockIdx.x=0`,

- schedule **warp1**: `gridDim.x=166656`, `blockDim.x=256`, `blockIdx.x=0`,

- ...

- schedule **warp7**: `gridDim.x=166656`, `blockDim.x=256`, `blockIdx.x=0`,

Note that these warps are only *scheduled,* not *dispatched* yet. They have to wait until resources are available that allows them to be dispatched.

9.2.12 Dispatch Units

Once the resources are available to dispatch a warp, the dispatcher units do so. This is where the dispatchers slap the `threadIdx.x`, `threadIdx.y`, and `threadIdx.z` on each thread before passing them down to the cores, DPUs, SFUs, or the LD/ST units. For example, when dispatching warp0, the following parameters would be passed onto the 32 cores that are responsible for executing warp0:

- `gridDim.x=166656`, `blockDim.x=256`, `blockIdx.x=0`, `threadIdx.x=0...32`.

Note the important concept that warps execute *serially,* which is an un-GPU-like phenomenon. If one warp hits a big wait because of a memory load, another warp starts executing. When all of the warps are done, a block commits its results and disappears. Therefore, all 8 of the warps in *Block0* would have to finish executing before *Block0* can commit its results and get removed from *SM0*'s execution queue. The weird serial nature of the way warps are scheduled has important implications for a programmer; although a programmer writes code with the assumption of each block executing independently, even warps must be assumed to execute independently. As we will see in Chapter 10, this will force us to use explicit synchronization for memory reads that occur at intra-warp boundaries. For now, the reader does not need to worry about this.

9.3 PARALLEL THREAD EXECUTION (PTX) DATA TYPES

In this section, we will look at the data types as they are defined in the Parallel Thread Execution (PTX) instruction set of the Nvidia GPUs. Understanding the assembly language of Nvidia (PTX) is important in getting a sense for how the GPU cores execute the instructions in a kernel. Remember that the first PTX was PTX 1.0, which was introduced in 2009. The newest PTX 5.0 was introduced in January 2017, which is only supported by Pascal GPUs. Most of the data types and arithmetic, logic, and floating point operations stayed the same while others were added to provide support for emerging applications such as deep learning. When I introduce each data type, I will provide a set of PTX instructions that use that data type. We will use this insight to improve our kernels in the following few chapters.

9.3.1 *INT8*: 8-bit Integer

These are the 8-bit integers as defined in PTX:

.u8   PTX type is the unsigned 8-bit integer (range 0 $\cdots$ 255)

.s8   PTX type is the signed 8-bit integer (range -128 $\cdots$ 127)

.b8   PTX type is the untyped 8-bit integer

Example PTX instructions for the INT8 data type are as follows:

```
add.u8 d, a, b;    // add unsigned 8-bit integers a to b, save result in d
```

Here are two new instructions that work on "vector" 8-bit data:

```
dp4a.u32.u32 d,b,a,c;    // d=c+four-way dot product of bytes of a,b
dp2a.lo.u32.u32 d,b,a,c; // d=c+two-way dot product of words of a, Low bytes of b
```

Note that these two instructions are only available in PTX ISA 5.0 (i.e., Pascal family), which allow the processing of four bytes in one clock cycle, or two 16-bit words. Because of this, the Pascal family can achieve 4× the performance in processing byte-size data, as long as the code is compiled to take advantage of these instructions. As you can see here, Nvidia's architecture design trend is to turn their integer cores into more like the MMX and SSE, and AVX units that the i7 CPUs have. With, for example, the Intel AVX instruction extensions, it is possible to process a 512-bit vector as either 8 64-bit numbers, 16 32-bit integers, 32 16-bit, or 64 8-bit integers. The dp4a and dp2a instructions resemble this a little bit. My guess is that in the Volta family (the next generation after Pascal), there will be a much wider set of these instructions, potentially applicable to other data types.

9.3.2 *INT16*: 16-bit Integer

These are the 16-bit integers as defined in PTX:

.u16   PTX type is the unsigned 16-bit integer (range 0 $\cdots$ 65536)

.s16   PTX type is the signed 16-bit integer (range -32768 $\cdots$ 32767)

.b16   PTX type is the untyped 16-bit integer

Example PTX instructions for the INT16 data type are as follows:

```
min.u16 d, a, b;       // store the minimum of unsigned int16 numbers a,b in d
min.s16 d, a, b;       // same as above, but assumes that a and b are signed int16
mul.u16.lo d, a, b;    // unsigned multiply 16-bit a,b and store low 16-bit in d
mul.u16.wide d, a, b;  // unsigned mult. 16-bit a,b and store 32-bit result in d
mad.hi.sat.s32 d,a,b,c;  // signed mul a,b. add c, saturate; store hi-32 bits in d
```

9.3.3 24-bit Integer

There was a 24-bit integer data type in the early days of PTX 1.0. This was necessary because supporting 32-bit native operations would mean that every 32-bit multiplication would be forced to save its results in a 64-bit destination. However, if you multiply two 24-bit numbers, you get a 48-bit result and if the numbers are small enough, the result may actually fit in 32-bits; based on this, 24-bit multiplication instructions allowed one to save either the upper or lower 32 bits of the result. That way, you can use the lower 32 bits if you know that your numbers are small to start with. Alternatively, you can use the upper 32 bits if you were storing fixed point numbers and the lower bits only mean more resolution and can be ignored. If you cannot live without all 48 bits of the result, you can always perform both of the multiplications and save both of the results for future use.

Example PTX instructions for the 24-bit data type are as follows:

```
mul24.hi.u32 d,a,b;        // unsigned multiply 24b a, b; save higher 32b in d
mul24.lo.s32 d,a,b;        // signed multiply 24b a, b; save lower 32b in d
mad24.hi.u32 d,a,b,c;      // multiply a, b and add c. Save the result in d
mad24.hi.sat.s32 d,a,b,c;  // saturating signed multiply a, b, add c, save in d
```

9.3.4 *INT32*: 32-bit Integer

These are the 32-bit integers as defined in PTX:

.u32 PTX type is the unsigned 32-bit integer (range $0 \cdots 2^{32}-1$)

.s32 PTX type is the signed 32-bit integer (range $-2^{31} \cdots 2^{31}-1$)

.b32 PTX type is the untyped 32-bit integer

Example PTX instructions for the INT32 data type are as follows:

```
rem.u32 d,a,b;        // d=remainder of integer division of u32-type a/b
mad.lo.u32 d,a,b,c;   // d=a*b+c (lower 32 bits)
abs.s32 d,a;          // d=absolute value of a. Only applies to signed types
popc.b32 d,a;         // d=population count (number of 1s) in a
add.sat.s32 d, a, b;  // saturating add s32 a, b and save result into d
ld.global.b32 f, [addr]; // load a 32-bit value from memory into f register
```

Note that a "saturating" addition avoids overflow by limiting the result to the MININT$\cdots$MAXINT range; it only applies to the s32 type. For example, adding 1 to the highest number $(2^{31}-1)$ would cause an overflow because the result (2^{31}) is an out-of-range s32 value. However, add.sat limits the result to MAXINT $(2^{31}-1)$ and voids overflow. This is perfect for digital signal processing applications, where a lot of filter coefficients and sampled voice or image data are being multiplied. The inaccuracy caused by the saturation is inaudible to the human ear, but avoiding overflow prevents the results from being completely wrong and meaningless and outputting white noise like garbage at the output of the filter.

9.3.5 Predicate Registers (32-bit)

This is the predicate type as defined in PTX:

.pred PTX type is a 32-bit predicate register

Example PTX instructions for the predicate registers are as follows:

```
    .reg .pred p,q,r;    // declare p, q, r as predicate registers
    setp.lt.s32 p, a,b;  // p= (a<b);
@p  add.s32 c,c,2;       // c+=2 if p is True (i.e., if a<b);
```

Here, the `@p` is the guard predicate, which executes the conditional add instruction based on the Boolean value of the `p` predicate register. The reverse of the predicate can also be used for conditional instructions, as follows:

```
      setp.lt.s32 p, a,b;    // p= (a<b);
@!p   bra OUT;               // branch if predicate p is false (i.e., a>=b)
      mul...
      ...
OUT:
```

9.3.6 *INT64*: 64-bit Integer

These are the 64-bit integers as defined in PTX:

.u64   PTX type is the unsigned 64-bit integer (range $0 \cdots 2^{64} - 1$)

.s64   PTX type is the signed 64-bit integer (range $-2^{63} \cdots 2^{63} - 1$)

.b64   PTX type is the untyped 64-bit integer

Example PTX instructions for the INT64 data type are as follows:

```
rem.s64 d.a.b;      // d=remainder of integer division s64-type a/b
abs.s64 d,a;        // d=absolute value of a. Only applies to signed types
clz.b64 d,a;        // d=leading zeros in the 64 bits of a
bfind.s64 d,a;      // d=position of the most significant non-sign bit
bfe.u64 d,a,b,c;    // d=bit field extract(a) at position b, length c
bfi.b64 f,a,b,c,d;  // f=aligned,inserted bitfield from a into b, start@d,length=d
```

9.3.7 128-bit Integer

Although there aren't any direct 128-bit instructions, extended addition and subtraction instructions allow 128-bit operations indirectly as exemplified below:

```
// first number=(a3,a2,a1,a0), second=(b3,b2,b1,b0), result=(d3,d2,d1,d0)
add.cc.u32 d0,a0,b0;    // add the lowest 32 bits. save carry in CC.CF
addc.cc.u32 d1,a1,b1;   // add-with-carry next 32 bits. save carry in CC.CF
addc.cc.u32 d2,a2,b2;   // add-with-carry next 32 bits. save carry in CC.CF
addc.cc.u32 d3,a3,b3;   // add-with-carry highest 32 bits. save carry in CC.CF
```

The `CC.CF` is the carry flag in the condition register, which allows the carry to be used in the second, third, and last additions to extend beyond 32-bits. You can do the same thing to do a 128-bit multiplication by using the `madc` instruction, which multiply-accumulates and uses the carry during the accumulation.

9.3.8 FP32: Single Precision Floating Point (`float`)

This is the single precision floating point as defined in PTX:

.f32   PTX type is the single precision fp (min$\approx 1.17 \times 10^{-38}$. max$\approx 3.4 \times 10^{+38}$)

With the .f32 PTX data type, the smallest representable number (at full resolution) is $\approx 1.17 \times 10^{-38}$ and the highest representable number (at full resolution) is $\approx 3.4 \times 10^{+38}$. This conforms to the IEEE 754 single-precision floating point standard, which is one of the most commonly used data types in any computer. Although the same format allows the representation of smaller numbers (*denormalized numbers*), the resolution (i.e., the number of mantissa bits) of these numbers is lower. Every floating point number consists of three fields:

- **sign bit** is a single-bit value, where 0 indicates positive and 1 indicates negative.

- **exponent** is an 8-bit value and determines the *range* of the number.

- **mantissa** is a 23-bit value and determines the *precision* of the number. The effective precision of a float is actually 24 bits because when a float number is stored, the leading {1.} of the mantissa is not stored because a normalized mantissa always leads with this, what is called *hidden 1*; this effectively corresponds to an additional bit in the resolution of the mantissa.

The idea behind a floating point format in general — as compared to a same size integer INT32 — is that we sacrifice *precision* to gain *range*. For example, comparing FP32 to INT32, while INT32 has a 32-bit fixed precision and a fixed range, FP32 only has a 24-bit effective precision, but allows us to represent significantly larger numbers, i.e., has a much wider range. Note that range also implies being able to represent significantly smaller numbers. Example PTX instructions for the FP32 data type are as follows:

```
copysign.f32 d,a,b;      // copy sign bit of a into b. return result as d
add.rn.ftz.f32 d,a,b;    // d=a+b. round to nearest even. flush to zero
mul.rz.sat.f32 d,a,b;    // d=a*b. round to zero. saturate
rcp.rn.f32 d,a;          // d=1/a
rcp.approx.f32 d,a;      // d=1/a (much faster)
sqrt.approx.f32 d,a;     // d=sqrt(a) approximate, but much faster
div.ftz.f32 d,a,b;       // d=a/b. Flush to zero
fma.rn d,a,b,c;          // d=a*b+c
```

9.3.9 FP64: Double Precision Floating Point (double)

This is the double precision floating point as defined in PTX:

.f64   PTX type is the single precision fp (min$\approx 2.2 \times 10^{-308}$. max$\approx 1.8 \times 10^{+308}$)

With the .f64 PTX data type, the smallest representable number (at full resolution) is $\approx 2.2 \times 10^{-308}$ and the highest representable number (also at full resolution) is $\approx 1.8 \times 10^{+308}$. This conforms to the IEEE 754 double-precision floating point definition. This standard also allows smaller numbers to be represented at lower resolution (termed *denormal numbers*). Double precision floating point numbers also consist of three fields:

- **sign bit** is a single-bit value, where 0 indicates positive and 1 indicates negative.

- **exponent** is 11 bits for the double data type.

- **mantissa** is 52 bits, corresponding to an effective 53-bit precision.

Double-precision numbers are primarily used for their improved precision in applications that are based on a significant number of cascaded (e.g., continuously accumulated) floating point operations, where each operation contributes a small error due to its limited precision.

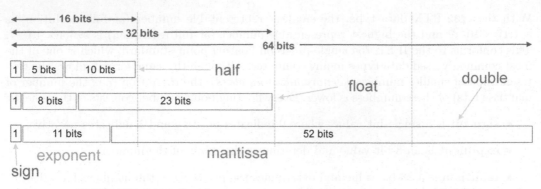

FIGURE 9.9 IEEE 754-2008 floating point standard and the supported floating point data types by CUDA. `half` data type is supported in Compute Capability 5.3 and above, while `float` has seen support from the first day of the introduction of CUDA. Support for `double` types started in Compute Capability 1.3.

As the accumulation of the numbers continue, the error grows, which effectively reduces the resolution of the result. Although using double precision does not prevent the accumulation of the error, it drastically reduced the ratio of the error in comparison to the result. Example PTX instructions for the FP64 data type are as follows:

```
fma.f64 d,a,b,c;        // d=a*b+c;
min.f64 d,a,b;          // d=min(a,b). Supports denormal numbers
sqrt.rnd.f64 d,a;       // d=sqrt(a)
```

9.3.10 FP16: Half Precision Floating Point (`half`)

Half precision floating point numbers were introduced with the IEEE754-2008 standard. PTX ISA 4.2 (Compute Capability 5.3) introduced the following instructions to support half-precision floating point in CUDA:

```
fma.ftz.f16 d,a,b,c;    // d=a*b+c (all half precision). ftz=flush to zero.
fma.ftz.f16x2 d,a,b,c;  // d=a*b+c (a, b, c are an array of two half-precision)
mov.b32 f, (h0,h1);     // pack h0,h1 (half's) into a f (float)
ld.global.b32 f, [addr]; // load a packed f16x2 into f (32-bit) from memory
cvt.rn.f16.f32 h,f;     // down-convert from float to half
```

Here, we see the ability of the GPU to do "packed" computations (i.e., two additions in one instruction). This is somehow similar to the `dp4a` instruction that allows the addition of 4 bytes in one instruction.

9.3.11 What is a FLOP?

Recall from Table 9.1 that we quantified the theoretical computational peak of multiple GPUs using the metric **G**iga **FL**oating point **OP**erations per **S**econd (GFLOPS). This begs the question: *What is a FLOP?* Considering the fact that a GPU's core (or the FPU inside the core of a CPU) is capable of a fused multiply accumulate (FMA) in a single instruction, what should we call a floating point "operation"? The answer is FMA.

In other words, if you executed 1 billion floating point additions, 1 billion multiplications, and 1 billion FMAs in a second, you computed 3 GFLOPS. With the FMA, you buy one (multiplication) and get one (addition) free! So, you don't get additional bonus points for executing these two operations in one instruction.

9.3.12 Fused Multiply-Accumulate (FMA) versus Multiply-Add (MAD)

The multiple-add instruction (`mad`) we saw before was introduced in PTX 1.0 and the newer fused multiply-accumulate (`fma`) instruction was only available with the later PTX versions; the FP64 version for `double` data types (`fma.f64`) was available in PTX1.4, the FP32 version for `float` data types (`fma.f32`) became available in PTX 2.0, while the FP16 version for the `half` data types (`fma.f16` and `fma.f16x2`) became available in PTX 4.2, as mentioned in Section 9.3.10. Also some `mad` instructions have exceptions, the difference is in the way the rounding is done. There are two possible ways of rounding for the general multiply-accumulate operations:

- **Double rounding**: the multiply-accumulate operation is computed as follows:

$$d \;=\; a \times b + c \;=\; \text{Round}(\text{MultiplyAndRound}(a, b) + c) \qquad (9.4)$$

- **Single rounding**: the "fused" multiply-accumulate operation is computed as follows:

$$d \;=\; a \times b + c \;=\; \text{Round}(\text{Multiply}(a, b) + c) \qquad (9.5)$$

The difference is that while the MultiplyAndRound operation rounds the resulting number to the precision of the operands, which reduces the intermediate resolution, the Multiply operation produces a result that has infinite resolution. Thus, the `fma` family operations prevent the accuracy loss twice. In modern CPUs and GPUs, `fma` is the only type of operation that makes sense to use, while the double rounding is somehow deprecated.

9.3.13 Quad and Octo Precision Floating Point

Figure 9.9 only shows the PTX-supported floating point data formats. However, there are other formats that are standard in IEEE 754-2008, although they are not supported in PTX. They are as follows:

- **Quadruple (Quad) Precision**: This format is called *decimal128* in the IEEE 754-2008 format. Its format is identical to the other floating point numbers; however, its precision and range are higher. It has a single sign bit, like the other formats, while it has a 15-bit exponent and a 112-bit mantissa, for a total of 128 bits.

- **Octuple (Octo) Precision**: This format is called *decimal256* in IEEE 754-2008 and has a single sign bit, 19 exponent bits, and 236 mantissa bits, for a total of 256 bits.

9.3.14 Pascal GP104 Engine SM Structure

The GTX 1080 uses a GP104 engine [32], which has a different SM structure than the GP100 shown in Figure 9.8. An interesting — and shockingly counterintuitive — property of the GP104 is that its performance in floating point operation using the `half` data types is 1/64 that of `float` operations. The reason lies in the design of the FPU inside the cores, which are not optimized for the `half` data types, while the performance of the GP100 is twice the `float` data types in GP100 because its DPU is responsible for the `half` data types, which is optimized for this type.

9.4   IMFLIPGC.CU: CORE-FRIENDLY IMFLIPG

We learned a great deal about the SM structure of different architecture families in this chapter, which should help us understand why our programs are performing poorly and how we can make them perform better. Let us see if we can apply some of the ideas we developed to the imflipG.cu to make the program use the core resources better, i.e., make it "core-friendly." The core-friendly version of the program is called imflipGCM.cu, which clearly contains its memory-friendly version, too. The only difference is the names of the kernels, as we will clarify in the following sections. In this section, our only goal is to utilize the *core* resources better, while using *memory* resources better is saved for Chapter 10. Here is what we will achieve in this chapter:

- We will make the Vflip() kernel in Code 6.7 core friendly. Its core-friendly versions are Vflip2() (Code 9.3), Vflip3() (Code 9.5), Vflip4() (Code 9.7), and Vflip5() (Code 9.9),

- We will make the Hflip() kernel (Code 6.8) core friendly. Its core-friendly versions are Hflip2() (Code 9.2), Hflip3() (Code 9.4), Hflip4() (Code 9.6), and Hflip5() (Code 9.8),

- We will make the PixCopy() kernel (Code 6.9) core friendly. Its core-friendly versions are PixCopy2() and PixCopy3() (both in Code 9.10),

The reason for the interleaved numbering is that whatever idea we can apply to Vflip2() (Code 9.3) can also be applied to design Hflip2() (Code 9.2); therefore, they are introduced sequentially. The imflipG.cu program does not contain a lot of core computations; aside from some exceptions, it contains a lot of data movement. Because of this, we would not expect a lot of kernel performance improvement by making it core friendly; however, even in this case, we will observe a significant performance improvement with the ideas we are introducing in this section. Generic data manipulation is a core operation and if we improve it we should expect an improvement. Additionally, generic kernel-based improvements — relating to passing arguments into the kernel — are covered in this section, too. We will use the experience we gain in this chapter to improve a much more core-intensive program, imedgeG.cu, in Section 9.5.

The main() function of the imflipGCM.cu program is shown in Code 9.1. The added functionality, as compared to imflipG.cu, is the introduction of the multi-dimensional variables as follows:

```
dim3 dimGrid2D(BlkPerRow, ip.Vpixels);
```

In this example, `dimGrid2D` is a 2D variable. We will use it when we are passing 2D block dimensions. Aside from that, the program runs the correct kernel version based on a cascaded set of `switch` statements. For example, the statement below

```
case 3: Hflip3<<<dimGrid2D,ThrPerBlk>>>(GPUCopyImg, GPUImg, IPH, RowBytes);
        strcpy(KernelName,"Hflip3:Each thread copies 1 pixel (using a 2D grid)");
```

executes when the user enters the following command line:

$ imflipGCM H 128 3

The `KernelName[]` variable is used to report what the kernel does; using this feature, this program can run multiple improved versions of the same kernel and report the results one after the other with nice descriptions for each kernel. This is useful in comparing the different versions of the kernels and their quantitative impact on performance.

CODE 9.1: imflipGCM.cu   main() ... {...

In imflipGCM.cu, main() is modified to allow specifying the kernel name at the command line.

```
#define CEIL(a,b)              ((a+b-1)/b)
#define SWAP(a,b,t)            t=b; b=a; a=t;
#define DATAMB(bytes)          (bytes/1024/1024)
#define DATABW(bytes,timems)   ((float)bytes/(timems * 1.024*1024.0*1024.0))
...
int main(int argc, char **argv)
{
  int KernelNum=1;   char KernelName[255];
   ...
  strcpy(ProgName, "imflipG");
  switch (argc){
  case 6:  KernelNum = atoi(argv[5]);
   ...
  default: printf("\n\nUs... [V/H/C/T] [ThrPerBlk] [Kernel=1-9]", ProgName);
           printf("\n\nExample: %s Astronaut.bmp Output.bmp V 128 2", ProgName);
  cudaEventRecord(time2, 0);  // Time stamp after the CPU --> GPU tfr is done
  RowBytes = (IPH * 3 + 3) & (~3);          RowInts = RowBytes / 4;
  BlkPerRow = CEIL(IPH,ThrPerBlk);          BlkPerRowInt = CEIL(RowInts, ThrPerBlk);
  BlkPerRowInt2 = CEIL(CEIL(RowInts,2), ThrPerBlk); NumBlocks = IPV*BlkPerRow;
  dim3 dimGrid2D(BlkPerRow,       ip.Vpixels);
  dim3 dimGrid2D2(CEIL(BlkPerRow,2), ip.Vpixels);
  dim3 dimGrid2D4(CEIL(BlkPerRow,4), ip.Vpixels);
  dim3 dimGrid2Dint(BlkPerRowInt, ip.Vpixels);
  dim3 dimGrid2Dint2(BlkPerRowInt2, ip.Vpixels);
  switch (Flip){
    case 'H': switch (KernelNum){
            case 1: Hflip<<<NumBlocks,ThrPerBlk>>>(...);
                    strcpy(KernelName,"Hflip:Each thread copies..."); break;
            case 2: Hflip2<<<NumBlocks,ThrPerBlk>>>(...);
                    strcpy(KernelName,"Hflip2..Uses pre-computed.."); break;
            case 3: Hflip3<<<dimGrid2D,ThrPerBlk>>>(...);
                     strcpy(KernelName, "Hflip3:...using a 2D grid"); break;
    case 'V': switch (KernelNum){
            case 1: Vflip<<<NumBlocks,ThrPerBlk>>>(...);
                    strcpy(KernelName,"Vflip:Each thread cop..."); break;
            case 2: Vflip2<<<NumBlocks,ThrPerBlk>>>(...);
                    strcpy(KernelName,"Vflip2:Each thread cop..."); break;
    case 'C': NumBlocks = CEIL(IMAGESIZE,ThrPerBlk);    NB2 = CEIL(NumBlocks,2);
            NB4 = CEIL(NumBlocks,4);                    NB8 = CEIL(NumBlocks,8);
            switch (KernelNum){
                case 1: PixCopy<<<NumBlocks,ThrPerBlk>>>(...);
                ...
  printf("------------------------------------------------------\n");
  printf("...", ProgName, ..., Flip, Thr..., KernelNum, Num..., BlkPerRow);
  printf("------------------------------------------------------\n");
  ...
}
```

CODE 9.2:   imflipGCM.cu   Hflip2() {...}

Hflip2() avoids computing `BlkPerRow` and `RowBytes` repeatedly.

```
// Improved Hflip() kernel that flips the given image horizontally
// BlkPerRow, RowBytes variables are passed, rather than calculated
__global__
void Hflip2(uch *ImgDst, uch *ImgSrc, ui Hpixels, ui BlkPerRow, ui RowBytes)
{
    ui ThrPerBlk = blockDim.x;
    ui MYbid = blockIdx.x;
    ui MYtid = threadIdx.x;
    ui MYgtid = ThrPerBlk * MYbid + MYtid;

    //ui BlkPerRow = CEIL(Hpixels,ThrPerBlk);
    //ui RowBytes = (Hpixels * 3 + 3) & (~3);
    ui MYrow = MYbid / BlkPerRow;
    ui MYcol = MYgtid - MYrow*BlkPerRow*ThrPerBlk;
    if (MYcol >= Hpixels) return;    // col out of range
    ui MYmirrorcol = Hpixels - 1 - MYcol;
    ui MYoffset = MYrow * RowBytes;
    ui MYsrcIndex = MYoffset + 3 * MYcol;
    ui MYdstIndex = MYoffset + 3 * MYmirrorcol;

    // swap pixels RGB @MYcol , @MYmirrorcol
    ImgDst[MYdstIndex] = ImgSrc[MYsrcIndex];
    ImgDst[MYdstIndex + 1] = ImgSrc[MYsrcIndex + 1];
    ImgDst[MYdstIndex + 2] = ImgSrc[MYsrcIndex + 2];
}
```

9.4.1 Hflip2(): Precomputing Kernel Parameters

Our first idea is very straightforward. It really has nothing to do with the cores. It has something to do with the efficiency in *passing function arguments* into a GPU kernel. Let us look at Code 6.8, where we introduced the Hflip() kernel), to see what the kernel does as its first few steps:

```
ui BlkPerRow = (Hpixels + ThrPerBlk - 1) / ThrPerBlk; // ceil
ui RowBytes = (Hpixels * 3 + 3) & (~3);
```

Note that the first line above is equivalent to `ui BlkPerRow = CEIL(Hpixels,ThrPerBlk);` This is a simple pre-computation of the `BlkPerRow` and `RowBytes` values, which can be computed from what is already passed into the kernel (`Hpixels`), as well as what can be obtained from the special registers (in this case, `ThrPerBlock` can be obtained from the special register `blockDim.x`). Although `ThrPerBlock` is needed later when `MYcol` is being computed, do we really need to compute `BlkPerRow` and `RowBytes` inside the kernel? Remember that anything we calculate inside the kernel is computed for ***every single thread***. What happens if we simply pass them as function arguments? This is possible because their values never change during the execution of any of the threads. So, instead of computing it millions of times, why not compute it once inside main() and pass it as a function argument? If we look at

TABLE 9.2 Comparison of kernel performances between (Hflip() and Hflip2()) as well as (Vflip() and HVflip2()).

Feature	Box II	Box III	Box IV	Box VII	Box VIII
GPU	K3000M	GTX 760	Titan Z	GTX 1070	Titan X
Engine	GK104	GK104	2xGK110	GP104-200	GP102-400
Cores	576	1152	2x2880	1920	3584
Compute Cap	3.0	3.0	3.5	6.1	6.1
GM BW GBps	89	192	336	256	480
Peak GFLOPS	753	2258	8122	5783	10157
DGFLOPS	31	94	2707	181	317
Kernel Performance: imflipGCM astronaut.bmp out.bmp H 128 1					
Hflip (ms)	20.12	6.73	4.17	2.15	1.40
GBps	11.82	35.35	57.02	110.78	169.5
Achieved (%)	(13%)	(18%)	(17%)	(43%)	(35%)
Kernel Performance: imflipGCM astronaut.bmp out.bmp H 128 2					
Hflip2 (ms)	17.23	5.85	3.63	1.98	1.30
GBps	13.81	40.69	65.54	119.85	182.34
Achieved (%)	(16%)	(21%)	(20%)	(47%)	(38%)
Improvement	14%	13%	13%	8%	7%
Kernel Performance: imflipGCM astronaut.bmp out.bmp V 128 1					
Vflip (ms)	20.02	6.69	4.11	2.12	1.40
GBps	11.88	35.56	57.83	112.19	169.5
Achieved (%)	(13%)	(19%)	(17%)	(44%)	(35%)
Kernel Performance: imflipGCM astronaut.bmp out.bmp V 128 2					
Vflip2 (ms)	17.23	5.84	3.67	1.96	1.30
GBps	13.81	40.71	64.85	121.63	182.34
Achieved (%)	(16%)	(21%)	(19%)	(48%)	(38%)
Improvement	14%	13%	11%	8%	7%

these computations carefully, we see the integer division, which is not something that you want to put inside every thread's computation, as we witnessed many times before. We also see two integer additions. Ironically, two integer additions can be more expensive than an addition followed by a multiplication in cases when the GPU cores' integer unit does not support a three-operand addition like $d = a + b + c$; PTX does not seem to have one.

Table 9.2 provides a comparison between the Hflip() and Hflip2() functions. Although some of the boxes are the same as the ones we saw previously, a new box is added (Box VII), which incorporates a Pascal series GTX Titan X GPU. So, this table includes two Kepler and two Pascal GPUs. No Maxwell is included, but you can expect the performance characteristics to be somewhere between these two families. Table 9.2 also provides a comparison between the Vflip() and Vflip2() functions. Because of the memory access patterns being very similar between Hflip() and Vflip() kernels, Table 9.2 shows an identical behavior for both kernels.

Code 9.2 shows the modified kernel, Hflip2(), in which the lines that are supposed to compute `BlkPerRow` and `RowBytes` are simply commented out and these two values are passed as arguments to the kernel, increasing the total number of arguments passed into the kernel to 5 (from 3).

Because the computation of these two values (`BlkPerRow` and `RowBytes`) depend on other values that do not change throughout the execution of the program once the user enters the command line parameters (`ThePerBlk` and `ip.Hpixels`), these values are readily calculated inside main() and passed into the kernel call as follows:

```
#define IPH        ip.Hpixels
main()
{
   ui         ThrPerBlk = 256, NumBlocks, NB2, NB4, NB8, GPUDataTransfer;
   ui         RowBytes, RowInts;
   ...
   RowBytes = (IPH * 3 + 3) & (~3);
   BlkPerRow = CEIL(IPH,ThrPerBlk);
   ...
   Hflip2<<<NumBlocks,ThrPerBlk>>>(GPUCopyImg,GPUImg,IPH,BlkPerRow,RowBytes);
   ...
```

Aside from this similarity, here is what we observe in Table 9.2:

- What we see is that our modified Hflip2() and Vflip2() kernels work ≈13% better on Kepler GPUs and ≈7% better on Pascal grade GPUs.

- The reason for the smaller relative improvement on Pascal GPUs seems to be the fact that the initial code must be working efficiently to begin with.

- The initial code saturates 35% of the bandwidth on Pascal GPUs; the reason for this must be related to the improvements in the architecture, as well as the new instructions that provide better data manipulation.

- We also have to remember another fact: despite the fact that we are passing 2 more arguments into the kernel is not degrading kernel performance. Intuitively, eliminating two lines of code is a good thing, but if increasing the number of arguments passed into the kernel by two counter-acts this benefit, we are back to square one. This doesn't seem to be happening.

9.4.2 Vflip2(): Precomputing Kernel Parameters

Code 9.3 shows the modified kernel Vflip2(), which is lacking two of the lines that compte the same `BlkPerRow` and `RowBytes` variables. The only small difference between the original `Vflip` kernel (Code 6.7) and this one is that `Vflip` had 4 arguments to begin with; in its modified version, `Vflip2`, this has increased to 6.

9.4.3 Computing Image Coordinates by a Thread

Before we look at our next kernel improvement, let us analyze the two lines of code below, which allow each thread to determine the x,y coordinates (`MYcol`, `MYrow`, respectively) it is responsible for processing:

```
ui MYrow = MYbid / BlkPerRow;
ui MYcol = MYgtid - MYrow*BlkPerRow*ThrPerBlk;
```

The bad news is that we cannot use the same trick as before by passing these values as function arguments; they change for every single thread and they are not fixed.

CODE 9.3:   imflipGCM.cu   Vflip2() {...}

Vflip2() avoids computing `BlkPerRow` and `RowBytes` repeatedly.

```
// Improved Vflip() kernel that flips the given image vertically
// BlkPerRow, RowBytes variables are passed, rather than calculated
__global__
void Vflip2(uch *ImgDst, uch *ImgSrc, ui Hpixels, ui Vpixels, ui BlkPerRow, ui
    RowBytes)
{
  ui ThrPerBlk = blockDim.x;
  ui MYbid = blockIdx.x;
  ui MYtid = threadIdx.x;
  ui MYgtid = ThrPerBlk * MYbid + MYtid;

  //ui BlkPerRow = CEIL(Hpixels,ThrPerBlk);
  //ui RowBytes = (Hpixels * 3 + 3) & (~3);
  ui MYrow = MYbid / BlkPerRow;
  ui MYcol = MYgtid - MYrow*BlkPerRow*ThrPerBlk;
  if (MYcol >= Hpixels) return;   // col out of range
  ui MYmirrorrow = Vpixels - 1 - MYrow;
  ui MYsrcOffset = MYrow      * RowBytes;
  ui MYdstOffset = MYmirrorrow * RowBytes;
  ui MYsrcIndex = MYsrcOffset + 3 * MYcol;
  ui MYdstIndex = MYdstOffset + 3 * MYcol;

  // swap pixels RGB @MYrow , @MYmirrorrow
  ImgDst[MYdstIndex] = ImgSrc[MYsrcIndex];
  ImgDst[MYdstIndex + 1] = ImgSrc[MYsrcIndex + 1];
  ImgDst[MYdstIndex + 2] = ImgSrc[MYsrcIndex + 2];
}
```

9.4.4 Block ID versus Image Row Mapping

The good news is that we are the ones that formulate the mapping between the x,y coordinates and the parameters of CUDA. Could we have done the mapping differently? Currently, in Hflip(), Hflip2(), Vflip(), and Vflip2(), we are using a 1D set of blocks, which contains a mixture of x and y coordinates of the image. What if we used a 2D set of blocks, where one of the dimensions of the block ID directly corresponded to one of the image coordinates? For example, what if the `MYrow` directly corresponded to the second dimension of the block ID as follows:

```
ui MYrow = blockID.y;
ui MYcol = MYbid*ThrPerBlk + MYtid;
```

This not only makes the computation of `MYrow` just a simple register `mov` operation, it makes the computation of `MYcol` extremely easy too. We know from Section 9.3.4 that computing `MYcol` is a single instruction `mad.lo.u32 d,a,b,c;` despite its complicated look. So, with this new index mapping, we converted the computation of the x,y coordinates to a mere two PTX instructions. It gets better ... we no longer need the `MYgtid` variable either.

CODE 9.4: imflipGCM.cu Hflip3() {...}

Hflip3() uses a 2D grid of blocks to make x,y image coordinate computation easier.

```
// Improved Hflip2() kernel using a 2D grid of blocks
__global__
void Hflip3(uch *ImgDst, uch *ImgSrc, ui Hpixels, ui RowBytes)
{
    ui ThrPerBlk = blockDim.x;
    ui MYbid = blockIdx.x;           ui MYtid = threadIdx.x;
    //ui MYgtid = ThrPerBlk * MYbid + MYtid;
    //ui BlkPerRow = CEIL(Hpixels,ThrPerBlk);
    //ui RowBytes = (Hpixels * 3 + 3) & (~3);
    //ui MYrow = MYbid / BlkPerRow;
    //ui MYcol = MYgtid - MYrow*BlkPerRow*ThrPerBlk;
    ui MYrow = blockIdx.y;                      ui MYcol = MYbid*ThrPerBlk + MYtid;
    if (MYcol >= Hpixels) return;              // col out of range
    ui MYmirrorcol = Hpixels - 1 - MYcol;    ui MYoffset = MYrow * RowBytes;
    ui MYsrcIndex = MYoffset + 3 * MYcol;
    ui MYdstIndex = MYoffset + 3 * MYmirrorcol;
    // swap pixels RGB @MYcol , @MYmirrorcol
    ImgDst[MYdstIndex] = ImgSrc[MYsrcIndex];
    ImgDst[MYdstIndex + 1] = ImgSrc[MYsrcIndex + 1];
    ImgDst[MYdstIndex + 2] = ImgSrc[MYsrcIndex + 2];
}
```

9.4.5   Hflip3(): Using a 2D Launch Grid

The only reason for the computation of MYgtid (i.e., the thread's global thread ID) was to be able to determine the image coordinates by relating the x,y coordinates to the MYgtid variable; this variable is not needed anymore. The modified version of the Hflip() that uses 2D block indexing is shown in Code 9.4, which is named Hflip3(). To launch this kernel, a 2D grid of blocks must be used in main() as follows:

```
dim3 dimGrid2D(BlkPerRow, ip.Vpixels);
...
case 3:Hflip3<<<dimGrid2D,ThrPerBlk>>>(GPUCopyImg,GPUImg,IPH,RowBytes);
       strcpy(KernelName,"Hflip3:Each thread copies 1 pixel (using a 2D grid)");
       break;
```

As we see from the code above, the y coordinate of the image has a one-on-one relationship with the blockIDx.y (i.e., the second dimension of the grid of blocks), which eliminates the need for each thread to compute the y image coordinate inside each kernel. Once the y coordinate is known, this makes it easier to compute the x coordinate also. This trick allows us to use the dimension computation hardware to get a *free integer division*! This is great considering the fact that we were getting a free for loop when we used the GPU internal hardware correctly, as we observed before. As we see in these examples, the trick with CUDA programming is to avoid *over-programming*; the more you use the internal GPU hardware to reduce the core instructions, the faster your programs will be.

CODE 9.5:   *imflipGCM.cu*   Vflip3() {...}

Vflip3() uses a 2D grid of blocks to make x,y image coordinate computation easier.

```
// Improved Vflip2() kernel that flips the given image vertically
// Grid is launched using 2D block numbers
__global__
void Vflip3(uch *ImgDst, uch *ImgSrc, ui Hpixels, ui Vpixels, ui RowBytes)
{
   ui ThrPerBlk = blockDim.x;
   ui MYbid = blockIdx.x;
   ui MYtid = threadIdx.x;
   //ui MYgtid = ThrPerBlk * MYbid + MYtid;

   //ui BlkPerRow = CEIL(Hpixels,ThrPerBlk);
   //ui RowBytes = (Hpixels * 3 + 3) & (~3);
   //ui MYrow = MYbid / BlkPerRow;
   //ui MYcol = MYgtid - MYrow*BlkPerRow*ThrPerBlk;
   ui MYrow = blockIdx.y;
   ui MYcol = MYbid*ThrPerBlk + MYtid;
   if (MYcol >= Hpixels) return;    // col out of range
   ui MYmirrorrow = Vpixels - 1 - MYrow;
   ui MYsrcOffset = MYrow     * RowBytes;
   ui MYdstOffset = MYmirrorrow * RowBytes;
   ui MYsrcIndex = MYsrcOffset + 3 * MYcol;
   ui MYdstIndex = MYdstOffset + 3 * MYcol;

   // swap pixels RGB @MYrow , @MYmirrorrow
   ImgDst[MYdstIndex] = ImgSrc[MYsrcIndex];
   ImgDst[MYdstIndex + 1] = ImgSrc[MYsrcIndex + 1];
   ImgDst[MYdstIndex + 2] = ImgSrc[MYsrcIndex + 2];
}
```

9.4.6   Vflip3(): Using a 2D Launch Grid

The Vflip3() kernel in Code 9.5 applies the same 2D indexing idea to the original Vflip() kernel. Table 9.3 compares the results of our improvements so far. Using 2D block indexing seems to be improving the performance by a mere 1% on desktop Kepler GPUs

TABLE 9.3   Kernel performances: Hflip(),⋯,Hflip3(), and Vflip(),⋯,Vflip3().

Kernel	Box II	Box III	Box IV	Box VII	Box VIII
Hflip (ms)	20.12	6.73	4.17	2.15	1.40
Hflip2 (ms)	17.23	5.85	3.63	1.98	1.30
Hflip3 (ms)	16.35	5.59	3.59	1.83	1.19
Vflip (ms)	20.02	6.69	4.11	2.12	1.40
Vflip2 (ms)	17.23	5.84	3.67	1.96	1.30
Vflip3 (ms)	16.40	5.62	3.65	1.87	1.19

(Boxes III, IV), a little healthier 5% on the mobile Kepler (Box II), but a nice 8% on the two Pascal GPUs (Boxes VII and VIII).

CODE 9.6:   imflipGCM.cu   Hflip4() {...}

Hflip4() computes two consecutive pixels, rather than one.

```
// Improved Hflip3() kernel that flips the given image horizontally
// Each kernel takes care of 2 consecutive pixels; half as many blocks are launched
__global__
void Hflip4(uch *ImgDst, uch *ImgSrc, ui Hpixels, ui RowBytes)
{
   ui ThrPerBlk = blockDim.x;
   ui MYbid = blockIdx.x;
   ui MYtid = threadIdx.x;
   //ui MYgtid = ThrPerBlk * MYbid + MYtid;

   ui MYrow = blockIdx.y;
   ui MYcol2 = (MYbid*ThrPerBlk + MYtid)*2;
   if (MYcol2 >= Hpixels) return;      // col (and col+1) are out of range
   ui MYmirrorcol = Hpixels - 1 - MYcol2;
   ui MYoffset = MYrow * RowBytes;
   ui MYsrcIndex = MYoffset + 3 * MYcol2;
   ui MYdstIndex = MYoffset + 3 * MYmirrorcol;

   // swap pixels RGB @MYcol , @MYmirrorcol
   ImgDst[MYdstIndex] = ImgSrc[MYsrcIndex];
   ImgDst[MYdstIndex + 1] = ImgSrc[MYsrcIndex + 1];
   ImgDst[MYdstIndex + 2] = ImgSrc[MYsrcIndex + 2];
   if ((MYcol2 + 1) >= Hpixels) return;     // only col+1 is out of range
   ImgDst[MYdstIndex - 3] = ImgSrc[MYsrcIndex + 3];
   ImgDst[MYdstIndex - 2] = ImgSrc[MYsrcIndex + 4];
   ImgDst[MYdstIndex - 1] = ImgSrc[MYsrcIndex + 5];
}
```

9.4.7   Hflip4(): Computing Two Consecutive Pixels

Considering the fact that the coordinate computation between any two adjacent pixels is almost identical, it is intuitive to think that we can possibly compute two pixels inside each kernel a lot cheaper than a single pixel (relatively, in terms of time per pixel). Modified Hflip4() kernel that achieves this goal is shown in Code 9.6. Let us analyze this kernel.

- The part that computes the pixel address is identical, with the small exception that it accounts for 2 pixels/thread when computing the column index. The row indexing is identical. Clearly, the main() is required to launch only half as many blocks. So, we seem to have saved a little bit of code; for the same number of C statements, we seemed to have computed the address for two pixels instead of one.

- The bottom part shows how we are writing the two pixels. We use the same address offset and write 2 pixels (6 bytes) that are consecutive. One big change is that we have the `if` statement after writing the first RGB to make sure that we are not going out of bounds in address range.

- Results of this kernel are shown in Table 9.4; this kernel gives us a worse performance than the previous Hflip3() kernel. Can this be true?

CODE 9.7:   imflipGCM.cu   Vflip4() {...}

Vflip4() computes two consecutive pixels, rather than one.

```
__global__
void Vflip4(uch *ImgDst, uch *ImgSrc, ui Hpixels, ui Vpixels, ui RowBytes)
{
    ui ThrPerBlk = blockDim.x;        ui MYbid = blockIdx.x;
    ui MYtid = threadIdx.x;           ui MYrow = blockIdx.y;
    ui MYcol2 = (MYbid*ThrPerBlk + MYtid)*2;
    if (MYcol2 >= Hpixels) return;       // col is out of range
    ui MYmirrorrow = Vpixels - 1 - MYrow;
    ui MYsrcOffset = MYrow     * RowBytes;
    ui MYdstOffset = MYmirrorrow * RowBytes;
    ui MYsrcIndex = MYsrcOffset + 3 * MYcol2;
    ui MYdstIndex = MYdstOffset + 3 * MYcol2;
    // swap pixels RGB @MYrow , @MYmirrorrow
    ImgDst[MYdstIndex] = ImgSrc[MYsrcIndex];
    ImgDst[MYdstIndex + 1] = ImgSrc[MYsrcIndex + 1];
    ImgDst[MYdstIndex + 2] = ImgSrc[MYsrcIndex + 2];
    if ((MYcol2+1) >= Hpixels) return;    // only col+1 is out of range
    ImgDst[MYdstIndex + 3] = ImgSrc[MYsrcIndex + 3];
    ImgDst[MYdstIndex + 4] = ImgSrc[MYsrcIndex + 4];
    ImgDst[MYdstIndex + 5] = ImgSrc[MYsrcIndex + 5];
}
```

9.4.8   Vflip4(): Computing Two Consecutive Pixels

Modified Vflip4() kernel is shown in Code 9.7, with its runtime results in Table 9.4; same thing! This kernel is slower than Vflip3(). Obviously, something is hurting the performance that we weren't suspecting. There is only one logical explanation: the `if` statement; while the

TABLE 9.4   Kernel performances: Hflip(),···,Hflip4(), and Vflip(),···,Vflip4().

Kernel	Box II	Box III	Box IV	Box VII	Box VIII
Hflip (ms)	20.12	6.73	4.17	2.15	1.40
Hflip2 (ms)	17.23	5.85	3.63	1.98	1.30
Hflip3 (ms)	16.35	5.59	3.59	1.83	1.19
Hflip4 (ms)	19.48	6.68	4.04	2.47	1.77
Vflip (ms)	20.02	6.69	4.11	2.12	1.40
Vflip2 (ms)	17.23	5.84	3.67	1.96	1.30
Vflip3 (ms)	16.40	5.62	3.65	1.87	1.19
Vflip4 (ms)	19.82	6.57	4.02	2.37	1.71

computation inside the if statement is harmless, the if statement itself is a huge problem. The performance penalty introduced by it totally negates the performance gain. The reason for this is *thread divergence*, which is due to the threads in a warp providing a different TRUE/FALSE answer to the same if statement; the divergence hurts the parallelism, because GPU does its best work when all 32 threads do exactly the same thing.

CODE 9.8:   imflipGCM.cu   Hflip5() {...}

Hflip5() computes four consecutive pixels, rather than one.

```
__global__
void Hflip5(uch *ImgDst, uch *ImgSrc, ui Hpixels, ui RowBytes)
{
    ui ThrPerBlk = blockDim.x;
    ui MYbid = blockIdx.x;
    ui MYtid = threadIdx.x;
    ui MYrow = blockIdx.y;
    ui MYcol4 = (MYbid*ThrPerBlk + MYtid) * 4;
    if (MYcol4 >= Hpixels) return;      // col (and col+1) are out of range
    ui MYmirrorcol = Hpixels - 1 - MYcol4;
    ui MYoffset = MYrow * RowBytes;
    ui MYsrcIndex = MYoffset + 3 * MYcol4;
    ui MYdstIndex = MYoffset + 3 * MYmirrorcol;
    // swap pixels RGB @MYcol , @MYmirrorcol
    for (ui a = 0; a<4; a++){
        ImgDst[MYdstIndex - a * 3] = ImgSrc[MYsrcIndex + a * 3];
        ImgDst[MYdstIndex - a * 3 + 1] = ImgSrc[MYsrcIndex + a * 3 + 1];
        ImgDst[MYdstIndex - a * 3 + 2] = ImgSrc[MYsrcIndex + a * 3 + 2];
        if ((MYcol4 + a + 1) >= Hpixels) return;   // next pixel is out of range
    }
}
```

9.4.9   Hflip5(): Computing Four Consecutive Pixels

I wouldn't be surprised if you are not convinced that the results are not accurate. So, let's double down on the idea. Why not compute four pixels instead of two? The Hflip5() kernel, shown in Code 9.8, does just that. In fact, let us even add a for loop at the end of the kernel to allow us to compute as many pixels as we want. For four pixels, this loop does four iterations. We are assuming that the image horizontal dimension is divisible by 4, which doesn't sound like an unrealistic assumption, although it restricts the application of imedgeGCM.cu to images with widths that are divisible by 4.

A careful analysis of Code 9.8 shows that the if statement at the end — which was at the center of our suspicions — has the potential to do more damage. Not only is it still there, but the statement inside now requires the computation of two integer additions, something we know that PTX does not support! The previous if statement was if((MYcol2+1)>= Hpixels) and this one is if((MYcol4+a+1)>= Hpixels).

We seem to be digging a bigger hole when we try to compute more pixels in each thread. Results in Table 9.5 show that the performance of Hflip5() is 50% lower than Hflip4() in Kepler GPUs and worse than half in Pascal GPUs. I guess Pascal family is great, although

TABLE 9.5 Kernel performances: Hflip(),···,Hflip5(), and Vflip(),···,Vflip5().

Kernel	Box II	Box III	Box IV	Box VII	Box VIII
Hflip (ms)	20.12	6.73	4.17	2.15	1.40
Hflip2 (ms)	17.23	5.85	3.63	1.98	1.30
Hflip3 (ms)	16.35	5.59	3.59	1.83	1.19
Hflip4 (ms)	19.48	6.68	4.04	2.47	1.77
Hflip5 (ms)	29.11	9.75	6.36	5.19	3.83
Vflip (ms)	20.02	6.69	4.11	2.12	1.40
Vflip2 (ms)	17.23	5.84	3.67	1.96	1.30
Vflip3 (ms)	16.40	5.62	3.65	1.87	1.19
Vflip4 (ms)	19.82	6.57	4.02	2.37	1.71
Vflip5 (ms)	29.13	9.75	6.35	5.23	3.90

the penalty for bad ideas seems to be much higher with it. As we witnessed many times before, this seems to be the trend with more advanced architectures.

CODE 9.9:   imflipGCM.cu   Vflip5() {...}

Vflip5() computes four consecutive pixels, rather than one.

```
// Improved Vflip3() kernel that flips the given image vertically
// Each kernel takes care of 4 consecutive pixels; 1/4 as many blocks are launched
__global__
void Vflip5(uch *ImgDst, uch *ImgSrc, ui Hpixels, ui Vpixels, ui RowBytes)
{
   ui ThrPerBlk = blockDim.x;       ui MYbid = blockIdx.x;
   ui MYtid = threadIdx.x;          ui MYrow = blockIdx.y;
   ui MYcol4 = (MYbid*ThrPerBlk + MYtid)*4;
   if (MYcol4 >= Hpixels) return;      // col is out of range
   ui MYmirrorrow = Vpixels - 1 - MYrow;
   ui MYsrcOffset = MYrow      * RowBytes;
   ui MYdstOffset = MYmirrorrow * RowBytes;
   ui MYsrcIndex = MYsrcOffset + 3 * MYcol4;
   ui MYdstIndex = MYdstOffset + 3 * MYcol4;
   // swap pixels RGB @MYrow , @MYmirrorrow
   for (ui a=0; a<4; a++){
      ImgDst[MYdstIndex + a * 3] = ImgSrc[MYsrcIndex + a * 3];
      ImgDst[MYdstIndex + a * 3 + 1] = ImgSrc[MYsrcIndex + a * 3 + 1];
      ImgDst[MYdstIndex + a * 3 + 2] = ImgSrc[MYsrcIndex + a * 3 + 2];
      if ((MYcol4 + a + 1) >= Hpixels) return;   // next pixel is out of range
   }
}
```

9.4.10 Vflip5(): Computing Four Consecutive Pixels

The Vflip5() kernel, shown in Code 9.9, performs as poorly as his sister, Hflip5(). What we do not know is what the bad idea was: **(1)** Was it the if statement? The answer is Yes. **(2)** Was it the for loop? The answer is Yes? **(3)** Was it the fact that we should never

TABLE 9.6   Kernel performances: PixCopy(), PixCopy2(), and PixCopy3().

Kernel		Box II	Box III	Box IV	Box VII	Box VIII
PixCopy	(ms)	22.76	7.33	4.05	2.33	1.84
PixCopy2	(ms)	15.81	5.48	3.56	1.81	1.16
PixCopy3	(ms)	14.01	5.13	3.23	1.56	1.05

attempt to compute multiple pixels in a kernel? The answer is absolutely No. We just have to do it right.

CODE 9.10:   imflipGCM.cu   PixCopy2(),PixCopy3() {...}

PixCopy2() copies 2 and PixCopy3() copies 4 consecutive pixels at a time.

```
// Improved PixCopy() that copies an image from one part of the
// GPU memory (ImgSrc) to another (ImgDst). Each thread copies 2 consecutive Bytes
__global__
void PixCopy2(uch *ImgDst, uch *ImgSrc, ui FS)
{
   ui ThrPerBlk = blockDim.x;         ui MYbid = blockIdx.x;
   ui MYtid = threadIdx.x;            ui MYgtid = ThrPerBlk * MYbid + MYtid;
   ui MYaddr = MYgtid * 2;
   if (MYaddr > FS) return;           // outside the allocated memory
   ImgDst[MYaddr] = ImgSrc[MYaddr];   // copy pixel
   if ((MYaddr + 1) > FS) return;     // outside the allocated memory
   ImgDst[MYaddr + 1] = ImgSrc[MYaddr + 1]; // copy consecutive pixel
}
// Improved PixCopy() that copies an image from one part of the
// GPU memory (ImgSrc) to another (ImgDst). Each thread copies 4 consecutive Bytes
__global__
void PixCopy3(uch *ImgDst, uch *ImgSrc, ui FS)
{
   ui ThrPerBlk = blockDim.x;         ui MYbid = blockIdx.x;
   ui MYtid = threadIdx.x;            ui MYgtid = ThrPerBlk * MYbid + MYtid;
   ui MYaddr = MYgtid * 4;
   for (ui a=0; a<4; a++){
      if ((MYaddr+a) > FS) return;
      ImgDst[MYaddr+a] = ImgSrc[MYaddr+a];
   }
}
```

9.4.11   PixCopy2(), PixCopy3(): Copying 2,4 Consecutive Pixels at a Time

My answer to "should we never process multiple pixels in a kernel?" was No. I said that we should do it right. Code 9.10 shows two new versions of the kernel PixCopy(), which copy two bytes within a kernel in PixCopy2() and four bytes in PixCopy3(). Performance results are shown in Table 9.6. We observe clearly that the more bytes we copy in the kernel, the better the result gets, even with the if statement right in the middle of the for loop. The reason is obvious if we compare this kernel to any of the previous ones: first of all, this kernel copies a byte, not a pixel, which makes the index computation logic much easier, eliminating a significant amount of overhead at the trailing end of the kernel code. Second, having to read so many bytes that are on consecutive addresses allows the memory controller to aggregate many consecutive bytes and issue them as much larger memory address region reads; we know that this makes the DRAM memory accesses much more efficient. Readers are encouraged to increase the amount of bytes being copied in each kernel to see where there will be an "inflection point" and the steady performance improvement with the increased amount of memory reads will slow down or stop.

9.5 IMEDGEGC.CU: CORE-FRIENDLY IMEDGEG

Now that we formulated a few good techniques to improve kernel performance, let us see if we can apply them to another program that we developed, imedgeGCM.cu, which includes core-friendly kernel versions of imedgeG.cu.

9.5.1 BWKernel2(): Using Precomputed Values and 2D Blocks

First, let us see if we can improve the performance of BWKernel() (Code 8.3) by using the simplest two tricks we learned: (1) passing more precomputed function arguments, and (2) 2D block addressing. This is done in BWKernel2() (Code 9.11). Table 9.7 compares the two kernels: while there is almost no performance difference in Pascal family, Kepler benefited from this idea in a healthy way. Why?

CODE 9.11: imedgeGCM.cu BWKernel2() {...}

BWKernel2() uses two of the highest impact improvements we utilized before in the flip kernels: (1) precomputed arguments and (2) launching a 2D grid of blocks.

```
// Improved BWKernel. Uses pre-computed values and 2D blocks.
__global__
void BWKernel2(double *ImgBW, uch *ImgGPU, ui Hpixels, ui RowBytes)
{
    ui ThrPerBlk = blockDim.x;      ui MYbid = blockIdx.x;
    ui MYtid = threadIdx.x;          double R, G, B;
    ui MYrow = blockIdx.y;           ui MYcol = MYbid*ThrPerBlk + MYtid;
    if (MYcol >= Hpixels) return;    // col out of range

    ui MYsrcIndex = MYrow * RowBytes + 3 * MYcol;
    ui MYpixIndex = MYrow * Hpixels + MYcol;

    B = (double)ImgGPU[MYsrcIndex];
    G = (double)ImgGPU[MYsrcIndex + 1];
    R = (double)ImgGPU[MYsrcIndex + 2];
    ImgBW[MYpixIndex] = (R + G + B) / 3.0;
}
```

Clearly, the architectural improvements in Pascal made some of our suggested techniques irrelevant to boost kernel performance because they were both attempting to cover for some hardware inefficiency (which doesn't exist anymore). This is very typical in GPU development in that it is hard to come up with performance improvement techniques that can be applied to many consecutive families. One example is atomic variables, which were

TABLE 9.7 Kernel performances: BWKernel() and BWKernel2().

Kernel	Box II	Box III	Box IV	Box VII	Box VIII
BWKernel (ms)	48.97	14.12	9.89	4.95	3.12
BWKernel2 (ms)	39.52	13.09	6.81	4.97	3.13

extremely slow to process in Fermi; however, they are orders-of-magnitude faster in Pascal. For this reason, during the development of this book, I intentionally stayed away from hard statements such as "this technique is great." Instead, I compared its impact on multiple families and observed which family benefited from it more. The reader should be aware that this trend will never stop. In the upcoming Volta family, many other hardware deficiencies will be addressed, potentially yielding very different results using our code.

The summary about all of the improvements I am suggesting in this book:

➤ *It will be very hard to find improvements that can survive the*
test of time for decades. GPU architecture development is very dynamic.

➤ *The user should be aware of this dynamism and try to come up with*
ideas that are widely applicable.

➤ *In GPU programming, "platform dependence" is not embarrassing...*
Low performance is a lot more embarrassing...
It is perfectly fine to write kernels that will only work in Pascal.

➤ *If platform-dependent (e.g., Pascal-only) code gets you a 10×*
speedup over a CPU, rather than 7×, you are a hero!
Nobody cares that it only works in Pascal. So, they should buy a Pascal!

9.5.2 GaussKernel2(): Using Precomputed Values and 2D Blocks

Although the results are nothing to brag about, Code 9.12 shows the GaussKernel2(), which uses the two simple tricks (precomputed argument passing and 2D blocks)) to improve GaussKernel() (Code 8.4). The comparison of the original and the improved kernel is shown in Table 9.8. This comparison shows no statistically meaningful improvement; it looks like GaussKernel2() eked out a ≈1% gain for some GPUs and nothing in others, as opposed to the much stronger gains we noticed with the kernels in imflipGCM.cu. Why? The answer is actually fairly clear: While the index computations, etc. were a significant portion of the computational load in the kernels of imflipGCM.cu (e.g., 10–15% of the computational time in the Vflip() kernel), it is a negligible percentage (something like ≈1%) in the GaussKernel2() due to the high computational intensity in this kernel. So, by improving a part of the code that is only 1% of our computational load can only give us a 1% performance improvement in the best case! This is what we did with GaussKernel2(). An important note to make is that the user should ignore the minor fluctuations that make it look like the performance actually went down in some GPUs. This is simply a measurement error. In general, it is expected that GaussKernel2() will have a minuscule performance improvement over GaussKernel(). My suggestion to developers would be to not even bother with incorporating such an improvement — if it is going to take too much development time — and stay with the heavy computation part to see if an improvement can be made there first.

TABLE 9.8 Kernel performances: GaussKernel() and GaussKernel2().

Kernel		Box II	Box III	Box IV	Box VII	Box VIII
GaussKernel	(ms)	183.95	149.79	35.39	30.40	33.45
GaussKernel2	(ms)	181.09	150.42	35.45	30.23	32.52

CODE 9.12:   imedgeGCM.cu   GaussKernel2() {...}
GaussKernel2() processes 1 pixel/kernel; precomputed values, 2D block indexing.

```
// Improved GaussKernel. Uses 2D blocks. Each kernel processes a single pixel
__global__
void GaussKernel2(double *ImgGauss, double *ImgBW, ui Hpixels, ui Vpixels)
{
   ui ThrPerBlk = blockDim.x;      ui MYbid = blockIdx.x;
   ui MYtid = threadIdx.x;         int row, col, indx, i, j;
   double G = 0.00;

   ui MYrow = blockIdx.y;          ui MYcol = MYbid*ThrPerBlk + MYtid;
   if (MYcol >= Hpixels) return;   // col out of range

   ui MYpixIndex = MYrow * Hpixels + MYcol;
   if ((MYrow<2) || (MYrow>Vpixels - 3) || (MYcol<2) || (MYcol>Hpixels - 3)){
      ImgGauss[MYpixIndex] = 0.0;
      return;
   }else{
      G = 0.0;
      for (i = -2; i <= 2; i++){
         for (j = -2; j <= 2; j++){
            row = MYrow + i;
            col = MYcol + j;
            indx = row*Hpixels + col;
            G += (ImgBW[indx] * Gauss[i + 2][j + 2]);
         }
      }
      ImgGauss[MYpixIndex] = G / 159.00;
   }
}
```

## CHAPTER 10

Understanding GPU Memory

REMEMBER from the previous chapters that we introduced terms such as *memory friendly* and *core friendly* and tried to make our programs one or the other. The reality is that they are not independent concepts. Consider our GaussKernel(), which is a *core-intensive* kernel. It is so core-intensive that it is pointless to try to make it memory friendly. As a quantitative example, assume that this kernel is spending 10% of its time in memory accesses and 90% of its time in core computations. Let us assume that you made the kernel much more memory friendly by making memory accesses 2× faster. Now, instead of memory and core taking 10+90 units of time, respectively, they will take 5+90 units of time; you just made your program 5% faster! Instead, if you tried to make the core accesses 2× faster, your program would take 10+45=55 units of time, which would make it 45% faster. So, does this mean that we should pick one or the other and not bother with the other one? Not really. Let us continue the same example. Assume that your memory+core time was 10+90 units and you applied tricks that could make core accesses 6× faster, which would drop your execution time to 10+15=25 units and make your kernel 4× faster overall. Now, assume that you can still apply the same memory-friendly techniques to this kernel and make memory accesses 2× faster, which would drop your execution time to 5+15=20 units. Now, instead of a puny 5% improvement, the same memory-friendly technique can make your program 20% faster. The moral of the story is that the reason for the initial improvement to look weak was because your core accesses were very inefficient and were masking the potential improvements due to memory friendliness. This is why memory and core optimizations should be viewed as a "co-optimization" problem, rather than individual — and unrelated — problems.

In this chapter, we will study the memory architecture of different Nvidia GPU families and improve our kernels so their access to the data in different memory regions is efficient, i.e., we will make them memory friendly. As I mentioned in the previous chapter, we will learn how to use a very important tool named *CUDA Occupancy Calculator* in this chapter, which will allow us to establish a formal methodology for determining kernel launch parameters to ensure optimum resource utilization inside the SMs of the GPU.

10.1 GLOBAL MEMORY

When you are purchasing a GPU for gaming or scientific computing, what do you look at? Possibly the model name of the GPU (e.g., GTX 1080), how much memory it has (e.g., 8 GB for the GTX 1080), and possibly the number of cores or the advertised GFLOPS/TFLOPS. More savvy buyers will also look at the *type* of memory the GPU has and maybe even specific L1$, L2$ parameters, etc. For example, while the GTX Titan X Pascal GPU has a

480 GBps memory bandwidth due to its GDDR5X memory type, the GTX 1070 only has a 256 GBps bandwidth, due to its lower-bandwidth GDDR5 memory type. In either case, which memory is being advertised? The answer is the *global memory* (abbreviated GM), which is the main memory of the GPU.

Starting with Section 3.5, we spent quite a bit of time talking about CPU memory and how *accessing consecutive large chunks* was the best way to read data from the CPU's main memory. The GPU memory is surprisingly similar and almost every bit of intuition you gained about how to efficiently access CPU memory will apply to GPU memory. The only big difference is that because a significantly higher number of cores — as compared to a CPU — need to be fed data simultaneously from the GPU memory, GDDR5 (and the newer GDDR5X) are designed to provide data to multiple sources a lot more efficiently.

10.2   L2 CACHE

L2\$ is where all of the data read from GM is cached. L2\$ is *coherent*, meaning that an address in L2\$ means exactly the same address to every core in the GPU. So far, every Nvidia GPU architecture we looked at had an L2\$ as its Last Level Cache (LLC): Fermi (Figure 9.2) had a 768 KB L2\$, while Kepler (Figure 9.4) had 1.5 MB. Maxwell (Figure 9.6) increased its L2\$ to 2 MB, while Pascal (Figure 9.8) enjoys a large 4 MB L2\$. While GK110, GM200, and GP100 represent the biggest architectures in their respective families, smaller (scaled-down) versions were released also; for example, although GTX 1070 is a Pascal family GPU, it only has a 2 MB L2\$ because this is what the GP104-200 engine includes.

Table 10.1 lists some example GPUs and their global memory and L2\$ sizes. Additionally, a new metric, *bandwidth per core*, is shown to demonstrate how much memory a GPU has, relative to its number of cores. Bandwidth per core was ≈0.1–0.17 GBps per core in the Kepler family and went down to about 0.08 in Maxwell and came back up to 0.13 GBps per core in Pascal. A comparison to two CPUs is shown below the thick line in Table 10.1. A CPU is designed to have nearly 50× more bandwidth allocated per core (e.g., 0.134 vs. 6.4 GBps per core). Similarly, a CPU-based PC enjoys nearly 1000× more main memory per core (e.g., 4.27 vs. 4096). Comparing the LLCs (which is L3\$ in the CPU), a CPU is, again, equipped with 2000× more (e.g., 1.14 vs. 2560). This shouldn't come as a surprise because the CPU architecture is significantly more sophisticated, allowing the CPU cores to do a lot more work. However, to deliver this performance the CPU cores need a lot more resources.

10.3   TEXTURE/L1 CACHE

In some generations, texture cache is separate from L1\$, but in Maxwell (Figure 9.6) and Pascal (Figure 9.8) they share the same cache area. In this book, we will not do any game design, so we are not interested in texture memory. However, the L1\$ in each SM works in exactly the same way a CPU L1\$ works. The data that the cores need immediately are cached in L1\$ by the hardware. The user has no say in which data gets thrown out or kept in cache. In this regard, L1\$ is called the *hardware cache*, or more clearly *hardware controlled cache*. L1\$ is *not coherent*, meaning that once the cores read some data from L2\$ and place it in their local L1\$, the address they use to refer to that data in their local L1\$ has no relationship to the other L1\$ memories in other SMs. In other words, L1\$ is strictly to improve access to specific data, rather than to share it with other L1\$ memories. In the Fermi and Kepler families, L1\$ is co-located with the shared memory, whereas in Maxwell and Pascal families, they are separate entities.

TABLE 10.1 Nvidia microarchitecture families and the size of global memory, L1\$, L2\$ and shared memory in each one of them.

Model (Engine)	#cores	Sh. Mem	L1\$	L2\$	Global Memory (MB/C)	(GBps/C)
		(KB/core)				
GTX550Ti (GF110)	192	64/32 (2.00 combined)		256/192 (1.33)	1024/192 (5.33)	98.5/192 (0.513)
GTX 760 (GK104)	1152	64/192 (0.33 combined)		768/1152 (0.67)	2048/1152 (1.78)	192/1152 (0.167)
Titan Z (2xGK110)	2x2880	64/192 (0.33 combined)		1536/2880 (0.53)	6144/2880 (2.13)	336/2880 (0.117)
Tesla K80 (2xGK210)	2x2496	112/192 (0.58 combined)		1536/2496 (0.62)	12288/2496 (4.92)	240/2496 (0.096)
GTX 980Ti (GM200)	2816	96/192 (0.50)		2048/2816 (0.73)	6144/2816 (2.19)	224/2816 (0.080)
GTX 1070 (GP104-200)	1920	96/128 (0.75)	48/128 (0.38)	2048/1920 (1.07)	8192/1920 (4.27)	256/1920 (0.134)
Titan X (GP102)	3584	64/64 (1.00)		4096/3584 (1.14)	12288/3584 (3.43)	480/3584 (0.134)
Xeon E5-2690 (Sandy Br EP)	8	**L3\$** (2560)	64/1 (64)	256/1 (256)	32768/8 (4096)	51.2/8 (6.400)
E5-2680v4 (Broadwell)	14	**L3\$** (2560)	64/1 (64)	256/1 (256)	262144/14 (18724)	76.8/14 (5.486)

Below the thick line, the same parameters are shown for two different CPUs. Note: /C means *per core*.

10.4 SHARED MEMORY

The most important type of memory in a GPU is the shared memory. Although this memory works similarly to L1\$, it is purely controlled by the programmer. The GPU hardware has no say in which data goes into shared memory. For this reason, shared memory is called *software cache* or *scratch pad memory*. The idea behind shared memory is that nobody knows which data elements a program will need — throughout its execution — better than the program developer. So, if the programmer is given the opportunity to cache certain data elements whenever necessary and evict them whenever they are not needed anymore, the efficiency of data caching can go up to 100%, as compared to even the best cache replacement algorithms, which achieve 80–90% efficiency.

10.4.1 Split versus Dedicated Shared Memory

As seen from Table 10.1, Fermi and Kepler families placed the L1\$ and shared memory in the same memory area, which is "split" between the two based on kernel demand as follows:

- Fermi : (Shared Memory, L1\$): (16 KB, 48 KB), (48 KB, 16 KB)

- Kepler: (Shared Memory, L1\$): (16 KB, 48 KB), (32 KB, 32 KB), (48 KB, 16 KB)

For example, if a kernel required 20 KB shared memory in Fermi, Fermi hardware would *automatically* split this memory as (shared memory=48 KB, L1$=16 KB), which is option 2. Alternatively, Kepler would split it as (shared memory=32 KB, L1$=32 KB), which is more efficient since it leaves more room for the hardware cache L1$. The decision as to when this split takes place is made at runtime by the streaming multiprocessors (SM) hardware. Because a different block can be running in each SM at different times, the split can be changed as new blocks are scheduled to run in the same SM.

While Kepler improved the L1$ and shared memory usage efficiency by introducing a (32 KB, 32 KB) split option in addition to the (16 KB, 48 KB) and (48 KB, 16 KB) that Fermi had, Nvidia decided to place the shared memory and L1$ in totally separate areas starting with the Maxwell generation. They decided that it is more efficient to have L1$ share the same area with texture memory, which is used in computer graphics operations. In these newer families, the shared memory area is fixed in size and is dedicated to this software cache duty.

10.4.2 Memory Resources Available Per Core

The amount of shared memory that is available per core kept getting progressively higher in the newer generations, as shown in Table 10.1; excluding the GTX550Ti, the amount of available shared memory per core increased from ≈0.20 KB/core in Kepler to 1.00 KB/core in Pascal. Note that *each block is limited to a maximum of 48 KB shared memory*, regardless of the amount of shared memory that an SM contains. This limitation existed since the first day of CUDA. That being the case, for Fermi and Kepler, because the shared memory and L1$ are shared, and the shared memory portion cannot be more than 48 KB, the GTX 760 cannot have a shared memory higher than 0.2 KB/core. A similar trend is visible in L2$ and memory bandwidth.

It is worthwhile commenting on why GTX 550Ti had a richer core, in terms of all allocated shared memory, L1$, and L2$, as compared to the next generation, Kepler. This is because there was a major jump in the number of cores in Kepler, as compared to Fermi. Such an increase in core count meant that the resources allocated for each core (shared memory, L1$, L2$) had to be reduced. This reduction was remedied partially by a more efficient SM design and advances in the memory controller and instruction set. Furthermore, in Maxwell and Pascal families, the ratio was improved and the priority shifted toward *making the cores richer*, rather than *increasing the number of cores*.

10.4.3 Using Shared Memory as Software Cache

Let us refresh our memory with Code 3.1, where using a localized memory array named `Buffer[]` made a big impact in program performance because we were fairly sure that this buffer was stored in L1$ or L2$ of the CPU's cores. However, this was nothing more than a "guess" or a "hope." We really had no control on where the CPU hardware would place the `Buffer[]` array because there are no explicit controls on cached data content in a CPU instruction set. Imagine what we could do if we could tell the hardware to keep certain data in the cache, so the access is much faster. Well, this is exactly what the shared memory does. It is a memory area that we know resides inside the SM and access to it is substantially faster than accessing the global memory. Furthermore, whatever is placed in shared memory stays in there until our code explicitly deletes it.

10.4.4 Allocating Shared Memory in an SM

Shared memory is an *SM-level resource*. Each SM has its own shared memory and every block requests a certain amount of it. As an example, assume that an SM (SM7) has a total of 64 KB for shared memory. Assume that two blocks, Block0 and Block16, are executing on SM7, which require a 20 KB shared memory each. Before the Giga Thread Scheduler can schedule Block0 and Block16 to execute on SM7, it would check the available shared memory. If no other block is scheduled and executing on SM7, the entire 64 KB shared memory area is available. When Block0 is scheduled, the SM7 hardware reduces the available shared memory area from 64 KB to 44 KB. This leaves room for happily scheduling Block16 also, which will reduce the available shared memory to 24 KB. If another block (say, Block 42) was to be scheduled, the available shared memory would go down to 4 KB, leaving no room for another block that is demanding 20 KB shared memory. However, assume that another kernel that demands only 2 KB shared memory has to be scheduled on SM7 next. This is not a problem because SM7 has 4 KB of free shared memory remaining. Indeed, two such blocks that require 2 KB of shared memory can be scheduled, increasing the total number of blocks executing on SM7 to 5.

Let us refresh our memory with the *Register File* limitations that we discussed in Section 9.2.4. Each kernel required a certain number of registers to operate, which meant that each block required a certain number of registers from the register file (RF). Scheduling each block meant that a portion of the RF would be dedicated to this newly scheduled block and would subtract away from the total available register count. We saw cases when a block could not be scheduled because there were no available registers in the RF. In this section, we just saw the second limitation that could prevent a block from being scheduled on an SM. Unfortunately, as far as these SM-level resources are concerned, the limitation that prevents a block from being scheduled is the *the most restrictive one*. In other words, it doesn't matter that you design a kernel so carefully to avoid using a large number of registers. Your efforts will be wasted if your blocks need a large amount of shared memory. Furthermore, making the kernels as small as possible to use virtually no registers or no shared memory is not a good strategy either; not only is there a limitation on the maximum number of blocks that can be scheduled on an SM (as detailed in Section 8.9.3), but also SM is designed to improve performance and it should be used for this purpose.

10.5 INSTRUCTION CACHE

The instruction cache and instruction buffer also reside in the SMs of the GPU. They cache the machine code instructions that are needed to run the kernels inside an SM. The programmer has no control on them and does not have to. As long as there is sufficient instruction cache, the size of this cache is somehow an irrelevant parameter from the programmer's standpoint.

10.6 CONSTANT MEMORY

The constant memory has existed from the very first days of CUDA. It is responsible for holding constant values that the threads need. There are a few major differences between regular memory and constant memory: (1) Constant memory is written only once and read many times by the kernels. (2) It is designed to provide the same constant value to multiple (e.g., 32) threads. It holds a single value, but makes 16 or 32 copies of it and replicates it; so, while L1\$, L2\$, and GM are 1-to-1 memory, constant memory is a 1-to-16 or 1-to-32 memory. Constant memory size has stayed at 64 KB from the introduction of

CUDA until now. However, the amount of *constant cache* has varied slightly. While constant memory is shared by the entire GPU, constant cache is local to the SMs. We will be using constant memory to speed up our kernels in this chapter.

10.7 IMFLIPGCM.CU: CORE AND MEMORY FRIENDLY IMFLIPG

In this section, we will use different types of memory and will study their impact on performance. In this section, let us add different kernel versions to imflipGCM.cu, which use the memory components we just introduced. Please note that using the memory sub-system of the GPU is tricky. So, there will be a lot of bad ideas we will try as well as good ideas. The idea is to see their impact and identify what is a good versus bad idea, rather than to continuously improve performance.

10.7.1 Hflip6(), Vflip6(): Using Shared Memory as Buffer

Code 10.1 shows two kernels that use shared memory to flip an image: Hflip6() and Vflip6() are the modified versions of the Hflip() and Vflip() kernels, respectively. In both kernels, the image is flipped by reading a pixel (3 bytes) from the GM into the shared memory first and written back to the GM (to its flipped location) from the shared memory. In both kernels, the shared memory is allocated using the following code:

```
__shared__ uch PixBuffer[3072]; // holds 3*1024 Bytes (1024 pixels).
```

which allocates 3027 elements of type unsigned char (uc). This is an allocation of 3072 Bytes total, allocated at the block level. So, any block that you launch running multiple threads like this will only be allocated single 3072 Byte shared memory area. If you, for example, launch your blocks with 128 threads, these 128 threads are allocated a total of 3072 Byte buffer area, corresponding to 3072/128=24 Bytes for each thread. The initial part of the kernel is identical to the original Hflip() and Vflip() kernels, however, the way that each thread addresses the shared memory depends on its tid. Because of the way the Hflip6() and Vflip6() kernels are written, they process only one pixel, i.e., 3 Bytes. So, if they are launched with 128 threads per block, they will only need 384 Bytes of shared memory, making the remaining 2688 Bytes of shared memory sit there idle during their execution. If we drop the allocated shared memory to 384 Bytes, then this kernel cannot be launched with more than 128 threads. So, there is an intricate formula in determining how much shared memory to declare.

Once the shared memory is declared, the SM allocates this much from its entire shared memory before launching a block. During its execution, these lines of code

```
PixBuffer[MYtid3]=ImgSrc[MYsrcIndex]; ...
```

copy pixels from GM (pointed to by ImgSrc) into shared memory (the PixBuffer array). These lines copy it back to GM (at its flipped position, using the pointer ImgDst)

```
ImgDst[MYdstIndex]=PixBuffer[MYtid3]; ...
```

The following line ensures that the reads into the shared memory by all of the threads in the block are completed before each thread is allowed to proceed.

```
__syncthreads();
```

TABLE 10.2 Kernel performances: Hflip() vs. Hflip6() and Vflip() vs. Vflip6().

Kernel	Box II	Box III	Box IV	Box VII	Box VIII
Hflip (ms)	20.12	6.73	4.17	2.15	1.40
Hflip6 (ms)	18.23	5.98	3.73	1.83	1.37
Vflip (ms)	20.02	6.69	4.11	2.12	1.40
Vflip6 (ms)	18.26	5.98	3.65	1.90	1.35

CODE 10.1:   imflipGCM.cu   Hflip6(), Vflip6() {...}

Hflip6() and Vflip6() use 3072 Bytes of shared memory.

```
// Each thread copies a pixel from GM into shared memory (PixBuffer[]) and back
__global__
void Hflip6(uch *ImgDst, uch *ImgSrc, ui Hpixels, ui RowBytes)
{
   __shared__ uch PixBuffer[3072]; // holds 3*1024 Bytes (1024 pixels).

   ui ThrPerBlk=blockDim.x;   ui MYbid=blockIdx.x;  ui MYtid=threadIdx.x;
   ui MYtid3=MYtid*3;         ui MYrow=blockIdx.y;  ui MYcol=MYbid*ThrPerBlk+MYtid;
   if(MYcol>=Hpixels) return;          ui MYmirrorcol=Hpixels-1-MYcol;
   ui MYoffset=MYrow*RowBytes;         ui MYsrcIndex=MYoffset+3*MYcol;
   ui MYdstIndex=MYoffset+3*MYmirrorcol;
   // swap pixels RGB @MYcol , @MYmirrorcol
   PixBuffer[MYtid3]=ImgSrc[MYsrcIndex]; PixBuffer[MYtid3+1]=ImgSrc[MYsrcIndex+1];
   PixBuffer[MYtid3+2]=ImgSrc[MYsrcIndex+2];
   __syncthreads();
   ImgDst[MYdstIndex]=PixBuffer[MYtid3]; ImgDst[MYdstIndex+1]=PixBuffer[MYtid3+1];
   ImgDst[MYdstIndex+2]=PixBuffer[MYtid3+2];
}
__global__
void Vflip6(uch *ImgDst, uch *ImgSrc, ui Hpixels, ui Vpixels, ui RowBytes)
{
   __shared__ uch PixBuffer[3072]; // holds 3*1024 Bytes (1024 pixels).
   ui ThrPerBlk=blockDim.x;   ui MYbid=blockIdx.x;  ui MYtid=threadIdx.x;
   ui MYtid3=MYtid*3;         ui MYrow=blockIdx.y;  ui MYcol=MYbid*ThrPerBlk+MYtid;
   if (MYcol >= Hpixels) return;       ui MYmirrorrow=Vpixels-1-MYrow;
   ui MYsrcOffset=MYrow*RowBytes;      ui MYdstOffset=MYmirrorrow*RowBytes;
   ui MYsrcIndex=MYsrcOffset+3*MYcol;  ui MYdstIndex=MYdstOffset+3*MYcol;
   // swap pixels RGB @MYrow , @MYmirrorrow
   PixBuffer[MYtid3]=ImgSrc[MYsrcIndex];  PixBuffer[MYtid3+1]=ImgSrc[MYsrcIndex+1];
   PixBuffer[MYtid3+2]=ImgSrc[MYsrcIndex+2];
   __syncthreads();
   ImgDst[MYdstIndex]=PixBuffer[MYtid3]; ImgDst[MYdstIndex+1]=PixBuffer[MYtid3+1];
   ImgDst[MYdstIndex+2]=PixBuffer[MYtid3+2];
}
```

Table 10.2 compares the Hflip() and Vflip() kernels to their shared memory versions, Hflip6() and Vflip6(). The small performance improvements are due to the previous improvements suggested in Chapter 9. So, it is safe to say that using the shared memory — in this specific kernel — made no difference. Now it is time to find out why.

10.7.2 Hflip7(): Consecutive Swap Operations in Shared Memory

In an attempt to improve on Hflip6(), we will come up with different theories about why the performance of this kernel is low and write a different version of the kernel to fix it. When we were moving pixels one pixel at a time, we could apply exactly the same ideas to the horizontal (e.g., Hflip6()) and vertical (e.g., Vflip6()) version of the kernel by simply exchanging x coordinates with y coordinates. However, flipping an image vertically versus horizontally corresponds to fundamentally different memory patterns. So, we will improve them separately in the rest of this chapter by prescribing more specific improvements for each kernel.

Let us start with Hflip6(). The first thing we suspect is that moving data one byte at a time is inefficient because the natural data size of a GPU is 32-bits (int). Hflip7() kernel is written in such a way that the memory transfers from the GM into the shared memory are all int type (32 bits at a time), as shown in Code 10.2. The idea is that 4 pixels take up 12 Bytes, which is 3 int's. So, we can read 3 int's (4 pixels) at a time from global memory and use the shared memory to swap these 4 pixels. When done, write them back to the GM in 3 int's. We will perform the swaps of each pixel's RGB bytes one byte at a time inside the shared memory. This idea will allow us to test what happens when we turn the global memory-to-shared memory transfers into the natural 32-bit size. The following lines read 3 int's from GM using a 32-bit int * type pointer named ImgSrc32:

```
PixBuffer[MYtid3]   = ImgSrc32[MYsrcIndex];
PixBuffer[MYtid3+1] = ImgSrc32[MYsrcIndex+1];
PixBuffer[MYtid3+2] = ImgSrc32[MYsrcIndex+2];
```

These lines correspond to nice, consecutive, clean, and fast GM read and should be extremely efficient. When written into the shared memory, they are stored in *Big Endian* notation, where the smaller addresses correspond to higher-valued bytes. Therefore, after this read, the shared memory holds the following three int's:

```
// PixBuffer: [B0 G0 R0 B1] [G1 R1 B2 G2] [R2 B3 G3 R3]
```

while our target in global memory is the following pattern:

```
// Our Target: [B3 G3 R3 B2] [G2 R2 B1 G1] [R1 B0 G0 R0]
```

Unfortunately, none of the bytes are where we want them to be; to put these 12 bytes in their desired form, 6 byte swap operations are needed as follows using a pointer that points to the shared memory as an int *:

```
// swap these 4 pixels inside Shared Memory
SwapPtr=(uch *)(&PixBuffer[MYtid3]);   //[B0 G0 R0 B1] [G1 R1 B2 G2] [R2 B3 G3 R3]
SWAP(SwapPtr[0], SwapPtr[9] , SwapB)   //[B3 G0 R0 B1] [G1 R1 B2 G2] [R2 B0 G3 R3]
SWAP(SwapPtr[1], SwapPtr[10], SwapB)   //[B3 G3 R0 B1] [G1 R1 B2 G2] [R2 B0 G0 R3]
SWAP(SwapPtr[2], SwapPtr[11], SwapB)   //[B3 G3 R3 B1] [G1 R1 B2 G2] [R2 B0 G0 R0]
SWAP(SwapPtr[3], SwapPtr[6] , SwapB)   //[B3 G3 R3 B2] [G1 R1 B1 G2] [R2 B0 G0 R0]
SWAP(SwapPtr[4], SwapPtr[7] , SwapB)   //[B3 G3 R3 B2] [G2 R1 B1 G1] [R2 B0 G0 R0]
SWAP(SwapPtr[5], SwapPtr[8] , SwapB)   //[B3 G3 R3 B2] [G2 R2 B1 G1] [R1 B0 G0 R0]
```

Flipped 4 pixels are written back to GM from shared memory as 3 consecutive int's again. Table 10.3 compares this kernel to the previous one; results are nothing to brag loudly about. Although we are getting closer, something is still not right.

TABLE 10.3 Kernel performances: Hflip(), Hflip6(), and Hflip7() using mars.bmp.

Kernel	Box II	Box III	Box IV	Box VII	Box VIII
Hflip (ms)	—	—	7.93	—	—
Hflip6 (ms)	—	—	7.15	—	—
Hflip7 (ms)	—	—	6.97	—	—

CODE 10.2: imflipGCM.cu   Hflip7() {...}

Hflip7() does the flipping using a massive set of byte swap operations.

```
// Each kernel uses Shared Memory (PixBuffer[]) to read in 12 Bytes (4 pixels) into
// Shared Mem. and flips 4 pixels inside Shared Mem. and writes to GM as 3 int's
// Horizontal resolution MUST BE A POWER OF 4.
__global__
void Hflip7(ui *ImgDst32, ui *ImgSrc32, ui RowInts)
{
    __shared__ ui PixBuffer[3072]; // holds 3*1024*4 Bytes (1024*4 pixels).

    ui ThrPerBlk=blockDim.x;               ui MYbid=blockIdx.x;
    ui MYtid=threadIdx.x;                  ui MYtid3=MYtid*3;
    ui MYrow=blockIdx.y;                   ui MYoffset=MYrow*RowInts;
    uch SwapB;                             uch *SwapPtr;
    ui MYcolIndex=(MYbid*ThrPerBlk+MYtid)*3;  if (MYcolIndex>=RowInts) return;
    ui MYmirrorcol=RowInts-1-MYcolIndex;   ui MYsrcIndex=MYoffset+MYcolIndex;
    ui MYdstIndex=MYoffset+MYmirrorcol-2;  // -2 is to copy 3 Bytes at a time
    // read 4 pixel blocks (12B = 3 int's) into Shared Memory
    // PixBuffer: [B0 G0 R0 B1] [G1 R1 B2 G2] [R2 B3 G3 R3]
    // Our Target: [B3 G3 R3 B2] [G2 R2 B1 G1] [R1 B0 G0 R0]
    PixBuffer[MYtid3] = ImgSrc32[MYsrcIndex];
    PixBuffer[MYtid3+1] = ImgSrc32[MYsrcIndex+1];
    PixBuffer[MYtid3+2] = ImgSrc32[MYsrcIndex+2];
    __syncthreads();
    // swap these 4 pixels inside Shared Memory
    SwapPtr=(uch *)(&PixBuffer[MYtid3]);  //[B0 G0 R0 B1] [G1 R1 B2 G2] [R2 B3 G3 R3]
    SWAP(SwapPtr[0], SwapPtr[9] , SwapB)  //[B3 G0 R0 B1] [G1 R1 B2 G2] [R2 B0 G3 R3]
    SWAP(SwapPtr[1], SwapPtr[10], SwapB)  //[B3 G3 R0 B1] [G1 R1 B2 G2] [R2 B0 G0 R3]
    SWAP(SwapPtr[2], SwapPtr[11], SwapB)  //[B3 G3 R3 B1] [G1 R1 B2 G2] [R2 B0 G0 R0]
    SWAP(SwapPtr[3], SwapPtr[6] , SwapB)  //[B3 G3 R3 B2] [G1 R1 B1 G2] [R2 B0 G0 R0]
    SWAP(SwapPtr[4], SwapPtr[7] , SwapB)  //[B3 G3 R3 B2] [G2 R1 B1 G1] [R2 B0 G0 R0]
    SWAP(SwapPtr[5], SwapPtr[8] , SwapB)  //[B3 G3 R3 B2] [G2 R2 B1 G1] [R1 B0 G0 R0]
    __syncthreads();
    //write the 4 pixels (3 int's) from Shared Memory into Global Memory
    ImgDst32[MYdstIndex] = PixBuffer[MYtid3];
    ImgDst32[MYdstIndex+1] = PixBuffer[MYtid3+1];
    ImgDst32[MYdstIndex+2] = PixBuffer[MYtid3+2];
}
```

10.7.3 Hflip8(): Using Registers to Swap Four Pixels

In formulating our next improvement, we will keep what is already working and improve what is not. The way we read 4 pixels in 3 natural-int-sized elements was perfect in Hflip7(). What was wrong in this kernel was that once inside shared memory, we were still manipulating the data in unsigned char size elements (bytes). Reading and writing such unnaturally sized elements from shared memory (or any type of memory) is very inefficient. In the Hflip7() kernel, every byte swap required 3 accesses to the shared memory, each of which is of unnatural unsigned char size.

In Hflip8(), shown in Code 10.3, our idea is to push the unnatural size accesses inside the cores because the cores are much more efficient in manipulating byte-sized data elements. For this, Hflip8() allocates 6 variables, A, B, C, D, E, F, all of which are the natural-sized unsigned int type (32 bits). Shared memory is not used at all; the only access to the global memory (GM) reads data from it in 32-bit sizes as follows:

```
ui A, B, C, D, E, F;
// read 4 pixel blocks (12B = 3 int's) into 3 long registers
A = ImgSrc32[MYsrcIndex];
B = ImgSrc32[MYsrcIndex + 1];
C = ImgSrc32[MYsrcIndex + 2];
```

This is the only access to GM before the flipped pixels are written back again to GM. An important note here is that the data in GM is stored in *Little Endian* format, contrary to shared memory. So, the goal of Hflip8() is to turn the following values in A, B, C

```
//NOW:          A=[B1,R0,G0,B0]  B=[G2,B2,R1,G1]   C=[R3,G3,B3,R2]
```

into the following ones in the D, E, F:

```
//OUR TARGET:   D=[B2,R3,G3,B3]  E=[G1,B1,R2,G2]   F=[R0,G0,B0,R1]
```

What makes this method efficient is that due to the requirement for a small number of variables in the kernel, all of these variables can be easily mapped to *core registers* by the compiler, making their manipulation extremely efficient. Furthermore, the core operations that are used in the byte manipulations is only shift, AND, OR operations, which are the fundamental operations of the ALU inside the cores and can be compiled into the fastest possible instructions by the compiler.

As an example, let us analyze the following C statement:

```
//NOW:          A=[B1,R0,G0,B0]  B=[G2,B2,R1,G1]    C=[R3,G3,B3,R2]
E = (B << 24) | (B >> 24) | ((A >> 8) & 0x00FF0000) | ((C << 8) & 0x0000FF00);
```

Here, there are 4 barrel shift/AND operations that result in the following 32-bit values:
(B<<24) = [G1, 0, 0, 0]   (B>>24) = [0, 0, 0, G2]
((A>>8) & 0x00FF0000) = ([0, B1, R0, G0] & [00, FF, 00, 00]) = [0, B1, 0, 0]
((C<<8) & 0x0000FF00) = ([G3, B3, R2, 0] & [00, 00, FF, 00]) = [0, 0, R2, 0]
When they are OR'ed, the following result is obtained: E = [G1, B1, R2, G2] which is listed as our goal in the code. The remaining manipulations are very similar by extracting different bytes from the initial 32-bit values. An important note here is that the barrel shift (i.e., shifting a 32-bit value by 0–31 times to the left or right) as well as the AND, OR instructions are native instructions of a GPU's integer unit.

TABLE 10.4 Kernel performances: Hflip6(), Hflip7(), Hflip8() using mars.bmp.

Kernel	Box II	Box III	Box IV	Box VII	Box VIII
Hflip (ms)	—	—	7.93	—	—
Hflip6 (ms)	—	—	7.15	—	—
Hflip7 (ms)	—	—	6.97	—	—
Hflip8 (ms)	—	—	3.88	—	—

CODE 10.3: imflipGCM.cu Hflip8() {...}

Hflip8() swaps 12 Bytes (4 pixels) using registers. This major shuffling is inside the core, therefore it is substantially more efficient than any other method that uses the shared memory.

```
// Swaps 12 bytes (4 pixels) using registers
__global__
void Hflip8(ui *ImgDst32, ui *ImgSrc32, ui RowInts)
{
    ui ThrPerBlk=blockDim.x;              ui MYbid=blockIdx.x;
    ui MYtid=threadIdx.x;                 ui MYrow=blockIdx.y;
    ui MYcolIndex=(MYbid*ThrPerBlk+MYtid)*3;  if(MYcolIndex>=RowInts) return;
    ui MYmirrorcol=RowInts-1-MYcolIndex;  ui MYoffset=MYrow*RowInts;
    ui MYsrcIndex=MYoffset+MYcolIndex;    ui A, B, C, D, E, F;
    ui MYdstIndex=MYoffset+MYmirrorcol-2; // -2 is to copy 3 Bytes at a time
    // read 4 pixel blocks (12B = 3 int's) into 3 long registers
    A = ImgSrc32[MYsrcIndex];
    B = ImgSrc32[MYsrcIndex + 1];
    C = ImgSrc32[MYsrcIndex + 2];
    // Do the shuffling using these registers
    //NOW:          A=[B1,R0,G0,B0]  B=[G2,B2,R1,G1]  C=[R3,G3,B3,R2]
    //OUR TARGET:   D=[B2,R3,G3,B3]  E=[G1,B1,R2,G2]  F=[R0,G0,B0,R1]
    // D=[B2,R3,G3,B3]
    D = (C >> 8) | ((B << 8) & 0xFF000000);
    // E=[G1,B1,R2,G2]
    E = (B << 24) | (B >> 24) | ((A >> 8) & 0x00FF0000) | ((C << 8) & 0x0000FF00);
    // F=[R0,G0,B0,R1]
    F=((A << 8) & 0xFFFF0000) | ((A >> 16) & 0x0000FF00) | ((B >> 8) & 0x000000FF);
    //write the 4 pixels (3 int's) from Shared Memory into Global Memory
    ImgDst32[MYdstIndex]=D;
    ImgDst32[MYdstIndex+1]=E;
    ImgDst32[MYdstIndex+2]=F;
}
```

Table 10.3 shows the results of Hflip8(). The significant improvement is due to the fact that although accessing memory using any format other than the natural 32-bit types causes significant inefficiencies, the same is not true to GPU cores, because they are designed to execute barrel shift, bitwise logical AND, OR instructions in a single cycle. As shown in Code 10.3, the unpleasant data access patterns are confined inside the GPU core instructions rather than being exposed as memory accesses.

CODE 10.4:   imflipGCM.cu   Vflip7() {...}

Vflip7() uses shared memory to copy 4 bytes (int) at a time.

```
// Improved Vflip6() kernel that uses shared memory to copy 4 B (int) at a time.
// It no longer worries about the pixel RGB boundaries
__global__
void Vflip7(ui *ImgDst32, ui *ImgSrc32, ui Vpixels, ui RowInts)
{
    __shared__ ui PixBuffer[1024];          // holds 1024 int = 4096B
    ui ThrPerBlk=blockDim.x;                ui MYbid=blockIdx.x;
    ui MYtid=threadIdx.x;                   ui MYrow=blockIdx.y;
    ui MYcolIndex=MYbid*ThrPerBlk+MYtid;    if (MYcolIndex>=RowInts) return;
    ui MYmirrorrow=Vpixels-1-MYrow;         ui MYsrcOffset=MYrow*RowInts;
    ui MYdstOffset=MYmirrorrow*RowInts;     ui MYsrcIndex=MYsrcOffset+MYcolIndex;
    ui MYdstIndex=MYdstOffset+MYcolIndex;
    // swap pixels RGB @MYrow , @MYmirrorrow
    PixBuffer[MYtid] = ImgSrc32[MYsrcIndex];
    __syncthreads();
    ImgDst32[MYdstIndex] = PixBuffer[MYtid];
}
```

10.7.4   Vflip7(): Copying 4 Bytes (int) at a Time

The next idea we will try is using natural size (32 bit) transfers for shared memory read-/writes. Vflip7() kernel, shown in Code 10.4, does this, which is an improvement over the Vflip6() kernel (Code 10.1) that used byte-by-byte transfers for shared memory. Our ability to apply this improvement is based on the fact that there is a major difference between horizontal and vertical flips; while horizontal flips have to worry about byte manipulations regardless of the size of the image, vertical flips can simply transfer an entire row (e.g., 7918 pixels, corresponding to 23,754 bytes) as a bulk transfer, which we know is very efficient as far as memory accesses are concerned. Considering the fact that the size of a BMP file is always a multiple of 32 bits (4 bytes), it is natural to have the kernels transfer one int at a time, as in Vflip7(). For this to work, the main() calls this kernel with unsigned int * arguments, instead of unsigned char * ones. The amount of shared memory allocated inside this kernel is for 1024 int elements (4096 bytes). Because each kernel transfers only a single int, a maximum of 1024 threads/block can be launched using Hflip7().

10.7.5   Aligned versus Unaligned Data Access in Memory

This section is a good place to mention the difference between *aligned data access* and *unaligned data access*. Remember the swap operations in Section 10.7.2:

```
SWAP(SwapPtr[0], SwapPtr[9] , SwapB)   //[B3 G0 R0 B1] [G1 R1 B2 G2] [R2 B0 G3 R3]
```

The last int contains [R2 B0 G3 R3], which is information relating to three different pixels. Because of the bytes of each pixel not aligning at 32-bit int boundaries, almost every pixel is guaranteed to require accesses to two int elements. So, it is not unreasonable to expect 2× the performance from its aligned version, Hflip8(), which we studied in Section 10.7.3.

TABLE 10.5 Kernel performances: Vflip(), Vflip6(), Vflip7(), and Vflip8().

Kernel	Box II	Box III	Box IV	Box VII	Box VIII
Vflip (ms)	20.02	6.69	4.11	2.12	1.40
Vflip6 (ms)	18.26	5.98	3.65	1.90	1.35
Vflip7 (ms)	9.43	3.28	2.08	1.28	0.82
Vflip8 (ms)	14.83	5.00	2.91	1.32	0.84

CODE 10.5:   imflipGCM.cu   Vflip8() {...}

Vflip8() uses shared memory to copy 8 bytes at a time.

```
__global__
void Vflip8(ui *ImgDst32, ui *ImgSrc32, ui Vpixels, ui RowInts)
{
    __shared__ ui PixBuffer[2048];              // holds 2048 int = 8192B

    ui ThrPerBlk=blockDim.x;                    ui MYbid=blockIdx.x;
    ui MYtid=threadIdx.x;                       ui MYtid2=MYtid*2;
    ui MYrow=blockIdx.y;
    ui MYcolIndex=(MYbid*ThrPerBlk+MYtid)*2;    if(MYcolIndex>=RowInts) return;
    ui MYmirrorrow=Vpixels-1-MYrow;             ui MYsrcOffset=MYrow*RowInts;
    ui MYdstOffset=MYmirrorrow*RowInts;         ui MYsrcIndex=MYsrcOffset+MYcolIndex;
    ui MYdstIndex=MYdstOffset+MYcolIndex;
    // swap pixels RGB @MYrow , @MYmirrorrow
    PixBuffer[MYtid2]=ImgSrc32[MYsrcIndex];
    if ((MYcolIndex+1)<RowInts)    PixBuffer[MYtid2+1]=ImgSrc32[MYsrcIndex+1];
    __syncthreads();
    ImgDst32[MYdstIndex]=PixBuffer[MYtid2];
    if ((MYcolIndex+1)<RowInts)    ImgDst32[MYdstIndex+1]=PixBuffer[MYtid2+1];
}
```

10.7.6 Vflip8(): Copying 8 Bytes at a Time

One reasonable question to ask next is whether copying 8 bytes at a time, rather than 4, could speed up the kernel. The Vflip8() kernel in Code 10.5 does that. Using this new kernel, the GPU only has to process half as many blocks as the Vflip7() kernel. Table 10.5 compares the runtime results of these two kernels and much like our previous attempts, Vflip8() loses. The reason is clear. Although a BMP file size is guaranteed to be a multiple 4 bytes, it is not guaranteed to be a multiple of 8. This forces the Vflip8() kernel to check to see whether it has exceeded the image boundary after copying every int. This check is very expensive, as the results in Table 10.5 show. Once the accesses are nicely aligned, the memory controller can neatly combine multiple consecutive read/write operations and the introduction of the if(...)   does more harm than the efficiency gain by copying two consecutive int elements. However, if we knew that the size of the image file was a multiple of 8 bytes, we could eliminate both of the if statements from the kernel and could expect some improvement over Vflip7(). This is a good exercise for the readers. One interesting observation from Table 10.5 is that the performance penalty due to the if statements is nearly negligible in Pascal GPUs. This is due to the architectural improvements in Pascal.

TABLE 10.6 Kernel performances: Vflip(), Vflip6(), Vflip7(), Vflip8(), and Vflip9().

Kernel	Box II	Box III	Box IV	Box VII	Box VIII
Vflip (ms)	20.02	6.69	4.11	2.12	1.40
Vflip6 (ms)	18.26	5.98	3.65	1.90	1.35
Vflip7 (ms)	9.43	3.28	2.08	1.28	0.82
Vflip8 (ms)	14.83	5.00	2.91	1.32	0.84
Vflip9 (ms)	6.76	2.61	1.70	1.27	0.82

CODE 10.6:   imflipGCM.cu   Vflip9() {...}

Vflip9() uses global memory only to copy 8 bytes at a time.

```
// Modified Vflip8() kernel that uses Global Memory only
// to copy 8 Bytes at a time (2 int). It does NOT use shared memory
__global__
void Vflip9(ui *ImgDst32, ui *ImgSrc32, ui Vpixels, ui RowInts)
{
   ui ThrPerBlk=blockDim.x;              ui MYbid=blockIdx.x;
   ui MYtid=threadIdx.x;                 ui MYrow=blockIdx.y;
   ui MYcolIndex=(MYbid*ThrPerBlk+MYtid)*2;  if(MYcolIndex>=RowInts) return;
   ui MYmirrorrow=Vpixels-1-MYrow;       ui MYsrcOffset=MYrow*RowInts;
   ui MYdstOffset=MYmirrorrow*RowInts;   ui MYsrcIndex=MYsrcOffset+MYcolIndex;
   ui MYdstIndex=MYdstOffset+MYcolIndex;

   // swap pixels RGB @MYrow , @MYmirrorrow
   ImgDst32[MYdstIndex]=ImgSrc32[MYsrcIndex];
   if((MYcolIndex+1)<RowInts) ImgDst32[MYdstIndex+1]=ImgSrc32[MYsrcIndex+1];
}
```

10.7.7 Vflip9(): Using Only Global Memory, 8 Bytes at a Time

Our next kernel Vflip9(), shown in Code 10.6, attempts to answer an obvious question:
What would happen if we didn't even try to use shared memory?
In other words, would it make a difference if we read from GM and wrote to GM and totally bypassed the shared memory? The answer is in Table 10.6: while the Pascal family — again — is immune to many of the inefficiencies related to the shared memory, Kepler family enjoys a major speed-up. For the Kepler family, this is the fastest vertical flip kernel, outperforming any of the previous ones we developed in Chapter 9. For Pascal, though, the results are completely different, making us conclude the following:

- Shared memory performance has been substantially improved in the Pascal family.

- Negative performance impact of if statements has been significantly reduced in Pascal.

- We also witnessed previously in this chapter that Pascal's ability to process byte-size elements has been drastically improved.

TABLE 10.7 Kernel performances: PixCopy(), PixCopy2(), ..., PixCopy5().

Kernel		Box II	Box III	Box IV	Box VII	Box VIII
PixCopy	(ms)	22.76	7.33	4.05	2.33	1.84
PixCopy2	(ms)	15.81	5.48	3.56	1.81	1.16
PixCopy3	(ms)	14.01	5.13	3.23	1.56	1.05
PixCopy4	(ms)	23.49	7.55	4.46	2.57	1.82
PixCopy5	(ms)	8.63	2.97	1.92	1.32	0.80

CODE 10.7:   imflipGCM.cu   PixCopy4(), PixCopy5() {...}

PixCopy4(), PixCopy5() copy 1 byte versus 4 bytes using shared memory.

```
// Uses shared memory as a temporary local buffer. Copies one byte at a time
__global__
void PixCopy4(uch *ImgDst, uch *ImgSrc, ui FS)
{
    __shared__ uch PixBuffer[1024];         // Shared Memory: holds 1024 Bytes.
    ui ThrPerBlk=blockDim.x;                ui MYbid=blockIdx.x;
    ui MYtid=threadIdx.x;                   ui MYgtid=ThrPerBlk*MYbid+MYtid;
    if(MYgtid > FS) return;                 // outside the allocated memory
    PixBuffer[MYtid] = ImgSrc[MYgtid];
    __syncthreads();
    ImgDst[MYgtid] = PixBuffer[MYtid];
}
__global__
void PixCopy5(ui *ImgDst32, ui *ImgSrc32, ui FS)
{
    __shared__ ui PixBuffer[1024];          // Shared Mem: holds 1024 int (4096 Bytes)
    ui ThrPerBlk=blockDim.x;                ui MYbid=blockIdx.x;
    ui MYtid=threadIdx.x;                   ui MYgtid=ThrPerBlk*MYbid+MYtid;
    if((MYgtid*4)>FS) return;               // outside the allocated memory
    PixBuffer[MYtid] = ImgSrc32[MYgtid];
    __syncthreads();
    ImgDst32[MYgtid] = PixBuffer[MYtid];
}
```

10.7.8 PixCopy4(), PixCopy5(): Copying One versus 4 Bytes Using Shared Memory

Code 10.7 shows two new versions of the original PixCopy() kernel: The PixCopy4() kernel uses shared memory to copy pixels one byte at a time, while PixCopy5() copies pixels one int at a time. Table 10.7 compares these two new kernels to all of its previous versions. Similar to many other results we have seen before, PixCopy4() does not show any performance gain, while PixCopy5() is ≈2× faster in Pascal and ≈2–3× faster in Kepler. Furthermore, compared to the previous versions of this kernel (PixCopy2() and PixCopy3()), only the PixCopy5() shows sufficient gain to become the fastest PixCopy() kernel so far.

TABLE 10.8   Kernel performances: PixCopy(), PixCopy4(), ..., PixCopy7().

Kernel		Box II	Box III	Box IV	Box VII	Box VIII
PixCopy	(ms)	22.76	7.33	4.05	2.33	1.84
PixCopy4	(ms)	23.49	7.55	4.46	2.57	1.82
PixCopy5	(ms)	8.63	2.97	1.92	1.32	0.80
PixCopy6	(ms)	6.44	2.29	1.59	1.27	0.79
PixCopy7	(ms)	4.54	1.58	1.19	0.69	0.44
	(GBps)	52.40	151	200	344	537
	(% BW)	(59%)	(79%)	(60%)	(>100%)	(>100%)

CODE 10.8:   imflipGCM.cu   PixCopy6(), PixCopy7() {...}

PixCopy6() copies 4 bytes, PixCopy7() copies 8 bytes using only global memory.

```
__global__
void PixCopy6(ui *ImgDst32, ui *ImgSrc32, ui FS)
{
    ui ThrPerBlk=blockDim.x;              ui MYbid=blockIdx.x;
    ui MYtid=threadIdx.x;                 ui MYgtid=ThrPerBlk*MYbid+MYtid;
    if((MYgtid*4)>FS) return;             // outside the allocated memory
    ImgDst32[MYgtid] = ImgSrc32[MYgtid];
}
__global__
void PixCopy7(ui *ImgDst32, ui *ImgSrc32, ui FS)
{
    ui ThrPerBlk=blockDim.x;              ui MYbid=blockIdx.x;
    ui MYtid=threadIdx.x;                 ui MYgtid=ThrPerBlk*MYbid+MYtid;
    if((MYgtid*4)>FS) return;             // outside the allocated memory
    ImgDst32[MYgtid] = ImgSrc32[MYgtid];
    MYgtid++;
    if ((MYgtid * 4) > FS) return;        // next 32 bits
    ImgDst32[MYgtid] = ImgSrc32[MYgtid];
}
```

10.7.9   PixCopy6(), PixCopy7(): Copying One/Two Integers Using Global Memory

Code 10.8 investigates how using only the GM would impact the performance. The new two kernels only use GM accesses and bypass the shared memory. PixCopy6() copies data one int at a time from a GM address to another GM address, while PixCopy6() copies data two int elements (8 bytes) at a time. Table 10.8 compares the performance of these two kernels to the previous two kernels that use shared memory. Results clearly show that using only GM results in a faster kernel, especially when we transfer a higher amount of data in each thread. In both the Kepler and Pascal families, the improvement over PixCopy5() is nearly 2×. At the bottom of Table 10.8, the relative bandwidth saturation is shown. While Kepler saturates only 60–80% of its performance, Pascal even exceeds the advertised maximum bandwidth!

10.8   IMEDGEGCM.CU: CORE- & MEMORY-FRIENDLY IMEDGEG

Using the experience we gained from the previous sections in this chapter, now it is time to attempt to improve a core-heavy program, imedgeG.cu. In Chapter 9, we only improved the GaussKernel() by applying two simple techniques. The resulting kernel, GaussKernel2() did not show any performance improvement in Table 9.8 because a substantial amount of time in this kernel is spent in heavy computations, rather than the index calculations that these two improvements target. In this section, we will only improve GaussKernel2() by designing its newer versions GaussKernel3()···GaussKernel8(). Because the SobelKernel() and ThresholdKernel() are also computation-intensive like the GaussKernel(), their improvement is left as an exercise for the reader.

10.8.1   BWKernel3(): Using Byte Manipulation to Extract RGB

Before we start designing versions of GaussKernel(), let us provide one exception. The BWKernel3() kernel (Code 10.9) is an improved version of BWKernel(). This time, let us look at the comparison in Table 10.9 before even we see the actual code. Although Pascal's (Boxes VII and VIII) are showing less of an improvement, they are still very healthy. Kepler's (Boxes II, III, IV) just seem to be loving any idea we can come up with. Now, let us analyze what is going on with BWKernel3():

- The key idea here is to access the data from memory in a size that is natural to the GPU architecture (32 bit int type).

- Because each pixel is made up of 3 bytes and the natural size is 4 bytes, why not use their product (12 bytes) to retrieve the pixels? In other words, every 3 int's will contain 4 pixels' RGB data. This is exactly the same idea we utilized in Hflip8().

- In the BWKernel3(), we are trying to sum up the RGB components (i.e., 3 consecutive bytes), so we will need to strip some of the bytes and shift some of them, etc. Again, the same idea in Hflip8().

- As seen in Code 10.9, a cascade of barrel shift (e.g., A >> 8, B >> 24) statements in C language can achieve the extraction of the bit fields we want to get the RGB values that are buried in these 12 bytes.

- The key to this technique's success is that barrel shift operations map naturally to the GPU core instructions and they are extremely fast.

- Note that for this technique to work, a pointer that is of type unsigned int * is used, thereby naturally addressing 32-bit values.

- Extracted 4 pixel RGB sums are saved in variables Pix1, Pix2, Pix3, and Pix4.

- Variables A, B, and C simply read three consecutive int's (12 bytes).

- Pixel 0's R0, G0, and B0 are buried in the first int, while one additional byte belongs to Pixel1, as shown in the comments of Code 10.9.

- (A & 0x000000FF) gets R0, ((A >> 8)& 0x000000FF) gets R1, ((A >> 16)& 0x000000FF) gets R2 as 32-bit values and their sum goes into Pix1, etc.

- From the standpoint of memory accesses, this is accessing 3 consecutive int and writing 4 consecutive double, both of which are very efficient.

- We are bypassing the shared memory and using global memory again.

TABLE 10.9 Kernel performances: BWKernel(), BWKernel2(), and BWKernel3().

Kernel		Box II	Box III	Box IV	Box VII	Box VIII
BWKernel	(ms)	48.97	14.12	9.89	4.95	3.12
BWKernel2	(ms)	39.52	13.09	6.81	4.97	3.13
BWKernel3	(ms)	22.81	6.86	4.34	3.07	2.33
	(GBps)	19.12	63.5	100	142	187
	(% BW)	(21%)	(33%)	(30%)	(55%)	(39%)

CODE 10.9:   imedgeGCM.cu   BWKernel3() {...}

BWKernel3() uses the byte-by-byte manipulation we saw before to extract the RGB values of pixels to calculate their B&W value.

```
// Improved BWKernel2. Calculates 4 pixels (3 int's) at a time
__global__
void BWKernel3(double *ImgBW, ui *ImgGPU32, ui Hpixels, ui RowInts)
{
    ui ThrPerBlk=blockDim.x;          ui MYbid=blockIdx.x;
    ui MYtid=threadIdx.x;             ui A, B, C;
    ui MYrow=blockIdx.y;              ui MYcol=MYbid*ThrPerBlk+MYtid;
    ui MYcolIndex=MYcol*3;            ui Pix1, Pix2, Pix3, Pix4;
    if(MYcolIndex>=RowInts) return;   ui MYoffset=MYrow*RowInts;
    ui MYsrcIndex=MYoffset+MYcolIndex; ui MYpixAddr=MYrow*Hpixels+MYcol*4;

    A = ImgGPU32[MYsrcIndex];          // A=[B1,R0,G0,B0]
    B = ImgGPU32[MYsrcIndex+1];        // B=[G2,B2,R1,G1]
    C = ImgGPU32[MYsrcIndex+2];        // C=[R3,G3,B3,R2]
    // Pix1 = R0+G0+B0;
    Pix1 = (A & 0x000000FF) + ((A >> 8) & 0x000000FF) + ((A >> 16) & 0x000000FF);
    // Pix2 = R1+G1+B1;
    Pix2 = ((A >> 24) & 0x000000FF) + (B & 0x000000FF) + ((B >> 8) & 0x000000FF);
    // Pix3 = R2+G2+B2;
    Pix3 = (C & 0x000000FF) + ((B >> 16) & 0x000000FF) + ((B >> 24) & 0x000000FF);
    // Pix4 = R3+G3+B3;
    Pix4=((C>>8) & 0x000000FF) + ((C>>16) & 0x000000FF) + ((C>>24) & 0x000000FF);
    ImgBW[MYpixAddr]     = (double)Pix1 * 0.33333333;
    ImgBW[MYpixAddr + 1] = (double)Pix2 * 0.33333333;
    ImgBW[MYpixAddr + 2] = (double)Pix3 * 0.33333333;
    ImgBW[MYpixAddr + 3] = (double)Pix4 * 0.33333333;
}
```

From Table 10.9, we see that BWKernel3() is nearly 2× as fast as our previous version BWKernel2() in Kepler's and a little less than that in Pascal GPUs. The memory access efficiency of Pascal is evident here too, with Pascal GPUs reaching a much higher percentage of their maximum bandwidth. One interesting note to make here is that BWKernel3() is a balanced core-memory kernel. This is demonstrated by how the Pascal GPUs reach ≈50% of their peak bandwidth with this kernel. The memory operations in this kernel include GM read/writes, while core operations include index computations, barrel shift, AND, OR operations, and `double` type multiplications.

10.8.2 GaussKernel3(): Using Constant Memory

Our next improvement is targeted toward improving the access to repeatedly used constant values by the threads. In the GaussKernel2() (Code 9.12), the nested inner loops, which determined the overall performance of the kernel, looked like this:

```
G=0.0;
for (i = -2; i <= 2; i++){
   for (j = -2; j <= 2; j++){
       row=MYrow+i;              col = MYcol + j;
       indx=row*Hpixels+col;     G+=(ImgBW[indx]*Gauss[i+2][j+2]);
   }
}
ImgGauss[MYpixIndex] = G / 159.00;
```

The big question is where is the Gauss[] array stored, which holds these *filter coefficients*. They are stored in a double array of elements that do not change their values throughout the execution of these kernels, as initially shown in GaussKernel() (Code 8.4):

```
__device__
double Gauss[5][5] = { { 2,    4,    5,   4,   2 },
                         . . .
```

The problem is that multiple threads want to access the same exact value. Intuitively, assume that we launched a block with 128 threads. In this scenario, if each thread is calculating a single pixel, it has to access all 25 constant values. This is a total of $128 \times 25 = 3200$ accesses to these constant values (Gauss[i+2][j+2]) to compute 128 pixels. Even the simplest thinking indicates that when all 128 threads are trying to read these 25 constants, on average, $\lceil 128 \div 25 \rceil = 6$ threads will want to access the same constant simultaneously. In other words, instead of an N:N pattern, where N threads access N values, there are lots of N:1 type accesses, where N threads want to read a single value.

The constant memory and constant cache are designed precisely for this purpose in all Nvidia hardware. In the GaussKernel3() (Code 10.10), the constant coefficient array is declared as follows. A total of $25 \times 8 = 400$ bytes of storage is needed for this constant array.

```
__constant__
double GaussC[5][5] = { { 2,    4,    5,   4,   2 },
                          . . .
```

10.8.3 Ways to Handle Constant Values

There are multiple ways to handle the constants in CUDA:

1. Declare an array or simple values as __constant__ and the compiler will figure out what to do with it. In this case, constant cache will be utilized rather than constant memory.

2. Explicitly transfer values into the constant memory, which is limited to 64 KB in all GPU architecture families that Nvidia has introduced p to Pascal. This is done by using the cudaMemcpyToSymbol() API.

TABLE 10.10 Kernel performances: GaussKernel(), GaussKernel2(), GaussKernel3()

Kernel		Box II	Box III	Box IV	Box VII	Box VIII
GaussKernel	(ms)	183.95	149.79	35.39	30.40	33.45
GaussKernel2	(ms)	181.09	150.42	35.45	30.23	32.52
GaussKernel3	(ms)	151.35	149.17	19.62	13.69	9.15

CODE 10.10:   imedgeGCM.cu   GaussKernel3() {...}

GaussKernel3() uses constant memory to access constant coefficients efficiently.

```
__constant__
double GaussC[5][5] = { { 2,    4,    5,  4,  2 },
                        { 4,    9,   12,  9,  4 },
                        { 5,   12,   15, 12,  5 },
                        { 4,    9,   12,  9,  4 },
                        { 2,    4,    5,  4,  2 } };
// Improved GaussKernel2. Uses constant memory to store filter coefficients
__global__
void GaussKernel3(double *ImgGauss, double *ImgBW, ui Hpixels, ui Vpixels)
{
   ui ThrPerBlk=blockDim.x;          ui MYbid=blockIdx.x;
   ui MYtid=threadIdx.x;             int row, col, indx, i, j;
   double G;                         ui MYrow=blockIdx.y;
   ui MYcol=MYbid*ThrPerBlk+MYtid;   if (MYcol>=Hpixels) return;
   ui MYpixIndex=MYrow*Hpixels+MYcol;
   if ((MYrow<2) || (MYrow>Vpixels - 3) || (MYcol<2) || (MYcol>Hpixels - 3)){
      ImgGauss[MYpixIndex] = 0.0;
      return;
   }else{
      G = 0.0;
      for (i = -2; i <= 2; i++){
        for (j = -2; j <= 2; j++){
           row = MYrow + i;
           col = MYcol + j;
           indx = row*Hpixels + col;
           G += (ImgBW[indx] * GaussC[i + 2][j + 2]); // use constant memory
        }
      }
      ImgGauss[MYpixIndex] = G / 159.00;
   }
}
```

Table 10.10 compares the GaussKernel3() (Code 10.10) to its previous versions. It is clear that the impact of using the constant memory varies significantly in Kepler versus Pascal, as well as high-end versus low-end GPUs in each generation. This fact must be due to the major changes Nvidia made in its memory controller hardware, including separating the shared memory from texture/L1$ memory in Pascal, as well as their improved memory sub-system in their high-end GPUs (e.g., Titan Z), making certain parts of the hardware more efficient (e.g., constant memory). It is hard, at this point, to draw more concrete conclusions. We will keep building different versions of the kernel and observe their performance impact.

10.8.4 GaussKernel4(): Buffering Neighbors of 1 Pixel in Shared Memory

Our next kernel design, GaussKernel4(), is shown in Code 10.11. In this kernel, the idea is to buffer all 5×5 neighbors of a pixel in shared memory. For each thread that is executing, there will be 25 `double` elements stored in an array named `Neighbors[]`. In this array, the neighbors of a pixel at (x,y) pixel coordinates are stored in such a way that for a given thread with an ID of `MYtid`, corresponding neighbor elements are stored as follows:

- `Neighbors[MYtid][2][0...5]` stores 5 neighboring pixels of the pixel in the same row, with coordinates (x-2,y), (x-1,y), (x,y), (x+1,y), and (x+2,y).

- `Neighbors[MYtid][0][0...5]` stores two rows above this original pixel, with coordinates (x-2,y-2), (x-1,y-2), (x,y-2), (x+1,y-2), and (x+2,y-2).

- `Neighbors[MYtid][i][j]` stores the neighbor at coordinate (x+j-2, y+i-2).

- The entire `Neighbors[]` array is stored in shared memory. The total required shared memory space for this array is $8 \times 25 \times 128 = 25,600$ bytes or 25 KB for a block using 128 threads.

From this computation it is clear that this is a very expensive way of using shared memory because remembering the rule that a block can only use a maximum of 48 KB shared memory (Section 10.4.2), this kernel is getting close to hitting that limit. There is another big problem: for GPUs that only have 48 KB maximum total shared memory (such as Fermi and Kepler), this kernel cannot even be launched more than once in any SM because the total requested shared memory will be 50 KB (higher than 48 KB) and will not fit in the same SM. However, from the purposes of observing the performance, this is a very good example of "what not to do" as far as shared memory is concerned.

In order to avoid using even more shared memory, the limit is defined in main() as follows, which does not let you launch GaussKernel4() with more than 128 threads. A similar *maximum threads per block* will be defined on all variants of this kernel, from GaussKernel5() through GaussKernel8().

```
#define MAXTHGAUSSKN4 128
...
main()
{
   if ((GaussKN == 4) && (ThrPerBlk>MAXTHGAUSSKN4)){
      printf("ThrPerBlk cannot be higher than %d in Gauss Kernel 4 ... Set to
          %d.\n", MAXTHGAUSSKN4, MAXTHGAUSSKN4);
      ThrPerBlk = MAXTHGAUSSKN4;
   }
   ...
```

Table 10.11 shows the results of GaussKernel4(), which is possibly the most interesting set of results we have ever seen so far. The K3000M and the almighty Titan Z show a 6–7× performance degradation, while the GTX760 shows a modest 1.6–1.7× degradation. The two Pascal GPUs show a degradation of 1.6–1.7×, which is very similar to the GTX760. At this point the reader should be puzzled, but not about the 6–7×; instead, what is more puzzling is the major difference between 1.6× and the 6–7×. Let us continue formulating different versions of the kernel and the answer should start being obvious. For now, it suffices to say that shared memory accesses are not as cheap as one might think. They are most definitely

TABLE 10.11 Kernel performances: GaussKernel(), ..., GaussKernel4().

Kernel		Box II	Box III	Box IV	Box VII	Box VIII
GaussKernel	(ms)	183.95	149.79	35.39	30.40	33.45
GaussKernel2	(ms)	181.09	150.42	35.45	30.23	32.52
GaussKernel3	(ms)	151.35	149.17	19.62	13.69	9.15
GaussKernel4	(ms)	851.73	243.17	140.76	22.98	14.83

better than global memory accesses, but if the read/write patterns are not friendly to the shared memory, performance will suffer greatly.

CODE 10.11:   imedgeGCM.cu   GaussKernel4() ... {...}

GaussKernel4() reads 5x5 neighbors of the "pixel to be computed" into shared memory; each kernel processes one pixel.

```
#define MAXTHGAUSSKN4 128
...
__global__
void GaussKernel4(double *ImgGauss, double *ImgBW, ui Hpixels, ui Vpixels)
{
    __shared__ double Neighbors[MAXTHGAUSSKN4][5][5]; // 5 horiz. 5 vert neighbors
    ui ThrPerBlk=blockDim.x;          ui MYbid=blockIdx.x;
    ui MYtid=threadIdx.x;             int row, col, indx, i, j;
    double G;                         ui MYrow=blockIdx.y;
    ui MYcol=MYbid*ThrPerBlk+MYtid;   if (MYcol>=Hpixels) return;
    ui MYpixIndex = MYrow * Hpixels + MYcol;
    if ((MYrow<2) || (MYrow>Vpixels - 3) || (MYcol<2) || (MYcol>Hpixels - 3)) {
        ImgGauss[MYpixIndex] = 0.0;
        return;
    }
    // Read from GM to Shared Memory
    for (i = 0; i < 5; i++) {
        for (j = 0; j < 5; j++) {
            row=MYrow+i-2;            col=MYcol+j-2;
            indx=row*Hpixels+col;     Neighbors[MYtid][i][j]=ImgBW[indx];
        }
    }
    __syncthreads();
    G = 0.0;
    for (i = 0; i < 5; i++) {
        for (j = 0; j < 5; j++) {
            G += (Neighbors[MYtid][i][j] * GaussC[i][j]);
        }
    }
    ImgGauss[MYpixIndex] = G / 159.00;
}
```

An interesting note about GaussKernel4(): assuming that we launch our blocks with 128 threads, and considering the shared memory size of 25 KB, each thread is responsible for reading 25,600÷128=200 bytes from GM into shared memory using a `for` loop above and uses this information later in a different `for` loop.

TABLE 10.12 Kernel performances: GaussKernel1(), ..., GaussKernel5().

Kernel		Box II	Box III	Box IV	Box VII	Box VIII
GaussKernel	(ms)	183.95	149.79	35.39	30.40	33.5
GaussKernel2	(ms)	181.09	150.42	35.45	30.23	32.5
GaussKernel3	(ms)	151	149	19.6	13.7	9.15
GaussKernel4	(ms)	852	243	141	23.0	14.8
GaussKernel5	(ms)	1378	—	121	33.9	21.8

10.8.5 GaussKernel5(): Buffering Neighbors of 4 Pixels in Shared Memory

There were two major problems with GaussKernel4(): (i) too many accesses to the shared memory, due to the fact that Neighbors[] array had many repeated entries, and (ii) shared memory was so large that we could not launch multiple blocks in an SM.

Let us attempt to solve the first problem and ignore the second problem in GaussKernel5() (Code 10.12). Instead of decreasing the size of the shared memory, let us increase it by storing 5×8 of its neighbors of each pixel. The extra 3 pixels will be on the same row with the original pixel, allowing 8 horizontal pixels to be shared among 4 neighboring pixels. This will require a total of $8 \times 5 \times 8 \times 128 = 40,960$ bytes = 40 KB in shared memory for 128 threads, which increases the shared memory profile by 15 KB. However, GaussKernel4() required 25 KB of shared memory to process one pixel per thread and now, GaussKernel5() requires only 10 KB to per pixel (40 KB for 4 pixels).

In summary, we improved the first problem by improving the per-thread usage efficiency of shared memory; however, this made the size problem much worse. Which one wins? In other words, which problem was a more urgent one to solve? The answer is in Table 10.12. When a kernel uses such a high amount of shared memory, the results are very erratic as shown in Table 10.12. Most of the really high numbers mean that there isn't even a point in comparing these numbers to any other numbers. To summarize: using such a high amount of shared memory is simply a bad idea. When we start using our tool *CUDA Occupancy Calculator*, we will actually see the impact of these large shared memory usage profiles and go through many "what if" scenarios in launching our blocks.

Every single one of these kernels is launched with the maximum 128 threads per block. An alternative question that comes to mind is whether we could have solved the shared memory size problem by launching blocks with only 32 or 64 threads. This would, of course, require us to use a smaller #define MAXTHGAUSSKN5 value. If we end up doing that, though, we run into another problem: using such smaller blocks is not efficient either, as we witnessed many times before. Do not forget the fact that we also have to keep an eye on the number of registers, as well as the total number of blocks that can be launched on each SM. For now, it suffices to say that once we learn how to use the CUDA Occupancy Calculator in Section 10.9, a lot of these questions will be answered.

The GaussKernel4() utilized two different for loops, one to read data from GM into shared memory and another to use this data in computations. In GaussKernel5(), assuming, again, that we launch our blocks with 128 threads, each thread is now responsible for bringing 40,960÷128=320 bytes from GM to shared memory, but to process 4 pixels, each of which accesses 200 bytes from the shared memory (total=800 bytes for each thread). Therefore, the balance in GaussKernel4() was: read 200 bytes/use 200 bytes, while the balance in GaussKernel5() is: read 320 bytes/use 800 bytes.

CODE 10.12:   imedgeGCM.cu   GaussKernel5() ... {...}

GaussKernel5() reads 5x5 neighbors of four consecutive pixels and takes advantage of the massive overlap among them; each kernel processes 4 pixels.

```
#define MAXTHGAUSSKN5 128
...
__global__
void GaussKernel5(double *ImgGauss, double *ImgBW, ui Hpixels, ui Vpixels)
{
   __shared__ double Neighbors[MAXTHGAUSSKN5][5][8]; // 8 horz, 5 vert neighbors
   ui ThrPerBlk=blockDim.x;            ui MYbid=blockIdx.x;
   ui MYtid=threadIdx.x;               int row,col,indx,i,j,k;
   double G;                           ui MYrow=blockIdx.y;
   ui MYcol=(MYbid*ThrPerBlk+MYtid)*4; if (MYcol>=Hpixels) return;
   ui MYpixIndex = MYrow * Hpixels + MYcol;
   if ((MYrow < 2) || (MYrow > Vpixels - 3)){ // Top and bottom two rows
      ImgGauss[MYpixIndex] = 0.0;       ImgGauss[MYpixIndex+1] = 0.0;
      ImgGauss[MYpixIndex+2] = 0.0;     ImgGauss[MYpixIndex+3] = 0.0;        return;
   }
   if (MYcol > Hpixels - 3) {  // Rightmost two columns
      ImgGauss[MYpixIndex] = 0.0;       ImgGauss[MYpixIndex + 1] = 0.0;      return;
   }
   if (MYcol < 2) {             // Leftmost two columns
      ImgGauss[MYpixIndex] = 0.0;       ImgGauss[MYpixIndex + 1] = 0.0;      return;
   }
   MYpixIndex += 2;                     MYcol += 2;   // Process 2 pix. shifted
   // Read from GM to Shared Memory
   for (i = 0; i < 5; i++){
      for (j = 0; j < 8; j++){
         row=MYrow+i-2;                 col=MYcol+j-2;
         indx=row*Hpixels+col;          Neighbors[MYtid][i][j]=ImgBW[indx];
      }
   }
   __syncthreads();
   for (k = 0; k < 4; k++){
      G = 0.000;
      for (i = 0; i < 5; i++){
         for (j = 0; j < 5; j++){
            G += (Neighbors[MYtid][i][j+k] * GaussC[i][j]);
         }
      }
      ImgGauss[MYpixIndex+k] = G / 159.00;
   }
}
```

TABLE 10.13 Kernel performances: GaussKernel3(), ..., GaussKernel6().

Kernel		Box II	Box III	Box IV	Box VII	Box VIII
GaussKernel3	(ms)	151	149	19.6	13.7	9.15
GaussKernel4	(ms)	852	243	141	23.0	14.8
GaussKernel5	(ms)	1378	—	121	33.9	21.8
GaussKernel6	(ms)	387	164	49.6	22.3	14.3

10.8.6   GaussKernel6(): Reading 5 Vertical Pixels into Shared Memory

For our next version, we will attack the "second problem," which is reducing the amount of elements that we are saving in the shared memory. At the same time, we will attack the "first problem," which is the existence of the repeated entries we are storing that do not have to be there. It turns out these two problems are related and can be solved together. If we simply reduce the number of repeated entries, we expect the amount of shared memory to go down as a consequence. So, let us start by storing only 5 vertical neighbors of each pixel; for a thread that is responsible for processing pixel coordinate (x,y), we will store pixels at coordinates (x,y), (x,y+1), (x,y+2), (x,y+3), and (x,y+4) in shared memory. This is a little different than what we did before when we stored pixels that were *two to the left* and *two to the right* of our pixel (x,y).

Here, we are storing only pixels to the right of our own (x,y). This means that when we compute (x,y), we have to write the result in (x+2,y). In other words, we will never compute the two pixels that are *on the left edge*. We also will not compute two pixels that are *on the right edge*. Using this technique, the only threads that we have to worry about writing to an illegal memory area are the *four threads at the very end of the row*. Others are guaranteed to be safe. The following lines allocate the shared memory:

```
#define MAXTHGAUSSKN67 1024
...
   __shared__ double Neighbors[MAXTHGAUSSKN67+4][5];
```

The reason for adding "+4" should be clear to the reader, because they store the 4 edge pixels, which are wasted but make the code a little cleaner. By eliminating such a large amount of space, we are increasing the maximum number of threads to 1024. With the 1024 set as the maximum number of threads, $8 \times 1028 \times 5 = 41,120$ bytes or ≈ 41 KB. The reality is that the Nvidia hardware will possibly allocate 48 KB for each block that is running this kernel, because 41 KB is simply an "ugly" number to an architecture that loves numbers like 32 or 48 ... This is still a very large amount of shared memory and will possibly put strain on shared memory resources. A wise thing would be to reduce the `#define MAXTHGAUSSKN67` down to 128, which will allocate 5280 bytes (≈ 6 KB) for each block, leaving room for each SM to run 8 blocks inside it (i.e., 48÷6=8 blocks). Or, on Pascal GPUs, which have 64 KB shared memory allocated to an SM, 10 blocks can be launched in each SM without hitting the shared memory limit.

Results of GaussKernel6() are shown in Table 10.13, which indicate that computing a single pixel per thread is a better idea than computing 4, because the additional `for` loops necessary mean that the loop is being done with internal kernel variables rather than the Nvidia hardware that gives us free `for` loops, etc.

CODE 10.13:   imedgeGCM.cu   GaussKernel6() ... {...}

GaussKernel6() reads 5 vertical pixels into shared memory.

```
#define MAXTHGAUSSKN67 1024
...
__global__
void GaussKernel6(double *ImgGauss, double *ImgBW, ui Hpixels, ui Vpixels)
{
   // 5 vertical neighbors for each pixel that is represented by a thread
   __shared__ double Neighbors[MAXTHGAUSSKN67+4][5];

   ui ThrPerBlk=blockDim.x;          ui MYbid=blockIdx.x;     ui MYtid=threadIdx.x;
   int indx, i, j;                   double G;                ui MYrow=blockIdx.y;
   ui MYcol=MYbid*ThrPerBlk+MYtid;   if(MYcol>=Hpixels) return;
   ui MYpixIndex=MYrow*Hpixels+MYcol;
   if ((MYrow<2) || (MYrow>Vpixels - 3) || (MYcol<2) || (MYcol>Hpixels - 3)) {
      ImgGauss[MYpixIndex]=0.0;       return;
   }
   ui IsEdgeThread=(MYtid==(ThrPerBlk-1));
   // Read from GM to Shared Memory. Each thread will read a single pixel
   indx = MYpixIndex-2*Hpixels-2; // start 2 rows above & 2 columns left
   if (!IsEdgeThread) {
      for (j = 0; j < 5; j++) {
         Neighbors[MYtid][j]=ImgBW[indx];
         indx += Hpixels; // Next iteration will read next row, same column
      }
   }else{
      for (j = 0; j < 5; j++) {
         Neighbors[MYtid][j]=ImgBW[indx];         Neighbors[MYtid+1][j]=ImgBW[indx+1];
         Neighbors[MYtid+2][j]=ImgBW[indx+2];  Neighbors[MYtid+3][j]=ImgBW[indx+3];
         Neighbors[MYtid+4][j]=ImgBW[indx+4];
         indx += Hpixels; // Next iteration will read next row, same column
      }
   }
   __syncthreads();
   G = 0.0;
   for (i = 0; i < 5; i++) {
      for (j = 0; j < 5; j++) {
         G += (Neighbors[MYtid+i][j] * GaussC[i][j]);
      }
   }
   ImgGauss[MYpixIndex] = G / 159.00;
}
```

10.8.7 GaussKernel7(): Eliminating the Need to Account for Edge Pixels

Note that the GaussKernel6() was tested with the `#define MAXTHGAUSSKN67 1024` maximum threads per block value. Although we commented on how reducing this down to 128 should help significantly, we never tested its results. This is left as an exercise to Section 10.9, when we study the CUDA Occupancy Calculator; we will find the best parameters for shared memory and more, not as a result of a "guess," but as informed careful calculations.

Before we do that, let us work on a few more improvement ideas. The next idea is GaussKernel7() (Code 10.14), which attempts to solve one small problem that existed in GaussKernel6(): having to check for "edge pixels" using the following lines of code:

```
ui IsEdgeThread=(MYtid==(ThrPerBlk-1));
// Read from GM to Shared Memory. Each thread will read a single pixel
indx = MYpixIndex-2*Hpixels-2; // start 2 rows above & 2 columns left
if (!IsEdgeThread) {
   ...
}else{
   ...
}
```

In this GaussKernel6() code above, you will notice that the code is trying to avoid going out of bounds in index computations. So, it populates the `Neighbors[][]` array differently based on whether a thread is an *edge thread* or not. These cases are designated by a Boolean variable `IsEdgeThread`, which is TRUE if the pixel is an edge thread.

The necessity to check for edge pixels is eliminated by two different facts. Because the following lines of code simply write 0.00 in the `GaussImage` for the two leftmost and two rightmost pixels

```
if ((MYrow<2) || (MYrow>Vpixels - 3) || (MYcol<2) || (MYcol>Hpixels - 3)) {
   ImgGauss[MYpixIndex] = 0.0;          return;
}
```

there is no reason to worry about the edge pixels as long as we store the pixel at coordinates (x-2,y-2) in shared memory. Because of this (−2, −2) "shifted storage," we can multiply this with the constants without shifting during the actual computation as follows:

```
for (i = 0; i < 5; i++) {
   for (j = 0; j < 5; j++) {
      G += (Neighbors[MYtid + i][j] * GaussC[i][j]);
   }
}
```

Indeed, this modification solves multiple problems:

- Edge checking is no longer necessary when reading from GM into shared memory,

- Shared memory profile is reduced by 8× for the same number of threads/block,

- Storing the additional 4 pixels in shared memory is no longer necessary,

- Because each block now computes only `ThrPerBlk-4` pixels, the edge detection, checking is reduced to a single line as follows:

```
if(MYtid>=ThrPerBlk-4) return; // Each block computes ThrPerBlk-4 pixels
```

TABLE 10.14 Kernel performances: GaussKernel3(), ..., GaussKernel7().

Kernel		Box II	Box III	Box IV	Box VII	Box VIII
GaussKernel3	(ms)	151	149	19.6	13.7	9.15
GaussKernel4	(ms)	852	243	141	22.98	14.83
GaussKernel5	(ms)	1378	—	121	33.9	21.8
GaussKernel6	(ms)	387	164	49.6	22.3	14.3
GaussKernel7	(ms)	270	154	35.6	19.9	12.5

CODE 10.14:   imedgeGCM.cu   GaussKernel7() ... {...}

GaussKernel7() eliminates the need for accounting for edge pixels.

```
#define MAXTHGAUSSKN67 1024
...
__global__
void GaussKernel7(double *ImgGauss, double *ImgBW, ui Hpixels, ui Vpixels)
{
   // 5 vertical neighbors for each pixel (read by each thread)
   __shared__ double Neighbors[MAXTHGAUSSKN67][5];

   ui ThrPerBlk=blockDim.x;              ui MYbid=blockIdx.x;
   ui MYtid=threadIdx.x;                 int indx, i, j;
   double G;                             ui MYrow = blockIdx.y;
   ui MYcol=MYbid*(ThrPerBlk-4)+MYtid;   if (MYcol >= Hpixels) return;
   ui MYpixIndex = MYrow * Hpixels + MYcol;
   if ((MYrow<2) || (MYrow>Vpixels - 3) || (MYcol<2) || (MYcol>Hpixels - 3)) {
      ImgGauss[MYpixIndex] = 0.0;         return;
   }
   // Read from GM to Shr Mem. Each threads 1 pixel and 5 neighboring rows
   // Each block reads ThrPerBlk pixels starting at (2 left) location
   indx = MYpixIndex - 2 * Hpixels - 2; // start 2 rows above & 2 columns left
   for (j = 0; j < 5; j++) {
      Neighbors[MYtid][j] = ImgBW[indx];
      indx += Hpixels; // Next iteration will read next row, same column
   }
   __syncthreads();
   if (MYtid >= ThrPerBlk - 4) return; // Each block computes ThrPerBlk-4 pixels
   G = 0.0;
   for (i = 0; i < 5; i++) {
      for (j = 0; j < 5; j++) {
         G += (Neighbors[MYtid + i][j] * GaussC[i][j]);
      }
   }
   ImgGauss[MYpixIndex] = G / 159.00;
}
```

Table 10.14 compares GaussKernel7() to GaussKernel6(). Results indicate that eliminating the necessity to compute edge pixels improves the performance in both Kepler and Pascal.

TABLE 10.15 Kernel performances: GaussKernel3(), ..., GaussKernel8().

Kernel		Box II	Box III	Box IV	Box VII	Box VIII
GaussKernel3	(ms)	151	149	19.6	13.7	9.15
GaussKernel4	(ms)	852	243	141	22.98	14.83
GaussKernel5	(ms)	1378	—	121	33.9	21.8
GaussKernel6	(ms)	387	164	49.6	22.3	14.3
GaussKernel7	(ms)	270	154	35.6	19.9	12.5
GaussKernel8	(ms)	272	190	29.4	24.0	15.1

10.8.8 GaussKernel8(): Computing 8 Vertical Pixels

Perhaps one last idea we can try is computing 8 vertical pixels in the same kernel. GaussKernel8() (Code 10.15) shows an implementation of this, which stores 12 vertical pixels to be able to compute 8 vertical pixels in the same kernel. The difference between this kernel and GaussKernel7() is that to compute a pixel, we need 4 vertical and 4 horizontal neighbors' values. Because we are using the 4 horizontal neighbors inside the kernel as plain reads from global memory, we need to store 4 vertical neighbors in shared memory (i.e., 1+4=5 total, as shown in GaussKernel7()). Following this logic, to compute 8 vertical pixels, we need to store 4+8=12 vertical pixels.

Table 10.15 compares GaussKernel8() to its previous versions. It is less efficient than the previous one because of the fact that to write the results of the 8 computed pixels, an additional for loop is needed at the end, which is shown below:

```
double G[8] = { 0.00, 0.00, 0.00, 0.00, 0.00, 0.00, 0.00, 0.00 };
...
// Write all computed pixels back to GM
for (j = 0; j < 8; j++) {
   ImgGauss[MYpixIndex] = G[j] / 159.00;
   MYpixIndex += Hpixels;
}
```

which requires the results of these 8 pixels to be saved in an array named G[] that takes up 8×8=64 bytes of storage. If the compiler allocates registers for this array, a valuable 128 registers are used in this kernel out of the available 255 maximum that is limited by Nvidia hardware; 64 double registers take up a 128 native 32 bit register allocation in the register file. If the compiler decides to place this array in the global memory, than this defeats the purpose of using the shared memory in the first place. As the reader can see from this example, trying to do more work in each kernel increases the resource profile of the kernel, which limits the number of blocks that can be launched that run this kernel, thereby limiting the performance. In a lot of the cases we investigated, the solutions that involved kernels with small resource profiles always won, although this is not a hard rule.

Trying to compute multiple pixels has other side effects, such as having to check whether we are computing the last few rows, as shown by the code below:

```
ui isLastBlockY=(blockIdx.y == (blockDim.y-1));
if (isLastBlockY) {
   indx=(Vpixels-2)*Hpixels + MYcol;
   ImgGauss[indx]=0.0;   ImgGauss[indx+Hpixels]=0.0; // last row-1, last row
}
```

CODE 10.15:   imedgeGCM.cu   GaussKernel8() ... {...}

GaussKernel8() computes 8 vertical pixels, rather than 1.

```
#define MAXTHGAUSSKN8 256
...
// Each block reads 12 rows; each thread computes 8 vertical pix
__global__
void GaussKernel8(double *ImgGauss, double *ImgBW, ui Hpixels, ui Vpixels)
{
  // 12 vertical neighbors are saved in ShMem and used to compute 8 vertical pixels
  // by each thread; Reads from 2 top and 2 bottom pixels are wasted.
  __shared__ double Neighbors[MAXTHGAUSSKN8][12];
  ui ThrPerBlk=blockDim.x;       ui MYbid=blockIdx.x;
  ui MYtid=threadIdx.x;          int indx, i,j,row;
  ui MYrow=blockIdx.y*8;         ui isLastBlockY=(blockIdx.y == (blockDim.y-1));
  ui MYcol=MYbid*(ThrPerBlk-4)+MYtid;   if(MYcol>=Hpixels) return;
  if(MYrow>=Vpixels) return;            ui MYpixIndex=MYrow*Hpixels+MYcol;
  double G[8] = { 0.00, 0.00, 0.00, 0.00, 0.00, 0.00, 0.00, 0.00 };
  if ((MYcol<2) || (MYcol>Hpixels - 3)) {  // first and last 2 columns
     ImgGauss[MYpixIndex] = 0.0;       return;
  }
  if (MYrow == 0) {
     ImgGauss[MYpixIndex]=0.0;  ImgGauss[MYpixIndex+Hpixels]=0.0; // row 0,1
  }
  if (isLastBlockY) {
     indx=(Vpixels-2)*Hpixels + MYcol;
     ImgGauss[indx]=0.0;   ImgGauss[indx+Hpixels]=0.0; // last row-1, last row
  }
  // Read from GM to Shared Memory; each thr. reads 1 pixel for 12 neighboring rows
  // Each thread reads 12 pixels, but will only compute 8
  for (indx = MYpixIndex, j = 0; j < 12; j++) {
     if ((MYrow+j) < Vpixels) {
        Neighbors[MYtid][j] = ImgBW[indx];
        indx += Hpixels; // Next iteration will read next row, same column
     }else    Neighbors[MYtid][j] = 0.00;
  }
  __syncthreads();
  if (MYtid >= ThrPerBlk - 4) return; // Each block computes ThrPerBlk-4 pixels
  for (row = 0; row < 8; row++) {
     for (i = 0; i < 5; i++) {
        for (j = 0; j < 5; j++) {
           G[row] += (Neighbors[MYtid + i][row+j] * GaussC[i][j]);
        }
     }
  }
  // Write all computed pixels back to GM
  for (j = 0; j < 8; j++) {
     ImgGauss[MYpixIndex] = G[j] / 159.00;
     MYpixIndex += Hpixels;
  }
}
```

10.9   CUDA OCCUPANCY CALCULATOR

In all of the examples in this chapter, although we observed some improvements in performance when we made changes to the code, we never really totally understood how these changes altered the way the GPU hardware is utilized at runtime. Furthermore, our analysis was a little shaky when comparing two different kernels. For example, in Table 10.15, when we were comparing different kernels, we jumped to the conclusion that GaussKernel8() was better than GaussKernel7(), when run on Box IV. Let us strictly focus on Box IV as an example in this subsection. Table 10.15 shows a runtime of 35.6 ms for GaussKernel7() and 29.4 ms for GaussKernel8(). So, why wouldn't we conclude that GaussKernel8() is better?

The question we didn't ask ourselves is this: we tested both of these kernels with 256 threads/block, and the same image astronaut.bmp. The problem is that we don't know if 256 threads/block is optimum for either one of the kernels. Additionally, we don't know if the size of the image (astronaut.bmp) will expose the best behavior of either kernel. Without thinking about this, it sounds fair — and logical — that if you are testing both kernels with the "same" this and "same" that, the result should be comparable. The fact is that *it is not*; because the "same" threads/block is not optimum for any of the kernels. This means that *you are comparing junk to junk*. A fair comparison should use the "same picture," but the "optimum parameters for each algorithm." This is what we will do in this section. We will try to find the best parameters for each kernel, by using the CUDA Occupancy Calculator.

To provide an analogy, if you were comparing the performance of a Ford Mustang to a GM Corvette, it would sound fair to race them both on the dirt road and with the speed restriction of less than 60 mph. Sixty mph is very low for a high-performance car. Corvette might be able to handle 60 mph much better than Mustang, but if you increased the speed limit to 100 mph, maybe Mustang will do much better. To phrase it differently, your comparison means nothing. A much more meaningful comparison would be to remove the speed limit and say "whoever reaches the first 5 mile mark wins." This will allow the driver of the car to drive it at the optimum speed that is best suited for each car and truly expose which car is better.

Let us prove how faulty our analysis was, when comparing the performances of GaussKernel7() and GaussKernel8(), by using the results shown in Table 10.15. The only reason we compared both of them using 256 threads/block was because GaussKernel8() is limited to 256 threads/block due of the definition,   #define MAXTHGAUSSKN8 256, in Code 10.15. Because GaussKernel8() is not even allowed to exceed 256 threads/block, we figured that a fair comparison should use 256 threads/block for both GaussKernel7() and GaussKernel8(). This logic has two flaws: First of all, GaussKernel7() is allowed to go to a maximum of 1024 threads per block because, as shown in Code 10.15, the limiting constant is different: #define MAXTHGAUSSKN67 1024. What if its performance is higher for 384, 512, 768, or 1024 threads/block? If this were true, we would have missed a parameter that would allow GaussKernel7() perform better; therefore, we would not have concluded that GaussKernel8() is better. Second, how can we even guarantee that 256 is the best threads/ block for even GaussKernel8() is 256? What if it is 192, 128, or even lower, such as 96, 64? If a lower value was the optimum one for GaussKernel8(), then we are truly *comparing junk to junk*. This is the general *parameter space exploration problem*, and the only solution is to exhaustively search for all possible parameters. To put it in more concrete mathematical terms, the comparison of GaussKernel7() and GaussKernel8() should include all possible combinations of parameters for them. GaussKernel8() has the following 6 options that make sense: {32, 64, 96, 128, 192, 256} and GaussKernel8() has the following 10–15 options that are practical: {32, 64, 96, 128, 192, 256, . . . , 1024}.

TABLE 10.16 Kernel performances for GaussKernel6(), GaussKernel7(), and GaussKernel8() for Box IV, under varying threads/block choices.

Threads per Block	Runtime (ms)		
	GaussKernel6	GaussKernel7	GaussKernel8
32	450	340	202
64	181	130	105
96	122	83	86
128	121	62	48
192	73	61	47
256	50	36	29
320	41	30	
384	35	25	
448	44	24	
512	35	30	
768	24	24	
1024	22	19	

10.9.1 Choosing the Optimum Threads/Block

Let us gain some intuition into the impact of the threads/block parameter on performance. Table 10.16 provides a more detailed version of Table 10.15 for only Box IV, which is the box with the GTX Titan Z GPU. These comparison results may change for different GPUs, sometimes in surprising ways, as we have seen over and over again in the past chapters; however, Table 10.16 should give us a good starting point. I only used rounded numbers because even this accuracy is more than sufficient to make the point. The highlighted values are for 256 threads/block (identical to what is reported in Table 10.15), which allows us to see what we might have missed by using the faulty logic to create Table 10.15. Here are the conclusions we can draw from Table 10.16:

- We definitely missed a better parameter value for GaussKernel7(): 1024 threads/block allows this kernel to perform much better than GaussKernel8().

- We can further conclude that GaussKernel7() is better than GaussKernel8(), because GaussKernel8() cannot use higher threads/block values, due to the limitation: #define MAXTHGAUSSKN8 256.

- The #define MAXTHGAUSSKN8 256 restriction in GaussKernel8() was not just some random number we picked; any value higher than 256 would have resulted in exceeding the maximum allowed shared memory in this kernel and the kernel would not have compiled. Because GaussKernel7() uses a lot less shared memory, we could have pushed it to 1024, without exceeding the 48 KB shared memory limitation of the GPU.

- The GaussKernel8() kernel in Section 10.8.8 achieved its performance by using a lot of shared memory, so it was destined to be uncompetitive against other kernels (e.g., GaussKernel7()) that are not so shared memory-hungry!

- GaussKernel7() seems to always perform better then GaussKernel6(), because both have the same restriction on the threads/block.

10.9.2 SM-Level Resource Limitations

The question is: *Is there a tool to let us determine the optimum threads/block to launch a kernel?* The answer is: Yes; it is the *CUDA Occupancy Calculator*. Let us start using it; it is a simple Excel sheet you download from Nvidia's website. The purpose of this Excel sheet is to calculate how "occupied" each streaming multiprocessor (SM) will be under different choices of shared memory, threads/block, and registers/kernel choices. The following are the three resources that can constrain how many threads you can launch in the entire SM:

- **Shared memory** usage is determined by the design of the kernel; if you design it to require a lot of shared memory (like our GaussKernel8()), this will end up being the limiting factor for how many blocks you can fit in an SM, because you will eventually run out of shared memory. Let us refresh our memory with the fact that shared memory is a *block-level* resource, which is not dependent on the number of threads/block. The maximum shared memory each block can use is 48 K for many of the newer Compute Capability levels. For example, if your blocks require 25 K, you cannot launch more than a single block, because 50 K for two blocks would make you go over the 48 K limit if two blocks were launched. Alternatively, if your blocks require 13 K, you can only launch 3 of them; each would require 13 K, with a total of 39 K for 3 of them. The fourth one would hit the shared memory limit and would not be able to reside in the same SM with the other three.

- **Number of registers** in a kernel depends on how many variables you use to design the kernel; using too many registers causes two problems: first, the maximum number of registers that can be used in a kernel is 255. If your kernels exceed this, it will cause a *register spill*, which is the scenario when the compiler is desperately using memory addresses to act as a register. Your performance will go down drastically if you have a register-spill situation; second, you will not be able to launch too many blocks without hitting the "total number of registers in a block" limitation, which is either 32 K or 64 K depending on the Compute Capability. Therefore, *register-hungry* kernels will limit the performance. For example, if your kernels require 240 registers, launching 512 threads/block will require $512 \times 240 = 122{,}880 = 120$ K registers inside the register file. However, even if your Compute Capability allows 64 K registers, this is still more than the 120 K you need. This means that you cannot launch more than 256 threads/block before hitting the register file limitation, which would require $256 \times 240 = 61{,}440 = 60$ K registers.

- **Threads/block** value determines how many blocks you can launch in an SM; if you use a small threads/block value, you will need to launch a lot of blocks and you will hit the "maximum blocks launched per SM" limitation, which is 16 for Kepler, and 32 for Maxwell and Pascal. This can become a limitation before the other two. Alternatively, if you launch a large threads/block value, the granularity of your blocks is too high and it will be difficult to fit a large number of them; you will potentially leave a wasted space.

- **Number of warps** per SM is really what it all boils down to. This is the only restriction that matters. In Compute Capability, this number is 64. Considering that each warp is 32 threads, having a "maximum 64 warps per SM" limitation translates to having a maximum of 2048 threads per block.

The goal of the CUDA Occupancy Calculator is to graphically demonstrate the interplay among these parameters, to allow the programmer to choose the best parameter set.

10.9.3 What is "Occupancy"?

The four limitations described in Section 10.9.2 must all be met. This means that the worst limitation ends up being the performance limiting factor and the other restrictions do not even matter. Every single one of these constraints can be converted to a "warp count" and, therefore, corresponds to an "occupancy." If your warp limitation is 64, then 64 warps per SM is considered 100% occupancy for that SM. From the standpoint of the SM, it doesn't matter who makes up 64 warps. For example, if you launch 32 threads/block, you are indeed launching 1 warp/block. In a Kepler GPU, you can launch a maximum of 32 blocks in an SM, therefore with such small blocks, you can launch 32 blocks, each running a single warp; so, you are running 32 warps. Considering that 64 warps is considered 100% occupancy, 32 warps means 50% occupancy. In other words, you are keeping that SM only 50% *stuffed with work*. Alternatively, if you launched your blocks with 64 threads/block (i.e., 2 warps/block), launching 32 blocks would get you to stuff the SM with 64 warps total, i.e., 100% occupancy. So, why wouldn't anyone do this?

I used nice and smooth numbers. Now, let us assume that you are launching your blocks with 640 threads/block (i.e., 10 warps/block). When you launched 6 of them, you are at 60 warps and left 4 wasted warps on the table. Therefore, your occupancy is $60 \div 64 \approx 94\%$; not that terrible, but not 100%. Things could have been a lot worse. What if each block requires 10 KB shared memory? We know that the maximum shared memory allowed for a block is 48 K; this would limit the number of blocks you can launch to 4. Say you launch your blocks with 64 threads (2 warps). Because each block has 2 warps, and you can launch a maximum of 4 blocks, you can launch a maximum of 8 warps; you are at $8 \div 64 = 12.5\%$ occupancy. Naturally, if this is the case, why not launch real big blocks, like 512 threads/block (16 warps/block); then, you will hit a total of 64 warps (100% occupancy).

10.9.4 CUDA Occupancy Calculator: Resource Computation

The types of "what if" scenarios we contemplated for resources in Section 10.9.3 are exactly what the CUDA Occupancy Calculator is supposed to compute. Figure 10.1 (see ebook for color version) provides a screen shot of the CUDA Occupancy Calculator Excel sheet, in which you plug in the CUDA Compute Capability (CC) and the amount of shared memory you have in your SMs in the green area (top). Based on your CC, it determines all of the restrictions for each SM on the bottom (gray area). The most important restriction that determines our occupancy is the *Max Warps per SM*, which is 64 for CC 3.0; we will specifically use the CC 3.0 case in our analyses in this section because this is what we compiled our program with so far.

The orange area in Figure 10.1 is where you plug in the kernel and block parameters; the case that is shown in Figure 10.1 is for launching 128 threads/block (i.e., 4 warps per block) and when each thread uses 16 registers. Furthermore, each block needs 8192 (8 K) shared memory. For this specific case, let us analyze the limitations that come from three different resources: **(1)** shared memory is limited to 48 K and each block uses 8 K of it. Therefore, beyond 6 blocks, I would hit the shared memory limit and would not be able to launch more blocks, **(2)** register file limitation for CC 3,0 is 64 K, therefore, I cannot launch more than $65,536 \div 16 = 4096$ threads. This will not be a restriction at all because before I hit the register wall, I will hit the "maximum number of threads I can launch in the SM is 2048" wall, **(3)** maximum number of warps is 64, so, if we launch 4 warps/block, we can launch a maximum of 16 blocks. It is clear that the strictest limitation is the shared memory limitation, which limits us to 6 blocks, with 4 warps in each block, with a total of 24 warps for the entire SM, translating to $24 \div 64 = 38\%$ occupancy.

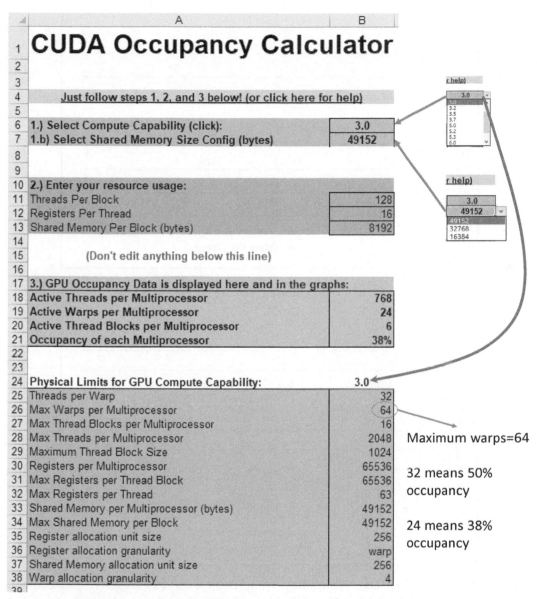

FIGURE 10.1 CUDA Occupancy Calculator: Choosing the Compute Capability, max. shared memory size, registers/kernel, and kernel shared memory usage. In this specific case, the occupancy is 24 warps per SM (out of a total of 64), translating to an occupancy of $24 \div 64 = 38\,\%$.

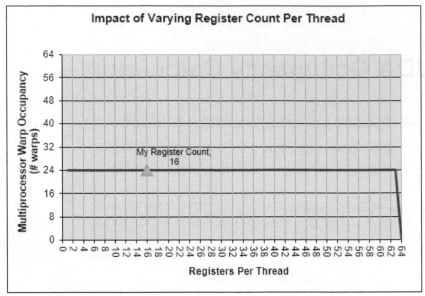

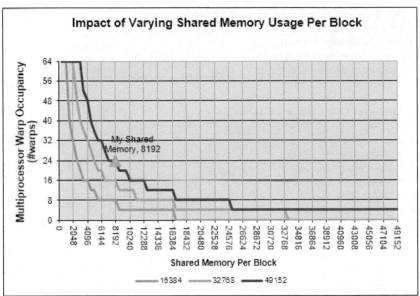

FIGURE 10.2 Analyzing the occupancy of a case with **(1)** registers/thread=16, **(2)** shared memory/kernel=8192 (8 KB), and **(3)** threads/block=128 (4 warps). CUDA Occupancy Calculator plots the occupancy when each kernel contains more registers (top) and as we launch more blocks (bottom), each requiring an additional 8 KB. With 8 KB/block, the limitation is 24 warps/SM; however, it would go up to 32 warps/block, if each block only required 6 KB of shared memory (6144 Bytes), as shown in the shared memory plot (below).

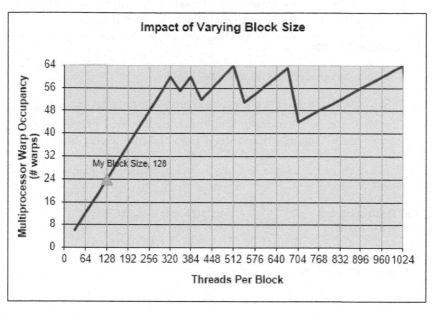

40	Allocated Resources		Per Block	Limit Per SM	= Allocatable Blocks Per SM
41	Warps	(Threads Per Block / Threads Per Warp)	4	64	16
42	Registers	(Warp limit per SM due to per-warp reg count)	4	128	32
43	Shared Memory (Bytes)		8192	49152	6
44	Note: SM is an abbreviation for (Streaming) Multiprocessor				
45					

46	Maximum Thread Blocks Per Multiprocessor	Blocks/SM	* Warps/Block	= Warps/SM
47	Limited by Max Warps or Max Blocks per Multiprocessor	16		
48	Limited by Registers per Multiprocessor	32		
49	**Limited by Shared Memory per Multiprocessor**	**6**	**4**	**24**

50	Note: Occupancy limiter is shown in orange	Physical Max Warps/SM = 64
51		Occupancy = 24 / 64 = 38%

FIGURE 10.3 Analyzing the occupancy of a case with (1) registers/thread=16, (2) shared memory/kernel=8192 (8 KB), and (3) threads/block=128 (4 warps). CUDA Occupancy Calculator plots the occupancy when we launch our blocks with more threads/block (top) and provides a summary of which one of the three resources will be exposed to the limitation before the others (bottom). In this specific case, the limited amount of shared memory (48 KB) limits the total number of blocks we can launch to 6. Alternatively, the number of registers or the maximum number of blocks per SM does not become a limitation.

10.9.5 Case Study: GaussKernel7()

Now that we saw an example of how the CUDA Occupancy Calculator can be used, we should apply it to the kernels we freshly developed in the previous sections. Let us start with GaussKernel7(). Here are the lines that declare the shared memory:

```
#define MAXTHGAUSSKN67 1024
...
    __shared__ double Neighbors[MAXTHGAUSSKN67][5];
```

This kernel is using a total of $1024 \times 8 \times 5 = 40,960$ Bytes (40 KB) shared memory; note: 8 is the sizeof(double). We are estimating that this kernel needs 16 registers per kernel. The number of registers can be roughly determined from the initial part of the kernel, which is shown below:

```
void GaussKernel7(...)
{
    ui ThrPerBlk=blockDim.x;        ui MYbid=blockIdx.x;
    ui MYtid=threadIdx.x;           int indx, i, j;
    double G;                       ui MYrow = blockIdx.y;
    ui MYcol=MYbid*(ThrPerBlk-4)+MYtid;   if (MYcol >= Hpixels) return;
    ui MYpixIndex = MYrow * Hpixels + MYcol;
    ...
```

This kernel requires nine 32-b registers and one 64-b register (i.e., double, which requires 2×32-b), totaling 11 32-b registers. Conservatively, it will need a few temporary registers to move the data around. Let us assume 5 temporary registers. So, 16 is not a bad estimate for the number of 32-b registers.

Remember that we ran and reported GaussKernel7() with 256 threads per block in Table 10.15. When we plug these parameters into the CUDA Occupancy Calculator, we get the plot in Figure 10.4 for the register-induced limitation (top) and shared memory-induced limitation (bottom). Undoubtedly, the shared memory is the worst of the two limitations. In fact, it is so bad that it doesn't even allow us to launch more than a single block in an SM. If we launch our blocks with 256 threads/block (i.e., 8 warps/block), our occupancy will always be 8 warps/SM. Because the SM is limited to 64 warps/SM, our occupancy is only $8 \div 64 \approx 13\%$, which is shown in Figure 10.5. As this example demonstrates, we designed our kernels without knowing the "occupancy" concept. Now we realize that we created a kernel that only keeps each SM 13% occupied, leaving 87% of the warps unused. There should be no doubt in the reader's mind at this point that we could have done better. Of course, if we launched the blocks with 1024 threads/block (i.e., 32 warps per block), then our occupancy would have been 50%, although we are still limited by the same 40 KB shared memory. By simply changing the threads/block parameter, we increased our occupancy from 13% to 50%. One would wonder how this translates to performance, when we have $4\times$ higher occupancy. The answer is in Table 10.16: our runtime goes from 36 down to 19 ms. In other words, it increased substantially (almost $2\times$ in this case). Although it is difficult to make generalizations such as "x amount of occupancy corresponds to y amount of performance increase," etc., one thing that is for sure: you will never know how high you can go unless you increase your occupancy as much as possible. Notice that even with 1024 threads/block, we are still at 50% occupancy. This begs the question about whether GaussKernel7() could have done much better. This requires the programmer to go back to the drawing board and design the kernel to require a little less shared memory.

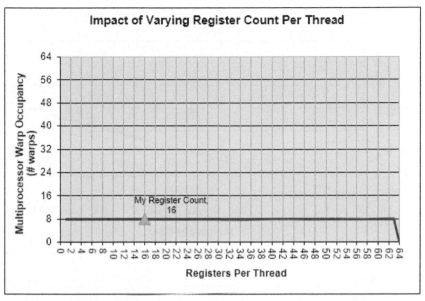

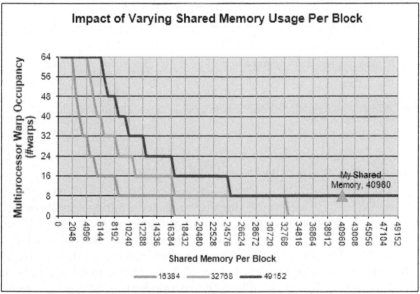

FIGURE 10.4 Analyzing the GaussKernel7(), which uses **(1)** registers/thread ≈ 16, **(2)** shared memory/kernel=40,960 (40 KB), and **(3)** threads/block=256. It is clear that the shared memory limitation does not allow us to launch more than a single block with 256 threads (8 warps). If you could reduce the shared memory down to 24 KB by redesigning your kernel, you could launch at least 2 blocks (16 warps, as shown in the plot) and double the occupancy.

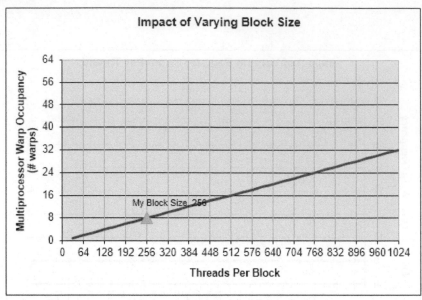

Allocated Resources		Per Block	Limit Per SM	= Allocatable Blocks Per SM
40 **Allocated Resources**		Per Block	Limit Per SM	Blocks Per SM
41 Warps	(Threads Per Block / Threads Per Warp)	8	64	8
42 Registers	(Warp limit per SM due to per-warp reg count)	8	128	16
43 Shared Memory (Bytes)		40960	49152	1

44 Note: SM is an abbreviation for (Streaming) Multiprocessor

46 **Maximum Thread Blocks Per Multiprocessor**	Blocks/SM	* Warps/Block	= Warps/SM
47 Limited by Max Warps or Max Blocks per Multiprocessor	8		
48 Limited by Registers per Multiprocessor	16		
49 **Limited by Shared Memory per Multiprocessor**	**1**	**8**	**8**

50 Note: Occupancy limiter is shown in orange

Physical Max Warps/SM = 64
Occupancy = 8 / 64 = 13%

FIGURE 10.5 Analyzing the GaussKernel7() with **(1)** registers/thread=16, **(2)** shared memory/kernel=40,960, and **(3)** threads/block=256.

For the GaussKernel7() case study, threads/block and limiting conditions are shown in Figure 10.5. The top plot shows that you could have gone up to 32 warps if you launched 1024 threads/block, although this would have made you hit the shared memory limitation, shown in Figure 10.4. The implication of the occupancy concept is far-reaching: maybe even a technically worse kernel can perform better, if it is not resource-hungry, because it can "occupy" the SM better. To phrase differently: writing kernels with a high resource profile has the danger of putting shackles on the kernel's ankles. Maybe it is better to design kernels with much smaller profiles, although they are not *seemingly* high performance. When a lot of them are running on the SM, at a much higher occupancy, maybe they will translate to an overall higher performance. This is the beauty of GPU programming. It is art as much as a science. You can be sure that the resource limitations will not go away even 10 generations from today, although you might have a little more shared memory in each, a little more registers, etc. So, this "resource-constrained thinking" will always be a part of GPU programming.

To provide an analogy: if you have two runners — one is a world-class marathon runner athlete and the other one is an ordinary guy like me — who will win the race? This is an

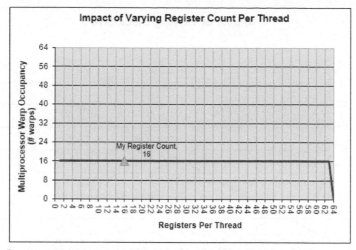

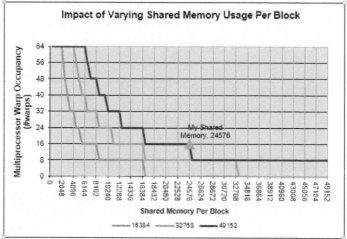

FIGURE 10.6 Analyzing the GaussKernel8() with **(1)** registers/thread=16, **(2)** shared memory/kernel=24,576, and **(3)** threads/block=256.

obvious answer. But, here is the second question: who will win the race, *if you put shackles on the marathon runner's feet?* This is what a technically excellent, but resource-heavy kernel is like. It is a marathon runner with shackles on his feet. You are destroying his performance due to resource limitations.

10.9.6 Case Study: GaussKernel8()

This is the case study for GaussKernel8(): 24576 KB shared memory per block, 256 threads per block, 16 registers per kernel. Register usage and shared memory usage are shown in Figure 10.6. Just because it uses 24 KB shared memory, it can squeeze two blocks into each SM; together, they will max out the shared memory at 48 KB, but it works! It is launched with 256 threads/block (8 warps/block) and two blocks occupy 16 warps in the SM. Note that even a 24.1 KB shared memory requirement would have halved the occupancy.

For the GaussKernel8() case study, threads/block and limiting conditions are shown in Figure 10.7. Although we can push the threads/block to much higher levels, the shared

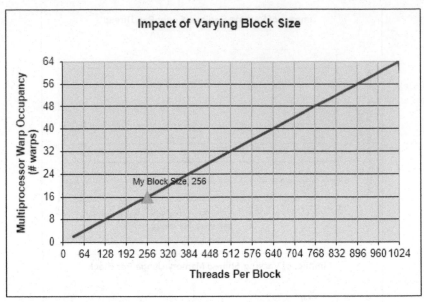

		Per Block	Limit Per SM	= Allocatable Blocks Per SM
40	**Allocated Resources**			
41	Warps (Threads Per Block / Threads Per Warp)	8	64	8
42	Registers (Warp limit per SM due to per-warp reg count)	8	128	16
43	Shared Memory (Bytes)	24576	49152	2
44	Note: SM is an abbreviation for (Streaming) Multiprocessor			
45				

	Maximum Thread Blocks Per Multiprocessor	Blocks/SM	* Warps/Block =	Warps/SM
46				
47	Limited by Max Warps or Max Blocks per Multiprocessor	8		
48	Limited by Registers per Multiprocessor	16		
49	**Limited by Shared Memory per Multiprocessor**	**2**	**8**	**16**

50 Note: Occupancy limiter is shown in orange   **Physical Max Warps/SM = 64**

51   **Occupancy = 16 / 64 = 25%**

FIGURE 10.7 Analyzing the GaussKernel8() with **(1)** registers/thread=16, **(2)** shared memory/kernel=24,576, and **(3)** threads/block=256.

memory limitation does not let us to do so. Because of the shared memory limitation, we are stuck at 25% occupancy (16 warps out of the max 64 warps). What we deduce from Figure 10.6 is that we can only reach 100% occupancy if our shared memory-per-kernel requirement is 6 KB per kernel.

The reader is now encouraged to go back to all of the Gauss kernels we designed and check their CUDA occupancy to see if we missed any potential excellent candidate that is low-performance but has a much lower occupancy limitation. It is likely that such a kernel can beat any of the others, because it can reach 100% occupancy.

## CHAPTER 11

CUDA Streams

I N THE entire rest of this book, we focused on improving the kernel execution time. In a CUDA program, first the data must be transferred from the CPU memory into the GPU memory; it is only when the data is in GPU memory that the GPU cores can access it and process it. When the kernel execution is done, the processed data must be transferred back into the CPU memory. One clear exception to this sequence of events is when we are using the GPU as a graphics machine; if the GPU is being used to render the graphics that are required for a computer game, the output of the processed data is the monitor (or multiple monitors), which is directly connected to the GPU card. So, there is no loss of time to transfer data back and forth between the CPU and the GPU.

However, the type of programs we are focusing on in this book are GPGPU applications, which use the GPU as a general purpose processor; these types of computations are, in the end, totally under the CPU control, which are originated by the CPU code. Therefore, the data must be shuttled back and forth between the CPU and GPU. What is worse, the data transfer rate between these two dear friends (CPU and GPU) is bottlenecked by the PCIe bus, which has a much lower rate than either the CPU-main memory bus or GPU-global memory bus (see Sections 4.6 and 10.1 for details).

As an example, what if we are performing edge detection on the astronaut.bmp file? Table 11.1 (top half) shows the run time results of this operation on four different GPUs (two Kepler and two Pascal), which is broken down into three different components for the execution time, as shown below:

- CPU→GPU transfers take, on average, 31% of the total execution time.

- Kernel execution inside the GPU takes, on average, 39% of the execution time.

- GPU→CPU transfers take, on average, 30% of the total execution time.

For the purposes of this discussion, the details for the Kepler versus Pascal are not important. It suffices to say that the CPU→GPU, kernel execution, and GPU→CPU components of the runtime are approximately a third of the total time. Taking the Pascal GTX 1070 as an example, we wait to transfer the entire 121 MB astronaut.bmp image into the GPU (which takes 25 ms); once this transfer is complete, we run the kernel and finish performing edge detection (which takes another 41 ms); once the kernel execution is complete, we transfer it back to the CPU (which is another 24 ms).

When we look at the results for the horizontal flip in Table 11.1 (bottom half), we see a different ratio for the execution time of operations; the CPU→GPU and GPU→CPU transfers are almost half of the total execution time (53% and 43%, respectively), while the kernel execution time is negligible (only 4% of the total).

The striking observation we make from these two cases is that most of the execution time goes to shuttling the data around, rather than doing the actual (and useful) work! Although

TABLE 11.1 Runtime for edge detection and horizontal flip for astronaut.bmp (in ms).

Operation	Task	Titan Z	K80	GTX 1070	Titan X	Avg %
Edge Detection	CPU→GPU tfer	37	46	25	32	31%
	Kernel execution	64	45	41	26	39%
	GPU→CPU tfer	42	18	24	51	30%
	Total	143	109	90	109	100%
Horizontal Flip	CPU→GPU tfer	39	48	24	32	53%
	Kernel execution	4	4	2	1	4%
	GPU→CPU tfer	43	17	24	34	43%
	Total	86	69	50	67	100%

Kernel execution times are lumped into a single number for clarity. GTX Titan Z and K80 use the Kepler architecture, while the GTX 1070 and the Titan X (Pascal) use the Pascal architecture.

our efforts to try to decrease the kernel execution time was well justified so far, Table 11.1 makes it clear that we cannot just worry about the kernel execution time when trying to improve our program's overall performance. The data transfer times must be considered as an integral part of the total runtime. So, here is our motivating question for this chapter: could we have started processing *a part of this image* once it was in GPU memory, so we wouldn't have to wait for the *entire* image to be transferred? The answer is most definitely YES; the idea is that the CPU⟷GPU data transfers can be overlapped with the kernel execution because they use two different pieces of hardware (PCI controller and the GPU cores) that can work independently and, more importantly, *concurrently*. Here is an analogy that helps us understand this concept:

ANALOGY 11.1: *Pipelining.*

Cocotown hosted an annual competition among two teams; the goal was to harvest 100 coconuts in the shortest amount of time. The first team did this in three separate steps: **(1)** Charlie went to the jungle, picked 100 coconuts, brought them to the harvesting area, **(2)** Katherine met Charlie at the harvesting area precisely when he was done, as they coordinated; Kenny started harvesting the coconuts and Charlie went home, and when Katherine finished, she went home, **(3)** Greg brought the harvested coconuts back to the competition area when Katherine was done, at which time the clock stopped. Each step took approximately 100 minutes and the first team was done in 300 minutes.

The second team had a completely different approach: **(1)** Cindy went and picked only 20 coconuts and brought them to Keith, **(2)** Keith immediately started harvesting the coconuts when he received them, and **(3)** Gina immediately brought the finished coconuts to the competition area. The team continued this cycle 5 times.

Although both Team 1 (Charlie, Katherine, Greg) and Team 2 (Cindy, Keith, Gina) were equally experienced farmers, the second team shocked everyone because they completed their harvesting in 140 minutes.

TABLE 11.2 Execution timeline for the second team in Analogy 11.1.

Time	Cindy	Keith	Gina
0–20	Bring 20 coconuts to Keith	—	—
20–40	Bring 20 more coconuts to Keith	Harvest 20 coconuts	—
40–60	Bring 20 more coconuts to Keith	Harvest 20 coconuts	Deliver 20 coconuts
60–80	Bring 20 more coconuts to Keith	Harvest 20 coconuts	Deliver 20 coconuts
80–100	Bring 20 more coconuts to Keith	Harvest 20 coconuts	Deliver 20 coconuts
100–120	—	Harvest 20 coconuts	Deliver 20 coconuts
120–140	—	—	Deliver 20 coconuts

Cindy brings 20 coconuts at a time from the jungle to Keith. Keith harvests the coconuts immediately when he receives them. When 20 of them are harvested, Gina delivers them from Keith's harvesting area to the competition desk. When the last 20 coconuts are delivered to the competition desk, the competition ends.

11.1 WHAT IS PIPELINING?

According to Analogy 11.1, why was the second team able to finish their harvesting in 140 minutes? Let us think. It is obvious when three different tasks are *serialized*; because each task takes 100 minutes, they take 300 minutes cumulatively when executed serially. However, if the three tasks can be executed concurrently, the formula changes; the second team was able to overlap the execution time of the two different transfers with the amount of time it takes to actually do the work. What Charlie and Cindy did in Analogy 11.1 corresponds to the CPU→GPU transfers, while Greg and Gina's job is analogous to GPU→CPU transfers. Clearly, Katherine and Keith had tasks that compare to kernel execution.

11.1.1 Execution Overlapping

Table 11.2 sheds light onto how the second team finished their job in 140 minutes. In the ideal theoretical case, if the three tasks could be perfectly overlapped, it would only take 100 minutes to finish all of them because all three of them are being executed concurrently. Let us use a notation to explain the run time of a serial operation (the case for Team 1):

- Task1 execution time $= T1 = 100$,

- Task2 execution time $= T2 = 100$,

- Task3 execution time $= T3 = 100$,

$$\text{Serialized runtime} = T1 + T2 + T3, \tag{11.1}$$
$$= 100 + 100 + 100 = 300.$$

If any of the executions can be overlapped, as shown in Table 11.1, we can use the following notation to explain the total runtime, by breaking the individual tasks (say, Task1, denoted as $T1$) into subtasks that can be partially overlapped with other subtasks (e.g., $T1a$, $T1b$, and $T1c$). Some of these subtasks might be serially executed (e.g., $T1a$) and some can be overlapped with other subtasks (e.g., $T1b$ can be overlapped with $T2a$). So, the total runtime is (the case for Team 2):

- Task1 $= 100 = 20$ non-overlapped, 20 overlapped with Task2, 60 fully overlapped,

  $T1 = T1a + T1b + T1c = 20 + 20 + 60$,

- Task2 = 100 = 20 overlapped w/ Task1, 60 fully overlapped, 20 overlapped w/ Task3,

  $T2 = T2a + T2b + T2c = 20 + 60 + 20,$

- Task3 = 100 = 60 fully overlapped, 20 overlapped with Task 2, 20 non-overlapped.

  $T3 = T3a + T3b + T3c = 60 + 20 + 20,$

- Total = T1a + (T1b||T2a) + (T1c||T2b||T3a) + (T2c||T3b) + T3c.

Pipelined runtime $= T1a + (T1b||T2a) + (T1c||T2b||T3a) + (T2c||T3b) + T3c,$ (11.2)

$$= 20 + (20||20) + (60||60||60) + (20||20) + 20 = 140.$$

11.1.2 Exposed versus Coalesced Runtime

So, we are using the parallel symbol (||) to denote the tasks that can be executed concurrently (i.e., in parallel). Because Keith has to wait for Cindy to bring in the first 20 coconuts and cannot start harvesting until this has been done, the runtime for this initial 20 coconuts is *exposed* in the total runtime (denoted as $20 + \cdots$). However, Cindy can bring in the next 20 coconuts while Keith is harvesting the first 20 coconuts. Therefore, these two operations can execute in parallel (denoted as 20||20). Unfortunately, Gina sits idle until Keith has processed the very first 20 coconuts Cindy brought. Therefore, the overlapping among all three tasks cannot happen until Keith finishes the first 20 coconuts. Once this is done, Cindy can bring in the third batch of 20 coconuts, while Keith is harvesting the second set of 20 coconuts and Gina starts delivering the first harvested 20 coconuts to the competition bench. This overlapping of all three tasks is only feasible for 60 minutes (denoted as 60||60||60). Unfortunately, after the 100th minute, Cindy has nothing else to do because she brought all 100 coconuts already. Therefore, between the 100–120 minutes, only Keith's and Gina's tasks can be overlapped (denoted as 20||20 again). In the very last 20 minutes of the entire operation, Gina is the only one that can bring back the final 20 coconuts to the competition bench, while Cindy and Keith sit idle (denoted as $\cdots + 20$).

As is evident from Table 11.2, the non-overlapped portion of Task 1 (20 minutes to transfer the very first 20 coconuts), as well as the non-overlapped portion of Task 3 (20 minutes to transfer the very last 20 coconuts) are fully exposed in the overall runtime. The two partially overlapped run times (20||20) and (20||20) allowed us to *coalesce* 20 minutes each, while the fully overlapped portion of all three tasks allowed us to coalesce 60 minutes from every task. We can do the math in the reverse order as follows:

- Fully serialized total run time would have been 300,

- (Task1||Task2) overlap coalesced 20 minutes, dropping the total run time to 280,

- (Task1||Task2||Task3) overlap coalesced 60 minutes from two tasks, exposing only 60 out of the 180 and saving 120; this dropped the total run time to 160,

- (Task2||Task3) overlap coalesced 20 minutes, dropping the total run time to 140.

What is described in Analogy 11.1 is called *pipelining*, in which a large task is broken down into concurrently executable smaller sub-tasks and executed in parallel. All modern microprocessors use this concept to save a significant amount of time by pipelining their fetch, decode, memory read, execute, and memory write operations. This pipelining concept is at the heart of how streaming works in a GPU.

11.2 MEMORY ALLOCATION

Assume that a computer has 8 GB of physical memory. You would have 8 GB of memory if you purchased your computer with an "8 GB" label on it. Does this mean that you cannot run any program that requires beyond this much memory? Let us ask the same question in a different context, traveling back 40–50 years in time: assume that an IBM mainframe computer is running 20 users, each with a demand for 1 MB. So, the total required memory is 20 MB. If you add the demand from the OS (say, 4 MB), a total of 24 MB memory is required. However, the computer only has 8 MB of physical memory. Does this mean that no more than a few users can use the mainframe? One obvious solution would be to increase the memory from 8 MB to 24 MB. However, this can be cost prohibitive. Engineers found a much more clever solution for this problem back then. Look at Analogy 11.2.

ANALOGY 11.2: *Physical versus Virtual Memory.*

Cocotown's biggest warehouse is *CocoStore*, which can store millions of coconuts, from hundreds of farmers. Because of such a high volume, they invented a highly sophisticated warehousing system, in which every farmer is only allowed to store their coconuts in large containers; 4096 (4 K) coconuts per container. CocoStore has sufficient physical warehouse storage for 20,480 (20 K) containers, with a total storage capacity of 80 M coconuts (i.e., 20 K×4 K = 80 M). These containers in the warehouse are numbered P0 through P20479 (P denoting *physical*).

If a farmer wants to store 40,960 coconuts, he or she brings in 10 containers and drops them off at CocoStore; CocoStore assigns this farmer a *virtual* warehouse number for each container, say, V10000 through V10009 (V denoting *virtual*). They do not want to assign a number that corresponds to a physical container number because they can move the containers at any point in time to make their storage management easier.

Because of these two different numbering schemes, the number they gave the farmer never changes; instead, the *mapping* from virtual to physical container numbers change. For example, if they placed V10000 in P20000 originally and decided to move the P20000 container into the container next to it (P20001), all they have to do is to change the mapping from V10000→P20000 to V10000→P20001; when the farmer comes to claim his container V10000, all they have to do is to look it up in the list; once they determine that the virtual container number V10000 corresponds to the physical container number P20001, they go grab it and give it to the farmer. The farmer is never aware of the physical container numbers; neither does he care, because all that is needed to access his container is the virtual number, V10000.

The conversion from virtual to physical addresses is called *address translation* and requires a translation list, which is kept in an easily accessible area in the company because it is accessed every single time a farmer needs a container. Because it is a tedious and time-consuming task, they hired a person, Taleb, who is responsible for maintaining and updating this list.

11.2.1 Physical versus Virtual Memory

The idea is that the amount of physical memory can be 8 MB; however, the *illusion* of a higher amount of *virtual* memory can be given to the users, even the OS by a clever trick: if we treated the required memory — by the users and even the OS — as a bunch of "pages," which are typically 4 KB in size, we could build the illusion that we have many more pages available than the actual physical memory has. If a user requires 1 MB, he or she requires 256 pages of storage. The OS needs 4 MB, i.e., 1024 pages. If we used the hard disk to store all of the pages that are much less frequently accessed, and say, we allocated 24 MB area on the disk as our virtual memory, we could store all of the pages there. Although a user needs 256 pages, the way a program works is that a page is used heavily before another page is needed; this *locality* of usage means that we can keep the pages that the user currently requires in physical memory, while the rest can stay on the disk. In this scheme, only the virtual addresses are given to any application that is running. They are not aware of the actual physical address their data is sitting in.

11.2.2   Physical to Virtual Address Translation

Although this expands the boundaries of the memory substantially, it comes at a price: a continuous *translation* from the virtual memory to the physical memory is needed every time a memory address is accessed. This is such an intensive process that most modern processors incorporate a *Translation Lookaside Buffer (TLB)* for this translation and provide direct hardware support. In our Analogy 11.2, Taleb is the person that is analogous to a TLB in a CPU. Giving the users only the virtual addresses allows the OS to keep only the frequently used pages in memory and keeps the infrequent pages in a slower storage area, such as disk, and *swaps* a page that is becoming high-demand. Although this swap takes a long time, common research shows that once a page is swapped in, it is used a lot before it has to be swapped back to disk; this quickly reduces the negative impact of the initial swap time. The malloc() function returns a virtual address, which can map to anywhere in physical memory, without our control.

11.2.3   Pinned Memory

A virtual address is only an "illusion" and cannot be used to access any data unless it is *translated* into a physical address; a physical address is an actual address that is used to access the data in DRAM main memory. So, when malloc() gives us a pointer, it is a virtual address, which must be translated into a physical address. This is not a big deal because the TLB in the CPU, coupled with the page table in the OS does this fairly easily. One interesting twist to this story is that it is actually possible for a program to request physical memory. Doing so completely eliminates any translation overhead because the hardware can directly access the data from the physical address that is readily available. In other words, you do not need to translate the physical addresses to physical addresses because the DRAM memory address is already there! What if there was a memory allocation function that allocated physical memory? Wouldn't it be much faster to access it? Absolutely! This type of memory is called *pinned memory.*

Something has to give though. What is the catch? Simple ... You would lose all of the flexibility that virtual addresses bring you. Furthermore, the amount of allocated physical memory would eat away from your available physical memory. To exemplify, if you have 8 GB physical memory and the OS is working with a 64 GB virtual address space, allocating 2 GB virtual memory would reduce your available virtual memory to 62 GB, while your available physical memory is still 8 GB. Alternatively, allocating 2 GB physical memory would reduce your available physical memory down to 6 GB; in this case, although you still have 64 GB virtual memory available, the OS has much less flexibility in mapping it because it is now working with only 6 GB physical memory. This would potentially require more frequent page swaps.

11.2.4 Allocating Pinned Memory with cudaMallocHost()

CUDA runtime library introduces the following API to allocate pinned memory:

```
void *p;
...
AllocErr=cudaMallocHost((void**)&p, IMAGESIZE);
   if (AllocErr == cudaErrorMemoryAllocation){
   ...
}
```

Here, the cudaMallocHost() function works very similar to malloc(), but allocates physical memory from the CPU memory. It returns a CUDA error code if it could not allocate the requested amount of physical memory. In case it is successful, it places the pointer to the physical memory address in the first argument (&p). Note that cudaMallocHost() is a CUDA API, which is designed with the sole purpose to facilitate fast CPU⟷GPU transfers over the PCIe bus. It might sound a little strange to the reader that CUDA (the GPU-side people) is being used to allocate CPU-side resources. Not really. This API communicates peacefully with the CPU side to allocate CPU-side resources for use strictly in GPU-side functionality (i.e., transferring data from the CPU to the GPU and vice versa).

11.3 FAST CPU⟷GPU DATA TRANSFERS

Once pinned memory is allocated using cudaMallocHost(), further CUDA APIs must be used that allow data transfers between CPU⟷GPU using pinned memory. Before we get into the details of this, let us look at the two different types of transfers that can happen between the CPU and the GPU: *synchronous* and *asynchronous*. Let us analyze these two and see their differences.

11.3.1 Synchronous Data Transfers

A *synchronous* transfer is shown below:

```
cudaMemcpy(GPUImg, TheImg, IMAGESIZE, cudaMemcpyHostToDevice);
```

The reader, at this point, might be analyzing every character in the code above trying to find any difference between this transfer and the ones we have seen in almost every code listing in the past GPU chapters. There isn't! What we saw over and over again in the previous code listings was all synchronous transfers. We just haven't seen anything else. In a synchronous transfer, the cudaMemcpy() API function is called and the code continues to the next line when cudaMemcpy() *finishes execution*. Until then, the execution "hangs" on that line and cannot continue. The bad news is that nothing below this line can be executed while the cudaMemcpy() is in progress, even if there is work that doesn't depend on this memory transfer.

11.3.2 Asynchronous Data Transfers

An *asynchronous* transfer is shown below:

```
cudaMemcpyAsync(GPUImg, TheImg, IMAGESIZE, cudaMemcpyHostToDevice, stream[0]);
```

Almost nothing is different, except that the API is named cudaMemcpyAsync() and it is associated with a *stream*. When the execution hits the line with cudaMemcpyAsync(), it doesn't actually *execute* this transfer; rather, it queues this transfer task in a stream (whose number is given in `stream[0]`) and immediately moves onto the next C command. While we knew that the data transfer was *complete* upon reaching the next line with a synchronous transfer, we cannot make any such assumption with an asynchronous transfer. The only assumption we can make is that "the transfer is scheduled and will be completed sooner or later." The good news is that we can move onto doing other useful work, such as some computations that do not depend on this data being transferred. While we are doing this extra work, the transfer continues in the background and will eventually be completed.

11.4   CUDA STREAMS

The idea behind a CUDA stream is that we queue up many operations into a stream and they are executed serially within that stream. We do this for multiple streams. Each stream executes everything we queued up serially; however, if there is any part between two streams that can be overlapped, the CUDA runtime engine automatically overlaps these operations. Therefore, although each stream is serially executed within itself, multiple streams can overlap execution.

11.4.1   CPU→GPU Transfer, Kernel Exec, GPU→CPU Transfer

To turn one of our older programs into its streamed version, let us examine the edge detection programs we reported in Table 11.1. The top half of Table 11.1 shows the results of edge detection, which requires three different tasks to be executed. Note that the kernel operations are considered one of those tasks and the total kernel execution time is lumped together into a single number; because all kernels use the same resources (GPU cores) and cannot be overlapped with each other, there is no reason to separate different kernel runtimes, such as SobelKernel(), ThresholdKernel(), etc. Rather, the CPU→GPU transfers, all of the lumped kernel operations, and the GPU→CPU transfers are considered as three operations with major prospects for execution overlapping. Let us take the Pascal GTX 1070 as an example. These three operation runtimes are

- CPU→GPU transfer takes 25 ms,

- Kernel execution takes 41 ms

  (including BWKernel(), GaussKernel(), SobelKernel(), and ThresholdKernel()),

- GPU→CPU transfer takes 24 ms.

- Total serialized (i.e., non-streamed) execution time (per Equation 11.1) is 90 ms.

These three operations can be partially overlapped. In the ideal — theoretical — case, if everything can be overlapped, we expect the runtime of the streamed version of the program to drop down to 41 ms. In other words, the CPU→GPU and GPU→CPU transfer times are coalesced completely. But, this is not really realistic. If we do something like what Team 2 did in Analogy 11.1, i.e., chop the big image into 10 chunks, each chunk will take 2.5 ms (i.e., 25/10) to transfer into the GPU and 2.4 ms (i.e., 24/10) to transfer out of the GPU. So, 2.5 ms of the CPU→GPU and 2.4 ms of the GPU→CPU time will be exposed; this means that our streamed runtime (from Equation 11.2) will be $(2.5 + (22.5||41||21.6) + 2.4)$, or $2.5 + 41 + 2.4 = 45.9$ ms.

11.4.2   Implementing Streaming in CUDA

Asynchronous transfers and pinned memory form the skeleton of CUDA streams. Whenever we want to stream our operations, we first allocate pinned memory. We, then, create multiple streams and break our large task into multiple chunks that can be executed independently. We assign each chunk into a different stream and let the CUDA streaming mechanism find the potential execution overlaps. This will automatically achieve the effect I described in Analogy 11.1.

Nvidia GPUs need only two different types of engines to implement streaming: (1) kernel execution engine, and (2) copy engine. Let's describe them in detail.

11.4.3   Copy Engine

The copy engine performs the tasks that were associated with Cindy and Gina in Analogy 11.1. Each stream has a copy engine. The purpose of this engine is to queue the incoming (CPU→GPU) and outgoing (GPU→CPU) operations and perform them when the PCIe bus is available. The queued transfer must be asynchronous; otherwise, the program would know when the transfer takes place and there would be no need for any sort of queuing. However, for asynchronous transfers like the one below:

```
// queue an asynchronous CPU --> GPU transfer in steam[0]
cudaMemcpyAsync(GPUImg, TheImg, IMAGESIZE, cudaMemcpyHostToDevice, stream[0]);
```

the copy engine of `stream[0]` immediately queues this host-to-device (CPU→GPU) memory copy request and the execution immediately moves onto the next line. The program has no idea about when the transfer will actually takes place. Although there are API functions to check the status of the queue, in general, the program does not need to be worried about it. It is guaranteed that whenever the PCIe bus is available for a transfer, this transfer will immediately initiate. Until then, it will sit in the queue, waiting for the right time to start the transfer. In the meantime, because the program has advanced to the next line, something on the CPU side can be executed, or something else can be queued up in `stream[0]`. Better yet, something else can be queued up *in a different stream*. This is how a streamed program keeps shoving the tasks into the queues of different streams and the streams execute concurrently.

11.4.4   Kernel Execution Engine

The kernel execution engine is Keith in Analogy 11.1. There is one per stream. Its job is to queue up the execution of different kernels for a given stream. Here is an example of this:

```
cudaStream_t stream[MAXSTREAMS];
...
BWKernel2S <<< dimGrid2DSm5, ThrPerBlk, 0, stream[i] >>> (...);
```

The parameters are not shown intentionally because they are not relevant. The important part is the inclusion of a stream ID, which is `stream[i]` in this case. This is a kernel call that is associated to a specific stream ID. If we omit the last two parameters as follows:

```
BWKernel2S <<< dimGrid2D, ThrPerBlk >>> (...);
```

this is exactly the same kernel launch, with the exception that it is assigned to the *default stream*. The default stream is a special stream in CUDA for unstreamed operations (pretty much everything we have done so far up until this chapter). The unstreamed kernel launches work exactly the way the streamed ones do; however, you cannot execute them in a streamed fashion and take advantage of the execution overlapping among different streams.

11.4.5 Concurrent Upstream and Downstream PCIe Transfers

There is a very important hardware concept that must be noted when it comes to the Copy engine. Not every Nvidia GPU is capable of transferring data from CPU→GPU and GPU→CPU *simultaneously*, although the PCIe bus is fully capable of simultaneous transfers. Some low-end home-grade Nvidia GPUs can only transfer data in one direction; while a CPU→GPU transfer is taking place, a GPU→CPU cannot take place simultaneously. In other words, although the runtimes of Task 1 and Task 2 can be overlapped (as well as Task 2 and Task 3), Task 1 and Task 3 cannot be overlapped. This feature of the GPU can be queried with the `GPUProp.deviceOverlap` as follows (which is reported by our program as follows):

```
cudaGetDeviceProperties(&GPUprop, 0);
// Shows whether the device can transfer in both directions simultaneously
 deviceOverlap = GPUprop.deviceOverlap;
 ...
printf("This device is %s capable of simultaneous CPU-to-GPU and GPU-to-CPU data
     transfers\n", deviceOverlap ? "" : "NOT");
```

So, what would happen to our expectations in this case if we are using a low end GPU, which cannot overlap incoming and outgoing data transfers (in which case GPUprop.deviceOverlap()=FALSE)? If the runtimes are 25, 40, and 30, assuming 10 chunks, 2.5 of the incoming transfer would be exposed, leaving 22.5 to overlap with the 40; only 22.5 of the 40 is overlapped, leaving another 17.5; this 17.5 can be overlapped with 30 partially; so, we could expect a runtime of: 2.5 + (22.5||22.5) + (17.5||30) = 2.5 + 22.5 + 30 = 55.

If the incoming and outgoing transfers could be performed concurrently (i.e., `GPUProp.deviceOverlap`= TRUE), then we would expect the streamed runtime to go down to 2.5 + (40||40||40) + 3 = 45.5. In this case, 2.5 is the non-overlapped portion of the incoming transfer and 3 is the non-overlapped version of the outgoing transfer.

Although this looks like the savings is minimal (i.e., 55 vs. 45.5), let us analyze a more dramatic case: If we are performing the horizontal flip, as shown on the bottom of Table 11.1, the CPU→GPU transfer time is 24, kernel execution time is 2, and the GPU→CPU transfer time is 24. The serialized version takes 24 + 2 + 24 = 50 ms, while the streamed version on a high-end GPU is expected to take 2.4+(21.6||2||21.6)+2.4=26.6 ms. However, on a low-end GPU that cannot simultaneously perform incoming and outgoing transfers, we expect the total runtime to be (24||2)+24=48 ms; in other words, only the kernel and one of the transfers can be overlapped and the incoming and outgoing transfers are serialized. It clearly doesn't sound impressive that the streaming allowed us to go from 50 to 48! However, on a high-end GPU, the savings is drastic, going from 50 to 26.6, which is almost a 2× improvement. It should be very clear to the reader at this point that the number of GPU cores is hardly the only parameter to look at when buying a GPU!

11.4.6 Creating CUDA Streams

For us to be able to use CUDA steams, we have to first create them. Creating CUDA streams is extremely easy using as shown below:

```
cudaStream_t stream[MAXSTREAMS];
...
if(NumberOfStreams != 0){
    for (i = 0; i < NumberOfStreams; i++) {
        chkCUDAErr(cudaStreamCreate(&stream[i]));
    }
}
```

11.4.7 Destroying CUDA Streams

Because CUDA streams are *resources*, much like memory, they have to be destroyed when we no longer need them. This is equally easy as shown below:

```
if (NumberOfStreams != 0) {
    for (i = 0; i < NumberOfStreams; i++) {
        chkCUDAErr(cudaStreamDestroy(stream[i]));
    }
}
```

11.4.8 Synchronizing CUDA Streams

Although we use the streams to queue things up and not worry about how CUDA finds the right time to execute them, there are some cases we intentionally want to block the execution of the program until a stream has completed all of the operations in its queue. This would be useful if we feel like there is nothing else to queue up in any stream before a given, say, stream 12, has completed. Using the code below, we can wait until a stream has completed its operations:

```
cudaStreamSynchronize(stream[i]);
```

In which case, we are telling the CUDA runtime: *do not do anything else until the stream with the ID contained in stream[i] has completed all of its copy and kernel operations.* This will completely execute everything that is in this stream's FIFO (First In First Out) buffer and the control will move onto the next line. If you wanted to synchronize every stream we created, we would use the following:

```
for (i = 0; i < NumberOfStreams; i++)
    cudaStreamSynchronize(stream[i]);
```

You would use something like this if you wanted to be sure that a large batch of things you queued up — in many streams — has all completed before moving onto another part of the program that will queue a bunch of other stream jobs.

11.5   IMGSTR.CU: STREAMING IMAGE PROCESSING

We will call the streaming version of edge detection and horizontal flip imGStr.cu. Here is the command line to execute it:

imGStr InputfileName OutputfileName [E/H] [NumberOfThreads] [NumberOfStreams]

This code is meant to perform streamed GPU image processing, hence its name imGStr. It can perform edge detection with the 'E' option and horizontal flip with the 'H' option. Most of the other options are identical to the previous code listings we have seen. The NumberOfStreams option determines how many streams will be launched inside the program; 0 denotes "synchronous execution without any streaming." Code 11.2 shows the part of the main() function that is responsible for creating and destroying the streams as well as reading the image. The command line option that determines how many streams will be launched is saved in variable NumberOfStreams.

11.5.1   Reading the Image into Pinned Memory

If it is 0, the image is read using the ReadBMPlin() function (shown in Code 11.1). Remember that this function returns a virtual address, which it gets from malloc(). If the number of streams is between 1 and 32, a different version of the same function, ReadBMPlinPINNED() is called, which returns a physical "pinned" memory address, which it receives from cudaMallocHost(), which is also shown in Code 11.1.

CODE 11.1:   imGStr.cu   ReadBMPlinPINNED() {...}

Reading the image into pinned memory.

```
uch *ReadBMPlin(char* fn)
{
   ...
   Img = (uch *)malloc(IMAGESIZE);
   ...
   return Img;
}

uch *ReadBMPlinPINNED(char* fn)
{
   static uch *Img;      void *p;
   cudaError_t AllocErr;
   ...
   AllocErr=cudaMallocHost((void**)&p, IMAGESIZE); // allocate pinned mem
   if (AllocErr == cudaErrorMemoryAllocation){
      Img=NULL;           return Img;
   }else                  Img=(uch *)p;
   ...
   return Img;
}
```

CODE 11.2: imGStr.cu   ... main() {...

First part of main() in imGStr.cu.

```
#define MAXSTREAMS    32
...
int main(int argc, char **argv)
{
   char          Operation = 'E';
   float         totalTime, Time12, Time23, Time34; // GPU code run times
   cudaError_t   cudaStatus;
   cudaEvent_t   time1, time2, time3, time4;
   int           deviceOverlap, SMcount;
   ul            ConstMem, GlobalMem;
   ui            NumberOfStreams=1,RowsPerStream;
   cudaStream_t  stream[MAXSTREAMS];
   ...
   if (NumberOfStreams > 32) {
      printf("Invalid NumberOfStreams (%u). Must be 0...32.\n", NumberOfStreams);
      ...
   }
   if (NumberOfStreams == 0) {
      TheImg=ReadBMPlin(InputFileName);   // Read the input image into a memory
         if(TheImg == NULL) { ...
   }else{
      TheImg=ReadBMPlinPINNED(InputFileName); // Read input img into a PINNED mem
         if(TheImg == NULL) { ...
   }
   ...
   cudaGetDeviceProperties(&GPUprop, 0);
   ...
   deviceOverlap=GPUprop.deviceOverlap; // bi-directional PCIe transfers?
   ConstMem    = (ul) GPUprop.totalConstMem;
   GlobalMem   = (ul) GPUprop.totalGlobalMem;
   // CREATE EVENTS
   cudaEventCreate(&time1); ...     cudaEventCreate(&time4);
   // CREATE STREAMS
   if(NumberOfStreams != 0){
      for (i = 0; i < NumberOfStreams; i++) {
         chkCUDAErr(cudaStreamCreate(&stream[i]));
      }
   }
    ...
   // Deallocate CPU, GPU memory
   cudaFree(GPUptr);
   // DESTROY EVENTS
   cudaEventDestroy(time1); ...     cudaEventDestroy(time4);
   // DESTROY STREAMS
   if (NumberOfStreams != 0) for(i=0; i<NumberOfStreams; i++)
       chkCUDAErr(cudaStreamDestroy(stream[i]));
   ...
}
```

CODE 11.3: imGStr.cu   ... main() {...
Second part of main() in imGStr.cu.

```
BlkPerRow = CEIL(IPH, ThrPerBlk);
RowsPerStream = ((NumberOfStreams == 0) ? IPV : CEIL(IPV, NumberOfStreams));
dim3 dimGrid2D(BlkPerRow, IPV);      dim3 dimGrid2DS(BlkPerRow, RowsPerStream);
dim3 dimGrid2DS1(BlkPerRow, 1);      dim3 dimGrid2DS2(Blk..., 2);
dim3 dimGrid2DS4(BlkPerRow, 4);      dim3 dimGrid2DS6(Blk..., 6);
dim3 dimGrid2DS10(BlkPerRow, 10);    dim3 dimGrid2DSm1(Blk..., Rows...eam-1);
dim3 dimGrid2DSm2(Blk..., Rows...eam-2); dim3 dimGrid2DSm3(Blk..., Rows...eam-3);
dim3 dimGrid2DSm4(Blk..., Rows...eam-4); dim3 dimGrid2DSm5(Blk..., Rows...eam-5);
dim3 dimGrid2DSm6(Blk..., Rows...eam-6); dim3 dimGrid2DSm10(Blk...,Row...eam-10);
uch *CPUstart, *GPUstart;
ui  StartByte, StartRow;
ui  RowsThisStream;
switch (NumberOfStreams) {
   case 0:  cudaMemcpy(GPUImg,TheImg,IMAGESIZE,cudaMemcpyHostToDevice);
         cudaEventRecord(time2, 0);   // Time stamp @ begin kernel exec
         switch(Operation){
            case 'E': BWKernel2S<<<dimGrid2D,ThrPerBlk>>>(GPUBWImg, ..., 0);
                  GaussKernel3S<<<dimGrid2D,ThrPerBlk>>>(GPUGaussImg, ..., 0);
                  SobelKernel2S<<<dimGrid2D,ThrPerBlk>>>(GPUGradient, ..., 0);
                  ThresholdKernel2S<<<dimGrid2D,ThrPerBlk>>>(GPUResult..., 0);
                  break;
            case 'H': Hflip3S<<<dimGrid2D,ThrPerBlk>>> (GPUResultImg, ..., 0);
                  break;
         }
         cudaEventRecord(time3, 0);   // Time stamp @end kernel exec
         cudaMemcpy(CopyImg,GPUResultImg,IMAGESIZE,cudaMemcpyDeviceToHost);
         break;
   case 1: cudaMemcpyAsync(GPUImg,TheImg,...,cudaMemcpyHostToDevice,stream[0]);
         cudaEventRecord(time2, 0);   // Time stamp @begin kernel exec
         switch(Operation) {
            case 'E': BWKernel2S<<<dimGrid2D,ThrPerBlk,0,stream[0]>>>(...);
                  GaussKernel3S<<<dimGrid2D,ThrPerBlk,0,stream[0]>>>(...);
                  SobelKernel2S<<<dimGrid2D,ThrPerBlk,0,stream[0]>>>(...);
                  ThresholdKernel2S<<<dimGrid2D,ThrPerBlk,0,stream[0]>>>(...);
                  break;
            case 'H': Hflip3S<<<dimGrid2D,ThrPerBlk,0,stream[0]>>>(...);
                  break;
         }
         cudaEventRecord(time3, 0);   // Time stamp @end kernel exec
         cudaMemcpyAsync(CopyImg,GPU...,cudaMemcpyDeviceToHost,stream[0]);
         break;
```

11.5.2 Synchronous versus Single Stream

The second part of the main() is shown in Code 11.3, where the variable NumberOfStreams is checked to determine whether the user is requesting synchronous operation (i.e., NumberOfStreams=0) or with a single stream (i.e., NumberOfStreams=1). In the former case, regular (virtual) memory is allocated for the image using the ReadBMPlin() function and the code is no different than its previous version, with the exception that a different version of the kernel is used; kernels with their name ending with 'S' (e.g., BWKernel2S()) are designed to work in a streamed environment. Even in the case of NumberOfStreams=0, the same kernel is launched. This will allow us to provide a fair performance comparison between the streamed and synchronous version of the same kernel.

As an example, the synchronous and single-streamed version of the horizontal flip operation are shown below. Here is the synchronous version:

```
case 0:  cudaMemcpy(GPUImg,TheImg,IMAGESIZE,cudaMemcpyHostToDevice);
         cudaEventRecord(time2, 0);   // Time stamp @ begin kernel exec
         switch(Operation){
            case 'E': BWKernel2S<<<dimGrid2D,ThrPerBlk>>>(GPUBWImg, ..., 0);
               ...
         }
             cudaMemcpy(CopyImg,GPUResultImg,IMAGESIZE,cudaMemcpyDeviceToHost);
```

and here is the single-stream version:

```
case 1: cudaMemcpyAsync(GPUImg,TheImg,...,cudaMemcpyHostToDevice,stream[0]);
        cudaEventRecord(time2, 0);   // Time stamp @ begin kernel exec
        switch(Operation){
           case 'E': BWKernel2S<<<dimGrid2D,ThrPerBlk,0,stream[0]>>>(...);
              ...
        }
            cudaMemcpyAsync(CopyImg,GPU...,cudaMemcpyDeviceToHost,stream[0]);
```

Note that when there is only a single stream — in the streaming version of the code — we can simply use stream[0]. In this case, if there is any improvement in performance, it will be due to using the pinned memory, not necessarily from the streaming effect. Because, obviously, the operation, again, serialized, because there is not other stream that can be used for execution overlapping. First, the CPU→GPU transfer finishes, then the kernel execution, then the GPU→CPU transfer. However, because of using pinned memory, the transfers over the PCIe bus are much faster, which speeds up the execution. This is the reason why the single-stream version of the program is analyzed thoroughly.

11.5.3   Multiple Streams

When there are two or more streams, we would like to divide the large image processing task into multiple tasks. The third part of the main() function is in Code 11.4. The idea behind sub-dividing a large image processing task into multiple streams is to have each stream process a set of *image rows*. So, processing each image row will be the unit computation. For example, if we are performing horizontal flipping on the astronaut.bmp file, which has 5376 rows) and using 4 streams, each stream is supposed to horizontally flip 5376/4 = 1344 rows. Because none of the rows depend on each other for processing, this task distribution is fairly straightforward. However, in the case of edge detection, things get a little bit more complicated. Performing edge detection on the astronaut.bmp image using 4 streams requires each stream to edge-detect 1344 rows, however, the top 2 rows of each chunk require the bottom 2 rows of the previous row for Gaussian filtering. Furthermore, the Sobel filter requires the bottom row of the previous chunk to compute its top row. A straightforward assignment of 1344 rows to each stream won't do the trick, because of this data dependence.

CODE 11.4: imGStr.cu ... main() {...

Third part of main() in imGStr.cu, where we queue up the operations in different streams' FIFOs.

```
default:  // Check to see if it is horizontal flip
    if (Operation == 'H') {
        for (i = 0; i < NumberOfStreams; i++) {
            StartRow = i*RowsPerStream;
            StartByte = StartRow*IPHB;
            CPUstart = TheImg + StartByte;
            GPUstart = GPUImg + StartByte;
            RowsThisStream = (i != (NumberOfStreams - 1)) ?
                    RowsPerStream : (IPV-(NumberOfStreams-1)*RowsPerStream);
            cudaMemcpyAsync(G...,RowsThisStream*IPHB,...ToDevice,stream[i]);
            cudaEventRecord(time2, 0);        // beginCPU --> GPU transfer
            Hflip3S<<<dimGrid2DS,ThrPerBlk,0,stream[i]>>>(...,StartRow);
            cudaEventRecord(time3, 0);     // end of kernel exec
            CPUstart = CopyImg + StartByte;
            GPUstart = GPUResultImg + StartByte;
            cudaMemcpyAsync(CPU...,RowsThisStream*IPHB,...ToHost,stream[i]);
        }
        break;
    }
    // If not horizontal flip, do edge detection (STREAMING)
    // Pre-process: 10 rows B&W,6 rows Gauss,4 rows Sobel,2 rows Thresh
    for (i = 0; i < (NumberOfStreams-1); i++) {
        StartRow = (i+1)*RowsPerStream-5;
        StartByte = StartRow*IPHB;
        CPUstart = TheImg + StartByte;
        GPUstart = GPUImg + StartByte;
        // Transfer 10 rows between chunk boundaries
        cudaMemcpy(GPUstart, CPUstart, 10*IPHB, cudaMemcpyHostToDevice);
        // Pre-process 10 rows for B&W
        BWKernel2S<<<dimGrid2DS10,ThrPerBlk>>>(...,StartRow);
        // Calc 6 rows of Gauss, starting @ the last 3 rows, except last
        StartRow += 2;
        GaussKernel3S<<<dimGrid2DS6,ThrPerBlk>>>(...,StartRow);
        // Calc 4 rows Sobel starting @last-1 row, except last
        StartRow ++;
        SobelKernel2S <<< dimGrid2DS4, ThrPerBlk >>> (...,StartRow);
        // Calc 2 rows Thr starting @last
    }
    cudaEventRecord(time2, 0);        // end of the pre-processing
    // Stream data from CPU --> GPU, B&W, Gaussian, Sobel
    for (i = 0; i < NumberOfStreams; i++) {
        if (i == 0) {
            RowsThisStream = RowsPerStream - 5;
        }else if (i == (NumberOfStreams - 1)) {
            RowsThisStream = IPV-(NumberOfStreams-1)*RowsPerStream-5;
        }else{
            RowsThisStream = RowsPerStream - 10;
        }
```

```
        StartRow = ((i == 0) ? 0 : i*RowsPerStream + 5);
        StartByte = StartRow*IPHB;
        CPUstart = TheImg + StartByte;
        GPUstart = GPUImg + StartByte;
        cudaMemcpyAsync(GPU...,RowsThisStream*IPHB,...ToDevice,stream[i]);
        if (i==0){
           BWKernel2S<<<dimGrid2DSm5,...,stream[i]>>>(...,StartRow);
           GaussKernel3S<<<dimGrid2DSm3,...,stream[i]>>>(...,StartRow);
           SobelKernel2S<<<dimGrid2DSm2,...,stream[i]>>>(...,StartRow);
        }else if (i == (NumberOfStreams - 1)) {
           BWKernel2S<<<dimGrid2DSm5,,,,,stream[i]>>>(...,StartRow);
           StartRow -= 2;
           GaussKernel3S<<<dimGrid2DSm3,,,,,stream[i]>>>(...,StartRow);
           StartRow--;
           SobelKernel2S<<<dimGrid2DSm2,,,,,stream[i]>>>(...,StartRow);
        }else {
           BWKernel2S<<<dimGrid2DSm10,,,,,stream[i]>>>(...,StartRow);
           StartRow -= 2;
           GaussKernel3S<<<dimGrid2DSm6,...,stream[i]>>> (...,StartRow);
           StartRow--;
           SobelKernel2S<<<dimGrid2DSm4,...,stream[i]>>>(...,StartRow);
        }
     }
     cudaEventRecord(time3, 0);      // end of BW+Gauss+Sobel
     // Streaming Threshold
     for (i = 0; i < NumberOfStreams; i++) {
        StartRow = i*RowsPerStream;
        ThresholdKernel2S<<<dimGrid2DS,,,,,stream[i]>>>(...);
     }
     // Stream data from GPU --> CPU
     for (i = 0; i < NumberOfStreams; i++) {
        StartRow = i*(RowsPerStream-5);
        StartByte = StartRow*IPHB;
        CPUstart = CopyImg + StartByte;
        GPUstart = GPUResultImg + StartByte;
        RowsThisStream = (i != (NumberOfStreams-1)) ? (RowsPerStream-5) :
              (IPV-(NumberOfStreams-1)*(RowsPerStream-5));
        cudaMemcpyAsync(CPU...,RowsThisStream*IPHB,...ToHost,stream[i]);
     }
  }
```

11.5.4 Data Dependence Across Multiple Streams

There are many ways to solve the problem with the data dependence across multiple streams. A good portion of Code 11.4 is dedicated to addressing this problem, which only exists in edge detection. Horizontal flip is immune to this problem because every row only requires that row; therefore, due to our unit computation being a *row*, this problem is naturally eliminated in streaming.

The solution used in Code 11.4 is fairly intuitive; assume that we are using 4 streams for edge detection and we would like to process 1344 rows using each stream. The following scenario emerges:

- For Stream 0, because the Gaussian filtered value of the *top two rows* will be 0.00, there is no data dependence.

- For Stream 0, the *bottom two rows* require the data from the *top two rows* of Stream 1.

- For Stream 1, the *top two rows* require the data from the *bottom two rows* of Stream 0.

- For Stream 1, the *bottom two rows* require the data from the *top two rows* of Stream 2.

- ...

It should be clear that the very first stream (i.e., Stream #0) and the very last stream (i.e., Stream #NumberOfStreams-1) only have one-sided dependence because their two edge rows will be set to 0.00 anyway. The "inside" streams have two-sided dependence because they must calculate their top and bottom two rows. In Code 11.4, a simple approach is taken: The crossing rows between any two streams is calculated in a non-streaming fashion at the very beginning; that way, when the streamed execution starts, the processed data is ready and the streams are only required to process the remaining rows.

11.5.4.1 Horizontal Flip: No Data Dependence

For the case of astronaut.bmp, the 5376 rows are processed as follows (assuming 4 streams and horizontal flip):

- Stream 0: Horizontal flip rows 0...1343

- Stream 1: Horizontal flip rows 1344...2687

- Stream 2: Horizontal flip rows 2688...4031

- Stream 3: Horizontal flip rows 4032...5375

In this case, each stream can transfer the data corresponding to the rows it is responsible for from CPU→GPU, process it, and transfer it back from GPU→CPU in a streaming fashion. The code looks like this:

```
if (Operation == 'H') {
   for (i = 0; i < NumberOfStreams; i++) {
      StartRow = i*RowsPerStream;
      StartByte = StartRow*IPHB;
      CPUstart = TheImg + StartByte;
      GPUstart = GPUImg + StartByte;
      RowsThisStream = (i != (NumberOfStreams - 1)) ?
        RowsPerStream : (IPV-(NumberOfStreams-1)*RowsPerStream);
      cudaMemcpyAsync(GPUstart,CPUstart,RowsThisStream*IPHB,
         cudaMemcpyHostToDevice,stream[i])
      cudaEventRecord(time2, 0);    // begin CPU --> GPU transfer
      Hflip3S<<<dimGrid2DS,ThrPerBlk,0,stream[i]>>>(GPUResultImg,
         GPUImg,IPH,IPV,IPHB,StartRow);
      cudaEventRecord(time3, 0);    // end of kernel exec
      CPUstart = CopyImg + StartByte;
      GPUstart = GPUResultImg + StartByte;
      cudaMemcpyAsync(CPUstart,GPUstart,RowsThisStream*IPHB,
         cudaMemcpyDeviceToHost,stream[i])
   }
}
```

11.5.4.2 Edge Detection: Data Dependence

Edge detection is more complicated because of data dependence among different chunks. The chunks are as follows:

- **Chunk0:** Rows 0...1343

  Rows 1339...1343 (last 5 rows) will be preprocessed synchronously

- **Chunk1:** Rows 1344...2687

  Rows 1344...1348 (first 5 rows) will be preprocessed synchronously

  Rows 2683...2687 (last 5 rows) will be preprocessed synchronously

- **Chunk2:** Rows 2688...4031

  Rows 2688...2692 (first 5 rows) will be preprocessed synchronously

  Rows 4027...4031 (last 5 rows) will be preprocessed synchronously

- **Chunk3:** Rows 4032...5375

  Rows 4032...4036 (first 5 rows) will be preprocessed synchronously

11.5.4.3 Preprocessing Overlapping Rows Synchronously

To handle the overlap, we will *preprocess* the last 5 rows of Chunk0 and the first 5 rows of Chunk1, which, in fact, corresponds to 10 consecutive rows. We will repeat this for the 10 overlapping rows between Chunk2 and Chunk3, and between Chunk3 and Chunk4. So, the following methodology is used (based on the same 4-stream example):

In a **synchronous** fashion, perform the following preprocessing:

- Transfer rows 1339–1348 from CPU→GPU,

- Run the BWKernel() for rows 1339–1348 (overlap is irrelevant for this kernel),

- Run the GaussKernel() for rows 1341–1346 (last 3 rows of Chunk0, first 3 of Chunk1),

- Run the SobelKernel() for rows 1342–1345 (last 2 rows of Chunk0, first 2 of Chunk1),

- Run the ThresholdKernel() for rows 1343–1344 (last row of Chunk0, first row Chunk1),

- Transfer rows 2683–2692 from CPU→GPU,

- Run the BWKernel() for rows 2683–2692,

- Run the GaussKernel() for rows 2685–2690,

- Run the SobelKernel() for rows 2686–2689,

- Run the ThresholdKernel() for rows 2687–2688,

- Transfer rows 4027–4036 from CPU→GPU,

- ...

The part of the code that performs the preprocessing is shown below:

```
// If not horizontal flip, do edge detection (STREAMING)
// Pre-process: 10 rows of B&W, 6 rows Gauss, 4 rows Sobel, 2 rows Threshold
for (i = 0; i < (NumberOfStreams-1); i++) {
  StartRow = (i+1)*RowsPerStream-5;
  StartByte = StartRow*IPHB;
  CPUstart = TheImg + StartByte;
  GPUstart = GPUImg + StartByte;
  // Transfer 10 rows between chunk boundaries
  cudaMemcpy(GPUstart,CPUstart,10*IPHB,cudaMemcpyHostToDevice);
  // Pre-process 10 rows for B&W
  BWKernel2S<<<dimGrid2DS10,ThrPerBlk>>>(GPUBWImg,GPUImg,IPH,IPV,IPHB,StartRow);
  // Calc 6 rows of Gauss, starting @ last 3 rows for every stream, except last
  StartRow += 2;
  GaussKernel3S<<<dimGrid2DS6,ThrPerBlk>>>(GPUGaussImg,GPUBWImg,
      IPH,IPV,StartRow);
  // Calc 4 rows of Sobel starting @last-1 row of every stream, except last
  StartRow ++;
  SobelKernel2S<<<dimGrid2DS4,ThrPerBlk>>>(GPUGradient,GPUTheta,GPUGaussImg,
      IPH,IPV,StartRow);
  // Calc 2 rows of Threshold starting @last row of every stream, except last
}
cudaEventRecord(time2, 0);      // the end of the pre-processing
```

11.5.4.4 Asynchronous Processing the Non-Overlapping Rows

Because the *preprocessing* part is synchronous, the execution will not reach this part until the preprocessing has been completed. Once it does, we know that the remaining rows can be queued up for asynchronous execution because any information they need from other chunks has already been processed and placed in GPU memory during synchronous execution. The remaining operations can, then, be queued up as follows:

In an **asynchronous** fashion, queue the following tasks:

- **Stream 0:** Transfer rows 0–1338 from CPU→GPU,

- Run the BWKernel() for rows 0–1338,

- Run the GaussKernel() for rows 0–1340,

- Run the SobelKernel() for rows 0–1341,

- Run the ThresholdKernel() for rows 0–1342,

- Transfer rows 0–1343 from GPU→CPU,

- **Stream 1:** Transfer rows 1349–2682 from CPU→GPU,

- Run the BWKernel() for rows 0–1338,

- Run the GaussKernel() for rows 0–1340,

- Run the SobelKernel() for rows 0–1341,

- Run the ThresholdKernel() for rows 0–1342,

- Transfer rows 1344–2687 from GPU→CPU,

- **Stream 2:** Transfer rows 2693–4026 from CPU→GPU,

- ...

- Transfer rows 2688–4031 from GPU→CPU,

- **Stream 3:** Transfer rows 4037–5375 from CPU→GPU,

- ...

- Transfer rows 4032–5375 from GPU→CPU,

Although it looks a little weird, the list above is fairly straightforward. As an example, between Chunk0 and Chunk1, the overlapping 10 rows were transferred and converted to B&W during the preprocessing, using BWKernel(). Because the next kernel (GaussKernel()) requires a 5x5 matrix, which accesses a 2 pixel-wide area on each pixel's surroundings, the 10 rows that are transferred can only be used to calculate the Gaussian version of the inner 6 rows because we lose 2 rows from the top and 2 from the bottom. Similarly, because Sobel filtering needs a one pixel-wide access on its surroundings, 6 rows of Gaussian can only be used to calculate 4 rows of Sobel. The threshold needs a one-pixel neighborhood access, which means that 2 rows of threshold can be calculated from 4 rows of Sobel (i.e., one row in one chunk, and another row in the adjacent chunk).

This means that when the preprocessing is complete, we have 10 rows of B&W, 6 rows of Gauss, 4 rows of Sobel, and 2 rows of threshold fully computed between any adjacent chunks. The goal of the asynchronous processing part is to compute the rest. In Code 11.4, in many places, the code uses a variable StartRow to either launch the appropriate kernels or perform data transfers. This is because all of the kernels are modified to accept a starting row number, as we will see very shortly. Additionally, the variable StartByte is used to determine the memory starting address that corresponds to this row (i.e., StartRow), as shown below:

```
StartRow = ((i == 0) ? 0 : i*RowsPerStream + 5);
StartByte = StartRow*IPHB;
CPUstart = TheImg + StartByte;
GPUstart = GPUImg + StartByte;
cudaMemcpyAsync(GPUstart, CPUstart, RowsThisStream * IPHB,
    cudaMemcpyHostToDevice, stream[i]);
if (i==0){
  BWKernel2S <<< dimGrid2DSm5, ThrPerBlk, 0, stream[i] >>> (GPUBWImg, GPUImg,
    IPH, IPV, IPHB, StartRow);
...
```

Note that, much like the synchronous version of the code, the GPU memory is allocated only once using a single bulk cudaMalloc() and different parts of this bulk memory area are used to store B&W, Gauss, etc. This is possible because the pinned memory is only a relevant concept when it comes to CPU memory allocation. When pinned memory is being used, the GPU memory allocation is exactly the same as before, using cudaMalloc(). For example, in the memory transfer cudaMemcpyAsync(GPUstart, CPUstart,...), GPUstart is no different than the GPU memory pointer we used in the synchronous version of the program; however, CPUstart is pointing to pinned memory.

CODE 11.5:   imGStr.cu   ... Hflip3S() {...}

Hflip3S() kernel, used in the streaming version of imGStr differs slightly from its synchronous version (Hflip3() in Code 9.4); it takes a row number as its argument, rather than the starting memory address of the entire image.

```
__global__
void Hflip3S(uch *ImgDst,uch *ImgSrc,ui Hpixels,ui Vpixels,ui RowBytes,ui StartRow)
{
    ui ThrPerBlk=blockDim.x;            ui MYbid=blockIdx.x;
    ui MYtid=threadIdx.x;               ui MYrow = StartRow + blockIdx.y;
    if (MYrow >= Vpixels) return;       // row out of range
    ui MYcol = MYbid*ThrPerBlk + MYtid;
    if (MYcol >= Hpixels) return;       // col out of range

    ui MYmirrorcol=Hpixels-1-MYcol;     ui MYoffset=MYrow*RowBytes;
    ui MYsrcIndex=MYoffset+3*MYcol;     ui MYdstIndex=MYoffset+3*MYmirrorcol;
    // swap pixels RGB @MYcol , @MYmirrorcol
    ImgDst[MYdstIndex] = ImgSrc[MYsrcIndex];
    ImgDst[MYdstIndex + 1] = ImgSrc[MYsrcIndex + 1];
    ImgDst[MYdstIndex + 2] = ImgSrc[MYsrcIndex + 2];
}
```

11.6 STREAMING HORIZONTAL FLIP KERNEL

Streaming horizontal flipper kernel, Hflip3S(), is shown in Code 11.5. This kernel is nearly identical to its synchronous version (Hflip3(), listed in Code 9.4) with one exception: because Hflip3() was not designed to flip "parts" of an image, it didn't have to accept a parameter to indicate which part of the image was being flipped. The redesigned kernel Hflip3S() accepts a new parameter, StartRow, which is the starting row of the image that needs to be flipped. An "ending row" does not need to be specified because the second dimension of the grid (blockIdx.y) automatically contains that information. So, the column number is calculated exactly the same way as Hflip3() as follows:

```
ui ThrPerBlk = blockDim.x;          ui MYbid = blockIdx.x;
ui MYtid = threadIdx.x;             ui MYrow = blockIdx.y;
if (MYrow >= Vpixels) return;       // row out of range
```

The older kernel Hflip3() computed the row number as follows:

```
ui MYrow = blockIdx.y;
```

Because as many kernels in the y dimension as the number of rows in the image were launched, an error checking was not necessary to see if the row number went out of range. It sufficed to use blockIdx.y to determine the row number that is being flipped. In the new kernel, Hflip3S(), the modification is simple by adding a starting row number as follows:

```
ui MYrow = StartRow + blockIdx.y;
if (MYcol >= Hpixels) return;       // col out of range
```

Without the error checking on the MYcol variable, we would have some issues when, for example, we have 5 streams; dividing 5376 by 5, we get CEIL(5376,5), which is 1076. So, the first chunk would be 1076 rows and the rest would be 1075 rows. Because we are launching the second dimension of every kernel with 1076 rows, the last kernel launch would go beyond the image and would error out. The error checking prevents that.

11.7 IMGSTR.CU: STREAMING EDGE DETECTION

The streaming version of the edge detection is implemented by making a similar modification to the four kernels that are used in edge detection as follows:

- BWKernel2S() (in Code 11.6) is the streaming version of BWKernel2() (in Code 9.11).

- GaussKernel3S() (in Code 11.7) is the streaming version of GaussKernel3() (in Code 10.10).

- SobelKernel2S() (in Code 11.8) is the streaming version of SobelKernel() (in Code 8.5).

- ThresholdKernel2S() (in Code 11.9) is the streaming version of ThresholdKernel() (in Code 8.6).

As you can see from the list above, not a lot of effort was put into using the latest and greatest versions of any kernel because the primary goal of this book, as always, is making the code *instructive*, before making it *efficient*. The goal of this chapter is to demonstrate streaming, rather than writing efficient kernels, so a simple version of each kernel is used. The reader can, certainly, experiment with more improved versions of each kernel.

CODE 11.6:   imGStr.cu   BWKernel2S() {...}

BWKernel2S() kernel processes blockIdx.y rows starting at StartRow.

```
void BWKernel2S(double *ImgBW, uch *ImgGPU, ui Hpixels, ui Vpixels, ui RowBytes,
    ui StartRow)
{
    ui ThrPerBlk = blockDim.x;          ui MYbid = blockIdx.x;
    ui MYtid = threadIdx.x;             ui R, G, B;
    ui MYrow = StartRow + blockIdx.y;   ui MYcol = MYbid*ThrPerBlk + MYtid;
    if (MYcol >= Hpixels) return;       if (MYrow >= Vpixels) return;
    ui MYsrcIndex = MYrow * RowBytes + 3 * MYcol;
    ui MYpixIndex = MYrow * Hpixels + MYcol;

    B = (ui)ImgGPU[MYsrcIndex];
    G = (ui)ImgGPU[MYsrcIndex + 1];
    R = (ui)ImgGPU[MYsrcIndex + 2];
    ImgBW[MYpixIndex] = (double)(R + G + B) * 0.333333;
}
```

CODE 11.7:   imGStr.cu   ... GaussKernel3S() {...}

GaussKernel3S() function processes blockIdx.y rows starting at StartRow.

```
__constant__
double GaussC[5][5] = { { 2,  4,  5,   4, 2 },
                        { 4,  9, 12,   9, 4 },
                        { 5, 12, 15,  12, 5 },
                        { 4,  9, 12,   9, 4 },
                        { 2,  4,  5,   4, 2 } };
__global__
void GaussKernel3S(double *ImgGauss,double *ImgBW,ui Hpixels,ui Vpixels,ui
    StartRow)
{
   ui ThrPerBlk = blockDim.x;
   ui MYbid = blockIdx.x;
   ui MYtid = threadIdx.x;
   int row, col, indx, i, j;
   double G;

   ui MYrow = StartRow+blockIdx.y;
   ui MYcol = MYbid*ThrPerBlk + MYtid;
   if (MYcol >= Hpixels) return;   // col out of range
   if (MYrow >= Vpixels) return;   // row out of range

   ui MYpixIndex = MYrow * Hpixels + MYcol;
   if ((MYrow<2) || (MYrow>Vpixels - 3) || (MYcol<2) || (MYcol>Hpixels - 3)) {
     ImgGauss[MYpixIndex] = 0.0;
     return;
   }else{
     G = 0.0;
     for (i = -2; i <= 2; i++) {
       for (j = -2; j <= 2; j++) {
          row = MYrow + i;
          col = MYcol + j;
          indx = row*Hpixels + col;
          G += (ImgBW[indx] * GaussC[i + 2][j + 2]); // use constant memory
       }
     }
     ImgGauss[MYpixIndex] = G * 0.0062893; // (1/159)=0.0062893
   }
}
```

CODE 11.8: imGStr.cu ... SobelKernel2S() {...}

SobelKernel2S() function processes `blockIdx.y` rows starting at `StartRow`.

```
__device__
double Gx[3][3] = {   { -1, 0, 1 },
                      { -2, 0, 2 },
                      { -1, 0, 1 } };
__device__
double Gy[3][3] = {   { -1, -2, -1 },
                      {  0,  0,  0 },
                      {  1,  2,  1 } };
__global__
void SobelKernel2S(double *ImgGrad, double *ImgTheta, double *ImgGauss, ui
   Hpixels, ui Vpixels, ui StartRow)
{
   ui ThrPerBlk = blockDim.x;
   ui MYbid = blockIdx.x;
   ui MYtid = threadIdx.x;
   int indx;
   double GX,GY;

   ui MYrow = StartRow + blockIdx.y;
   ui MYcol = MYbid*ThrPerBlk + MYtid;
   if (MYcol >= Hpixels) return;    // col out of range
   if (MYrow >= Vpixels) return;    // row out of range

   ui MYpixIndex = MYrow * Hpixels + MYcol;
   if ((MYrow<1) || (MYrow>Vpixels - 2) || (MYcol<1) || (MYcol>Hpixels - 2)){
      ImgGrad[MYpixIndex] = 0.0;
      ImgTheta[MYpixIndex] = 0.0;
      return;
   }else{
      indx=(MYrow-1)*Hpixels + MYcol-1;
      GX = (-ImgGauss[indx-1]+ImgGauss[indx+1]);
      GY = (-ImgGauss[indx-1]-2*ImgGauss[indx]-ImgGauss[indx+1]);

      indx+=Hpixels;
      GX += (-2*ImgGauss[indx-1]+2*ImgGauss[indx+1]);

      indx+=Hpixels;
      GX += (-ImgGauss[indx-1]+ImgGauss[indx+1]);
      GY += (ImgGauss[indx-1]+2*ImgGauss[indx]+ImgGauss[indx+1]);
      ImgGrad[MYpixIndex] = sqrt(GX*GX + GY*GY);
      ImgTheta[MYpixIndex] = atan(GX / GY)*57.2957795; // 180.0/PI = 57.2957795;
   }
}
```

CODE 11.9:   imGStr.cu   ThresholdKernel2S() {...}

ThresholdKernel2S() function processes `blockIdx.y` rows starting at `StartRow`.

```
__global__
void ThresholdKernel2S(uch *ImgResult, double *ImgGrad, double *ImgTheta, ui
   Hpixels, ui Vpixels, ui RowBytes, ui ThreshLo, ui ThreshHi, ui StartRow)
{
   ui ThrPerBlk = blockDim.x;                ui MYbid = blockIdx.x;
   ui MYtid = threadIdx.x;
   ui MYrow = StartRow + blockIdx.y;         if(MYrow >= Vpixels) return;
   ui MYcol = MYbid*ThrPerBlk + MYtid;       if(MYcol >= Hpixels) return;
   unsigned char PIXVAL;                     double L, H, G, T;

   ui ResultIndx = MYrow * RowBytes + 3 * MYcol;
   ui MYpixIndex = MYrow * Hpixels + MYcol;
   if ((MYrow<1) || (MYrow>Vpixels - 2) || (MYcol<1) || (MYcol>Hpixels - 2)){
      ImgResult[ResultIndx]=ImgResult[ResultIndx+1]=ImgResult[ResultIndx+2]=NOEDGE;
      return;
   }else{
      L=(double)ThreshLo;      H=(double)ThreshHi;        G=ImgGrad[MYpixIndex];
      PIXVAL = NOEDGE;
      if (G <= L){                      // no edge
         PIXVAL = NOEDGE;
      }else if (G >= H){                // edge
         PIXVAL = EDGE;
      }else{
         T = ImgTheta[MYpixIndex];
         if ((T<-67.5) || (T>67.5)){
            // Look at left and right: [row][col-1] and [row][col+1]
            PIXVAL = ((ImgGrad[MYpixIndex - 1]>H) || (ImgGrad[MYpixIndex + 1]>H)) ?
               EDGE : NOEDGE;
         }
         else if ((T >= -22.5) && (T <= 22.5)){
            // Look at top and bottom: [row-1][col] and [row+1][col]
            PIXVAL = ((ImgGrad[MYpixIndex - Hpixels]>H) || (ImgGrad[MYpixIndex +
               Hpixels]>H)) ? EDGE : NOEDGE;
         }
         else if ((T>22.5) && (T <= 67.5)){
            // Look at upper right, lower left: [row-1][col+1] and [row+1][col-1]
            PIXVAL = ((ImgGrad[MYpixIndex - Hpixels + 1]>H) || (ImgGrad[MYpixIndex +
               Hpixels - 1]>H)) ? EDGE : NOEDGE;
         }
         else if ((T >= -67.5) && (T<-22.5)){
            // Look at upper left, lower right: [row-1][col-1] and [row+1][col+1]
            PIXVAL = ((ImgGrad[MYpixIndex - Hpixels - 1]>H) || (ImgGrad[MYpixIndex +
               Hpixels + 1]>H)) ? EDGE : NOEDGE;
         }
      }
      ImgResult[ResultIndx]=ImgResult[ResultIndx+1]=ImgResult[ResultIndx+2]=PIXVAL;
   }
}
```

TABLE 11.3 Streaming performance results (in **ms**) for imGStr, on the astronaut.bmp image.

Operation	# Streams	Titan Z	K80	GTX 1070	Titan X
Edge	SYNCH	143 (37+64+42)	109 (46+45+18)	90 (25+41+24)	109 (32+26+51)
	1	103	68	59	70
	2	92	66	53	60
	3	81	69	51	56
	4	79	56	50	54
	5	75	60	50	54
	6	73	66	50	53
	7	88	55	50	53
	8	82	65	47	51
Hflip	SYNCH	86 (39+4+43)	69 (48+4+17)	50 (24+2+24)	67 (32+1+34)
	1	44	30	22	42
	2	44	23	19	33
	3	44	22	19	30
	4	44	24	18	28
	5	44	23	18	28
	6	44	21	18	27
	7	44	23	18	27
	8	44	21	17	26

Synchronous results are repeated from Table 11.1, where the three numbers (e.g., 37+64+42) denote the CPU→GPU transfer time, kernel execution time, and GPU→CPU transfer time, respectively.

11.8 PERFORMANCE COMPARISON: IMGSTR.CU

Table 11.3 provides runtime results for imGStr; synchronous (i.e., NumberOfStreams=0) versus asynchronous (1–8 streams). The bottom half lists horizontal flip results (using the Hflip3S() kernel in Code 11.5), while the top half of the table tabulates the results for edge detection (using the kernels in Code 11.6, 11.7, 11.8, and 11.9). As you may have guessed, Table 11.3 repeats the results from Table 11.1, which only shows synchronous versions of the same edge detection and horizontal flip operations.

11.8.1 Synchronous versus Asynchronous Results

Let us take GTX 1070 as an example to evaluate these results. The synchronous version of the program took 90, where the CPU→GPU transfer time was 25 ms, kernel execution was 41 ms, and the GPU→CPU transfer time was 24 ms. From our detailed discussion in Section 11.4.1, we calculated a potential runtime of 45.9 ms when we used 10 streams. With 8 streams, we should possibly expect a little bit worse performance. What we see in Table 11.3 is almost the book definition of what we described for 8 streams.

11.8.2   Randomness in the Results

Although the perfect streaming example we saw in the previous subsection (for 8 streams) might give us the false sense of "perfectness," it is just by luck that the numbers ended up being so perfect. At runtime, there will be some degree of randomness built into the result. Because when we queue up the operations into 8 streams in a FIFO (First In First Out) fashion, we have no idea which stream will execute its individual operations *in comparison to the other streams*. The only thing we are sure about is that each stream will execute its own queued operations in the order they were queued.

11.8.3   Optimum Queuing

Although it sounds like the performance of our code depends completely on random factors, this is not true. There is a way to queue up the operations in the streams to achieve optimum performance. We will demonstrate in the next section that the way the operations were queued in Code 11.4 is not optimum. To understand what can go wrong, let us quickly look at a hypothetical scenario for the GTX 1070; assume that 41 ms kernel execution time (in Table 11.3) is broken down to four kernels as follows:

- BWKernel()=4 ms (for simplicity, assume ≈1 ms for each chunk, for four streams),

- GaussKernel()=11 ms (≈3 ms for four streams),

- SobelKernel()=23 ms (≈5 ms for four streams),

- ThresholdKernel()=3 ms (≈1 ms for four streams).

These numbers should be fairly accurate because they are based on the results we have observed in the previous chapters for this GPU. Based on our Code 11.4, each stream's FIFO is filled with the operations shown in Section 11.5.4.4, which is summarized below:

- **Stream 0:** CPU→GPU [6 ms] ⟹ BWKernel() [1 ms] ⟹ GaussKernel() [3 ms]

- ⟹SobelKernel() [5 ms] ⟹ ThresholdKernel() [1 ms] ⟹ GPU→CPU [6 ms]

- **Stream 1:** CPU→GPU [6 ms] ⟹ BWKernel() [1 ms] ⟹ GaussKernel() [3 ms]

- ⟹SobelKernel() [5 ms] ⟹ ThresholdKernel() [1 ms] ⟹ GPU→CPU [6 ms]

- **Stream 2:** CPU→GPU [6 ms] ⟹ BWKernel() [1 ms] ⟹ GaussKernel() [3 ms]

- ⟹SobelKernel() [5 ms] ⟹ ThresholdKernel() [1 ms] ⟹ GPU→CPU [6 ms]

- **Stream 3:** CPU→GPU [6 ms] ⟹ BWKernel() [1 ms] ⟹ GaussKernel() [3 ms]

- ⟹SobelKernel() [5 ms] ⟹ ThresholdKernel() [1 ms] ⟹ GPU→CPU [6 ms]

Minor details, such as the fact that the first and the last streams will have slightly different runtimes, are ignored for simplicity. We are assuming that the FIFOs of all four streams are filled with the above operations and the question, now, is which ones will get executed in which order? To repeat one more time, we are *guaranteed* that the order of operations in Stream 0 are strictly what is shown above. For example, the ThresholdKernel() in Stream 0 cannot execute before the data has been transferred (CPU→GPU), and BWKernel(), GaussKernel(), and SobelKernel() have all been executed after this transfer. It is only then that the ThresholdKernel() can execute, followed by the GPU→CPU transfer. This strict ordering is true for all of the streams within themselves; however, there is absolutely no guarantee that SobelKernel() of Stream 1 will run anytime we can guess *in relationship to Stream 0*, or any other stream for that matter. Let us now go through a few runtime scenarios. For readability, the names of the kernels will be shortened (e.g., **BW** instead of BWKernel()); additionally, the stream names are shortened (e.g., **S0** to denote Stream 0).

11.8.4 Best Case Streaming Results

First, the good news: At runtime, here is a perfect scenario for us:

- **Copy Engine:** CPU→GPU [S0] [6 ms] $\Longrightarrow$ CPU→GPU [S1] [6 ms]

  $\Longrightarrow$ CPU→GPU [S2] [6 ms] $\Longrightarrow$ CPU→GPU [S3] [6 ms]

- **Kernel Engine:** BW [S0] [4 ms] $\Longrightarrow$ Gauss [S0] [11 ms] $\Longrightarrow$ Sobel [S0] [23 ms]

  $\Longrightarrow$ Thresh [S0] [3 ms] $\Longrightarrow$ BW [S1] [4 ms] $\Longrightarrow$ Gauss [S1] [11 ms]

  $\Longrightarrow$ Sobel [S1] [23 ms] $\Longrightarrow$ Thresh [S1] [3 ms] $\Longrightarrow$ BW [S2] [4 ms]

  $\Longrightarrow$ Gauss [S2] [11 ms] $\Longrightarrow$ Sobel [S2] [23 ms] $\Longrightarrow$ Thresh [S2] [3 ms]

  $\Longrightarrow$ BW [S3] [4 ms] $\Longrightarrow$ Gauss [S3] [11 ms] $\Longrightarrow$ Sobel [S3] [23 ms]

  $\Longrightarrow$ Thresh [S3] [3 ms]

- **Copy Engine:** GPU→CPU [S0] [6 ms] $\Longrightarrow$ GPU→CPU [S1] [6 ms]

  $\Longrightarrow$ GPU→CPU [S2] [6 ms] $\Longrightarrow$ GPU→CPU [S3] [6 ms]

This scenario is good news for us because the following sequence of events would unfold at runtime (time values are all given in ms):

- **Time 0–6 (Copy Engine):** The GPU cores, therefore the kernel engine, cannot do anything until the data for the first chunk has been copied into the GPU memory. Assume that the copy engine carries out the CPU→GPU [S0] data transfer [6 ms].

- **Time 6–7 (Kernel Engine):** Up until the CPU→GPU transfer [S0] was complete, kernel engine was blocked, unable to find anything to do. At Time=6, it can proceed; because all of the other streams are waiting for their data to be transferred, the only available option to the kernel engine is to launch the **BW** [S0] kernel [1 ms].

- **Time 6–12 (Copy Engine):** Once the copy engine is done with the first CPU→GPU transfer [S0], it is now looking for an opportunity to initiate another transfer. At this point, the copy engine has 3 options; initiating a CPU→GPU copy on **S1**, **S2**, or **S3**. We will assume that it initiates S1 next [6 ms].

- **Time 7–10 (Kernel Engine):** Once the kernel engine is done with **BW** [S0], it is now looking at its options. It can either launch the execution of **Gauss** [S0], or launch **BW** [S1]. At this point, because S2 and S3 haven't even transferred their data in the CPU→GPU direction, they cannot provide kernel alternatives to the kernel engine. Let us assume that the kernel engine picks **Gauss** [S0] [3 ms].

- **Time 10–16 (Kernel Engine):** Assume that the kernel engine picks **Sobel** [S0] [5 ms] and **Thresh** [S0] [1 ms] next.

- **Time 12–18 (Copy Engine):** At Time=12, the copy engine has two options; it can initiate a CPU→GPU transfer for either **S2** or **S3**. Let us assume that it initiates CPU→GPU [S2] [6 ms].

- **Time 16–22 (Copy Engine):** At Time=16, because S0 has finished all of its kernels, it is now ready for its GPU→CPU transfer [S0] [6 ms], which the copy engine can initiate.

- **Time 18–24 (Copy Engine):** If this is a GPU that supports bi-directional PCI transfers simultaneously (i.e., `GPUProp.deviceOverlap`=TRUE, as detailed in Section 11.4.5), it can also initiate the CPU→GPU transfer [S3], which is the final CPU→GPU transfer. Assuming `GPUProp.deviceOverlap`=TRUE, all CPU→GPU transfers are done at Time=24, as well as the CPU→GPU transfer [S0]. Note that because the kernel execution is still continuing non-stop, the CPU→GPU transfer time [S0] is 100% coalesced. Furthermore, the GPU→CPU transfer times for S1, S2, and S3 are also 100% coalesced, which totals 18 ms. Only the CPU→GPU transfer time [S0] is completely exposed (6 ms).

- **Time 16–26 (Kernel Engine):** Because we are studying the perfect scenario, let us assume that the kernel engine launches and finishes **BW** [S1] [1 ms], **Gauss** [S1] [3 ms], **Sobel** [S1] [5 ms], and **Thresh** [S1] [1 ms] next. We are at Time=26, at which point the kernel execution for S1 is complete.

- **Time 24–32 (Copy Engine):** Notice that the copy engine has to sit idle between Time 24–26, because no stream has data to transfer. At Time=26, because the kernel execution of S1 is complete, the copy engine can initiate its GPU→CPU transfer [S1] [6 ms]. Because kernel execution will continue during this interval, this transfer time is also 100% coalesced.

- **Time 26–36 (Kernel Engine):** Let us assume the best case scenario again: **BW** [S2] [1 ms], **Gauss** [S2] [3 ms], **Sobel** [S2] [5 ms], and **Thresh** [S2] [1 ms] next. We are at Time=36, at which point the kernel execution for S2 is complete.

- **Time 36–42 (Copy Engine):** Copy engine sits idle between Time 32–36 and initiates GPU→CPU transfer [S2] [6 ms], which will be 100% coalesced again.

- **Time 36–46 (Kernel Engine):** The next best case scenario is **BW** [S3] [1 ms], $\cdots$, **Thresh** [S2] [1 ms] next. We are at Time=46, at which point the kernel execution for S3 is complete. Indeed, all four streams have completed their kernel execution at this point. The only remaining work is the GPU→CPU transfer [S3].

- **Time 46–52 (Copy Engine):** Copy engine sits idle between Time 42–46 and initiates GPU→CPU transfer [S3] [6 ms], which will be 100% exposed, because there is no concurrent kernel execution that can be launched.

A total of 52 ms will be required for four streams, which is the sum of all kernel executions (40 ms based on our approximations), and the exposed CPU→GPU transfer (6 ms), as well as the exposed CPU→GPU transfer (6 ms).

11.8.5 Worst Case Streaming Results

What we saw in Section 11.8.4 was an overly optimistic scenario, where everything went right. What if it doesn't? Now, the bad news: Assume the following scheduling at runtime (and assume that `GPUProp.deviceOverlap`=TRUE):

- **Time 0–24 (Copy Engine):** We know that the copy engine will complete all incoming transfers until Time=24.

- **Time 0–9 (Kernel Engine):** Kernel engine launches and finishes BW [S0] [1 ms], Gauss [S0] [3 ms], Sobel [S0] [5 ms].

- **Time 9–18 (Kernel Engine):** Kernel engine launches and finishes BW [S1] [1 ms], Gauss [S1] [3 ms], Sobel [S1] [5ms].

- **Time 18–27 (Kernel Engine):** Kernel engine launches and finishes BW [S2] [1 ms], Gauss [S2] [3 ms], Sobel [S2] [5 ms].

- **Time 27–36 (Kernel Engine):** Kernel engine launches and finishes BW [S3] [1 ms], Gauss [S3] [3 ms], Sobel [S3] [5 ms].

- **Time 36–37 (Kernel Engine):** Kernel engine executes Thresh [S0] [1 ms].

- **Time 37–43 (Copy Engine):** Until Time=37, no stream had any GPU→CPU transfer available. At Time=37, S0 is the first stream that has it available, so the copy engine can schedule it. A large portion of the transfer time is exposed.

- **Time 37–40 (Kernel Engine):** Kernel engine executes Thresh [S1] [1 ms], Thresh [S2] [1 ms], and Thresh [S3] [1 ms]. Because of the concurrent GPU→CPU transfers, these execution times are 100% coalesced.

- **Time 43–49 (Copy Engine):** After the previous transfer, now the copy engine can initiate the next GPU→CPU transfer [S1] [6 ms]. This transfer time is 100% exposed, along with the next two transfer times.

- **Time 49–55 (Copy Engine):** The next GPU→CPU transfer [S2] [6 ms].

- **Time 55–61 (Copy Engine):** The next GPU→CPU transfer [S3] [6 ms].

In this worst case scenario, the runtime is 61 ms, which is not too much worse than the best case result we analyzed, which was 52 ms. However, this 15% performance penalty could have been prevented by a carefull scheduling of events.

11.9 NVIDIA VISUAL PROFILER: NVVP

One question that comes to mind is whether there is a tool that we can use to *visualize* the timeline (i.e., schedule of events) that we tabulated in Section 11.8.3. After all, it is a simple timing chart that will make it clear how the events happen one after the other. Yes, there is such a tool. It is called the *Nvidia Visual Profiler* (nvvp).

11.9.1 Installing nvvp and nvprof

There is no specific installation needed for the nvvp. It automatically gets installed when you install the CUDA toolkit, both in Windows and Unix. It is also available in command line version (nvprof), which can collect data and import it to nvvp. The goal of either program is to run a given CUDA code and continuously log the CPU←→GPU data transfers as well as kernel launches. Once the execution completes, the results are viewed (i.e., visualized) in a graphics format. There is definitely a computation l overhead that the profiler introduces, while performing its profiling; however, this is taken into account and shown to the user, which will allow the user to estimate the results without the overhead.

FIGURE 11.1 Nvidia visual profiler.

11.9.2 Using nvvp

To run the profiler, follow these steps:

- In Windows, double-click the executable for nvvp. In Unix, type: nvvp &

- In either case, you will have the screen in Figure 11.1.

- Once in the Profiler, go to File → New Session. Click Next.

- In the "Create New Session: Executable Properties" window, type the name of the CUDA executable you want to profile under "File" and fill in the command line arguments if the program needs it (for example, our imGStr program will need the arguments we discussed in Section 11.5).

- In the "Profiling Options" window, you can change the profiling options. Usually, the default settings suffice. You can learn more about each *settings* options through the following link [20].

- The Nvidia Visual Profiler main window is divided into different sections. The timeline section represents the timing results of your program. The upper part is the CPU timeline while the lower part is the GPU timeline. You can learn more about the timeline options, through the following link [21].

- Click Finish to close the "Profiling Options" window. The profiler will run the code and will save the results for visualization. After this step, you are ready to view the timing of events.

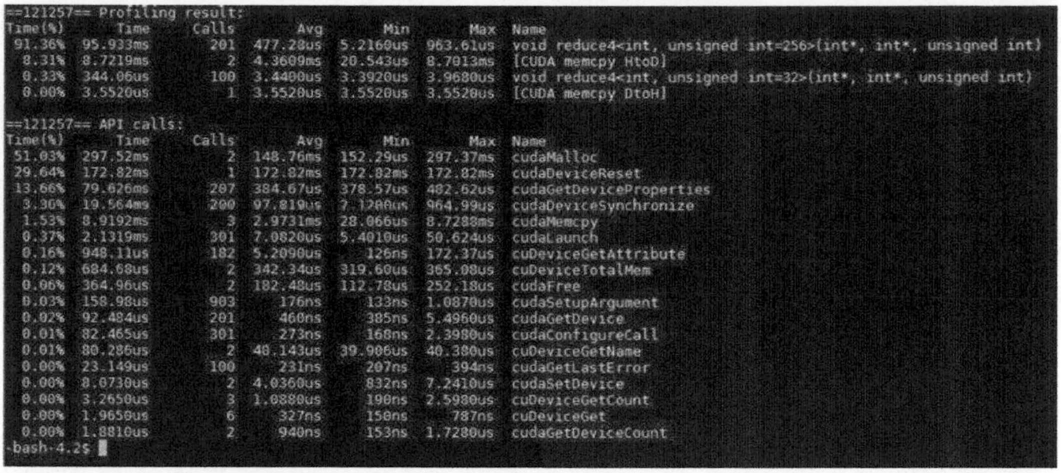

```
==121257== Profiling result:
Time(%)      Time     Calls       Avg       Min       Max  Name
 91.36%   95.933ms       201   477.28us   5.2160us  963.61us  void reduce4<int, unsigned int=256>(int*, int*, unsigned int)
  8.31%   8.7219ms         2   4.3609ms  20.543us  8.7013ms  [CUDA memcpy HtoD]
  0.33%   344.06us       100   3.4400us  3.3920us  3.9680us  void reduce4<int, unsigned int=32>(int*, int*, unsigned int)
  0.00%   3.5520us         1   3.5520us  3.5520us  3.5520us  [CUDA memcpy DtoH]

==121257== API calls:
Time(%)      Time     Calls       Avg       Min       Max  Name
 51.03%   297.52ms         2   148.76ms  152.29us  297.37us  cudaMalloc
 29.64%   172.82ms         1   172.82ms  172.82ms  172.82ms  cudaDeviceReset
 13.66%   79.626ms       207   384.67us  378.57us  482.62us  cudaGetDeviceProperties
  3.30%   19.564ms       200   97.819us  7.1700us  964.99us  cudaDeviceSynchronize
  1.53%   8.9192ms         3   2.9731ms  28.066us  8.7288ms  cudaMemcpy
  0.37%   2.1319ms       301   7.0820us  5.4010us  50.624us  cudaLaunch
  0.16%   948.11us       182   5.2090us     126ns  172.37us  cuDeviceGetAttribute
  0.12%   684.68us         2   342.34us  319.60us  365.08us  cuDeviceTotalMem
  0.06%   364.96us         2   182.48us  112.78us  252.18us  cudaFree
  0.03%   158.98us       903      176ns     133ns  1.0870us  cudaSetupArgument
  0.02%   92.484us       201      460ns     385ns  5.4960us  cudaGetDevice
  0.01%   82.465us       301      273ns     168ns  2.3980us  cudaConfigureCall
  0.01%   80.286us         2   40.143us  39.906us  40.380us  cuDeviceGetName
  0.00%   23.149us       100      231ns     207ns     394ns  cudaGetLastError
  0.00%   8.0730us         2   4.0360us     832ns  7.2410us  cudaSetDevice
  0.00%   3.2650us         3   1.0880us     190ns  2.5980us  cuDeviceGetCount
  0.00%   1.9650us         6      327ns     150ns     787ns  cuDeviceGet
  0.00%   1.8810us         2      940ns     153ns  1.7280us  cudaGetDeviceCount
-bash-4.2$
```

FIGURE 11.2 Nvidia profiler, command line version.

11.9.3   Using nvprof

You can use the command line profiler, nvprof, to collect the data, and then import the result into the Visual Profiler. Here is an example command line to run it:

$ nvprof -o exportedResults.nvprof my_cuda_program my_args

This command will run your program and save the profiling results to exported Results.nvprof. You may want to add the `--analysis-metrics` flag to nvprof as well, which will provide extra info for analysis in the profiler. Note: it will cause your program to be run several times, which may take a while. If you run nvprof without saving the output to a file, you will get an output like Figure 11.2. To see the first portion as a timeline, including some more details about resources and throughput, try the `--print-gpu-trace` flag. Using this method, you may not always need the GUI. See the following link for reference [19].

11.9.4   imGStr Synchronous and Single-Stream Results

We are now ready to run out imGStr and see the results in the profiler. Figure 11.3 depicts the synchronous (top) and single-stream (bottom) results of imGStr, executed on a K80 GPU. The only difference between the two figures is the fact that the execution time drops from 109 ms (synchronous) down to 70 ms (single stream), as previously shown in Table 11.3. What nvvp shows us is what we expected: the six different pieces of the execution (CPU→GPU transfer ⇒ **BW** ⇒ · · · ⇒ **Thresh** ⇒ GPU→CPU transfer) are executed one after the other, in the same exact order, in both cases.

The reason for the 39 ms time savings is due to the fact that the CPU⟷GPU transfers are substantially faster when pinned memory is used. Per Table 11.3 we know that the kernel execution is 26 ms and will not change during asynchronous execution. This means that the transfers took a total of 70−26=44 ms, instead of the previous total of 32+51=83 ms. In other words, the speed at which we transfer the data between CPU⟷GPU nearly doubled. A similar trend is evident in almost all GPUs. This reduction in transfer time is the fact that the OS doesn't have to be bothered with *page swaps* during the transfers because the reads from the CPU memory are referencing physical memory, instead of virtual memory.

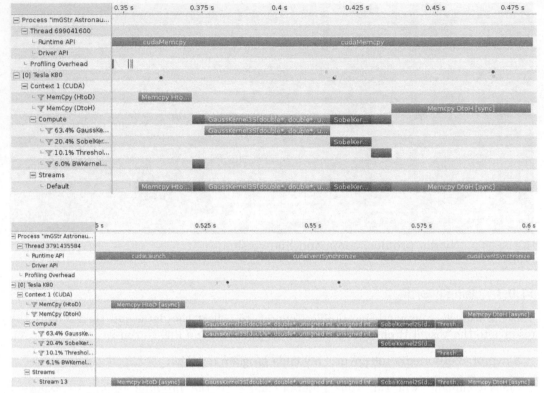

FIGURE 11.3 Nvidia NVVP results with no streaming and using a single stream, on the K80 GPU.

11.9.5 imGStr 2- and 4-Stream Results

Figure 11.4 depicts the 2-stream and 4-stream results of imGStr, on a K80 GPU. When more than one stream is involved, the copy engine and kernel engine have some flexibility in choosing from the available operations. Unfortunately, what we witness in Figure 11.4 is that something close to what we described as "the worst case scenario," described in Section 11.8.5 happens. While the CPU→GPU transfers are 100% coalesced, excepting the first one, the GPU→CPU transfers are only partially coalesced.

It looks like the final ThresholdKernel() batch is preventing the GPU→CPU transfers to be initiated. Instead of providing what has to be done to remedy the situation, I will leave it up to the reader to try a few different things. Here is a list of ideas:

- Look at the order in which we launched the kernels. Would it make a difference had we placed the ThresholdKernel() launches right after the SobelKernel() launches?

- Would it be a better idea to merge all four kernels into a single piece and launch just a single kernel?

- Can we use cudaStreamSynchronize() to somehow manually control when a stream starts and when it ends?

- Can we launch different versions of the same kernel to achieve better occupancy in the SMs?

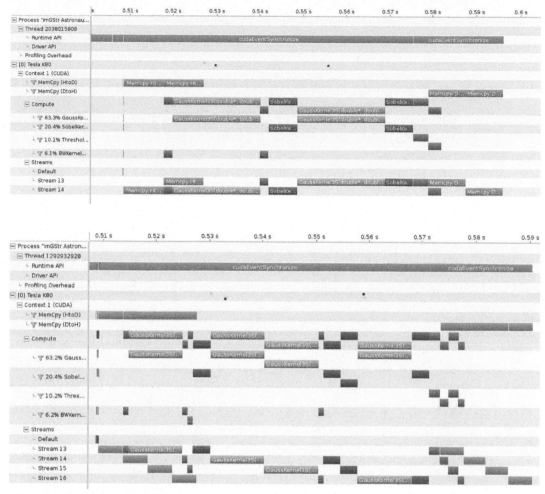

FIGURE 11.4 Nvidia NVVP results with 2 and 4 streams, on the K80 GPU.

There are so many things to try that some of these ideas will eventually lead to the most efficient answer. Thank goodness for the Visual Profiler; with it, you can see what kind of a runtime execution pattern is being induced for these different ideas.

PART III

More To Know

## CHAPTER 12

CUDA Libraries

Mohamadhadi Habibzadeh

University at Albany, SUNY

Omid Rajabi Shishvan

University at Albany, SUNY

Tolga Soyata

University at Albany, SUNY

I N the previous chapters of this book, we learned how to create a CUDA program without the help of any "prepackaged" library, like the ones we will see in this chapter. This was intentional; a deeper understanding of the inner-workings of the GPU can only be gained — and appreciated — when you create a program with the primitives that allow you to get close to the metal (i.e., the GPU cores). Surely, assembly language is a little too low-level, but the libraries we will see in this chapter barely require you to know how the GPU works; so, they are too high-level. The clear choice for the right "level" was plain and simple CUDA, which we based our GPU programming on up to this point in the book. In real life, however, when you are developing GPU programs, it is tedious to have to build everything from scratch. For example, there is no way you can build and optimize a matrix multiplication code the way Nvidia engineers can; because they spend weeks, months optimizing it. Because of this, CUDA programmers typically use high level libraries, such as cuBLAS, cuFFT, etc. and use CUDA itself as a *glue* language, to make everything work together. Every now and then, you will find something that there is no library for; well, this is when you go back to good-and-old CUDA. Aside from that, there is nothing wrong with using the libraries, especially because they are provided free of charge.

12.1 cuBLAS

The roots of Basic Linear Algebra Subprograms (BLAS) go back to the late 1970s and was initially written in Fortran. Note: The exciting programming language of the late 1950s, Formula Translation (Fortran), provided a way for programmers to do scientific computation without requiring assembly language. Having BLAS on top of that was a Godsend.

12.1.1 BLAS Levels

cuBLAS is an implementation of BLAS on the CUDA architecture. It can be downloaded from Nvidia's website royalty free. cuBLAS APIs provide support for vector and matrix algebraic operations such as addition, multiplication, etc., allowing developers to accelerate their programs easily. Because vector operations are significantly less compute-intensive than matrix operations, BLAS comes in three different flavors.

The BLAS operations can be categorized into the following:

- **BLAS Level 1** operations include algebraic operations on vectors. For example, adding two vectors or dot multiplication is considered a level 1 BLAS operation.

  BLAS Level 1 vector-vector operations have the generic form:

$$\mathbf{y} \leftarrow \alpha \cdot \mathbf{x} + \mathbf{y}$$

$$\mathbf{z} \leftarrow \alpha \cdot \mathbf{x} + \mathbf{y}$$

  where $\mathbf{x}$, $\mathbf{y}$, and $\mathbf{z}$ are vectors and α is a scalar. Shown above are the different flavors of BLAS Level 1, allowing two or three vectors.

- **BLAS Level 2** operations include matrix-vector operations. For example, multiplying a matrix by a vector is a BLAS Level 2 operation. These matrix-vector operations have the generic form:

$$\mathbf{y} \leftarrow \alpha \cdot \mathbf{A} \cdot \mathbf{x} + \beta \cdot \mathbf{y}$$

$$\mathbf{z} \leftarrow \alpha \cdot \mathbf{A} \cdot \mathbf{x} + \beta \cdot \mathbf{y}$$

  where $\mathbf{x}$, $\mathbf{y}$, and $\mathbf{z}$ are vectors, α and β are scalars, and $\mathbf{A}$ is a matrix. The above two forms represent the variants of BLAS Level 2, where the result is written in one of the source vectors (top) versus into a totally new vector (bottom).

- **BLAS Level 3** operations are matrix-matrix operations such as addition and multiplication of two matrices. Matrix-matrix operations are also called GEMM (General Matrix-Matrix Multiply). These matrix-vector operations have the generic form:

$$\mathbf{C} \leftarrow \alpha \cdot \mathbf{A} \cdot \mathbf{B} + \beta \cdot \mathbf{C}$$

$$\mathbf{D} \leftarrow \alpha \cdot \mathbf{A} \cdot \mathbf{B} + \beta \cdot \mathbf{C}$$

  where $\mathbf{A}$, $\mathbf{B}$, and $\mathbf{C}$ are matrices and α and β are scalars. The above two forms represent the variants of BLAS Level 3, where the result is written in one of the source matrices (top) versus into a totally new matrix (bottom).

12.1.2 cuBLAS Datatypes

Every cuBLAS API function comes in four different data types; single precision floating point (**S**), double precision floating point (**D**), complex single precision floating point (**C**), and complex double precision floating point (**Z**). Therefore, the GEMM APIs take the following names:

- **SGEMM**: Single Precision General Matrix-Matrix Multiply

- **DGEMM**: Single Precision General Matrix-Matrix Multiply

- **CGEMM**: Complex Single Precision General Matrix-Matrix Multiply

- **ZGEMM**: Complex Double Precision General Matrix-Matrix Multiply

12.1.3 Installing cuBLAS

CUDA BLAS toolkit 10 (latest version) is available for free download from Nvidia website. All CUDA BLAS versions below 4 use the legacy APIs. Starting with version 4 (introduced with CUDA 6.0), new APIs are implemented which provide additional services and cleaner interface. For a complete list of changes, refer to the Nvidia website: http://docs.nvidia.com/cuda/cublas/index.html#axzz4reNqlFkR [3]. New and legacy APIs can be used by including the related header files:

```
// Use new APIs
#include "cublas_v2.h"

// Or use Legacy APIs
// #include "cublas.h"
```

However, the use of legacy APIs is not recommended by the Nvidia. Creating cuBLAS programs is similar to other CUDA implementations. Typically, every cuBLAS code can be implemented in the following six stages.

12.1.4 Variable Declaration and Initialization

The host must first declare the input and output variables of the program. Furthermore, enough memory space must be allocated in the host memory for each variable. Typically, the host also needs to initialize the input variables. In some programs, however, the inputs might be initialized by the device. The code segment below shows the declaration, memory allocation, and initialization of input variables.

```
// Number of rows
#define M 6
// Number of columns
#define N 5
// Converting column-major to row-major format
#define IDX2C(i,j,ld) (((j)*(ld))+(i))

int main( void )
{
   int i, j;
    float* a = 0;
    // Allocate memory in host memory for matrix a
    a = (float *)malloc (M * N * sizeof (*a));
    // Check for errors
    if (!a)
      printf ("host memory allocation failed"); return EXIT_FAILURE;
    // Initialize matrix a on host
    for (j = 1; j <= N; j++)
      for (i = 1; i <= M; i++)
        a[IDX2F(i,j,M)] = (float)((i-1) * M + j);
} // end main
```

Defining IDX2C is necessary because cuBLAS supports column-major storage by default, which guarantees compatibility with Fortran. However, C and C++ use row-major storage. A conversion mechanism is, therefore, necessary by using the IDX2C macro.

12.1.5   Device Memory Allocation

Similar to other CUDA programs, enough space for the inputs and/or outputs must be allocated in device memory. In the example below, we allocate space for an M by N matrix. The matrix will hold the values of matrix a defined by the host in host memory.

```
float* devPtrA;
cublasStatus_t stat;

cudaStat = cudaMalloc ((void**)&devPtrA, M*N*sizeof(*a));

if (cudaStat != cudaSuccess){
   printf ("device memory allocation failed");
   return EXIT_FAILURE;
}
```

12.1.6   Creating Context

All calls to cuBLAS APIs require a *handle* to a context. Specifying handles for every context allows the user to have multithreaded host applications running on a single or multiple GPUs. The runtime library automatically dispatches the work to devices. Although it is possible that multiple host threads share the same handle, this is not recommended as it may lead to synchronization issues (for example, for deleting the handle). The code segment below shows how a handle to a context can be created.

```
cublasStatus_t stat;
cublasHandle_t handle;

stat = cublasCreate(&handle);
if (stat != CUBLAS_STATUS_SUCCESS){
   printf ("CUBLAS initialization failed\n");
   return EXIT_FAILURE;
}
```

12.1.7   Transferring Data to the Device

Data must be transferred to the device. The usual `cudaMalloc` and `cudaMemcpy` APIs can be used for this purpose. Alternatively, equivalent cuBLAS APIs exist that automatically allocate memory and transfer the data to the device, combining the two steps into one. Therefore, they are preferred to normal APIs. In the code segment below, we use cuBLAS APIs.

```
stat = cublasSetMatrix (M, N, sizeof(*a), a, M, devPtrA, M);
if (stat != CUBLAS_STATUS_SUCCESS){
   printf ("data download failed");
   cudaFree(devPtrA);
   cublasDestroy(handle);
   return EXIT_FAILURE;
}
```

The cuBLAS API `cublasSetMatrix` header is shown below:

```
cublasStatus_t
cublasSetMatrix(int rows, int cols, int elemSize, const void *A, int lda, void *B,
    int ldb)
```

This API allocates a `rows` × `cols` matrix in the device memory. It assumes that each element is of size `elemSize`. A pointer to the beginning of allocated space is copied in variable B. The API then transfers the context of matrix A (from host memory) to matrix B. `lda` and `ldb` represent the leading dimensions of matrices A and B, respectively. In other words, these variables determine the number of rows in A to B.

`cublasSetMatrix` API assumes that matrices are in column-major format.

12.1.8   Calling cuBLAS Functions

cuBLAS functions can be called to perform various operations on the data. In this example we use `cublasSscal`. The code segment below shows the usage of this function.

```
static __inline__ void
modify (cublasHandle_t handle, float *m, int ldm, int n, int p, int q, float
    alpha, float beta)
{
    cublasSscal (handle, n-p, &alpha, &m[IDX2C(p,q,ldm)], ldm);
    cublasSscal (handle, ldm-p, &beta, &m[IDX2C(p,q,ldm)], 1);
} // end modify

int main (void)
{
    ...
    modify (handle, devPtrA, M, N, 1, 2, 16.0f, 12.0f);
    ...
} // end main
```

Function `cublasSscal` is defined as:

```
cublasStatus_t
cublasSscal (cublasHandle_t handle, int n, const float *alpha, float *x, int incx)
```

where `handle` is a pointer to a context. This function multiplies each element of vector x (in device memory) by the scalar `alpha`. The length of the vector is determined by `n` and `incx` determines the stride size.

12.1.9 Transfer Data Back to the Host

Transferring data back to the host memory can be done through CUDA APIs (`cudaMemcpy`). Alternatively, cuBLAS APIs can be used.

```
stat = cublasGetMatrix (M, N, sizeof(*a), devPtrA, M, a, M);
if (stat != CUBLAS_STATUS_SUCCESS){
    printf ("data upload failed");
    cudaFree (devPtrA);
    cublasDestroy(handle);
    return EXIT_FAILURE;
}
```

The usage format of `cublasGetMatrix` is shown below:

```
cublasStatus_t
cublasGetMatrix(int rows, int cols, int elemSize, const void *A, int lda, void *B,
    int ldb)
```

12.1.10 Deallocating Memory

In the end, the allocated memory must be freed to prevent memory leakage problems as shown below.

```
cudaFree (devPtrA);
cublasDestroy(handle);
free(a);
return EXIT_SUCCESS;
```

12.1.11 Example cuBLAS Program: Matrix Scalar

The code below uses cuBLAS to multiply every element of a matrix by a scalar:

```
//Example 2. Application Using C and CUBLAS: 0-based indexing
#include <stdio.h>
#include <stdlib.h>
#include <math.h>
#include <cuda_runtime.h>
#include "cublas_v2.h"

#define M 6
#define N 5
#define IDX2C(i,j,ld) (((j)*(ld))+(i))

static __inline__ void
modify (cublasHandle_t handle, float *m, int ldm, int n, int p, int q, float
    alpha, float beta)
{
    cublasSscal (handle, n-p, &alpha, &m[IDX2C(p,q,ldm)], ldm);
    cublasSscal (handle, ldm-p, &beta, &m[IDX2C(p,q,ldm)], 1);
}
```

```
int main (void)
{
  cudaError_t cudaStat;
  cublasStatus_t stat;
  cublasHandle_t handle;
  int i, j;                  float* devPtrA;          float* a = 0;

  a = (float *)malloc (M * N * sizeof (*a));
  if (!a)       printf ("host memory allocation failed"); return EXIT_FAILURE;

  for (j = 0; j < N; j++)
    for (i = 0; i < M; i++)
      a[IDX2C(i,j,M)] = (float)(i * M + j + 1);

  cudaStat = cudaMalloc ((void**)&devPtrA, M*N*sizeof(*a));
  if (cudaStat != cudaSuccess){
    printf ("device memory allocation failed");
    return EXIT_FAILURE;
  }

  stat = cublasCreate(&handle);
  if (stat != CUBLAS_STATUS_SUCCESS){
    printf ("CUBLAS initialization failed\n");
    return EXIT_FAILURE;
  }

  stat = cublasSetMatrix (M, N, sizeof(*a), a, M, devPtrA, M);
  if (stat != CUBLAS_STATUS_SUCCESS){
    printf ("data download failed");
    cudaFree (devPtrA);
    cublasDestroy(handle);
    return EXIT_FAILURE;
  }

  modify (handle, devPtrA, M, N, 1, 2, 16.0f, 12.0f);
  stat = cublasGetMatrix (M, N, sizeof(*a), devPtrA, M, a, M);
  if (stat != CUBLAS_STATUS_SUCCESS){
    printf ("data upload failed"); cudaFree (devPtrA);
    cublasDestroy(handle);
    return EXIT_FAILURE;
  }

  cudaFree (devPtrA);
  cublasDestroy(handle);

  for (j = 0; j < N; j++){
    for (i = 0; i < M; i++) printf ("%7.0f", a[IDX2C(i,j,M)]);
    printf ("\n");
  } // for( j )
  free(a);
  return EXIT_SUCCESS;
} // end main
```

12.2 CUFFT

cuFFT is the CUDA Fast Fourier Transform (FFT) API library, which allows working in the frequency domain by computing the frequency components of images or audio signals. In digital signal processing (DSP), the *convolution* operation in the time domain corresponds to the *Fourier transform* in the frequency domain. Using FFT allows filters to be built, which work on the *frequency domain*, rather than the *time domain*. Note that the FFT operations involve complex numbers; therefore, they are much slower than using the time domain. However, for the right applications, they can vastly simplify the necessary operations.

12.2.1 cuFFT Library Characteristics

cuFFT consists of two libraries

- **cuFFT** library is designed to perform FFT computation on Nvidia GPUs.

- **cuFFTW** library is an easy porting tool for FFTW (C Library) users to Nvidia GPUs.

The cuFFT library provides support for a variety of FFT inputs and outputs such as:

- 1D, 2D, 3D transforms

- Complex and real value inputs and outputs

    Real to complex

    Complex to complex

    Complex to real

- Transforms in half-precision, single-precision, and double-precision

The cuFFT library has the following characteristics:

- Its computational complexity is $O(n \cdot \log n)$

- It is highly optimized for input sizes in the following form: $2^a \times 3^b \times 5^c \times 7^d$

- Allows execution of transforms over multiple GPUs

12.2.2 A Sample Complex-to-Complex Transform

The following code is an example of a cuFFT-based code that performs a 3D complex-to-complex transform:

```
#define NX64
#define NY64
#define NZ128

cufftHandleplan;
cufftComplex*data1, *data2;

cudaMalloc((void**)&data1,sizeof(cufftComplex)*NX*NY*NZ);
cudaMalloc((void**)&data2,sizeof(cufftComplex)*NX*NY*NZ);
```

```
/* Create a 3D FFT plan. */
cufftPlan3d(&plan, NX, NY, NZ, CUFFT_C2C);

/* Transform the first signal in place. */
cufftExecC2C(plan, data1, data1, CUFFT_FORWARD);

/* Transform the second signal using the same plan. */
cufftExecC2C(plan, data2, data2, CUFFT_FORWARD);

/* Destroy the cuFFT plan. */
cufftDestroy(plan);
cudaFree(data1); cudaFree(data2);
```

12.2.3   A Sample Real-to-Complex Transform

This is another sample code that provides a 1D real-to-complex transform:

```
#define NX 256
#define BATCH 1

cufftHandle plan;
cufftComplex *data;
cudaMalloc((void**)&data, sizeof(cufftComplex)*(NX/2+1)*BATCH);
if (cudaGetLastError() != cudaSuccess){
   fprintf(stderr, "Cuda error: Failed to allocate\n");
   return;
}

if (cufftPlan1d(&plan, NX, CUFFT_R2C, BATCH) != CUFFT_SUCCESS){
   fprintf(stderr, "CUFFT error: Plan creation failed");
   return;
}

...

/* Use the CUFFT plan to transform the signal in place. */
if (cufftExecR2C(plan, (cufftReal*)data, data) != CUFFT_SUCCESS){
   fprintf(stderr, "CUFFT error: ExecC2C Forward failed");
   return;
}

if (cudaDeviceSynchronize() != cudaSuccess){
   fprintf(stderr, "Cuda error: Failed to synchronize\n");
   return;
}

...

cufftDestroy(plan);
cudaFree(data);
```

12.3 NVIDIA PERFORMANCE PRIMITIVES (NPP)

Nvidia performance primitives library (NPP) is a library for imaging and video processing functions. There are many different types of functions implemented in NPP. The following list shows some of the higher level classes of functions and their subclasses which are implemented in NPP.

- Image Arithmetic and Logic Operations

  Arithmetic operations: Add, Mul, Sub, Div, AbsDiff, Sqr, Sqrt, Ln, Exp, etc.

  Logical operations: And, Or, Xor, RShift, LShift, Not, etc.

- Image Color Conversion

  Color model conversion, color sampling formant conversion, color gamma correction, color processing, complement color key.

- Image Compression

- Image Filters

  1D linear filter, 1D window sum, convolution, 2D fixed linear filters, rank filters, fixed filters

- Image Geometry

  Resize, remap, rotate, mirror, affine transform, perspective transform

- Image Morphological

  Dilation, erode

- Image Statistic and Linear

  Sum, min, max, mean, Mean_StdDev, norms, DotProd, integral, histogram, error, etc.

- Image Support and Data Exchange

  Set, copy, convert, scale, transpose, etc.

- Image Threshold and Compare

- Signal

  Not working on a picture, but rather a "signal"

  Many subclasses, e.g., all the arithmetic and logical operations

  Many more such as Cauchy, Cubrt, Arctan, etc.

Following is a sample code that implements box filter using NPP:

```
// declare a host image object for an 8-bit grayscale image
npp::ImageCPU_8u_C1 oHostSrc;
// load gray-scale image from disk
npp::loadImage(sFilename, oHostSrc);
// declare a device image and copy construct from the host image,
// i.e. upload host to device
npp::ImageNPP_8u_C1 oDeviceSrc(oHostSrc);
// create struct with box-filter mask size
```

```
NppiSize oMaskSize = {5, 5};
// create struct with ROI size given the current mask
NppiSize oSizeROI = {oDeviceSrc.width() - oMaskSize.width + 1, oDeviceSrc.height()
    - oMaskSize.height + 1};
// allocate device image of appropriately reduced size
npp::ImageNPP_8u_C1 oDeviceDst(oSizeROI.width, oSizeROI.height);
// set anchor point inside the mask to (0, 0)
NppiPoint oAnchor = {0, 0};
// run box filter
NppStatus eStatusNPP;
eStatusNPP = nppiFilterBox_8u_C1R(oDeviceSrc.data(), oDeviceSrc.pitch(),
                                  oDeviceDst.data(), oDeviceDst.pitch(),
                                  oSizeROI, oMaskSize, oAnchor);
NPP_ASSERT(NPP_NO_ERROR == eStatusNPP);
// declare a host image for the result
npp::ImageCPU_8u_C1 oHostDst(oDeviceDst.size());
// and copy the device result data into it
oDeviceDst.copyTo(oHostDst.data(), oHostDst.pitch());
saveImage(sResultFilename, oHostDst);
std::cout << "Saved image: " << sResultFilename << std::endl;
```

12.4 THRUST LIBRARY

cuBLAS APIs provide a convenient way for the developer to accelerate matrix and vector operations. Similarly, the Thrust library includes GPU-based massively paralleled implementation of some useful algorithms, allowing developers to achieve $10\times$ and $100\times$ performance boost for operations such as sorting and reduction. This library also includes definitions for data structure (such as device vector, host vector, device and host pointers, etc.). These data structures facilitate the implementation of the code. Thrust implementation is based on C++ templates, and is not compatible with C. Thrust header files are typically included in the CUDA toolkit. Therefore, separate installation and setup is not required. As Thrust is a C++ based library, it provides C++ vector types that are either defined on the host or are directly allocated on the device. This makes the program development much easier, as the developer is not required to use CUDA APIs such as cudaMalloc or cudaMemcpy to transfer data to and from the device.

The example below shows how vectors are defined on the host side and the device side using the Thrust library as opposed to conventional memory allocation.

```
//Creating the host vector;
//Thrust:
thrust::host_vector<int> host_vec(1000);
//C:
int* host_vec_c = new int[1000];

//Creating the vector on the device and copying the host vector to the device
//Thrust:
thrust::device_vector<int> device_vec = host_vec;
//C:
int* device_vec_c;
cudaMalloc((void**)&device_vec_c, 1000 * sizeof(int));
cudaMemcpy(device_vec_c, host_vec_c, 1000 * sizeof(int), cudaMemcpyHostToDevice);
```

You can see how you can create a vector with length of 16 and assign random numbers to it using the Thrust library on the host side. Then copy this vector to the device side and fill all the values with "10."

```
#include <thrust/host_vector.h>
#include <thrust/device_vector.h>
#include <thrust/fill.h>

int main(void)
{
   //Creating the host vector;
   thrust::host_vector<int> host_vec(16);
   //Assigning random numbers to the host vector
   thrust::generate(host_vec.begin(), host_vec.end(), rand);

   //Creating the devce vector
   thrust::device_vector<int> device_vec = host_vec;
   //Filling the copy with values of 10
   thrust::fill(device_vec.begin(), device_vec.end(), 10);

   return 0;
}
```

Many algorithms are implemented efficiently in the Thrust library. Some of these algorithms are

- **Reductions:** Reduces a vector to single value. Examples are max, min, sum, etc. For example, the following code computes the sum of numbers from 1 to 100:

  ```
  #include <thrust/host_vector.h>
  #include <thrust/device_vector.h>
  #include <thrust/reduce.h>

  int main(void)
  {
     thrust::host_vector<int> host_vec(100);
     for (int i = 0; i < 100; i++)
        host_vec[i] = i+1;

     thrust::device_vector<int> device_vec = host_vec;

     //reduce sums up all the values in a vector and return the final value
     int x = thrust::reduce(device_vec.begin(), device_vec.end(), 0,
        thrust::plus<int>());
     return 0;
  }
  ```

- **Transformations:** Refers to algorithms that operate on each element of a vector at a time such as fill, sequence, replace, and transform (transform applies a function to each element of a vector and writes the result in that element. For example, it can negate all elements).

- **Searching:** Refers to finding strings in a given text.

```
#include <thrust/find.h>

thrust::device_vector<int> device_vec;
thrust::device_vector<int>::iterator iter;
...
thrust::find(device_vec.begin(), device_vec.end(), 3);
```

- **Sorting:** Refers to sorting a list of strings, integers, etc.

```
#include <thrust/sort.h>

thrust::device_vector<int> device_vec;
...
thrust::sort(device_vec.begin(), device_vec.end());
```

## CHAPTER 13

Introduction to OpenCL

Chase Conklin
University of Rochester

Tolga Soyata
University at Albany, SUNY

I N this chapter, we will be familiarized with OpenCL, which is the most popular GPU programming language, excluding CUDA. This chapter is designed to show how OpenCL simplifies writing multiplatform parallel programs. From previous chapters, we have become familiar with programs such as imflip and imedge. Though OpenCL and CUDA both exist to write highly parallel code, you will quickly see how the approach for writing programs in each differs.

13.1 WHAT IS OpenCL?

OpenCL was released in 2009 by the Khronos Group as a framework for writing parallel programs on many different platforms. Unlike CUDA, which only runs on Nvidia GPUs, OpenCL code is capable of running on CPUs, GPUs, and other devices such as field programmable gate arrays (FPGAs) and digital signal processors (DSPs), as long as the device supports OpenCL. Also unlike CUDA, OpenCL kernels are compiled at runtime, which is how it is able to easily work on many different platforms.

13.1.1 Multiplatform

OpenCL supports many different devices, but it is up to the device manufacturer to implement the drivers that allow OpenCL to work on their devices. These different implementations are known as *platforms*. Depending on your hardware, your computer may have multiple OpenCL platforms available; for example one from Intel to run on the integrated graphics, and one from Nvidia to run on their discrete graphics.

OpenCL considers devices to be in one of three categories: **(1)** CPU, **(2)** GPU, and **(3)** accelerator. Of the three, only the accelerator should be unfamiliar. Hardware accelerators include FPGAs and DSPs, or devices such as Intel's Xeon Phi (See Section 3.9 for a detailed introduction to Xeon Phi).

13.1.2 Queue-Based

Unlike CUDA, which operates on either synchronous blocking calls, or can operate asynchronously using streams, OpenCL's execution is queue-based, where all commands are dispatched to a command queue, and execute once they reach the head of the queue.

As mentioned previously, OpenCL can support multiple different devices. In fact, it can support running on multiple devices *simultaneously*. To do this, one only needs to create a queue for each device.

13.2 IMAGE FLIP KERNEL IN OPENCL

Let's look at an OpenCL kernel. The kernel shown below is for the imflip program, and it horizontally flips an image. For simplicity, each work-item (the analog of threads in CUDA) will be responsible for one pixel, swapping the red, green, and blue components.

```
__kernel void
hflip(
   __global const unsigned char *input,
   __global unsigned char *output,
   const int M,
   const int N)
{
   int idx = get_global_id(0);

   int row = idx / N;
   int col = (idx % N) * 3;

   int start = row*N*3;     // first byte of row in 1D array
   int end = start + N*3 - 1; // last byte of row in 1D array

   if (idx >= M * N) return;

   output[start+col] = input[end-col-2];
   output[start+col+1] = input[end-col-1];
   output[start+col+2] = input[end-col];

   output[end-col-2] = input[start+col];
   output[end-col-1] = input[start+col+1];
   output[end-col] = input[start+col+2];
}
```

The `__global__` identifier is gone, and is replaced with `__kernel`. Pointers to global memory must be prefaced with the `__global` identifier, and other values should be declared with `const`. Read-only memory should also be declared with `const` so that attempts to modify it will raise reasonable errors rather than crash the program. CUDA's `threadIdx` and `blockIdx` are replaced by a call to `get_global_id()`. The argument passed to `get_global_id()` determines which dimension the id is in. For example, if this was a 2D kernel, we could write

```
int row = get_global_id(0);
int col = get_global_id(1) * 3;
```

Otherwise, this OpenCL kernel is very similar to the comparable CUDA kernel. Table 13.1 shows some of the comparable CUDA and OpenCL terms.

TABLE 13.1 Comparable terms for CUDA and OpenCL.

CUDA	OpenCL
Thread	Work-Item
Thread Block	Work-Group
Shared Memory	Local Memory
Local Memory	Private Memory
__global__ function	__kernel function
__device__ function	Implicit
__shared__ variable	__local variable
gridDim	get_num_groups()
blockDim	get_local_size()
blockIdx	get_group_id()
threadIdx	get_local_id()
__syncthreads()	barrier()

13.3   RUNNING OUR KERNEL

Let's look at what is needed to run an OpenCL kernel. Any OpenCL program will need to include the OpenCL headers. On Linux systems, this will require including the `<CL/cl.h>` header file. Apple's OSX, which implements its own version of OpenCL, names the headers differently, and instead includes `<OpenCL/opencl.h>`. For programs that may need to be compiled in OSX or on Linux, use of preprocessor conditionals can selectively include the correct header as follows:

```
#ifdef __APPLE__
#include <OpenCL/opencl.h>
#else
#include <CL/cl.h>
#endif
```

13.3.1 Selecting a Device

Our program will have the user select the compute device they wish to run their program on. To do this, we will define a convenience function that will make the necessary OpenCL calls.

```
cl_device_id
selectDevice(void)
{
   int i, choice = -1;
   char * value;
   size_t valueSize;

   cl_uint deviceCount;
   cl_device_id * devices, selected;

   clGetDeviceIDs(NULL, CL_DEVICE_TYPE_ALL, 0, NULL, & deviceCount);
   devices = (cl_device_id *) malloc(sizeof(cl_device_id) * deviceCount);
   assert(devices);
   clGetDeviceIDs(NULL, CL_DEVICE_TYPE_ALL, deviceCount, devices, NULL);

   for (i = 0; i < deviceCount; i ++) {
      clGetDeviceInfo(devices[i], CL_DEVICE_NAME, 0, NULL, & valueSize);
      value = (char *) malloc(valueSize);
      clGetDeviceInfo(devices[i], CL_DEVICE_NAME, valueSize, value, NULL);
      printf("[%d]: %s\n", i, value);
      free(value);
   }

   while (choice < 0 || choice >= deviceCount) {
      printf("Select Device: ");
      scanf("%d", & choice);
   }
   selected = devices[choice];
   free(devices);
   return selected;
}
```

This function returns a `cl_device_id`, which is an OpenCL type that refers to a specific device. We call `clGetDeviceIDs` once to get the number of available devices (stored into `deviceCount`). We then allocate an array to hold each device id, and call `clGetDeviceIDs()` again to populate our newly allocated array with device ids.

The function `clGetDeviceIds` has a signature as shown below:

```
cl_int
clGetDeviceIDs(
   cl_platform_id platform,
   cl_device_type device_type,
   cl_uint num_entries,
   cl_device_id *devices,
   cl_uint *num_devices)
```

The first argument, `platform`, allows specification of an OpenCL platform; for simplicity we will ignore this argument and pass `NULL`.

The second argument, `device_type`, allows filtering of which OpenCL devices to show. It accepts `CL_DEVICE_TYPE_CPU` (only the CPU), `CL_DEVICE_TYPE_GPU` (any GPUs), `CL_DEVICE_TYPE_ACCELERATOR` (such as Xeon Phi), `CL_DEVICE_TYPE_DEFAULT` (the default CL device in the system), and `CL_DEVICE_TYPE_ALL` (all available OpenCL devices).

The third argument, `num_entries`, specifies the number of device ids that may be placed in the array specified by `devices`. If `devices` is not `NULL`, this must be greater than 0.

The fourth argument, `devices`, is a pointer to a space of memory that will be filled with device ids. If this value is `NULL`, it is ignored.

The final argument, `num_devices`, is the number of devices that match `device_type`. This argument is ignored if it is `NULL`.

13.3.2 Running the Kernel

Before we can run our kernel, we need to set up the world. This involves creating a compute context, creating a command queue, loading our kernel into OpenCL, building the OpenCL program, and setting up our kernel invocation. While this sounds like a significant amount of work, it is not as difficult as it sounds.

13.3.2.1 Creating a Compute Context

Using the device id provided by `selectDevice`, we create a compute context by calling `clCreateContext`.

```
cl_context context = clCreateContext(0, 1, &device_id, NULL, NULL, &err);
if (!context){
    printf("Error: Failed to create a compute context!\n");
    return EXIT_FAILURE;
}
```

13.3.2.2 Creating a Command Queue

Using our newly created context, we can now create a command queue. The command queue is used to schedule tasks such as kernels and memory transfers on our device.

```
cl_command_queue commands = clCreateCommandQueue(context, device_id,
    CL_QUEUE_PROFILING_ENABLE, &err);
if (!commands){
    printf("Error: Failed to create a command queue, named commands!\n");
    return EXIT_FAILURE;
}
```

13.3.2.3 Loading Kernel File

Unlike CUDA, where kernels are defined in the same code as the rest of the C/C++ functions, in OpenCL, kernels are written into a string, which is then loaded via an OpenCL API call. This example has the kernels written in another file, imflip.cl. Because the contents of this file need to be a string for the OpenCL API call to work, we need a function to read the kernel file.

```c
char *kernelSource(const char *kernel_file) {
  FILE *fp;
  char * source_str;
  size_t source_size, program_size;

  fp = fopen(kernel_file, "rb");
  if (!fp){
      printf("Failed to load kernel from %s\n", kernel_file);
      exit(1);
  }
  fseek(fp, 0, SEEK_END);
  program_size = ftell(fp);
  rewind(fp);
  source_str = (char*)malloc(program_size + 1);
  source_str[program_size] = '\0';
  fread(source_str, sizeof(char), program_size, fp);
  fclose(fp);
  return source_str;
}
```

Now that we can read the file, we need to load it into OpenCL.

```c
char *KernelSource = kernelSource("imflip.cl");
cl_program program = clCreateProgramWithSource(context, 1,
   (const char **) &KernelSource, NULL, &err);
if (!program){
    printf("Error: Failed to create compute program!\n");
    return EXIT_FAILURE;
}

err = clBuildProgram(program, 0, NULL, NULL, NULL, NULL);
if (err != CL_SUCCESS){
    size_t len;
    char buffer[2048];

    printf("Error: Failed to build program executable!\n");
    clGetProgramBuildInfo(program, device_id, CL_PROGRAM_BUILD_LOG, sizeof(buffer),
        buffer, &len);
    printf("%s\n", buffer);
    exit(1);
}
```

The OpenCL function `clCreateProgramWithSource` accepts a context and an array of strings. Because we loaded the entirety of the imflip.cl into one string, we set the length of the string array to 1, and pass a pointer to the source string. We pass the next parameter as `NULL`, which indicates that the source string is 0 terminated.

We now need to create our kernel. This involves telling OpenCL which function in the loaded program we want to use.

```
cl_kernel kernel = clCreateKernel(program, "hflip", &err);
if (!kernel || err != CL_SUCCESS){
    printf("Error: Failed to create compute kernel!\n");
    exit(1);
}
```

13.3.2.4 Setting Up Kernel Invocation

Before we can run our kernel, we need to allocate device memory and transfer our image to it. Allocating device memory in OpenCL is accomplished with the `clCreateBuffer()` function. Assume that `TotalSize` is the number of bytes in our image.

```
cl_mem input = clCreateBuffer(context, CL_MEM_READ_ONLY, TotalSize, NULL, NULL);
cl_mem output = clCreateBuffer(context, CL_MEM_WRITE_ONLY, TotalSize, NULL, NULL);
if (!input || !output){
    printf("Error: Failed to allocate device memory!\n");
    exit(1);
}
```

This is similar to allocating memory in CUDA, with the exception that we can set the read/write permissions of the memory. These flags give the OpenCL more information on how the memory will be used, which allows it to better optimize performance. Since our input image will only be read and our output image will only be written to, we use `CL_MEM_READ_ONLY` and `CL_MEM_WRITE_ONLY`, respectively.

Having allocated memory on our device, we can transfer our image to the device.

```
err = clEnqueueWriteBuffer(commands, input, CL_TRUE, 0, TotalSize, CPU_InputArray,
    0, NULL, NULL);
if (err != CL_SUCCESS){
    printf("Error: Failed to write to source array!\n");
    exit(1);
}
```

The OpenCL function `clEnqueueWriteBuffer()` enqueues the transfer onto the command queue. By passing `CL_TRUE` as the third argument, we ensured that `clEnqueueWriteBuffer()` will block until the transfer is complete, though in other applications, it may be possible to schedule a transfer, perform other useful work in the shadow of the transfer, then execute the kernel.

Now that we have the data on the device, we can finally run our kernel! The first step to doing that is to set the kernel arguments. Unlike in CUDA, these are not set during the kernel call (since to CUDA, the kernel is just another function), but using the OpenCL API.

```
err = 0;
err = clSetKernelArg(kernel, 0, sizeof(cl_mem), &input);
err |= clSetKernelArg(kernel, 1, sizeof(cl_mem), &output);
err |= clSetKernelArg(kernel, 2, sizeof(int), &M);
err |= clSetKernelArg(kernel, 3, sizeof(int), &N);
if (err != CL_SUCCESS){
    printf("Error: Failed to set kernel arguments! %d\n", err);
    exit(1);
}
```

Now, we need to get our work group sizes. This example uses a 1D work-group. Because setting the size of the work-groups is non-trivial, we place the logic for it into another function, getWorkGroupSizes().

```
void getWorkGroupSizes(cl_device_id device_id, cl_kernel kernel, size_t * local,
    size_t * global, size_t desired_global) {
  int err;
  size_t max_work_group_size;
  clGetDeviceInfo(device_id, CL_DEVICE_MAX_WORK_GROUP_SIZE, sizeof(size_t),
      &max_work_group_size, NULL);
  // Get the maximum work group size for executing the kernel on the device
  err = clGetKernelWorkGroupInfo(kernel, device_id, CL_KERNEL_WORK_GROUP_SIZE,
      sizeof(size_t), local, NULL);
  if (err != CL_SUCCESS){
    printf("Error: Failed to retrieve kernel work group info! %d\n", err);
    exit(1);
  }
  *global = desired_global + (desired_global % max_work_group_size);
}
```

This function (getWorkGroupSizes) takes a device, kernel, and desired global work-group size, and outputs a local and global work-group size compatible with the requested values. The call to clGetKernelWorkGroupInfo() determines the maximum local work-group size for the device. Because the global work-group size must be an integer multiple of the local work-group size, we "round" the size up. Finally, we can execute our kernel!

```
getWorkGroupSizes(device_id, kernel, &local, &global, M * N);

err = clEnqueueNDRangeKernel(commands, kernel, 1, NULL, &global, &local, 0, NULL,
    NULL);
if (err){
    printf("Error: Failed to execute kernel!\n");
    return EXIT_FAILURE;
}

clFinish(commands);
```

We call the getWorkGroupSizes() function which we just defined, then we enqueue our kernel to the command queue. We then call clFinish(), which is similar to

TABLE 13.2 Runtimes for imflip, in ms.

Device	Runtime
Intel i7-3820QM CPU @ 2.70GHz	148.6
Intel i7-6700K CPU @ 4.00GHz	158.1
Intel HD Graphics 4000	959.0
Nvidia GeForce 650M	63.9
AMD Radeon R9 M395X	5.5

`cudaDeviceSynchronize()`. Having run our kernel, we can enqueue a transfer back from the device of our flipped image.

```
err = clEnqueueReadBuffer( commands, output, CL_TRUE, 0, TotalSize,
    CPU_OutputArray, 0, NULL, NULL );
if (err != CL_SUCCESS){
    printf("Error: Failed to read output array! %d\n", err);
    exit(1);
}
```

Finally, we need to clean up the OpenCL runtime.

```
clReleaseMemObject(input);
clReleaseMemObject(output);
clReleaseProgram(program);
clReleaseKernel(kernel);
clReleaseCommandQueue(commands);
clReleaseContext(context);
```

13.3.3 Runtimes of Our OpenCL Program

Let's look at the performance of our OpenCL program, imflip.cl, on Astronaut.bmp, as tabulated in Table 13.2.

- Unsurprisingly, discrete GPUs had the lowest runtimes, as their GDDR5 memory has a much higher bandwidth than the DDR3 memory used by the other three devices.

- However, the integrated graphics was slower than the CPU! This is because the CPU has more limited memory access patterns which allows for better caching and higher-bandwidth accesses from main memory.

- Also interesting is that the 6700K CPU ran slower than the 3820QM CPU, despite being newer and having a higher clock rate. Note: These runtimes were the average of multiple runs, as to avoid variance.

13.4 EDGE DETECTION IN OpenCL

Often, real applications will require multiple kernels to perform their required task. Let us look at an edge detection program imedge. Here is the Gaussian kernel:

```
__constant float Gauss[5][5] = {
    { 2,  4,  5,  4,  2 },
    { 4,  9, 12,  9,  4 },
    { 5, 12, 15, 12,  5 },
    { 4,  9, 12,  9,  4 },
    { 2,  4,  5,  4,  2 } };

__constant float Gx[3][3] = {
    { -1, 0, 1 },
    { -2, 0, 2 },
    { -1, 0, 1 } };

__constant float Gy[3][3] = {
    { -1, -2, -1 },
    {  0,  0,  0 },
    {  1,  2,  1 } };


__kernel void gaussian_filter(
    __global float * output,
    __global float * input,
    __local float * working,
    const int M,
    const int N)
{
    int idx = get_global_id(0);           int local_id = get_local_id(0);
    int local_size = get_local_size(0);
    int row = idx / N;                     int col = idx % N;
    int local_row = 2;                     int local_col = local_id + 2;
    int i, j;                              int local_idx, gidx;
    float G;

    if (idx >= M * N) return;
    if ((row<2) || (row > (M - 3)) || (col < 2) || (col > N - 3)) {
        output[idx] = input[idx];
        return;
    }

    // Check if work-group size is worthwhile for local memory
    if (local_size >= 64) {
        if (local_id == 0) {
            // Left side
            for (i = 0; i < 5; i ++) {
                local_idx = i * (local_size + 4) + local_col - 2;
                working[local_idx] = input[(row+i) * N + col - 2];
                local_idx = i * (local_size + 4) + local_col - 1;
                working[local_idx] = input[(row+i) * N + col - 1];
            }
        } else if (local_id == local_size - 1) {
```

```
        for (i = 0; i < 5; i ++) {
            local_idx = i * (local_size + 4) + local_col + 2;
            working[local_idx] = input[(row+i) * N + col + 2];
            local_idx = i * (local_size + 4) + local_col + 1;
            working[local_idx] = input[(row+i) * N + col + 1];
        }
    }
    for (i = 0; i < 5; i ++) {
        local_idx = i * (local_size + 4) + local_col;
        working[local_idx] = input[(row+i) * N + col];
    }
    barrier(CLK_LOCAL_MEM_FENCE);
    G=0.0;
    for(i=-2; i<=2; i++){
        for(j=-2; j<=2; j++){
            gidx = (local_row+i) * (local_size + 4) + local_col+j;
            G += working[gidx] * Gauss[i+2][j+2];
        }
    }
  } else {
    G=0.0;
    for(i=-2; i<=2; i++){
        for(j=-2; j<=2; j++){
            gidx = (row+i) * N + col+j;
            G += input[gidx] * Gauss[i+2][j+2];
        }
    }
  }
  output[idx] = (G/159.0);
}
```

And, here is the Sobel kernel:

```
__kernel void sobel(
    __global unsigned char * output,
    __global float * input,
    __local float * working,
    const int M,
    const int N)
{
    int idx = get_global_id(0);
    int local_id = get_local_id(0);
    int local_size = get_local_size(0);
    int row = idx / N;
    int col = idx % N;
    int local_row = 1;
    int local_col = local_id + 1;
    int local_idx;
    float GX,GY;
    int gidx;
    int i, j;

    // this thread ID is out of bounds
    if (idx >= M * N) return;
```

```
// row, column out of bounds
if ((row<1) || (row > (M - 2)) || (col < 1) || (col > N - 2)) {
    output[idx] = 255;
    return;
}

// Check if work-group size is worthwhile for local memory
if (local_size >= 64) {
    if (local_id == 0) {
        // Left side
        for (i = 0; i < 3; i ++) {
            local_idx = i * (local_size + 2) + local_col - 1;
            working[local_idx] = input[(row+i) * N + col - 1];
        }
    } else if (local_id == local_size - 1) {
        for (i = 0; i < 3; i ++) {
            local_idx = i * (local_size + 2) + local_col + 1;
            working[local_idx] = input[(row+i) * N + col + 1];
        }
    }
    for (i = 0; i < 3; i ++) {
        local_idx = i * (local_size + 2) + local_col;
        working[local_idx] = input[(row+i) * N + col];
    }
    barrier(CLK_LOCAL_MEM_FENCE);
    // calculate Gx and Gy
    GX=0.0; GY=0.0;
    for(i=-1; i<=1; i++){
        for(j=-1; j<=1; j++){
            gidx = (local_row+i) * (local_size + 2) + local_col+j;
            GX += working[gidx] * Gx[i+1][j+1];
            GY += working[gidx] * Gy[i+1][j+1];
        }
    }
} else {
    GX=0.0; GY=0.0;
    for(i=-1; i<=1; i++){
        for(j=-1; j<=1; j++){
            gidx = (row+i) * N + col+j;
            GX += input[gidx] * Gx[i+1][j+1];
            GY += input[gidx] * Gy[i+1][j+1];
        }
    }
}
float THRESHOLD = 64.0;
output[idx] = sqrt(GX*GX+GY*GY) < THRESHOLD ? 255 : 0;
}
```

Our imedge.cl file starts with three array definitions. Note how they have been declared to be __constant, which specifies them as read-only. Not only does this provide some guarantees that they will not be erroneously modified, but may also allow the implementation to optimize by placing this data in a constant cache. We also made use of __local memory, which places the data in a special cache with L1 access times. Local memory is used in a manner similar to `__shared__` memory in CUDA. In our kernels, we first load the data from global memory into shared memory. Because we have a 1D arrangement of threads, we need to load the pixels in the same column as the pixel that the initial thread works on. The edges need to load the values next to them as well. Were the threads dispatched in a 2D arrangement, we could have made better use of thread locality with respect to local memory; however, running with a 2D arrangement of threads runs significantly slower than a 1D arrangement.

After loading our data into local memory, we use a barrier to ensure that the local memory is fully populated before any threads proceed. Unlike with CUDA, the argument `CLK_LOCAL_MEM_FENCE` is passed to the barrier call. This argument tells the barrier to synchronize only the work-items in the current work-group before continuing, while a different argument `CLK_GLOBAL_MEM_FENCE` would synchronize all work-items in all work-groups for the current kernel. Since we are working with local memory, `CLK_LOCAL_MEM_FENCE` will be sufficient.

The host-side code for imedge is similar to imflip.

As before, we select a device, create a context, a command queue, and compile our program. Creating each kernel is the same as before as well, except we need to do it once per kernel.

```
gauss = clCreateKernel(program, "gaussian_filter", &err);
if (!gauss || err != CL_SUCCESS){
    printf("Error: Failed to create compute kernel (gaussian_filter)!\n");
    exit(1);
}

sobel = clCreateKernel(program, "sobel", &err);
if (!sobel || err != CL_SUCCESS){
    printf("Error: Failed to create compute kernel (sobel)!\n");
    exit(1);
}
```

Setting the arguments is also just as before.

We now need to allocate some buffers to hold our input, output, and some intermediate steps.

As before, the input buffer only needs to be read, so we mark it with `CL_MEM_READ_ONLY`. Additionally, our output (on the device) is write only, since our kernels will only write to this location; as such, we mark it with `CL_MEM_WRITE_ONLY`. The buffers that contain our image run through the Gaussian filter will be written to and read. Thus, it must be marked as `CL_MEM_READ_WRITE`.

```
input = clCreateBuffer(context, CL_MEM_READ_ONLY, input_size, NULL, NULL);
output = clCreateBuffer(context, CL_MEM_WRITE_ONLY, output_size, NULL, NULL);
blurred = clCreateBuffer(context, CL_MEM_READ_WRITE, input_size, NULL, NULL);
if (!input || !output || !blurred){
    printf("Error: Failed to allocate device memory!\n");
    exit(1);
}
```

Let's look at how we run each of these kernels.

```
// Set the kernel arguments.
err = 0;
err = clSetKernelArg(gauss, 0, sizeof(cl_mem), &blurred);
err |= clSetKernelArg(gauss, 1, sizeof(cl_mem), &input);
err |= clSetKernelArg(gauss, 2, ((gauss_local + 4) * 5) * sizeof(float), NULL);
err |= clSetKernelArg(gauss, 3, sizeof(int), &M);
err |= clSetKernelArg(gauss, 4, sizeof(int), &N);
if (err!=CL_SUCCESS){ printf("Failed to set kernel args! %d\n", err); exit(1);}
// Set the arguments to our sobel kernel
err = 0;
err = clSetKernelArg(sobel, 0, sizeof(cl_mem), &output);
err |= clSetKernelArg(sobel, 1, sizeof(cl_mem), &blurred);
err |= clSetKernelArg(sobel, 2, ((sobel_local + 2) * 3) * sizeof(float), NULL);
err |= clSetKernelArg(sobel, 3, sizeof(int), &M);
err |= clSetKernelArg(sobel, 4, sizeof(int), &N);
if (err!=CL_SUCCESS){ printf("Failed to set kernel args! %d\n", err); exit(1);}
// Get the work-group sizes
getWorkGroupSizes(device_id, gauss, &gauss_local, &gauss_global, M * N);
getWorkGroupSizes(device_id, sobel, &sobel_local, &sobel_global, M * N);
// Write our data set into the input array in device memory
//
err = clEnqueueWriteBuffer(commands, input, CL_TRUE, 0, input_size,
    CPU_InputArray, 0, NULL, & xin);
if (err!=CL_SUCCESS){ printf("Failed to write to source array\n"); exit(1);}
// Execute the kernel over the entire range of our 1d input data set
// using the maximum number of work group items for this device
err = clEnqueueNDRangeKernel(commands, gauss, 1, NULL, &gauss_global,
    &gauss_local, 0, NULL, &event1);
if (err){ printf("Error: Failed to execute kernel! (%d)\n", err); return
    EXIT_FAILURE; }
err = clEnqueueNDRangeKernel(commands, sobel, 1, NULL, &sobel_global,
    &sobel_local, 0, NULL, &event2);
if (err){
   printf("Error: Failed to execute kernel! (%d)\n", err);
   return EXIT_FAILURE;
}
err = clEnqueueNDRangeKernel(commands, edge, 1, NULL, &edge_global, &edge_local,
    0, NULL, &event3);
if (err){
   printf("Error: Failed to execute kernel! (%d)\n", err);
   return EXIT_FAILURE;
}
// Read back the results from the device to verify the output
err = clEnqueueReadBuffer(commands, output, CL_TRUE, 0, output_size,
    CPU_OutputArray, 0, NULL, & xout );
if (err != CL_SUCCESS){
   printf("Error: Failed to read output array! %d\n", err);
   exit(1);
}
```

TABLE 13.3 Runtimes for imedge, in ms.

Device	Tfr In	Tfr Out	Gauss	Sobel
Intel i7-3820QM CPU	59.86	5.943	210.26	135.27
Intel i7-6700K CPU	48.16	5.806	166.54	102.45
Nvidia GeForce 650M	51.95	12.74	119.99	60.02
AMD Radeon R9 M395X	16.84	10.31	5.316	3.70

First, we set the arguments to our kernel. This is similar to how arguments were set in imflip, but with one notable exception.

```
clSetKernelArg(gauss, 2, ((gauss_local + 4) * 5) * sizeof(float), NULL);
```

This line creates an array that will be the local memory used by our kernel. Unlike other arguments passed to the kernel, note that the address of the value is NULL.

Next, we determine the size of our work group. Note that this is done once per kernel, as this allows us to achieve maximum occupancy for each kernel.

We then enqueue the transfer, each kernel, and the transfer back. Because we do not allow OpenCL to execute commands in the queue out of order, we can be assured that each kernel will have the necessary data ready before starting.

How does our program perform? As before, we will test it using Astronaut.bmp. Results are reported in Table 13.3.

There are a few points to note here. The first is that there is still a memory transfer penalty on the CPU. This is because OpenCL has allocated the buffer in a different region of memory than was allocated by malloc(). Because of this, using the CPU is not a way to avoid transfer penalties, as this must be incurred regardless of the device.

Using local memory can greatly increase the performance of kernels, but this benefit can be offset by the penalty of synchronizing threads. To get the most performance gain, ensure that as much data is shared as possible so that each initial access to global memory that places the data in local memory offsets what would initially be multiple accesses to global memory.

## CHAPTER 14

Other GPU Programming Languages

Sam Miller

University of Rochester

Andrew Boggio-Dandry

University at Albany, SUNY

Tolga Soyata

University at Albany, SUNY

I N this chapter, we will briefly look at GPU programming languages other than OpenCL and CUDA. Additionally, we will investigate some of the common APIs, such as OpenGL, OpenGL ES, OpenCV, and Apple's Metal API. Although these APIs are not programming languages, they transform an existing language into a much more practical one.

14.1 GPU PROGRAMMING WITH PYTHON

Python has the ability to write GPU code through two powerful libraries, `PyOpenCL` and `PyCUDA`. Both of these libraries are written by the same author, and have similar functionality [31]. They closely follow the OpenCL and CUDA APIs, respectively, and abstract away much of the boilerplate code often needed to write GPU code. The base library itself is written in C++ so that Python can stay out of the way for maximum performance. Another extremely popular Python library, `Numpy`, is used throughout the community for numeric and scientific specific Python operations, and provides a basic, yet flexible numeric array type that prioritizes speed when compared to the standard Python list. Both `PyOpenCL` and `PyCUDA` create a `Numpy`-like array that make operations with these device arrays very similar to working with standard `Numpy` arrays, but handled on a CUDA or OpenCL device. Transferring data, error checking, profiling, and more are all wrapped in convenient Python methods to dovetail nicely with your existing Python code.

Both GPU libraries extensively use template code generation methods that can allow for simple creation of element-wise kernels as well as using parallel primitives like scan, reduce, and stream compaction. PyOpenCl also has the functionality to be combined with OpenGL and clBLAS libraries. All of this together makes it extremely easy to prototype code ideas or even speed up existing projects. Further information for working with PyOpenCL or PyCUDA can be found at https://documen.tician.de/pyopencl/ [26] and https://documen.tician.de/pycuda/ [25].

14.1.1 PyOpenCL Version of imflip

To show how simple it is to use PyOpenCL I will present two different programs. The first will be the Python version of the OpenCL imflip program shown earlier in the OpenCL section. Much of the syntax is very similar to the OpenCL API. The listing below gives the entire source code required to accomplish the same task as the OpenCL C version of imflip.

Before getting into the OpenCL portion of the code, the image is loaded into memory using the OpenCV Python library. The size of the image, rows, and columns are all extracted to get the proper sizing for the kernel later on in the code.

```
image = cv2.imread(sys.argv[1], cv2.CV_LOAD_IMAGE_COLOR)
rows, cols = image.shape[:2]
input_image = image.flatten()
flipped_image = np.empty_like(input_image)
```

This portion of the code (the listing below) allows the user to select which platform and device to choose. On an Apple device, there is only one platform available, since Apple writes the drivers and OpenCL library for its own devices. If on a Windows or Linux computer, there is the possibility for multiple platforms. For example, if you have an Nvidia GPU and an Intel CPU, you can download and install the OpenCL library for both devices. Each vendor is responsible for writing an OpenCL driver that properly takes your kernel and writes intermediate level language specific to the device (CPU or GPU). In the case of AMD, their OpenCL version can talk to both the CPU and GPU using the same drivers.

By using the get_devices method, PyOpenCL mirrors the OpenCL API calls to get devices of a certain type. In this version of the code, the device defaults to the CPU if nothing is specified. Note that this version does not do any extensive error checking to make sure that devices exist and that there is only one platform and one type of each device. Finally, the queue is created using the selected device. Note also that the properties include turning on profiling to allow timing of the kernel or any data transfers.

```
platform = cl.get_platforms()[0]
gpu_device = platform.get_devices(cl.device_type.GPU)
cpu_device = platform.get_devices(cl.device_type.CPU)
accel_device = platform.get_devices(cl.device_type.ACCELERATOR)

if cl_device_type == 'gpu':
    dev = gpu_device
elif cl_device_type == 'accelerator':
    dev = accel_device
else:
    dev = cpu_device
ctx = cl.Context(dev)

queue =
    cl.CommandQueue(ctx,properties=cl.command_queue_properties.PROFILING_ENABLE)
```

```
#Pyopencl version of imflip.cl
#
from __future__ import absolute_import, print_function
import numpy as np
import pyopencl as cl
import cv2
import sys

#Get the desired OpenCL device type (cpu, gpu, or accelerator)
cl_device_type = sys.argv[-1].lower()

image = cv2.imread(sys.argv[1], cv2.CV_LOAD_IMAGE_COLOR)
rows, cols = image.shape[:2]
input_image = image.flatten()
flipped_image = np.empty_like(input_image)

platform = cl.get_platforms()[0]
gpu_device = platform.get_devices(cl.device_type.GPU)
cpu_device = platform.get_devices(cl.device_type.CPU)
accel_device = platform.get_devices(cl.device_type.ACCELERATOR)

if cl_device_type == 'gpu':
    dev = gpu_device
elif cl_device_type == 'accelerator':
    dev = accel_device
else:
    dev = cpu_device
ctx = cl.Context(dev)

queue =
    cl.CommandQueue(ctx,properties=cl.command_queue_properties.PROFILING_ENABLE)

mf = cl.mem_flags

# Allocate memory on the device
image_device = cl.Buffer(ctx, mf.READ_ONLY | mf.COPY_HOST_PTR, hostbuf=input_image)
flipped_image_device = cl.Buffer(ctx, mf.WRITE_ONLY, input_image.nbytes)

with open("imflip.cl", "rb") as kernel_file:
    prg = cl.Program(ctx, kernel_file.read()).build()

exec_evt = prg.hflip(queue, input_image.shape, None, image_device,
    flipped_image_device, np.int32(rows), np.int32(cols))
exec_evt.wait()
kernel_run_time = 1e-9 * (exec_evt.profile.end - exec_evt.profile.start)

cl.enqueue_copy(queue, flipped_image, flipped_image_device)

flipped_image = np.reshape(flipped_image, (rows, cols, 3))
cv2.imwrite(sys.argv[2], flipped_image)

print("Kernel Runtime:", kernel_run_time)
```

The listing below shows how the memory is allocated on the device. These method calls the same syntax as the OpenCL C version.

```
mf = cl.mem_flags

# Allocate memory on the device
image_device = cl.Buffer(ctx, mf.READ_ONLY | mf.COPY_HOST_PTR, hostbuf=input_image)
flipped_image_device = cl.Buffer(ctx, mf.WRITE_ONLY, input_image.nbytes)
```

This portion of the code reads in the imflip.cl source code file and builds a kernel off of it. It then calls the kernel function `hflip()` with the same arguments as the original OpenCL C version. An event `exec_evt` is returned from the kernel function call. This is used to profile the runtime of the kernel. Note that `exec_evt.wait()` method ensures that the kernel has finished running before any timing values are used. The extraction of the runtime is then a simple arithmetic operation.

```
with open("imflip.cl", "rb") as kernel_file:
    prg = cl.Program(ctx, kernel_file.read()).build()

exec_evt = prg.hflip(queue, input_image.shape, None, image_device,
    flipped_image_device, np.int32(rows), np.int32(cols))
exec_evt.wait()
kernel_run_time = 1e-9 * (exec_evt.profile.end - exec_evt.profile.start)
```

Finally, the listing below shows how the flipped image buffer is copied back to the host. The array is reshaped to fit the OpenCV data format before writing it to disk as a BMP file.

```
cl.enqueue_copy(queue, flipped_image, flipped_image_device)

flipped_image = np.reshape(flipped_image, (rows, cols, 3))
cv2.imwrite(sys.argv[2], flipped_image)

print("Kernel Runtime:", kernel_run_time)
```

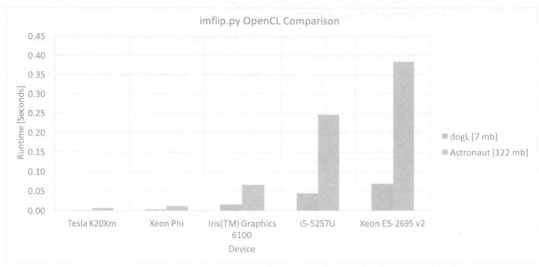

FIGURE 14.1   imflip.py kernel runtimes on different devices.

The results of running this imflip.py program are shown in Figure 14.1. Two different image sizes were tested on 5 different devices. Both the Tesla and Xeon Phi vastly outperformed the other devices as expected due to their high-end specs. The Intel Iris is a respectable middle-of-the-line integrated GPU found on most MacBook Pros. Surprisingly the Intel i5 CPU performed better than the server Xeon E5 CPU. The disparity between CPUs and GPUs could also be lessened if more effort was spent on optimization. One of the strong points of OpenCL is that it can be run on such a wide variety of devices. This high level of flexibility also comes with a slight caveat. Even though your OpenCL kernels will run just about everywhere, you will still need to think about device specific optimization, especially when running on CPUs, where memory layout is vastly different then GPUs. This leads us into the next section, where some of these hurdles are overcome by meta-templating and runtime code generation handled by PyOpenCL.

14.1.2 PyOpenCL Element-Wise Kernel

One of the unique and very convenient options within PyOpenCL and PyCUDA is the ability to quickly write kernels to perform element-wise operations. An element-wise operation is an operation that is performed on every single element with a vector or in the case of Python, a Numpy array container. An example of how to use this within PyOpenCL is shown in the listing below:

```python
#Elementwise operations in PyOpenCL
#
from __future__ import absolute_import, print_function
import numpy as np
import pyopencl as cl
import pyopencl.array
from pyopencl.elementwise import ElementwiseKernel

ctx = cl.create_some_context()
queue = cl.CommandQueue(ctx)

# Create random number array to represent a function f
f_cpu = np.random.rand(50000).astype(np.float32)

# Create a dx value
dx = 2.0

# Create a device array to hold f and transfer
f_device = cl.array.to_device(queue, f_cpu)

# Create an array to hold derivative value of the same
# size as f
dfdx_device = cl.array.empty_like(f_device)

# Create elementwise math kernel to compute the derivative
deriv_kernel = ElementwiseKernel(context=ctx,
                    arguments="float dx, float *f, float *dfdx",
                    operation="dfdx[i] = (f[i+1] - f[i])/dx",
                    name="deriv")
# Run the kernel
kernel_run_event = deriv_kernel(dx, f_device, dfdx_device)

# Get the derivative back to the device
dfdx_cpu = dfdx_device.get()
```

As in any OpenCL code, a platform and an appropriate queue needs to be set up. PyOpenCL has the convenience method create_some_context(), highlighted in the listing below, that will query the user at runtime to select which platform and device to run the code on. This is not the only way to select devices, as shown in the imflip.py example, where you can specify which device or types of device to choose.

```python
#Setting up a context and queue
#
ctx = cl.create_some_context()
queue = cl.CommandQueue(ctx)
```

To create an array of numbers to compute a derivative for an array, filled with random numbers, is created with standard Numpy methods on the CPU. A corresponding device array, f_device, is created using the to_device() method. This will create a device array exactly like the CPU version and transfer the contents to the device. An empty device array, dfdx_device, is created to hold the derivative values that the kernel will populate later in the code.

```python
#Creating device arrays
#
# Create random number array to represent a function f
f_cpu = np.random.rand(50000).astype(np.float32)

# Create a dx value
dx = 2.0

# Create a device array to hold f and transfer
f_device = cl.array.to_device(queue, f_cpu)

# Create an array to hold derivative value of the same
# size as f
dfdx_device = cl.array.empty_like(f_device)
```

Once the device arrays have been set up, the kernel can be created. The ElementiseKernel() method takes a set number of arguments, formatted in C style, and an operation, from which it creates a kernel behind the scenes to perform the operations. The real power in this portion of the code comes about because PyOpenCL uses meta-templating methods to create and analyze the kernel all in real time. This allows you to simply give it the basic operation, in this case a simple finite difference operation, and it will adapt it to your data. This runtime code generation and automatic tuning does a lot of work for you so that you simply have to give it the basic operation to perform. This is especially powerful when you wish to get optimal kernel performance on different devices.

```python
#Create kernel, run, and sending data back to the CPU
# Create elementwise math kernel to compute the derivative
#
deriv_kernel = ElementwiseKernel(context=ctx,
                    arguments="float dx, float *f, float *dfdx",
                    operation="dfdx[i] = (f[i+1] - f[i])/dx",
                    name="deriv")
# Run the kernel
kernel_run_event = deriv_kernel(dx, f_device, dfdx_device)

# Get the derivative back to the device
dfdx_cpu = dfdx_device.get()
```

Finally, once the element-wise kernel has been created and run, the data is copied back to the CPU using the device array method get(). Compare this with the imflip.py example where the data is transferred back to the CPU using cl.enqueue_copy(queue, output_image, res_g). Although both do essentially the same thing, the device array version adds yet another level of convenience.

14.2   OPENGL

Open Graphics Library (OpenGL) is a library of API functions that substantially simplify performing 2D and 3D computer graphics operations. The functionality included in OpenGL goes far beyond rotating objects, scaling them, etc. It also includes z-buffer functionality to turn a 3D image into a 2D projection of it to make it suitable for display in computer monitors, as well as compute the lighting effects of multiple lighting sources. This API is used to accelerate graphics computing by interfacing with a hardware accelerator.

OpenGL was introduced in the early 1990s by Silicon Graphics Inc. (SGI), which is when a *graphics accelerator* was vector units, built into a CPU or maybe other specialized chips, designed strictly to accelerate graphics. OpenGL is used extensively by Computer-Aided Graphics (CAD) applications; for example, the AutoCAD application, which is the de facto mechanical drawing tool among architects, to designers, and more, requires that a hardware graphics accelerator is available in the PC that is using this application. Furthermore, visualization programs, such as flight simulators, and more general information visualization tools (e.g., a tornado's travel pattern) use it to speed up the rate of information refresh (e.g., to visualize in real time). OpenGL is managed by the non-profit technology consortium Khronos Group, the same group that manages OpenCL.

The latest OpenGL 4.0 specification can be found on Khronos' website: https://www.khronos.org/registry/OpenGL/specs/gl/glspec40.core.pdf [23].

It is fairly common for OpenGL programmers to use higher level languages that build on top of OpenGL. These libraries are

- **OpenGL Utility Library (GLU):** A deprecated (as of 2009) library that provided higher-level functionality on top of OpenGL, such as support for spheres, disks, and cylinders.

- **OpenGL Utility Toolkit (GLUT):** Exposes a library of utilities to the programmer, who is writing an OpenGL-based program. These library functions are typically OS-specific, such as window definition, creation, control, and keyboard/mouse control. Without this add-on library, OpenGL alone is too low-level to program windows-based programs comfortably.

Installing `GLUT` on your computer is easy. A free version, `freeglut` is available on this website: http://freeglut.sourceforge.net/ [6].

Once you install it, you will have to include the following lines in your C code (this is just an example installation in Windows on an older version of MS Visual Studio):

```
#define WIN32_LEAN_AND_MEAN
#include <Windows.h>
#include <gl/gl.h>
#include <gl/glu.h>
#define FREEGLUT_STATIC
#include <gl/glut.h>
```

14.3   OPENGL ES: OPENGL FOR EMBEDDED SYSTEMS

OpenGL for Embedded Systems (OpenGL ES) was developed to enable 2D and 3D computer games on mobile devices (e.g., smartphones, tablet, and PDAs) and computer consoles that have limited computational power. It does not support GLU or GLUT. It is currently managed by the Khronos Group. It is royalty-free, just like OpenGL and OpenCL. It is expected that Vulkan will replace OpenGL ES. OpenGL ES, partly owing to its age, is the most widely deployed 3D graphics platform in history. More information about OpenGL ES can be found on Khronos' website: https://www.khronos.org/opengles/ [24].

14.4 VULKAN

Vulkan is another language by Khronos group, which intends to offer a more balanced CPU/GPU usage balance for the high-performance 3D graphics and generic computations. It was introduced in early 2016, which is when almost any commercial CPU included an integrated GPU. For example, Apple's A10 processor includes six CPU and 12 GPU cores, Intel's i5, i7 processors include four CPU cores and tens of GPU cores, AMD's APUs include multiple CPU and multiple GPU cores. Vulkan is similar to Direct 3D 12, which spreads the compute tasks among CPU and GPUs in a balanced manner, while staying loyal to its predecessor API, OpenGL. More information about Vulkan can be obtained on Khronos's website: https://www.khronos.org/registry/vulkan/specs/1.0/html/vkspec.html [27].

14.5 MICROSOFT'S HIGH-LEVEL SHADING LANGUAGE (HLSL)

In this book, we investigated GPU programming with the intention to perform generic scientific computations at a much higher performance than that is achievable by using only CPUs. In that sense, this is a GPGPU programming book, using a GPU's high-intensity computational capabilities on scientific computations using highly structured data, like an image or a mesh of electrons, or a volume of fluid elements. Remember that the emergence of the GPUs owes to computer games, not these *collateral* applications that need high performance computing; therefore, from day one, GPUs were designed with the capability to perform transformations, such as converting a computed 3D image into a 2D image that is seen on the computer monitor, or a TV (see Chapter 6 for a simple example). Because of this fact, each GPU is equipped with capabilities to perform *computer graphics* functions, such as dealing with image vertices, image geometry, putting texture on objects, etc. There are GPU programming languages to allow you to write a computer game; for example, Microsoft's High-Level Shading Language (HLSL) is one such language.

14.5.1 Shading

The first step in creating computer-generated visual effects is to design and build 3D objects using a set of points in the 3D space (see Chapter 6). These objects are then combined together with actual images taken by a camera to create a realistic *scene*. Raw 3D models, however, fail to blend well with the rest of the image. The result is a conspicuous 3D object that seems completely disconnected from its surrounding environment. This makes the entire image unrealistic and hence curtails the applicability of computer-generated visual effects in their target industries such as movie and video game production. This problem has been addressed by introducing shading programs. These programs analyze the surface of a 3D object and apply shading and light reflection effects in accordance with lighting characteristics of the scene with the ultimate aim to better blend in the objects and create more believable images.

Shading can be performed either *off-line* or *on-line*. The former is used in movie production, while the latter is typically employed in video game industries. OpenGL and Microsoft Direct3D are the examples of on-line shading tools. Although the initial shading tools were hard-coded in the GPU architecture, it was soon realized that a software-based assembly solution can improve the flexibility and customizability of the features. The complexity of assembly language, however, soon fueled the demand for a high-level shading language. Nvidia introduced their own language Nvidia Cg (C for graphics) and Microsoft introduced High Level Shading Language (HLSL). Both languages use the same syntax, however, they are branded with two different names for commercial reasons. Unlike Cg, however, HLSL can only compile shaders for DirectX and is *not* compatible with OpenGL (not surprisingly, because Direct 3D is a competitor of OpenGL). Cg is compatible with both OpenGL and DirectX.

14.5.2 Microsoft HLSL

The HLSL was developed by Microsoft for use with their Direct3D 9 Application Programming Interface (API) and became the required shading language for the unified shader model in Direct3D 10 and higher. HLSL was developed to evolve the coding of shaders from assembly language to a higher level language. The language works in the Direct3D pipeline and includes a built-in constructor, vector and matrix as native data types, swizzling and masking operators, and a large set of intrinsic functions, but does not have pointers, bitwise operations, function variables, or recursive functions. HLSL programs generally take one of following five forms:

1. *Pixel Shaders* calculate effects on a per-pixel basis.

2. *Vertex Shaders* add special effects to objects in a 3D environment by performing mathematical operations on the objects' vertex data. Note that the term *vertices* is plural for vertex.

3. *Geometry Shaders* accept a primitive as input and output zero or more primitives. Interestingly, a geometry shader can transform the given primitive to completely different primitives, even generating many more vertices than original input.

4. *Compute Shaders* are used for massively parallel GPGPU algorithms or to accelerate parts of game rendering. This is pretty much what we studied throughout this entire book.

5. *Tessellation Shaders* decompose surfaces, or patches, into smaller surfaces, as explained in Section 6.1.2.

14.6 APPLE'S METAL API

The Metal API (debuted with iOS 8, in 2014) was developed by Apple as part of their graphics API, which combines functionality found in OpenCL and OpenGL. Metal is based on the C++ 14 specification, but the C++ standard library is not to be used; instead, Metal has its own standard library. Metal also includes vectors and matrices as native data types, and does not include pointers. Metal is similar to OpenGL ES as it is a low-level language designed for interacting with 3D graphics hardware. Apple designed Metal specifically for their hardware to integrate with their OS and software. As a result, Metal can provide up to 10x the number of draw calls compared to OpenGL ES.

More information about the Metal language can be found on Apple's developer website: https://developer.apple.com/metal/ [1].

14.7 APPLE'S SWIFT PROGRAMMING LANGUAGE

Apple's Swift programming language is the result of substantial research, conducted by Apple, on programming languages. It was introduced in 2014 as a proprietary programming language and later became open source with Version 2.2. This is Apple's most recent programming language to program on all Apple platforms, such as iOS, macOS, watchOS, and tvOS. It can take advantage of the CPU and GPU capabilities of Apple platforms simultaneously.

More information about the Swift programming language can be found on Apple's website: https://developer.apple.com/swift/ [2].

14.8 OPENCV

The OpenCV library is one of the most important open source API libraries that is available today [22]. It is completely royalty-free and includes many image processing APIs, as well as high-end APIs such as face recognition. Later versions (OpenCV 3.3) include deep learning APIs. More information about OpenCV can be obtained on their website: http://opencv.org/ [22].

14.8.1 Installing OpenCV and Face Recognition

For a comprehensive tutorial for installing OpenCV and face recognition applications, refer to the following articles: [35, 28].

14.8.2 Mobile-Cloudlet-Cloud Real-Time Face Recognition

For an interesting application of real-time face recognition on mobile-cloudlet-cloud devices, see the following research, conducted over the past five years by the author [33, 36–38]. It describes a platform where the mobile phones can perform real-time face recognition using the cloud, which includes rich GPU resources.

14.8.3 Acceleration as a Service (AXaas)

For an interesting research, performed by the author, which suggests to sell GPU acceleration as a *rentable commodity* over cell phone usage, refer to the following articles: [34, 35]. In these works, heavy usage of OpenCV was made.

Deep Learning Using CUDA

Omid Rajabi Shishvan

University at Albany, SUNY

Tolga Soyata

University at Albany, SUNY

I N this chapter, we will study how GPUs can be used in deep learning. Deep learning is an emerging machine intelligence algorithm based on artificial neural networks (ANNs). ANNs were proposed to be computational models of neurological systems; they were designed to "learn" performing a certain task by mimicking the way a brain learns.

15.1 ARTIFICIAL NEURAL NETWORKS (ANNS)

ANNs are formed by multiple layers of "neurons" connected to each other, which create a network that takes input data, processes the data in each layer, and creates an output in the final layer. A common structure of a neural network is shown in Figure 15.1.

15.1.1 Neurons

Neurons are the building blocks of ANNs. The structure of a neuron is shown in Figure 15.2, in which the neuron is shown to take multiple inputs and create an output by passing a weighted sum of the inputs through an activation function. These input values may come from other neurons in the previous layer or from the primary inputs of a system.

15.1.2 Activation Functions

Activation functions in a neuron are implemented to introduce non-linearity to the network. If each neuron just output a linear combination of its inputs, the overall output of the network would be a linear combination of the inputs, which is not a desirable outcome for ANNs. To capture the nonlinear relation in input data, activation functions are used. Common activation functions in neural networks are shown in Table 15.1.

15.2 FULLY CONNECTED NEURAL NETWORKS

The ANNs, in which the connections between their neurons do not make a cycle, are called *Feed-Forward Neural Networks*. A common architecture for this type of ANN is when all of the neurons in one layer are connected to all of the neurons in the previous layer and the next layer; this makes a *fully connected* neural network. Figure 15.1 depicts a *fully connected feed-forward neural network*.

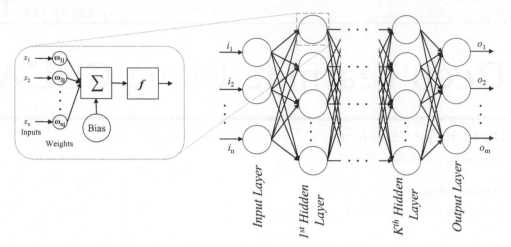

FIGURE 15.1 Generalized architecture of a fully connected artificial neural network with n inputs, k hidden layers, and m outputs.

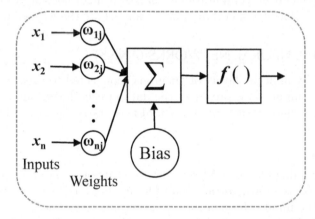

FIGURE 15.2 Inner structure of a neuron used in ANNs. ω_{ij} are the weights by which inputs to the neuron $(x_1, x_2, ..., x_n)$ are multiplied before they are summed. "Bias" is a value by which this sum is augmented, and $f()$ is the activation function, which is used to introduce a non-linear component to the output.

Fully connected networks work fine for shallower networks, but they show some problems when more hidden layers are added to their structure. One problem is saturating activation functions such as sigmoid function or hyperbolic tangent that are used in feedforward neural networks. In addition to saturating activation functions, the number of network parameters grows rapidly by adding extra hidden layers to them, which introduces computational complexities for training the network that is extra sensitive to even the smallest changes in the input, while those changes might not be important; this is called the *overfitting* problem. These two problems are addressed in other deep architectures such as *convolutional neural networks*.

TABLE 15.1 Common activation functions used in neurons to introduce a non-linear component to the final output.

Name	Equation
Identity	$f(x) = x$
Step Function	$f(x) = \begin{cases} 0 & \text{for } x < 0 \\ 1 & \text{for } x \geqslant 0 \end{cases}$
Logistic (Sigmoid)	$f(x) = \dfrac{1}{1 + e^{-x}}$
Hyperbolic Tangent	$f(x) = \dfrac{e^x - e^{-x}}{e^x + e^{-x}}$
Rectifier Linear Unit (ReLU)	$f(x) = \begin{cases} 0 & \text{for } x < 0 \\ x & \text{for } x \geqslant 0 \end{cases}$
Parametric ReLU	$f(\alpha, x) = \begin{cases} \alpha x & \text{for } x < 0 \\ x & \text{for } x \geqslant 0 \end{cases}$
Radial Basis Function	$\phi(x) = e^{-\beta \|x - \mu\|^2}$
Exponential Linear Unit (ELU)	$f(\alpha, x) = \begin{cases} \alpha(e^x - 1) & \text{for } x < 0 \\ x & \text{for } x \geqslant 0 \end{cases}$

15.3 DEEP NETWORKS/CONVOLUTIONAL NEURAL NETWORKS

Although the concept of deep neural networks has been around for a long time, they have gained popularity in recent years (since 2006) by using the computational power of GPUs. A form of the deep neural networks is convolutional neural networks (CNNs), which can take multidimensional data as input. Their main feature is that their layers are not fully connected and each neuron in a layer is only connected to some neurons in its *neighborhood* in the previous layers; this helps them identify *local* features and use almost unrestricted activation functions in their neurons. These differences with fully connected networks make training possible in deep convolutional neural networks. CNNs have proved themselves as a great candidate for image classification applications.

Each CNN is built by multiple different types of layers that are connected sequentially. Assuming that the network takes 2D images as input, the main types of layers used in a CNN are as follows:

- **Convolution Layer:** A 2D layer where each pixel (neuron) in the layer takes an area of pixels (for example, a 3×3 square area) from the previous layer, adds each pixel by a weight and sums those values to produce its output. Convolutional layers are capable of finding local features of the input data and passing those features to the following layers in the network.

- **Pooling Layer:** Also referred to as *downsampling layer*, is a filtering layer with multiple filtering options, maxpooling being the most popular one. Maxpooling takes a region in the previous layer (for example, a 2×2 region) and takes the maximum value of those pixels to assign to its output. The size of the region is called a stride (in this case, stride is equal to 2). This means that if a 6×6 input is given to a pooling layer with maxpooling function of size 2×2, it will create a 3×3 output.

- **Activation Layer:** Activation layer is a layer that each neuron takes one input and passes it through the activation function to produce its output. In a sense the activation functions of a neuron are decoupled from summing the inputs into a new layer. The activation function that is mostly used in CNNs is ReLU, since other activation functions tend to saturate due to having hard limits.

- **Fully Connected Layer:** Up to the late stages of CNNs, only convolutional, activation, and pooling layers are used which has detected the presence of local features in an input. At the end, there is a need to gather all the local data and have a global network to make a final decision on the input. This is where a few layers of fully connected networks are useful to "mix" the results from different local feature detectors into a final global result.

- **Softmax Layer:** Softmax layer is usually the last layer of a convolutional neural network that normalizes the output. Assume that the network is supposed to do image classification between n objects that are mutually exclusive. This means that we can use softmax function (Equation 15.1) to create n outputs that represent the probability that a given input is classified as a certain output. Note that softmax function creates outputs that are always summed to 1 and that softmax function is differentiable that helps with the training of the network.

$$f(v_i) = \frac{e^{v_i}}{\sum_j e^{v_j}} \tag{15.1}$$

15.4 TRAINING A NETWORK

ANNs are used for purposes such as regression or classification. Like other machine intelligence algorithms, ANNs need to be trained on a set of inputs. Training a network means that the internal parameters (including the weights that are assigned to the input values of neurons or the biases of the neurons) have to be adjusted to minimize the error in the performance of the network. Training a neural network is done through a process called "backpropagation." The network is initialized with random parameters and an input is given to the network. The input passes through all the layers and the error that is shown in the output is calculated. Through backpropagation, the amount of error that each parameter contributes to the whole outcome is calculated and those parameters are changed in small steps to reduce the error. After altering the parameters, a new input is passed through the network and by backpropagation, the parameters are changed again to reduce the error. This process is continued until there is minimal error.

15.5 CUDNN LIBRARY FOR DEEP LEARNING

cuDNN is a library that provides support for implementing deep networks on GPUs. Although using a GPU instead of a CPU already provides significant speedup for training a network, using cuDNN improves that speedup further.

cuDNN can be downloaded from `https://developer.nvidia.com/cudnn` [4]. After downloading the cuDNN files, make sure to follow the installation manual to fully install the library. The library is a host-callable C language API and like cuBLAS, it requires that the input and output data be resident on the GPU. The operations that are widely used in CNNs (and are optimized for both forward and backward passes) in cuDNN are

- Convolution

- Pooling

- Softmax

- Neuron activation functions
  - Rectified linear (ReLU)
  - Sigmoid
  - Hyperbolic tangent (tanh)
  - Exponential linear unit (ELU)

- Tensor transformation functions

15.5.1 Creating a Layer

Different types of layers can be described by different `struct` statement as shown below. Note that this is not the only way of implementing these layers; you can a single `struct` that describes multiple different layers.

```
struct Conv_Layer
{
    int inputs, outputs, kernelSize;
    int inputWidth, inputHeight, outputWidth, outputHeight;
    std::vector<float> convV;
    std::vector<float> biasV;
    ...
};

struct Maxpool_Layer
{
    int size, stride;
    ...
};

struct Fully_Connected_Layer
{
    int inputs, outputs;
    std::vector<float> neuronsV;
    std::vector<float> biasV;
    ...
};
```

In the layer descriptions the data and values related to that specific layer are contained. For example, a convolution layer needs to store the size of the input and output, the number of input and outputs (as you might want a convolution layer to create more than one output or take input from multiple inputs), and vectors of the biases and weights for the neurons. Depending on your application, you might want to implement functions that read these values from a pretrained network (that is saved to a file) or write the trained values to a new file at the end of the training phase.

15.5.2 Creating a Network

The network can be created through a struct. Setting up the tensors that are used to pass the data from one layer to the next one can be achieved here.

```
struct My_Network
{
    cudnnTensorDescriptor_t dataTensorDesc, convTensorDesc;
    cudnnConvolutionDescriptor_t convDesc;
    cudnnActivationDescriptor_t lastLayerActDesc;
    cudnnFilterDescriptor_t filterDesc;
    cudnnPoolingDescriptor_t poolDesc;

    void createHandles()
    {
        //General tensors and layers used in the network.
        //These need to be initialized by a descriptor.
        cudnnCreateTensorDescriptor(&dataTensorDesc);
        cudnnCreateTensorDescriptor(&convTensorDesc);
        cudnnCreateConvolutionDescriptor(&convDesc);
        cudnnCreateActivationDescriptor(&lastLayerActDesc);
        cudnnCreateFilterDescriptor(&filterDesc);
        cudnnCreatePoolingDescriptor(&poolDesc);
    }

    void destroyHandles()
    {
        cudnnDestroyTensorDescriptor(&dataTensorDesc);
        cudnnDestroyTensorDescriptor(&convTensorDesc);
        cudnnDestroyConvolutionDescriptor(&convDesc);
        cudnnDestroyActivationDescriptor(&lastLayerActDesc);
        cudnnDestroyFilterDescriptor(&filterDesc);
        cudnnDestroyPoolingDescriptor(&poolDesc);
    }
    ...
};
```

This is where the tensors and layers are described by descriptors and created.

15.5.3 Forward Propagation

Either to use or to train a network, you have to pass the data through the network layers. Some of the methods used in forward propagation for a convolution network are shown below:

```
convoluteForward(...)
{
    cudnnSetTensor4dDescriptor(dataTensorDesc, ...);
    cudnnSetFilter4dDescriptor(filterDesc, ...);
    cudnnSetConvolution2dDescriptor(convDesc, ...);
    cudnnConvolutionForward(...);
}
```

Note that all these functions take multiple inputs that are not shown here. These inputs vary based on the function, but common inputs are the cuDNN handle, descriptors of the input and output tensors, size of the data, and data types.

15.5.4 Backpropagation

For the training phase, backpropagation is done to adjust the weights and parameters in the network.

```
cudnnActivationBackward(...)
cudnnPoolingBackward(...)
cudnnConvolutionBackwardBias(...)
cudnnConvolutionBackwardFilter(...)
```

15.5.5 Using cuBLAS in the Network

The fully connected layers of the network are usually implemented using the cuBLAS library rather than the cuDNN library. This is due to the fact that the fully connected layers just need linear algebra and simple matrix multiplication that is supported in the cuBLAS library. A fully connected forward propagation is shown below:

```
fullyConenctedForward(...)
{
    ...
    cublasSgemv(...);
    ...
}
```

Multiple input arguments for `cublasSgemv` are the cuBLAS handle, source data, destination data, dimension of the data, etc.

15.6   KERAS

As shown in the previous sections, creating a CNN by using cuDNN is a time-consuming and confusing task. For the purpose of quickly creating and testing a prototype, it is not a good solution to use only cuDNN. There are many deep learning frameworks that take advantage of the GPU processing power and the cuDNN library to provide easy-to-develop networks that offer an acceptable performance. Frameworks such as Caffe, TensorFlow, Theano, Torch, and Microsoft Cognitive Toolkit (CNTK) are used to implement deep neural networks easily and achieve high performance.

We provide an example from the *Keras* framework on how to create a neural network. Keras is a Python library for deep learning; it can run on top of TensorFlow, CNTK, or Theano. Keras keeps all components of a network discretely, which are easy to add or remove. Keras is completely Python native, so there is no need for external file formats.

Keras provides support for different types of layers, even more than the layers that have been introduced previously in this chapter. It even provides the ability to define and write your own layer structure in the network. It also provides a variety of loss functions and performance metrics along many other supporting tools such as pre-existing widely used datasets, visualization support, and optimizers. A sample Keras code below shows the backbone of creating a simple network:

```python
from keras.models import Sequential
from keras.layers import Dense, Activation, Conv2D, MaxPooling2D
from keras import losses

model = Sequential()

model.add(Dense(units=..., input_dim=...))
model.add(Activation('relu'))
model.add(Conv2D(..., activation='relu'))
model.add(MaxPooling2D(pool_size=(2, 2)))
model.add(Dense(..., activation='softmax'))

model.compile(loss=losses.mean_squared_error,
            optimizer='sgd',
            metrics=['accuracy'])

model.fit([training data input], [training data output],
        batch_size=...,
        epochs=...)

score = model.evaluate([test data input], [test data output])
```

In the code above, a *sequential* network is created, meaning that the layers are connected through a linear stack. By using the add method, different layers are created and connected to each other in the network. There are a variety of network layers implemented in Keras and they take input arguments such as the dimensionality of the output, type of activation, the dimensionality of the input, etc.

The compile method is used before the training phase which configures the learning process. It takes an optimizer, such as stochastic gradient descent (sgd), a loss function, and a metric to set up the network.

After compiling, `fit` method is used to train the network. It takes the training input and output data, the size that the training data need to be chopped in the training process, input and output validation data if there are any available, etc.

The final stage is evaluating the network by using the `evaluate` method that takes input and output test data and checks the performance of the network on these data.

Other useful methods include `predict` that processes a given input and generates an output, `get_layer` that returns a layer in the network, and `train_on_batch` and `test_on_batch` that train and test the network on only one batch of input data.

Note that if Keras is running on the TensorFlow or CNTK backends, it automatically runs on the GPU if any GPU is detected. If the backend is Theano, there are multiple methods to use the GPU. One way is manually setting the device of the Theano configuration, as follows:

```
import theano
theano.config.device = 'gpu'
theano.config.floatX = 'float32'
```

Bibliography

[1] Apple: Metal Programming Language. https://developer.apple.com/metal/.

[2] Apple: Swift Programming Language. https://developer.apple.com/swift/.

[3] cuBLAS: CUDA Implementation of BLAS. http://docs.nvidia.com/cuda/cublas/index.html#axzz4reNqlFkR.

[4] cuDNN: Nvidia Deep Learning Library. https://developer.nvidia.com/cudnn.

[5] Cygwin Project. http://www.cygwin.com/.

[6] Free GLUT. http://freeglut.sourceforge.net/.

[7] INTEL DX79SR Motherboard. http://ark.intel.com/products/65143/Intel-Desktop-Board-DX79SR.

[8] INTEL i7-3820 4 Core Processor. https://ark.intel.com/products/63698/Intel-Core-i7-3820-Processor-10M-Cache-up-to-3_80-GHz.

[9] INTEL i7-4770K Quad Core Processor. http://ark.intel.com/products/75123/Intel-Core-i7-4770K-Processor-8M-Cache-up-to-3_90-GHz.

[10] INTEL i7-5930K Six Core Processor. https://ark.intel.com/products/82931/Intel-Core-i7-5930K-Processor-15M-Cache-up-to-3_70-GHz.

[11] INTEL i7-5960X 8 Core Extreme Processor. http://ark.intel.com/products/82930/Intel-Core-i7-5960X-Processor-Extreme-Edition-20M-Cache-up-to-3_50-GHz.

[12] INTEL Xeon E5-2680v4 Processor. https://ark.intel.com/products/91754/Intel-Xeon-Processor-E5-2680-v4-35M-Cache-2_40-GHz.

[13] INTEL Xeon E5-2690 8-Core Processor. https://ark.intel.com/products/64596/Intel-Xeon-Processor-E5-2690-20M-Cache-2_90-GHz-8_00-GTs-Intel-QPI.

[14] INTEL Xeon E7-8870 Processor. http://ark.intel.com/products/53580/Intel-Xeon-Processor-E7-8870-30M-Cache-2_40-GHz-6_40-GTs-Intel-QPI.

[15] INTEL Xeon W-3690 Six Core Processor. https://ark.intel.com/products/52586/Intel-Xeon-Processor-W3690-12M-Cache-3_46-GHz-6_40-GTs-Intel-QPI.

[16] Mobility Meets Performance: NvidiaOptimus Technology. http://www.nvidia.com/object/optimus_technology.html.

[17] Notepad++ Editor. https://notepad-plus-plus.org/.

[18] Nvidia CUDA Installation Guide for Mac OSX. http://docs.nvidia.com/cuda/pdf/CUDA_Installation_Guide_Mac.pdf.

[19] Nvidia Profiler Metrics Reference. http://docs.nvidia.com/cuda/profiler-users-guide/index.html#metrics-reference.

[20] Nvidia Visual Profiler User's Guide: Settings Options. http://docs.nvidia.com/cuda/profiler-users-guide/index.html#settings-view.

[21] Nvidia Visual Profiler User's Guide: Timeline Options. http://docs.nvidia.com/cuda/profiler-users-guide/index.html#timeline-view.

[22] OpenCV Library. http://opencv.org/.

[23] OpenGL 4.0 Specification. https://www.khronos.org/registry/OpenGL/specs/gl/glspec40.core.pdf.

[24] OpenGL ES. https://www.khronos.org/opengles/.

[25] PyCUDA Reference. https://documen.tician.de/pycuda/.

[26] PyOpenCL Reference. https://documen.tician.de/pyopencl/.

[27] Vulkan 1.0.59 Specification. https://www.khronos.org/registry/vulkan/specs/1.0/html/vkspec.html.

[28] A. Alling, N. Powers, and T. Soyata. Face Recognition: A Tutorial on Computational Aspects. In *Emerging research surrounding power consumption and performance issues in utility computing*, chapter 20, pages 405–425. IGI Global, 2016.

[29] Susan Blackmore. *Consciousness: an introduction*. Routledge, 2013.

[30] Daniel Kahneman. *Thinking, fast and slow*. Macmillan, 2011.

[31] Andreas Klckner, Nicolas Pinto, Yunsup Lee, B. Catanzaro, Paul Ivanov, and Ahmed Fasih. PyCUDA and PyOpenCL: A Scripting-Based Approach to GPU Run-Time Code Generation. *Parallel Computing*, 38(3):157–174, 2012.

[32] Nvidia. GTX 1080 White Paper. http://international.download.nvidia.com/geforce-com/international/pdfs/GeForce_GTX_1080_Whitepaper_FINAL.pdf.

[33] N. Powers, A. Alling, K. Osolinsky, T. Soyata, M. Zhu, H. Wang, H. Ba, W. Heinzelman, J. Shi, and M. Kwon. The Cloudlet Accelerator: Bringing Mobile-Cloud Face Recognition into Real-Time. In *Globecom Workshops (GC Wkshps)*, pages 1–7, San Diego, CA, Dec 2015.

[34] N. Powers and T. Soyata. AXaaS (Acceleration as a Service): Can the Telecom Service Provider Rent a Cloudlet? In *Proceedings of the 4th IEEE International Conference on Cloud Networking (CNET)*, pages 232–238, Niagara Falls, Canada, Oct 2015.

[35] N. Powers and T. Soyata. Selling FLOPs: Telecom Service Providers Can Rent a Cloudlet via Acceleration as a Service (AXaaS). In T. Soyata, editor, *Enabling real-time mobile cloud computing through emerging technologies*, chapter 6, pages 182–212. IGI Global, 2015.

[36] T. Soyata, H. Ba, W. Heinzelman, M. Kwon, and J. Shi. Accelerating Mobile Cloud Computing: A Survey. In H. T. Mouftah and B. Kantarci, editors, *Communication infrastructures for cloud computing*, chapter 8, pages 175–197. IGI Global, Sep 2013.

[37] T. Soyata, R. Muraleedharan, S. Ames, J. H. Langdon, C. Funai, M. Kwon, and W. B. Heinzelman. COMBAT: mobile Cloud-based cOmpute/coMmunications infrastructure for BATtlefield applications. In *Proceedings of SPIE*, volume 8403, pages 84030K–84030K, May 2012.

[38] T. Soyata, R. Muraleedharan, C. Funai, M. Kwon, and W. Heinzelman. Cloud-Vision: Real-Time Face Recognition Using a Mobile-Cloudlet-Cloud Acceleration Architecture. In *Proceedings of the 17th IEEE Symposium on Computers and Communications (ISCC)*, pages 59–66, Cappadocia, Turkey, Jul 2012.

[39] Jane Vanderkooi. *Your inner engine: An introductory course on human metabolism*. CreateSpace, 2014.

Index

__constant__, 321
__device__, 321
__global__, 166
__shared__, 309, 311, 314
__synchthreads(), 311, 314

Advanced Micro Devices (AMD), 3, 142
Apple's Metal API, 422
Apple's Swift Prog. Language, 422
Arithmetic Logic Unit (ALU), 5, 93
asynchronous data transfer, 351
ATI Technologies, 141

barrier synchronization, 128
Bitmap (BMP), 8, 38

C/T (cores/threads) notation, 24
cache memory (L1$ L2$ L3$), 24, 54, 67,
    82, 229, 304
color to B&W conversion, 113, 242
compile time, 36, 53, 253
compiler, 53
Compute-Unified Device Architecture
    (CUDA), 142
Constant Cache, 278, 307
core-intensive, 7
CPU cores, 4
CPU load, 33
cubin, 255
cuBLAS, 383
CUDA Compute Capability, 175, 206
CUDA cores, 227, 275
CUDA grid, 189, 257
CUDA kernel launch, 193
CUDA keywords, 170
CUDA occupancy calculator, 333
CUDA stream, 352
CUDA thread, 191
CUDA toolkit, 171
cuDNN, 428
cuFFT, 389
Cygwin, 13, 27, 182

data compression, 37

data splitting, 35
deep learning, 425
direct memory access (DMA), 58, 79
double precision floating point, 283
double precision units (DPU), 275
DRAM access patterns, 66
DRAM interface standards, 81
dynamic link library (DLL), 164, 255
dynamic random access memory (DRAM),
    66, 81

Eclipse IDE, 181
execution overlapping, 347
exposed versus coalesced memory access,
    348

Fermi architecture, 207, 231, 263
floating unit (FPU), 93
fused multiply-accumulate, 285

Gaussian filter, 109
gdb, 20
Giga bits per second (Gbps), 9, 250
Giga thread scheduler, 227, 258
Giga-floating-point-operations (GFLOPS),
    272, 284
global memory, 207, 303
GPGPU, 140
GPU graphics driver, 255
GPU kernel execution, 157

half precision floating point, 284
hard disk drive (HDD), 8
hardware versus software threads, 62
heap, 74

image edge detection, 107
image flipping, 27, 147
image rotation, 83
image thresholding, 110
ImageMagick, 39
In-Order Core (inO), 55
integrated develop. environ. (IDE), 13
INTEL Atom CPU, 56

INTEL x64 64-bit instruction set, 164
INTEL Xeon CPU, 4, 55
intermediate representation (IR), 164
International Business Machines (IBM), 3

Kepler architecture, 207, 232, 265
Keras, 432
kilo bytes (KB), 151

Makefile, 13
many integrated core (MIC), 56, 76
Maxwell architecture, 207, 232, 267
Mega bytes (MB), 151
memory address linearization, 163
memory bandwidth, 25, 119, 201, 207
memory leak, 11
memory-intensive, 7, 64
Microsoft HLSL, 422
Microsoft visual studio, 171
multithreaded, 4, 37
mutex, 129

network interface card (NIC), 9
Notepad++, 13
nvcc compiler, 144
NVIDIA Nsight, 171
Nvidia optimus technology, 179
Nvidia performance primitives (NPP), 391
Nvidia visual profiler: nvvp, nvprof, 375
NVLink, 233

old school debugging, 21, 214
OpenCL, 143, 397
OpenCV, 39, 423
OpenGL, 419
out-of-order core (OoO), 55

parallel programming, 3, 27
parallel thread execution (PTX), 164, 255,
        279
parallelization overhead, 89, 145
Pascal architecture, 207, 233, 270, 285
PCI express (PCIe) bus, 79, 200, 354
pinned memory, 350
pipelining, 346
POSIX-compliant, 27
precomputation, 133
predicate registers, 281
process memory map, 74
program performance, 5, 82
Pthreads, 27

PyCUDA, 413
PyOpenCL, 413

red green blue (RGB), 113
resource contention, 7
run time, 36, 53, 257

serial programming, 3
shared memory, 229, 278, 305
shared resources, 6
single precision floating point, 282
single-threaded, 3
Sobel gradient operator, 110
solid state disk (SSD), 8, 79
special function units (SFU), 276
ssh, 182
stack, 74
static random access memory (SRAM), 81
streaming multiprocessor (SM), 225

task granularity, 133
task parallelization, 186
task splitting, 35
tesselation, 139
texture mapping, 139
thin versus thick threads, 57
thread block, 175, 186, 190, 259
thread handle, 43, 59
thread ID, 3
thread launch/execution, 34, 43, 59
thread status, 60
throughput, 200
Thrust library, 393
tid, 3, 36, 66
time stamping, 112, 152
transparent scalability, 261

universal serial bus (USB), 53
Unix commands, 16

valgrind, 22
virtual memory, 349
Volta architecture, 207
Vulkan, 421

warp, 146, 186, 192, 278
Windows task manager, 34
wrapper function, 156

X11 forwarding, 182
Xcode, 15, 180
Xeon Phi, 56, 76